D0909674

Gastrointestinal Physiology II

Publisher's Note

The *International Review of Physiology* remains a major force in the education of established scientists and advanced students of physiology throughout the world. It continues to present accurate, timely, and thorough reviews of key topics by distinguished authors charged with the responsibility of selecting and critically analyzing new facts and concepts important to the progress of physiology from the mass of information in their respective fields.

Following the successful format established by the earlier volumes in this series, new volumes of the *International Review of Physiology* will concentrate on current developments in neurophysiology and cardiovascular, respiratory, gastrointestinal, endocrine, kidney and urinary tract, environmental, and reproductive physiology. New volumes on a given subject generally appear at two-year intervals, or according to the demand created by new developments in the field. The scope of the series is flexible, however, so that future volumes may cover areas not included earlier.

University Park Press is honored to continue publication of the *International Review of Physiology* under its sole sponsorship beginning with Volume 9. The following is a list of volumes published and currently in preparation for the series:

Volume 1: **CARDIOVASCULAR PHYSIOLOGY** (A. C. Guyton and C. E. Jones)
Volume 2: **RESPIRATORY PHYSIOLOGY** (J. G. Widdicombe)
Volume 3: **NEUROPHYSIOLOGY** (C. C. Hunt)
Volume 4: **GASTROINTESTINAL PHYSIOLOGY** (E. D. Jacobson and L. L. Shanbour)
Volume 5: **ENDOCRINE PHYSIOLOGY** (S. M. McCann)
Volume 6: **KIDNEY AND URINARY TRACT PHYSIOLOGY** (K. Thurau)
Volume 7: **ENVIRONMENTAL PHYSIOLOGY** (D. Robertshaw)
Volume 8: **REPRODUCTIVE PHYSIOLOGY** (R. O. Greep)
Volume 9: **CARDIOVASCULAR PHYSIOLOGY II** (A. C. Guyton and A. W. Cowley, Jr.)
Volume 10: **NEUROPHYSIOLOGY II** (R. Porter)
Volume 11: **KIDNEY AND URINARY TRACT PHYSIOLOGY II** (K. Thurau)
Volume 12: **GASTROINTESTINAL PHYSIOLOGY II** (R. K. Crane)
Volume 13: **REPRODUCTIVE PHYSIOLOGY II** (R. O. Greep)

(Series numbers for the following volumes will be assigned in order of publication)

RESPIRATORY PHYSIOLOGY II (J. G. Widdicombe)
ENDOCRINE PHYSIOLOGY II (S. M. McCann)
ENVIRONMENTAL PHYSIOLOGY II (D. Robertshaw)

Consultant Editor: Arthur C. Guyton, M.D., Department of Physiology and Biophysics, University of Mississippi Medical Center

INTERNATIONAL
REVIEW OF PHYSIOLOGY

Volume 12

Gastrointestinal Physiology II

Edited by

Robert K. Crane, Ph.D.
Department of Physiology
College of Medicine
and
Dentistry of New Jersey
Rutgers Medical School

UNIVERSITY PARK PRESS
Baltimore • London • Tokyo

UNIVERSITY PARK PRESS
International Publishers in Science and Medicine
Chamber of Commerce Building
Baltimore, Maryland 21202

Typeset by The Composing Room of Michigan, Inc.

Manufactured in the United States of America by Universal Lithographers, Inc., and The Optic Bindery Incorporated

Library of Congress Cataloging in Publication Data
Main entry under title:

Gastrointestinal physiology II.

(International review of physiology; v. 12)
Includes index.
1. Alimentary canal. 2. Digestive organs.
I. Crane, Robert K. II. Series. [DNLM: 1. Gastrointestinal system—Physiology—Periodicals. W1 IN834F v. 12 etc.]
QP1.P62 vol. 12 [QP145] 599'.01'08s [599'.01'3]
ISBN 0-8391-1061-8 76-54750

Consultant Editor's Note

In 1974 the first series of the *International Review of Physiology* appeared. This new review was launched in response to unfulfilled needs in the field of physiological science, most importantly the need for an in-depth review written especially for teachers and students of physiology throughout the world. It was not without trepidation that this publishing venture was begun, but its early success seems to assure its future. Therefore, we need to repeat here the philosophy, the goals, and the concept of the *International Review of Physiology.*

The *International Review of Physiology* has the same goals as all other reviews for accuracy, timeliness, and completeness, but it also has policies that we hope and believe engender still other important qualities often missing in reviews, the qualities of critical evaluation, integration, and instructiveness. To achieve these goals, the format allows publication of approximately 2,500 pages per series, divided into eight subspecialty volumes, each organized by experts in their respective fields. This extensiveness of coverage allows consideration of each subject in great depth. And, to make the review as timely as possible, a new series of all eight volumes is published approximately every two years, giving a cycle time that will keep the articles current.

Yet, perhaps the greatest hope that this new review will achieve its goals lies in its editorial policies. A simple but firm request is made to each author that he utilize his expertise and his judgment to sift from the mass of biennial publications those new facts and concepts that are important to the progress of physiology; that he make a conscious effort not to write a review consisting of an annotated list of references; and that the important material that he does choose be presented in thoughtful and logical exposition, complete enough to convey full understanding, as well as woven into context with previously established physiological principles. Hopefully, these processes will continue to bring to the reader each two years a treatise that he will use not merely as a reference in his own personal field but also as an exercise in refreshing and modernizing his whole store of physiological knowledge.

A. C. Guyton

Contents

Preface ix

1
Gastrointestinal Circulation 1
J. Svanik and O. Lundgren

2
Gastrointestinal Motility 35
E. Atanassova and M. Papasova

3
Gastrointestinal Hormones 71
S. R. Bloom

4
Morphology and Physiology of Salivary Myoepithelial Cells 105
J. A. Young and E. W. van Lennep

5
Gastric Secretion 127
G. Sachs, J. G. Spenney, and W. S. Rehm

6
The Exocrine Pancreas 173
H. Sarles

7
Biliary Secretion and Motility 223
A. Gerolami and J.-C. Sarles

8
Intestinal Secretion 257
T. R. Hendrix and H. T. Paulk

9
Mechanisms Underlying the Absorption of Water and Ions 285
H. J. Binder

10
Digestion and Absorption of Lipids 305
B. Borgström

11
Digestion and Absorption: Water-soluble Organics 325
R. K. Crane

Index 367

Preface

Editing this volume was a joy and an awakening. I can lay no claim to personal expertise concerning the broad span of gastrointestinal physiology represented in this volume. However, I was responsible, with guidance from many friends of acknowledged expertise, for the assemblage of authors. I followed the concept of the originators of this series and obtained the broadest possible representation of sound points of view by asking for advice with such questions as, "Whom, among those who disagree with you, do you most admire?" Then, acting on the advice obtained in answer to my questions, I sought from each author an expression of his point of view rather than a compilation of facts.

Some of the areas of gastrointestinal physiology are now so fast moving, and so much has been done in the interim since the writing of the preceding volume, that limitations of space required several of the reviews to take a form that lies somewhere between a point of view and a compilation of data. Happily, one or two of the chapters are quite freely provocative and reflect true daring on the part of the authors. All the reviews are at least cogent, informative, and interesting.

I thank the authors for the pleasure it gave me to be involved with them as the editor of this volume. I also thank my secretary, Miss Helen Sedlowski, not just for her competence and diligence, but even more for her forbearance.

R. K. Crane

International Review of Physiology
Gastrointestinal Physiology II, Volume 12
Edited by Robert K. Crane
Copyright 1977 University Park Press Baltimore

1
Gastrointestinal Circulation

J. SVANVIK AND O. LUNDGREN

Department of Physiology and Surgery III, University of Göteborg
Göteborg, Sweden

FUNCTIONAL ANATOMY
OF GASTROINTESTINAL VASCULAR BEDS 2

CURRENT METHODS OF
STUDYING GASTROINTESTINAL CIRCULATION 4
Morphological Methods 4
Studies of Total Organ Blood Flow 4
Direct Measurements of Volume Blood Flow 4
Indirect Methods of Studying Gross Flow 4
Studies of Parallel-coupled Vascular Sections 5
In Vivo Microscopy 5
Clearance Methods 5
Isotope and Microsphere Fractionation Techniques 5
Inert Gas Wash-out Methods 5
Local Tissue Monitoring of Indicator Dilution Curves 6
Absorption of Gas from Gastrointestinal Lumen 6
Studies of Consecutive Vascular Sections 7
Plethysmographic and Gravimetric Methods 7
Resistance Measurements in Severed Small Intestine 7
Methods for Determining Capillary Permeability 7

GASTRIC CIRCULATION 8
Vascular Dimensions of the Stomach 8
Parallel-coupled Vascular Circuits 8
Consecutive Vascular Sections 8

The literature search for this review article was ended in September 1975. The authors are grateful to Drs. G. Bohlen, J. Sara, R. Gore, P. Guth, G. Ross, and E. Smith for preprints of their manuscripts in press. The work described herein as emanating from this laboratory was supported by grants from the Swedish Medical Research Council (14X-2855).

Adrenergic Control of Gastric Circulation 9
Gastric Circulation and Function 9

INTESTINAL CIRCULATION 10
Vascular Dimensions of the Small Intestine 10
Parallel-coupled Vascular Circuits 10
Series-coupled Vascular Sections 12
Adrenergic Control of Intestinal Circulation 15
Local and Hormonal Control of Intestinal Blood Flow 17
Intestinal Blood Flow and Function 19

COLONIC CIRCULATION 23

PANCREATIC CIRCULATION 24
Vascular Dimensions of the Pancreas 24
Adrenergic Control of Pancreatic Circulation 24
Pancreatic Blood Flow and Exocrine Secretion 25
Pancreatic Blood Flow and Endocrine Secretion 26

FUNCTIONAL ANATOMY OF GASTROINTESTINAL VASCULAR BEDS

The functions of the gastrointestinal tract are complex and so, indeed, is its vascular supply (Figure 1). The gross vascular morphology of the alimentary canal is characterized by a system of vascular arcades. This also is true for the intramural vessels. The arterial systems of the intestine and of the stomach form a close-meshed network of small interconnected arteries in the submucosa from which arterioles pass further into the different morphological and functional units of the organ wall.

Most anatomical studies of the intramural vasculature of the gastrointestinal tract indicate that it consists of a number of parallel-coupled vascular circuits (1–4), supplying the different wall layers. The degree of vascularization differs in the various layers apparently depending on functional demands. Thus, the vascularization of the mucosa is much denser than that of the muscularis, probably reflecting the great metabolic needs of the active transport mechanisms of the gastrointestinal epithelium.

The question of the possible existence of arteriovenous shunts in the gastrointestinal tract has been much debated, as reviewed by Grayson (5) in the earlier edition of this series of reviews. The shunts have been demonstrated by anatomists but their qualitative and quantitative physiological functions are obscure. Unfortunately, the terms shunt and shunting are often misused. Originally, the term shunt denoted a vessel forming a non-nutritional capillary bypass. However,

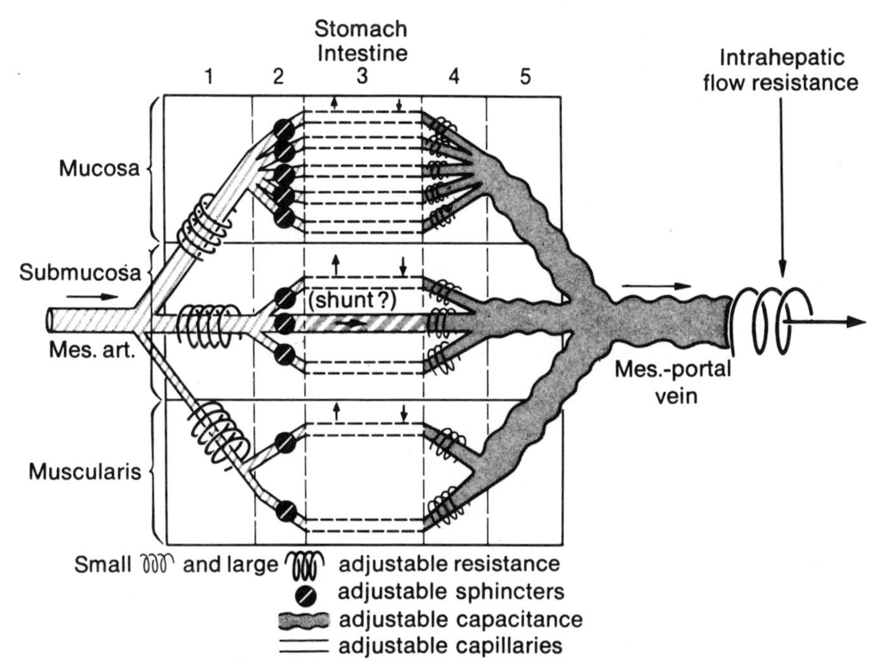

Figure 1. Schematic illustration of the different parallel-coupled vascular circuits of the stomach and the intestines with their consecutive vascular sections: *1*, precapillary resistance vessels; *2*, precapillary sphincters; *3*, exchange vessels; *4*, postcapillary resistance vessels; and *5*, capacitance vessels. (From Folkow (7), by courtesy of the Williams & Wilkins Co., Baltimore.)

the word is often wrongly used to denote arteriovenous connections constituted by wide capillaries in which transmural exchange occurs freely. Moreover, true shunting through non-nutritional channels should be clearly distinguished from "functional" shunting of blood due to an uneven capillary perfusion.

Each parallel-coupled vascular bed is made up of several anatomically and functionally different series-coupled (consecutive) vascular sections (see Figure 1) (6, 7). The *precapillary resistance section*, morphologically corresponding to the arteriole and metarteriole, is the main determinant of the blood supply. The *precapillary sphincters* are located proximal to capillaries. This section usually does not contribute much to total regional vascular resistance, but it is of paramount importance as regards the control of the numbers of perfused capillaries and, hence, the mean blood-tissue diffusion distance and the time available for transcapillary exchange. It should be underlined that the term precapillary sphincter, as used by us, is a functional concept not necessarily corresponding to an anatomical sphincter of the type observed in the mesentery.

The capillaries or the *exchange vessels* constitute the key section of any circulation. It is across their thin endothelial lining that the all important exchange takes place between intra- and extravascular compartments.

Distal to the capillaries are the venules and veins, sometimes called the *postcapillary resistance* section. The tonus of the smooth muscle cells of this section is one of the main determinants of mean hydrostatic capillary pressure and, thus, the rate of hydrodynamic fluid exchange across the capillary wall. The last mentioned section overlaps anatomically with the *capacitance section*, composed of the venous compartment as a whole. Changes in the tonus of the capacitance vessels may markedly alter regional blood volume without significantly influencing regional vascular resistance to blood flow.

The functional anatomy described above constitutes the conceptual framework of this review as regards the alimentary canal proper, and the function and regulation of the gastrointestinal circulation are discussed in these terms. Blood flow in the pancreas seems to be rather homogeneous (see below) and this vascular bed therefore lacks parallel-coupled vascular circuits. Special vascular morphological arrangements, as, for example, the vessels of the small intestinal mucosa, are discussed in connection with the organ concerned.

CURRENT METHODS OF STUDYING GASTROINTESTINAL CIRCULATION

Morphological Methods

Intravascular injections of different light or x-ray opaque material combined with light microscopy and microradiography have provided interesting information regarding the vascular anatomy of the gastrointestinal organs. Recently these techniques have been improved by use of silicon rubber (4) or methacrylic methyl ester resin, the latter in combination with scanning electron microscopy (8).

Studies of Total Organ Blood Flow

Direct Measurements of Volume Blood Flow There exist a number of well known methods for the study of total blood flow of an organ, and they need not be commented upon here. Techniques involving arterial cannulations with an extracorporal circuit with or without a pump should, however, be used with great caution, since the blood cells in such situations release vasoactive agents and tend to aggregate. This leads to unphysiological changes of vascular tone and a deterioration of the vascular bed of the organ studied (9).

Indirect Methods of Studying Gross Flow An indicator dilution method to measure superior mesenteric artery blood flow, based on the concepts of Stewart-Hamilton, was recently developed for the conscious man (10). The method included catheterizations of the superior mesenteric artery and of the superior mesenteric vein via the umbilical vein. This method also enabled a determination of blood volume between the place of injection and collection to be made.

Strandell and co-workers (11) described another technique for the study of blood flow in the splanchnic region of awake humans. Reopening the umbilical vein, Strandell et al. (11) could introduce up to three catheters via this route. By infusing ^{133}Xe dissolved in saline solution through one catheter and collecting it at another "downstream," blood flow was calculated.

Studies of Parallel-coupled Vascular Sections

In Vivo Microscopy During the last decade there has been a rapid development of a variety of techniques for quantifying microscopic observations of the vasculature. Intravascular pressure and red cell velocity, for example, can be measured continuously. The recording of pressures at various points in a vascular circuit provides an indirect measure of flow resistance. Most of these studies have for technical reasons been performed on the mesentery. Recently, however, some interesting studies on the small intestine were reported (12).

Clearance Methods Jacobson and co-workers (13) introduced the aminopyrine clearance technique for estimation of gastric mucosal blood flow in unanesthetized animals. At the pH of plasma the weak base, aminopyrine, is un-ionized and lipid soluble and easily crosses cell membranes. In the acid environment of the gastric juice, aminopyrine becomes ionized and water soluble. This prevents back diffusion into blood and aminopyrine therefore tends to accumulate in the gastric juice. This method has recently been modified using other weak bases such as a radioactive aniline (14) and neutral red (15). Tague and Jacobson (16) described the use of low concentrations of [^{14}C]aminopyrine. It should be possible to use the two latter methods (15, 16) in man for estimation of gastric mucosal blood flow.

Isotope and Microsphere Fractionation Techniques Sapirstein (17) proposed that intravenously injected ^{42}K and ^{86}Rb were distributed to an organ or tissue in proportion to its blood flow. These tracers are water soluble and pore restricted in their transcapillary passage so only a fraction of the injected tracer is taken up by a tissue during the first tracer passage. This is the great weakness of this method, since it is assumed that the fraction of tissue uptake is the same for *all* tissues of the body. The technique should therefore be used with great caution. Furthermore, only one determination can be performed on each animal. The method has been used for studies of the flow distribution within the gastrointestinal tract (18, 19) and within the intestinal wall (20–22).

A theoretically more attractive method is that of studying the tissue distribution of radioactively labeled microspheres usually injected into the left ventricle of the heart. The use of spheres labeled with different isotopes permits more than one determination in each animal. It should, however, be emphasized that what is studied with microspheres is the distribution of corpuscular elements and not whole blood. The method has been used to measure intramural flow distribution within the stomach (20) and the small intestine (23–25).

Inert Gas Wash-out Methods Kety (26) outlined one theory for these techniques based on the use of gases that are lipophilic and metabolically inert. The elimination of such a gas from a tissue, after local or intra-arterial injection of the gas dissolved in saline solution, is, according to Kety's calculations, limited by blood flow. ^{85}Kr and ^{133}Xe, radioactive isotopes of noble gases, are two of the most commonly used tracers. Recording the gas wash-out from an organ or tissue that is homogeneously perfused by blood gives a first order exponential ("monoexponential") elimination curve. Knowing the tissue to blood partition coefficient of the gas, blood flow can be calculated. If the tracer is injected into the arterial blood supply of a heterogeneously perfused organ, a multiexponential disappearance curve is recorded. It is always possible to resolve such a curve into 2–5 components, but great care should be taken when testing if these different components have any biological significance. Another way to treat externally monitored wash-out curves was proposed by Zierler (27). No assumptions were made concerning the shape of the elimination curve according to this approach.

Methods based on these principles have been used for the study of the stomach (28–30), the small intestine (31–35), the colon (36, 37), and the pancreas (38–40). Repeated determinations can be done in the same animal and recently the technique has been adapted to man (34, 37).

An elegant variation of the inert gas wash-out technique is the hydrogen gas method introduced by Aukland and co-workers (41) and used on pancreas by Aune and Semb (42). This technique involves addition of hydrogen gas to the inspiratory air and subsequent recording of the tissue desaturation of the gas by implanted detectors.

Local Tissue Monitoring of Indicator Dilution Curves Biber et al. (43) developed a method for the study of the blood circulation of the small intestinal mucosa based on the indicator dilution principle. The theoretical background for using external detectors to record indicator dilution curves from a tissue surface was outlined by Zierler (27) and Wolgast (44). The technique involves an injection of an intravascular β-radioactive tracer into the superior mesenteric artery and its detection with sensoring devices placed in the gut lumen. Knowing the amount of tracer injected and the flow in the superior mesenteric artery, it is possible to estimate regional blood flow from the height of the dilution curve and the regional blood content from the area under the curve. The tissue volume monitored by the detector depends on the energy level of the β-radiation and, by use of the proper tracer, the villous plasma flow and plasma volume could be studied separately and quantitatively. It is also possible to make repeated injections of tracer to follow transient hemodynamic adjustments. The method has been used to study mucosal hemodynamics of the small intestine in the cat (45–47).

Absorption of Gas from Gastrointestinal Lumen Considering the fast diffusion rate of lipophilic gases, it is tempting to assume that the rate of gastrointestinal absorption of these gases is blood flow limited. Forster (48) and Coburn (49) proposed the use of carbon monoxide (CO) for such studies and provided experimental evidence to indicate that the rate of CO disappearance was proportional to blood flow at a lumen pCO above 400 mm Hg.

A similar technique was used by Levitt and Levitt (50), who followed the relative rates of absorption of different inert gases from the gastrointestinal tract in conscious rats. The data obtained were fitted to a series of models for the interaction between perfusion and diffusion, and one of these seemed to predict the absorption rates of the gases from all organs investigated. Using this model it was possible to calculate mucosal blood flow from the absorption rates of the gases.

According to the countercurrent hypothesis (see below) gases absorbed from the small intestine are trapped in the countercurrent exchanger and delayed in their transport from the intestine into the blood (31, 51, 52). Hence, only "effective" blood flow is measured. This applied in particular to the Levitt method but probably also to the use of CO.

Studies of Consecutive Vascular Sections

Plethysmographic and Gravimetric Methods It is possible to follow quantitatively and continuously the reactions of the different series-coupled vascular sections of the gastrointestinal tract by recording changes in tissue volume (53) or tissue weight (54). These are usually recorded together with arterial inflow pressure, venous outflow pressure, and total blood flow. Since changes in tissue volume (or weight) depend on alterations of regional blood content and/or transcapillary fluid exchange, these two quantities must be separated. This is usually possible from the volume (or weight) recordings alone, since changes in the two parameters often have quite different time courses (cf. Wallentin (55)). Hence, these methods permit the continuous recording of the blood flow (resistance vessels), net capillary fluid exchange (exchange vessels), and changes of regional blood content (capacitance vessels). It is also possible to measure changes in capillary surface area available for transcapillary exchange by determining the capillary filtration coefficient (CFC). This parameter is a measure of the outward capillary filtration induced by a sudden increase of mean capillary hydrostatic pressure.

Resistance Measurements in Severed Small Intestine In an attempt to study the consecutive vascular sections of the rat small intestine Henrich and Singbartl (56) perfused the intestine during constant flow conditions and recorded perfusion pressure in the intact gut and after cutting the intestine at the antimesenteric border or at the mesenteric border. Apart from the fact that this technique is clearly "unphysiological," it is not obvious which part of the vascular tree was excluded by cutting at the antimesenteric border. Furthermore, the tissue was perfused with a Tyrode's solution containing neither albumin nor red cells so that the vascular smooth muscles were probably completely relaxed. Hence, conclusions drawn from experiments using this method are open to question.

Methods for Determining Capillary Permeability It is possible to estimate the capillary permeability for ^{86}Rb and ^{42}K by measuring the blood to tissue extraction of these radioactive isotopes as outlined by Renkin (57). These tracers were selected because they were quickly taken up by the cells, reducing their interstitial concentration and, hence, the tracer "back diffusion" from tissue to blood. From the arteriovenous concentration difference and blood flow a clearance value was calculated, called the PS product (permeability X surface area). The value of PS provided a semiquantitative measure of the tone of the precapillary sphincters. The method has been used on the small intestine (58, 59).

Another approach to the measurement of capillary permeability was utilized on the stomach by Alvarez and Yudilevich (60), who adapted the so-called single injection technique for that organ. Diffusable and intravascular tracers were injected intra-arterially and the venous indicator dilution curves were recorded. From these a PS value was calculated.

GASTRIC CIRCULATION

Vascular Dimensions of the Stomach

Parallel-coupled Vascular Circuits The values reported for "resting" total blood flow level in the stomach vary greatly among different authors, ranging from 10 to 55 ml/minX100 g as reviewed by Jacobson (61). The large scatter is no doubt explained by differences in the type of anesthesia, the extent of surgical intervention, the use of fasting or nonfasting animals, etc. As regards the intramural distribution of gastric blood flow, a major portion is distributed to the mucosa according to experiments by Delaney and Grim (20), using the microsphere method. These workers estimated that, "at rest," about 70% of gastric blood flow was delivered to the mucosa while the submucosa and muscularis received about 10% and 20%, respectively. This uneven flow distribution reflects varying blood flow rates in the different gastric wall layers. Thus, flow is 3–5 times greater in the mucosa than in the submucosa when expressed per unit weight of tissue (20, 21). In the vagally innervated stomach of the anesthetized dog, "at rest," mucosal blood flow measured 80–120 ml/minX100 g of mucosal tissue (20, 21, 28–30) while the muscularis flow amounted to about 25 ml/minX100 g of muscularis tissue (20, 21).

Consecutive Vascular Sections The resistance vessels are commented upon above. As regards the exchange vessels of the stomach our knowledge is scant. Indirect evidence suggests a fairly high permeability to protein since the collected lymph contained roughly half the concentration present in the plasma (62). On the other hand, horseradish peroxidase, a molecule of the same size as albumin, apparently did not leave the gastric capillaries to any large extent in the nonsecreting stomach (63). More quantitative results were obtained in studies of capillary permeability determining the capillary filtration coefficient (CFC) (64, 65) or using the single injection technique (60). CFC measured "at rest" was around 0.04 ml/minXmm HgX100 g. This value is similar to that reported for the "resting" intestine by Haglund and Lundgren (66), suggesting that the capillary surface area in the stomach may be 5–10 times larger than in skeletal muscle. A high capillary permeability, particularly in the mucosa, was also revealed by the single injection technique (60).

Adrenergic Control of Gastric Circulation

The effects on the consecutive vascular sections of stimulating the sympathetic vasoconstrictor fibers were investigated by Jansson et al. (64). They found that the neurally induced response pattern was identical from a qualitative point of view to that of the intestine, but that quantitative differences existed between the two organs. There was a drastic reduction of blood flow and a constriction of the capacitance vessels upon stimulation. However, when the stimulation was prolonged, blood flow again increased and within 3–4 min reached a new steady state flow level. This phenomenon has been named "autoregulatory escape from the vasoconstrictor fiber influence" (see below). Apart from the phasic decrease obtained initially when blood contents were expelled, the tissue volume recorded by the plethysmograph remained at a constant level throughout the stimulation period, indicating that mean hydrostatic capillary pressure was not significantly altered as compared with control. CFC was markedly reduced, however, reflecting a diminished surface area for transcapillary exchange. Quantitatively, blood flow was reduced to 30–40% of control upon stimulating the splanchnic nerves at 10 Hz during a gastrin pentapeptide-induced secretion (67). As regards the parallel-coupled vascular circuits, several observations, made during electrical stimulation of the vasoconstrictor fibers or during infusions of adrenergic drugs, clearly indicate that the mucosal vessels are constricted by adrenergic stimuli (13, 21, 67–70). The vasoconstriction elicited by an adrenergic influence seems to be relatively greater in the mucosa than in the rest of the gastric wall (21, 67), suggesting that blood is redistributed away from the mucosa. Hence, Grund et al. (67) reported that mucosal blood was reduced to one-third of control during maximal sympathetic vasoconstrictor activation, while non-mucosal blood flow was not significantly changed by this procedure.

Gastric Circulation and Function

The aminopyrine method for the study of gastric mucosal blood flow (13) represented a breakthrough for gastric physiology. Following this, an intense research has flowered on the question of the relationship between gastric blood flow and secretion. These investigations strongly suggest that an increase of gastric acid secretion is accompanied by a corresponding augmentation of gastric mucosal blood flow. An increase in blood flow by itself, however, does not induce secretion. A close correlation between flow and acid secretion was demonstrated by the original work by Jacobson and co-workers (71, 72) and has lately been confirmed by various workers inducing net secretion by vagal stimulation (68, 69, 73), by pentagastrin (74–78), or by histamine (76, 78, 79). Inhibition of gastric acid secretion by pharmacological means at the glandular cell level was accompanied by a reduction of mucosal blood flow (80). Induction of vasodilatation by isoprenaline in the unstimulated stomach did not produce any secretion (70). However, reducing blood flow by adrenergic stimuli decreased acid secretion owing, in all probability, to the lowered rate of mucosal blood flow (67, 68, 70, 75).

In contrast to the studies referred to above on acid secretion, Reed and Sanders (69) measured the secretion of pepsin induced by vagal stimulation and found a comparatively weak correlation with mucosal blood flow. These authors were unable to demonstrate any decrease of pepsin secretion during nervous vasoconstriction although the acid output was markedly inhibited. These findings obviously suggest that pepsin secretion is largely blood flow independent, which is a surprising observation in view of the fact that protein secretion in the pancreas is a highly oxygen-consuming process.

There are few reports on the effects of secreting agents on the consecutive vascular sections of the gastric vascular bed except for the investigation on the resistance vessels reported above. The changes of the capacitance vessels induced by pentagastrin or histamine, for example, are unknown. Stimulation of the high threshold vagal fibers (i.e., the nerve fibers eliciting secretion and "receptive relaxation" of the stomach) produced an increase of the capillary filtration coefficient that may indicate that vagal stimulation augments the number of perfused capillaries as well as size of the capillary "pores" (64). Bradykinin infusions, on the other hand, had apparently no effect on capillary permeability (65).

One report has been published on the effects of gastric motility on the regional blood flow. Semba et al. (81) increased motility by vagal stimulation and recorded blood flow in the gastroepiploic vein. Blood flow was sometimes increased and sometimes decreased by stimulation, and the authors discussed three different "patterns" with respect to the flow effects of motility. Unfortunately, the study was performed with unphysiologically high rates of vagal stimulation ($>$ 10 Hz).

INTESTINAL CIRCULATION

Vascular Dimensions of the Small Intestine

Parallel-coupled Vascular Circuits Grim and Lindseth (23) were the first to study intramural flow distribution in the small bowel. They used ^{24}Na-labeled glass spheres but since then a number of studies have been performed utilizing different techniques. Table 1 summarizes some results obtained on dogs and cats. It is quite clear from this table that a major portion of total blood flow is distributed to the mucosa-submucosa. In this tissue compartment, blood flow, expressed per unit weight tissue, is considerably greater than in the muscularis, reflecting the higher metabolic demands of active transport in the epithelium as compared with motility, as well as the fact that blood is the transport vehicle for

Table 1. Mean values reported for blood flow and flow distribution in the small intestine of dogs and cats measured with different techniques[a]

Method	Mucosa	Villi	Submucosa	Mucosa-submucosa	Muscularis	Total flow	Reference
Dog							
Microspheres	56 (62)		93 (20)	62 (82)	44 (18)	57	23
^{133}Xe wash-out				27 (88)	6.7 (12)	22	35
Cat							
Microspheres	132 (58)		217 (33)	154 (91)	16 (9)	85	24
^{85}Kr wash-out				40 (65)	12 (35)	20	33
Indicator dilution		48 (24)				20	45

[a]Flow values are expressed in ml/min×100 g tissue. Blood flow distribution, given in parentheses, is expressed in percent of total intestinal blood flow.

most absorbed solutes. Similar results have recently been obtained on the human gut (34).

It is a striking observation in Table 1 that when microspheres are used a large portion of these are trapped in the submucosa, a small compartment consisting of connective tissue with a very scarce vascularization of its own. This finding was discussed at length by Greenway and Murthy (24), who proposed that it could be explained by the nutritional vascular beds of the submucosa and mucosa being coupled in series. There are few, if any, anatomical studies which substantiate this conclusion (1–4). Furthermore, the interpretation of Greenway and Murthy would imply that the inner diameter of the vessels in the submucosa would be less than 14 μm, which was the size of the injected microspheres (24). This is a surprisingly small diameter (12, 82).

However, there exist at least two alternative explanations. First, the microspheres are supposed to behave as blood cells and reflect in essence red cell flow. Jodal and Lundgren (83) provided experimental support for the existence of plasma skimming in the intestine by finding the mucosa to be perfused with blood having a hematocrit half that of arterial blood. The localization of the "high-hematocrit" vessels is not known, but they must be located in the deeper layers of the mucosa or in the submucosa. Hence, one explanation of the results of Greenway and Murthy may be that the microspheres were trapped in "high-hematocrit" vessels in the submucosa. Second, it is well known that the microspheres tend to form aggregates, and in a study of the kidney Källskog et al. (84) demonstrated that these aggregates adhered to the walls of larger vessels such as the renal interlobular arteries. Such an adhesion to the walls of the intestinal submucosal vessels may possibly explain the above mentioned findings. To test this hypothesis it would be interesting to see the outcome of experiments on the small intestine in which microspheres of very small size (1–4 μm alouter diameter) were used. This crucial test of the microsphere technique has yet to be performed.

The values given in Table 1 represent "resting" flow values. The potential blood flow range was investigated in the cat by Kampp and Lundgren (33) and by Biber et al. (45). These experiments demonstrated that the blood flow capacity of the gut is very high indeed. Total venous outflow could be increased 8- to 10-fold from "resting" levels to flow values above 200 ml/min$\times$100 g by means of potent vasodilator drugs. Following intense vasodilatation induced by isoprenaline, villous blood flow reached 350 ml/min$\times$100 g of villous tissue. In this situation more than 90% of total blood flow was diverted to the mucosa-submucosa and almost half of total flow was passing through the villi. In this context it should be pointed out that during the functional hyperemia of the gut seen after a meal, intestinal blood flow never increases more than 200% above control (see below).

Series-coupled Vascular Sections The exchange vessels represent the consecutive vascular section that, for technical reasons, is more difficult to study quantitatively than the resistance vessels discussed above. Thus, our knowledge of capillaries is comparatively scant in most organs because capillaries have been studied in great detail only in certain easily accessible specialized tissues such as the mesentery, which has been investigated by several research groups. However, while strictly speaking the mesentery belongs to the gastrointestinal tract, it is of little functional importance and studies of the mesenteric microcirculation are usually carried out as investigations of general capillary physiology. For this reason, discussion of the rapidly expanding literature on the mesentery would carry the review too far afield. It is, indeed, sometimes claimed that the results from investigations of the mesenteric capillaries may be applicable also to the intestinal wall. However, such an extrapolation should be done with due caution if for no other reason than the fact that the intestinal wall, consisting of different tissues, is a much more complex organ.

The determination of the CFC and the ^{86}Rb clearance provides a semiquantitative measure of capillary function (see above). When measuring these parameters in the intestine (53, 58), values are obtained that are approximately 10 times higher than those for the skeletal muscle capillaries. This difference is explained by at least two facts. First, the intestine is more densely vascularized than skeletal muscle and second, the pore area per unit capillary surface area is greater in the intestine.

A bold attempt to study the permeability of the intestinal capillaries has been performed by Casley-Smith and co-workers (85) using quantitative electron microscopy. They found that the calculated capillary permeability for water and ^{86}Rb measured with the histological technique was several orders of magnitude larger than that measured with physiological techniques. This observation was taken as evidence for so-called "tunnel capillaries" (86) in which the endothelial wall is extremely permeable to most solutes and serves only to contain blood cells and macromolecules within the vessels.

Only a few of the calculations made in Casley-Smith's study can be compared with measurements made with other techniques. Capillary length and capillary surface area in the intestinal villi were estimated by Biber et al. (45) using an indicator dilution technique and, as is evident from Table 2, the values obtained in the two studies are very similar. However, the flow value calculated by Casley-Smith et al. in the villous tissue is far too low. Using an approach proposed by Intaglietta and Plomb (86), the authors (85) arrive at a flow of 2

Table 2. Capillary length and surface area in the villi as measured by Biber et al. (45) and Casley-Smith et al. (85)[a]

	Biber et al.	Casley-Smith et al.
Capillary length, km/100 g	98	103
Capillary surface area, m^2/100 g	2.45	1.9

[a]Values are given per 100 g of villous tissue.

ml/min×100 g of villous tissue. Assuming that a "resting" blood flow prevailed in these cats, villous blood flow was at least 20 times larger (see Table 1). Furthermore, using the morphological data on fenestrations provided by Casley-Smith (mean diameter = 531 Å; mean depth = 358 Å), it is easy to calculate the time for diffusion for ferritin across these "holes" assuming a restricted diffusion coefficient for ferritin of 0.0085×10^{-5} cm^2 s^{-1} (1/10 of that of serum albumin). With certain assumptions one can then calculate that a 60% diffusion equilibrium would be reached across the endothelium in 10^{-5} s (87). This value may be compared to the transcapillary diffusion time measured by Palade and co-workers (88) on electron micrographs, which was 3–4 min for ferritin. These results obtained by the Palade and the Casley-Smith groups are grossly contradictory and further experiments are needed to resolve the discrepancy.

Johnson and Richardson (89) investigated the effects of venous outflow pressure on the forces involved in the control of transcapillary fluid movement in the canine small bowel. By measuring tissue weight, lymph flow, lymph concentration of protein, and interstitial pressure, these authors were able to estimate interstitial oncotic and hydrostatic pressures. Using earlier estimations of mean capillary hydrostatic pressure obtained during identical experimental conditions, the authors reconstructed the forces in the Starling equilibrium at varying venous outflow pressures. From these considerations Johnson and Richardson concluded that the isogravimetric situation, obtained 10–20 min after an increase of venous pressure, was secondary to a decrease of interstitial oncotic pressure. This is at variance with the views of Wallentin (55), who concluded from plethysmographic experiments that an increasing interstitial pressure counterbalanced the increase of mean capillary hydrostatic pressure caused by the venous pressure increase. No such change was observed by Johnson and Richardson but, unfortunately, most techniques for measuring interstitial pressure are open to criticism. One difficulty with both studies discussed is that they treat the intestine as being a homogeneous tissue while, in reality, it consists of at least three well defined compartments in which the Starling equilibrium may be set at different levels. This is nicely demonstrated by the measurements of Bohlen and Gore (90). They found a mean capillary hydrostatic pressure around 30 mm Hg in the muscularis but a pressure of only 14 mm Hg in the villi.

The precapillary sphincters control the number of capillaries open to blood perfusion. The indicator dilution method was used by Biber et al. (45) to investigate the precapillary sphincter section of the mucosal vascular circuit. It was found that "at rest" only 30–40% of the capillaries in the villi were open for perfusion, while vasodilator drugs (45) and a reduced arterial inflow pressure (46) tended to relax all capillary sphincters and, hence, all the capillaries were perfused by blood.

The capacitance function of the mucosal vasculature was also investigated in the study of Biber et al. (45). The mucosal blood content ranged between 2 and 3 ml/100 g mucosa during rest and increased to about 5 ml/100 g of mucosa during hyperemia or when venous outflow pressure was raised to 25 mm Hg. This small range of mucosal blood content reflects the fact that a considerable part of the total intestinal blood volume is actually contained in the mesenteric veins.

Adrenergic Control of Intestinal Circulation

The extrinsic nervous control of the intestinal vasculature is exerted only by sympathetic vasoconstrictor fibers, since no parasympathetic vasodilator outflow to the small bowel exists (91, 92). The effects of an activation of the regional sympathetic vasoconstrictor fibers on the intestinal vascular bed were studied in detail by Folkow and co-workers (93–95). It was observed that a continuous graded activation of the splanchnic nerves produced constriction of resistance and capacitance vessels as well as precapillary sphincters. Within 2–4 min of the onset of constrictor fiber stimulation, intestinal blood flow increased ("autoregulatory escape from vasoconstrictor fiber influence") reaching a new steady state level moderately below control, while the neurogenic effect on the capacitance vessels and on the precapillary sphincters remained largely unaltered throughout the stimulation period. Mean capillary hydrostatic pressure remained at the control level during the steady state phase of vasoconstriction. Even though blood flow was only slightly reduced during the steady state phase, a pronounced reactive hyperemia was regularly seen upon cessation of the constrictor fiber stimulation.

The autoregulatory escape from vasoconstrictor fiber influence has lately been studied thoroughly and several mechanisms have been proposed to explain this increase of blood flow. It should be underlined, however, that some observations made by Henrich and Lutz (96) seemed to suggest that the autoregulatory escape was more or less absent when the infused dose of norepinephrine was slowly increased, simulating the changes occurring during "physiological" conditions. Similarly, one would assume that the rate of firing in the splanchnic nerves seldom increases from 0 to 8 Hz within a second or two (the experimental range used) in the intact organism. Hence, "autoregulatory escape" is in a certain sense an experimental artifact, and the steady state phase of vasoconstriction may represent a more "physiological" hemodynamic situation.

The original idea proposed by Folkow et al. (93–95) was that the escape was, at least in part, secondary to the opening of submucosal shunts. However, their further studies using ^{86}Rb clearance (58) excluded the possibility that true shunts were involved. The same has been demonstrated more recently (97). Wallentin (98), summarizing the work performed by Folkow and co-workers, concluded that the escape was due to a "physiological" shunting of blood through capillaries, as a consequence of intramural redistribution of blood flow. This view was supported by the investigations of Shepherd et al. (59), who observed a decline of oxygen consumption when stimulating the vasoconstrictor fibers during constant flow conditions. This was interpreted as being the result of a closing of some of the precapillary sphincters. A redistribution of blood flow was also inferred by Richardson and Johnson (99) from the studies of red blood cell velocity in the mesentery during intravascular norephinephrine infusions.

Folkow and co-workers (94, 95), using a rather crude method of India ink injections, obtained evidence that seemed to suggest a redistribution of blood flow mainly from the mucosa towards the submucosa. However, Svanvik (47) proposed that the redistribution of blood flow occurred mainly within the mucosa. This hypothesis, which is based on studies of villous and mucosal flow performed with an indicator dilution technique, implies that the vessels supplying the intestinal crypts, being less sensitive to vasodilating metabolites, are dominated by sympathetic fibers. The tone of villous arterial vessels, on the other hand, is largely determined by the chemical environment surrounding them as they pass between the crypts. Upon nerve stimulation both types of vessels constrict, the escape being explained by the dilatation of the villous vessels when metabolites accumulate in the crypt region. Experimental evidence for such a differentiated response to nervous stimulation was reported by Svanvik (47), who also showed that the rate of ^{85}Kr absorption was not altered by nervous vasoconstriction.

The redistribution hypothesis, as originally proposed by Folkow et al. (93–95), was challenged by Ross (22), who used a tracer fractionation method and failed to obtain any evidence for a decreased mucosal and increased submucosal blood flow, at least during norepinephrine infusion. These findings are, however, consistent with Svanvik's hypothesis. Ross concluded from his experiments that the "autoregulatory escape" was due to an opening of vessels previously constricted by norepinephrine or under vasoconstrictor fiber influence. This explanation was first proposed by Richardson and Johnson (101) comparing the autoregulatory escape with intestinal autoregulation.

Three explanations for such a reopening of previously closed vessels have been proposed. One hypothesis is based on findings reported by Fara and Ross (102, 103). They showed that isolated strips of arteries "escaped" when exposed to norepinephrine. This was demonstrated for several vessels even from vascular beds that do not show any autoregulatory escape from vasoconstrictor fiber stimulation in vivo (e.g., femoral artery) (103). From these experiments, Fara and Ross concluded that the escape may be due to an inherent property of the vascular smooth muscle cells. In line with this, Ross (104) recently proposed that the autoregulatory escape was due to a fading of propagated electrical activity of the vascular smooth muscle as a response to continuous stimuli, i.e., to a mechanism similar to the more general phenomenon of adaptation. Henrich (105) also arrived at the conclusion that the escape from adrenergic influence was, at least in part, a "myogenic" phenomenon but dismissed any involvement of "transmitter release and activation."

Microscopic observations made by Guth and co-workers (106, 107) on the intramural gastric and intestinal vessels were taken as evidence for a behavior of the arterioles in the muscularis and in the submucosa similar to that observed on isolated vessels by Fara and Ross. However, in the studies of Guth et al. one cannot exclude the influence of, for example, tissue metabolites on the vascular smooth muscles.

A second mechanism for autoregulatory escape, proposed by Ross (108) and by Swan and Reynolds (109), is that it is in part mediated via a stimulation of adrenergic β-receptors. This seems to be true for the escape seen during intravascular infusions of norepinephrine. The nervously elicited escape, however, was not influenced significantly by β-blocking drugs (108). This observation strongly suggests that so-called escape reactions seen after different vasoconstrictor agents such as norepinephrine, epinephrine, angiotensin, and prostaglandins (110, 111) are induced by different mechanisms.

Finally, accumulation of metabolites was inferred by Shepherd, Granger, and co-workers (59, 112) and by Svanvik (51) as the causative agent. The first mentioned authors suggested that metabolites, and low tissue pO_2 in particular, were more or less fully responsible for the autoregulatory escape from adrenergic influence. Svanvik (51) discussed the possibility that metabolites, accumulating in the crypt part of the mucosa, were regulating villous blood flow (see above).

In summary, the available experimental evidence strongly suggests that the autoregulatory escape from adrenergic influence is attributable to a relaxation of vessels previously constricted by the adrenergic agent. This vasodilatation is not of the same magnitude in all parts of the intestinal wall and, therefore, the adrenergic agent induces a certain redistribution of intestinal blood flow. The cause of this response is probably multifactorial and it seems probable that the mechanisms underlying a so-called escape reaction to one agent in one vascular bed are not necessarily identical with those causing a similar escape to the same agent in another vascular circuit. Furthermore, the mechanisms of vasoconstrictor escapes induced by different vasoconstrictor substances in one and the same vascular bed may not be identical.

Local and Hormonal Control of Intestinal Blood Flow

The intake of food leads to an increased blood flow in the splanchnic region amounting to 100–200% above control. This has been shown by a large number of workers (for references see Biber (113)), most recently by Norryd, who recorded blood flow in the superior mesenteric artery in the intact human (114). The mechanisms causing the functional hyperemia in the small bowel were until recently largely unknown. Most of the precapillary resistance to flow is situated in the vessels upstream to the mucosa, to judge from intravascular pressure measurements (12, 115), and the absorbing cells of the intestinal epithelium are therefore located relatively far away from these resistance vessels. Hence, the controlling mechanism must by necessity be organized in another manner than, for example, in the skeletal muscle, where locally produced tissue metabolites can exert a direct control of vascular smooth muscle tone due to the close anatomical relationship between the vessels and the skeletal muscle cells. In the intestine two other types of controlling mechanisms that may be involved in its functional hyperemia have been demonstrated. Fara and co-workers (116) showed in an extensive study that the intraduodenal instillation of corn oil, -phenylalanine, or hydrochloric acid, induced after a short latency, a selective increase of pancreatic and jejunal blood flow amounting to 30–40%, due probably to the physiological release of secretin and cholecystokinin (CCK). They also demonstrated that the vasodilatation was not mediated via adrenergic or cholinergic vascular receptors and suggested that the hyperemia was secondary to an increased metabolism. In a subsequent study Fara (117) obtained evidence of a direct effect of secretin and CCK on the vascular smooth muscle cells.

Another mechanism that may be of importance in explaining the functional hyperemia of the gut was proposed by Biber and co-workers (118). They demonstrated that slight mechanical stimulation of the mucosa of a denervated intestinal segment could increase gut blood flow more than 2-fold. This vascular response was not mediated via adrenergic or cholinergic receptors but was blocked by tetrodotoxin and 5-HT receptor blocking agents. From these observations it was concluded that the vascular response was elicited via an intramural nervous reflex arch involving 5-HT receptors (118, 119). The proposal was further substantiated by the finding that one could apparently stimulate the same reflex by applying an electrical field across the intestinal wall (120). The same reflex arch may possibly be involved also in the intestinal vasodilatation observed by Chou et al. (121) when hypertonic glucose was introduced into the intestinal lumen.

The hemodynamic adjustment of the intestinal vascular bed during its functional hyperemia has been the subject of several recent studies. In the investigations of Grim and Lindseth (23) flow distribution was measured in the jejunum and ileum following a meal. Using radioactively labeled glass microspheres they observed a slight blood flow redistribution from the mucosa to the muscularis after feeding. In most other studies, one or several of the underlying mechanisms proposed above have been studied. Thus, Fara and Madden (122) investigated the effects of physiological doses of secretin and CCK on intramural flow distribution in the feline small bowel using carbon microspheres. They observed that CCK increased mucosal blood flow and decreased submucosal blood flow while secretin induced a significant redistribution of blood away from the jejunal mucosa to the submucosa. The latter finding is at variance with the findings of Biber (123), who studied the rate of ^{85}Kr absorption and found that the increase of blood flow induced by CCK, secretin, and electrical field stimulation augmented the rate of krypton absorption. These findings led Biber to conclude that all agents tested increased blood flow in the "absorptive parts" of the mucosa. The discrepancy in results regarding secretin between Fara (122) and Biber (123) may be explained by the inadequacies of the microsphere technique (see above) and/or by the fact that the rate of ^{85}Kr absorption is probably not determined solely by the volume flow of blood, as is discussed below.

The reactions of the consecutive vascular sections during vasodilatation induced by physiological amounts of CCK and secretin and by field stimulation were investigated by Biber et al. (124). The vasodilator effects were mainly on the precapillary resistance vessels while the dilatation of the capacitance vessels was small and possibly only a passive effect of an increased transmural pressure. The CFC increased, reflecting a relaxation of the precapillary sphincters and probably also an increase of the permeability of the intestinal capillaries.

The influsion of pharmacological doses of secretin and CCK increased intestinal blood flow 2–3 times. However, the effects of CCK were only transient, since this drug also caused a strong contraction of the muscular layer and a decrease in total intestinal blood flow below control (125).

The intestine autoregulates its blood flow, i.e., flow tends to remain constant despite variations in arterial inflow pressure. Usually, this phenomenon is discussed with reference to total intestinal blood flow. In a recent report Lundgren and Svanvik (46) measured villous and mucosal flows (126) when changing perfusion pressure. They demonstrated that the autoregulatory ability of the villous vessels was indeed very great, plasma flow in the villi staying more or less constant although perfusion pressure was lowered from 100 to 30 mmg Hg.

Intestinal Blood Flow and Function

Motility is a local mechanism that may interfere mechanically with intestinal blood flow. The results of the studies of the relationship between motility and blood flow are, however, conflicting, as discussed by Jacobson et al. (127). For example, Sidky and Bean (128) and Semba and co-workers (129) reported clear effects on arterial inflow and venous outflow during motility, while Ziegler et al. (130) as well as Geber (131) were unable to detect any effect of motility on flow. It is striking that the two former groups of researchers used isolated segments and in the case of Semba et al. they were quite short (5–12.5 cm), while the two latter groups recorded flow in the whole intestine. This difference in experimental design suggests that blood flow in the vessels passing through the muscular layer was restricted at the site of the motility. This would be an important factor in short segments. However, in long intestinal segments blood flow through a part with no motility could have supplied the whole mucosa with blood via the rich anastomosing vascular network in the submucosa. This conclusion is substantiated by the observation of Kewenter (132) that when large enough doses of acetylcholine were given intra-arterially, motility interfered mechanically with blood flow. From this discussion one may infer that motility does not influence intestinal blood flow to any large extent during "physiological" conditions. However, in certain experimental situations and under the influence of drugs, motility may indeed interfere with flow.

Intestinal secretion is, according to current beliefs, due to active epithelial transport of various solutes. Transcapillary filtration is not considered to contribute to any measurable extent to the production of succus entericus, not even in pathological states such as cholera (133). Hence, blood flow is believed to be of importance for secretion as a deliverer of nutrition and of raw material.

As regards intestinal absorption, its relationship to blood flow is more complex. In an extensive series of experiments, Winne and co-workers have studied this relationship for passively (134–141) as well as actively (142–148) absorbed solutes. The main finding as regards passively absorbed solutes can be summarized as follows: the absorption rate for lipid-soluble compounds varies greatly with blood flow in a more or less linear fashion, while absorption for water-soluble compounds such as urea and erythrose is more or less independent of flow rate. These observations reflect the permeability characteristics of the epithelial lining of the intestinal tract, which is only slightly permeable to water-soluble compounds of molecular weight greater than 100, while lipid-soluble solutes easily traverse the gut epithelium. Hence, for most water-soluble substances (but not water itself) the passage across the epithelium represents the rate-limiting absorptive step. The data from Winne's experiments were fitted into different theoretical models, developed by Winne, with 2–4 different compartments (135).

Another model for the interaction between blood flow and passively absorbed solutes, particularly gases, was proposed by Levitt and Levitt (50). As described above, they followed the relative rates of absorption of different inert gases from the small and large bowel and from the stomach in the awake rat. The results obtained were fitted to a series of different models for the interaction between perfusion and diffusion, and one of these seemed to predict the absorption rates of the gases from all organs investigated. The proposed model implies that there exist two functionally distinct blood supplies. One flow is supposedly close to the lumen, and here the disappearance of the gases is flow limited. The second flow is further away from the lumen, and here rate of gas absorption is diffusion limited. In the small intestine the former flow region dominates, the absorption of gases being almost completely flow limited. As in most models, this model probably represents an oversimplification and the reliability of the proposed model would have been strengthened had the authors demonstrated that the model accurately predicted the results in a hemodynamic situation other than the "resting" state, in, for example, a vasodilatation.

The Levitt model for flow-absorption interaction is rather simple. A much more complicated model was proposed by Lundgren and co-workers (31, 51, 149) who have provided experimental evidence for the existence of a countercurrent exchanger in the intestinal mucosa. The anatomical basis for this mechanism in the cat is shown in Figure 2. Arterial vessels emerge towards the mucosa from the submucosal vascular network, each villus usually being supplied by a single vessel. This arterial vessel (diameter around 20 μm) runs in the central villous core without branching. It loses its muscular coat at the villous base. Close to the villus tip it arborizes into a dense, subepithelial capillary network.

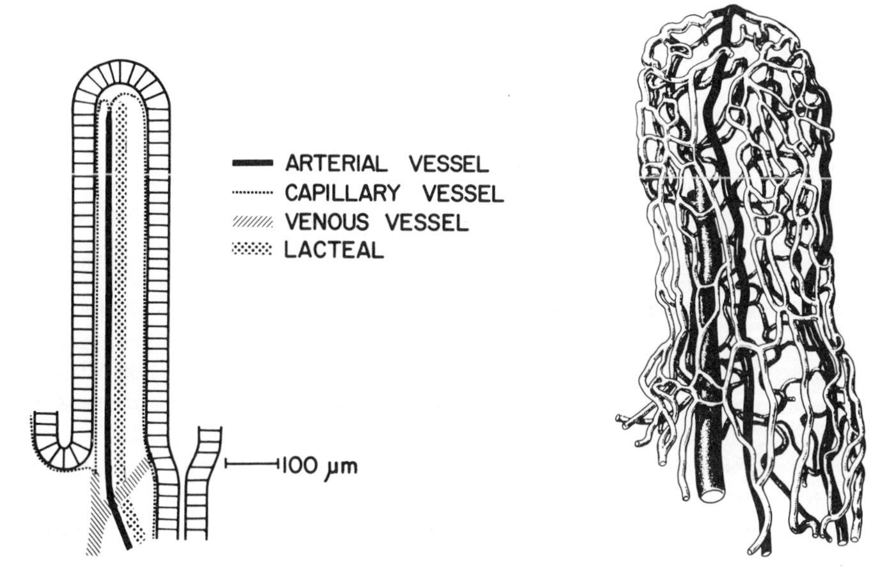

Figure 2. *Left panel,* schematic drawing of the vascular anatomy of a cat villus. The capillary vessels of the figure denote a dense subepithelial capillary network. Note that the ascending arterial vessel and descending capillary network and venous vessel form hairpin vascular loops. *Right panel,* the vascular anatomy of a human villus. (From Spanner (190), by courtesy of Akademische Verlagsgesellschaft, Leipzig.)

The capillaries collect into veins at the villous base. The fact that the vascular arrangement in the human villus is largely the same is also shown in Figure 2.

It is clear from the anatomical description above and from Figure 2 that the main direction of blood flow in the subepithelial capillary network must be opposite to that of the central arterial vessel. Thus, the anatomical prerequisites exist for an intestinal countercurrent exchanger. From a functional point of view it is, however, more interesting to know if the transit time of blood in the hairpin vascular loops of the villus is long enough to allow any significant diffusion of solutes across the 20 μm long intervascular distance. Mean transit time of plasma in the villi is 4–8 s at "rest" and about 1 s at intense vasodilatation (45). Assuming free diffusion in water, it can be calculated that a 75% concentration equilibrium is reached across a 20 μm distance in about 0.1 s for most physiological solutes having a molecular weight less than 1,000. Thus, diffusion across this distance is a fast process compared with the villous transit time of plasma and a considerable intervascular "cross diffusion" of substances may occur, provided that a concentration gradient exists and that vascular permeability to the solute is high. It seems therefore a priori reasonable to assume that some solutes, particularly lipid-soluble solutes, "cross diffuse" rapidly from one limb to the other and, hence, are easily "trapped" in the exchanger.

The experimental support for this hypothesis has been summarized in the theses by Jodal (149), Lundgren (31), and Svanvik (51). The functional implications of the countercurrent exchanger for easily diffusible solutes is shown in Figure 3. It is clear from the left part of Figure 3 that the exchanger may hinder net blood transport of absorbed, easily diffusible solutes. Hence, at first sight such a mechanism may seem unfavorable for the intestinal absorption. The studies of Jodal and Lundgren strongly suggest, however, that the most important function of the exchanger is to act as a countercurrent multiplier creating the hyperosmolar tissue compartment necessary for water absorption (150, 151). This hypothesis was most recently substantiated by the finding that there exists an osmolality gradient in the villous interstitial space, the villous tip having an osmolality of around 1200 mosm/kg of H_2O during absorption from an isotonic saline solution containing glucose (152). A theoretical treatment of the intestinal countercurrent exchanger was recently published (153).

The presence of an intestinal countercurrent exchanger was questioned by Levitt and co-workers (154, 155). Their criticism was partly based on experimental observations made in the rabbit, in which carbon monoxide uptake from the intestine was studied and compared with the absorption rate of other gases (155). This study was based on two assumptions. First, it assumed the Levitt model (50) described above to be true. Second, carbon monoxide was assumed to be so tightly bound to hemoglobin as not to be "trapped" in the exchanger. This assumption may not be true since mean transit time in the intestinal villi and in the mucosa is comparatively long (45).

The discussion so far has mainly centered around the role of blood flow for "passively" absorbed solutes. As regards "actively" transported substances, blood flow not only constitutes the transport vehicle but also delivers the nutrients for epithelial cell metabolism. This complex relationship between flow and "active" absorption is reflected by the varying results obtained by

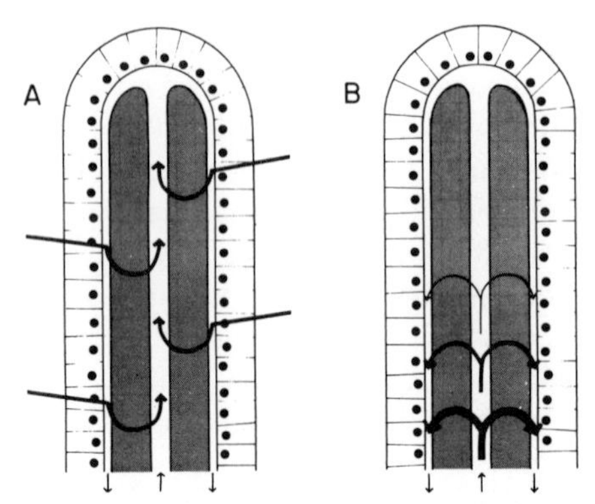

Figure 3. The functional implications of the intestinal countercurrent exchanger schematically illustrated. The intervascular distance is greatly exaggerated for the sake of clarity. (From Lundgren (31), by courtesy of Acta Physiol. Scand.)

different researchers (146–148, 156–158). In fact, in the same laboratory the absorption rate of L-phenylalanine and 3-*O*-methylglucose was affected differently by flow depending on whether flow was increased from a low level or decreased from a high one (146, 147). This observation was apparently explained by the fact that when the experiment was started with a very low blood flow, the intestinal epithelial cells were irreversibly damaged because of tissue hypoxia (148). In this context it should be pointed out that, according to the countercurrent hypothesis, a relative tissue hypoxia exists normally at the villous tips, since oxygen is "shunted" extravascularly in the way depicted in the right panel of Figure 3 (158).

COLONIC CIRCULATION

The circulation of blood through the colon has not received much attention and there are no obvious reasons for this neglect. In fact, it seems rather surprising considering the fairly common pathological conditions of this organ, such as ulcerative colitis, which may have a vascular pathogenesis.

"Resting" colonic blood flow is somewhat lower than that of the small intestine, as has been demonstrated in the dog (160), cat (161), and man (37). The flow distribution between the parallel-coupled vascular sections is largely the same as in the stomach and small bowel, i.e., a major portion of blood flow is diverted to the mucosa (37). The capillary filtration coefficient and the regional blood volume are of similar magnitude in the colon and in the small intestine (161, 162).

The nervous control of the colonic vasculature is very complex, since there are at least four anatomically well defined nerves supplying this organ. Electrical stimulation of the vagal nerves induced no direct effect on the colonic blood flow, while the other parasympathetic supply, the plevic nerve, induced a characteristic response (161). Initially, a pronounced transient vasodilatation was observed, followed by a fluctuating blood flow increase. The latter flow pattern seemed to be associated with the secretion of mucus, since atropine abolished both the secretory response and the blood flow augmentation (161, 162). The initial transient blood flow increase was atropine resistant (161). The observed blood flow increases upon pelvic nerve stimulation were confined to the mucosa, particularly to its most superficial part (36). Bradykinin has been proposed to be a transmitter substance in the parasympathetic outflow to the colon. Intravascular infusions of bradykinin elicited a vascular response that was rather similar to that produced by the pelvic nerves (163, 164).

The response pattern of the series-coupled vascular sections, observed on electrical stimulation of the sympathetic nervous outflow (splanchnic and lumbar colonic nerves) (162), was largely the same as that described above for the stomach and the small intestine. As regards the parallel-coupled vascular sections, it was noted that the mucosal vessels became constricted during nervous activation (36). Hence, the sympathetic response of the mucosal vessels in the colon resembled that observed in the stomach (see above).

The effects of colonic motility on blood flow were investigated by Semba and Fujii (165).

PANCREATIC CIRCULATION

Vascular Dimensions of the Pancreas

The "resting" state of the pancreas is difficult to define, since it has both exocrine and endocrine functions. Thus, its rate of blood flow is dependent on its exocrine secretion, but flow is also related to the glucose level in the blood (166). There is some recent evidence to indicate that the vascular supplies of the different pancreatic tissues are coupled in series in such a way that the exocrine tissue receives its blood supply to a great degree from capillaries that first have passed the islets of Langerhans (8). Usually a "resting" condition implies the fasting state of an animal. In this condition, most estimations of pancreatic blood flow are in the range of 50–130 ml/min×100 g of tissue in the dog, when determined with wash-out methods (39, 40, 42). The recorded elimination curves were usually monoexponential, implying that blood flow in the pancreatic tissue was homogeneous. Direct measurements in cats and dogs have given somewhat lower values in the range of 25–40 ml/min×100 g (167). In these studies it was also demonstrated that the blood flow could be increased 10 times by maximal vasodilatation, indicating a high resting vascular tone of the preparation.

The influence of anesthesia on pancreatic blood flow was demonstrated using the hydrogen gas method which in the conscious dog gave a mean flow value of about 75 ml/min×100 g, whereas in anesthetized preparations this fell to about 40 ml/min×100 g tissue (42). This effect of barbiturate was also corroborated by recent studies in the rat (168).

There are few studies concerning the capillary function in the pancreas. Eliassen and co-workers (167) estimated the resting capillary pressure in the pancreas to be 9–11 mm Hg by the Pappenheimer method. They also measured the CFC in the organ during "rest" and maximal vasodilatation. During "rest" it measured 0.1–0.3 ml/min×mm Hg×100 g, which is 10 times that of skeletal muscle. During maximal vasodilatation induced by intra-articular bradykinin, CFC was estimated to 1.3–1.6 ml/min×mm Hg×100 g, a value almost as large as the glomerular filtration when expressed in a similar fashion. These values clearly demonstrate the high hydraulic conductivity of the fenestrated capillaries in the pancreas.

Adrenergic Control of Pancreatic Circulation

Histochemical and electron microscopic studies of the ganglia and nerves within the pancreas revealed a dense network of cholinergic nerves supplying the acini, the ducts, and the islets of Langerhans (169). No evidence of a cholinergic vasomotor innervation was found (170), while an adrenergic innervation of the vascular smooth muscle was evident (171–173).

Recently Barlow and co-workers (174) investigated the effect of electrical stimulation of the splanchnic nerves on pancreatic blood flow and exocrine secretion. A few minutes of electrical stimulation of the splanchnic nerves were found to reduce pancreatic blood flow as well as its rate of secretion. Upon cessation of stimulation the blood flow increased above the resting level and remained elevated for several minutes. An α-receptor blocking agent blocked the vasoconstrictor phase, while a β-receptor blocking substance abolished the poststimulatory vasodilatation. The latter findings were in agreement with those of an earlier investigation by Barlow et al. (175) on the vascular effects of injected epinephrine and norepinephrine. Barlow and co-workers also demonstrated that when the vasoconstrictor response to splanchnic nerve stimulation was abolished by phenoxybenzamine, nerve stimulation still reduced the rate of secretion suggesting that this inhibitory effect was in part due to a direct nervous action on the secretory cells. The secretion of amylase induced by splanchnic stimulation was abolished by atropine, suggesting the occurrence of cholinergic fibers in the splanchnic nerves.

Marliss and co-workers (176) found a reduction of pancreatic blood flow when stimulating the nerves along the superior pancreatico-duodenal artery for periods of 10 min. Blood flow, however, tended to return to the prestimulatory control level during the stimulation period, suggesting the presence of an "autoregulatory escape from vasoconstrictor fiber influence" also in the pancreatic vascular bed. However, it may also be explained by the unphysiological stimulation frequency used (40 Hz) causing a failure of transmitter release. This suggestion is strengthened by the results reported by Rappaport et al. (177),who did not report any vascular escape during 30-min-long infusions of norepinephrine. Barlow et al. (174), on the other hand, demonstrated an escape when stimulating the sympathetic vasoconstrictor fibers at 2 Hz for about 3 min (see Figure 1 in Barlow's paper).

The effects of intravascularly injected adrenergic drugs on pancreatic blood flow were reported by Gorczynski et al. (178) and Takeuchi et al. (179).

Pancreatic Blood Flow and Exocrine Secretion

There is convincing evidence that both secretin and cholecystokinin (CCK) increase pancreatic blood flow (42, 179–183). It was, however, also demonstrated that this hyperemia was transient (42) and despite a continuous infusion of the drugs, blood flow returned to pre-infusion levels after 20 min in the face of an ongoing external secretion. Vaysse and co-workers (183) found that the secretory response to the hormones was much larger during constant pressure perfusion than when the arterial perfusion pressure was allowed to decrease upon vasodilatation. These observations could not be ascribed to any lack of oxygen during the constant flow experiments since the authors were unable to reveal any constant relationship between secretion rate and oxygen consumption. It was recently suggested that the hormonally induced secretory response, particularly that of secretin, not only was caused by an "active" secretion but also to some extent by a transcapillary filtration of fluid into the pancreatic acini (184). Such a filtration may be induced by an increased mean capillary hydrostatic pressure resulting from a dilatation of the precapillary resistance vessels. This speculative hypothesis would also explain the recent observations that secretion induced by secretin was augmented by infusions of papaverin, which is known to dilate precapillary vessels and thereby increase capillary pressure (185).

Electrical stimulation of the vagal nerve fibers to the pancreas induces an increase of venous outflow from the gland, indicating a fall of the vascular resistance (181, 186–188). Since the nerves innervating the blood vessels seem to be exclusively adrenergic, it is possible that the vagal effects on blood flow are mainly secondary to other effects in the gland. Such effects may be metabolic alterations in the secretory cells and/or the release of kinin-forming enzyme to the plasma to cause vasodilatation in the gland via bradykinin. This latter mechanism was proposed by Hilton and Jones (187), who found kallikrein in the pancreatic juice of the cat as well as a release of kinin-forming enzyme into the oxygenated Locke's solution perfusing the gland. This release was found to be increased by acetylcholine or CCK, substances which also increased blood flow through the organ in situ, suggesting the kallikrein mechanism to be a link between secretion and blood flow. Since the hyperemic response to acetylcholine (185) and to vagal stimulation (187) was abolished in the cat by atropine, a cholinergic receptor seemed to be involved. On the other hand, stimulation of the vagal nerve in the pig caused an intense atropine-resistant vasodilatation preceding the secretion in the pancreas (181).

Pancreatic Blood Flow and Endocrine Secretion

Fujita and Murakami (8), studying vascular casts of the microvasculature with a scanning electron microscope, described a portal system in the pancreas of the monkey. The intralobular arteries were found to branch into arterioles supplying the islets of Langerhans with capillaries that first passed the α-cells, then the β-cells, and then further into the exocrine tissue. According to these authors, it seemed that the exocrine pancreas received its blood supply to a great degree, if not exclusively, from the capillary network of the islets. The arrangement was referred to as a "insulo-acinar portal system" and it suggests interesting, but as yet unknown, connections between endocrine function and pancreatic blood flow.

A number of studies fail to show any consistent relationship between pancreatic blood flow and secretion of insulin and glucagon. Hence, Rappaport and co-workers (189) reported a concomitant increase of pancreatic blood flow and insulin output after intravenous glucose infusions. They also showed a parallel reduction of blood flow and insulin output during intravenous norepinephrine infusions, while the insulin output did not decrease when the blood flow was reduced by partially occluding the artery (177). Kaneto and co-workers (188) demonstrated that electrical stimulation of the dorsal vagus nerve of the dog increased blood flow as well as the output of insulin and glucagon. Atropine abolished all these effects. Glucagon was also released upon activation of the sympathetic fibers, concomitant with a reduction of blood flow (176).

ACKNOWLEDGMENTS

The authors are grateful to Doctors G. Bohlen, J. Fara, R. Gore, P. Guth, G. Ross, and E. Smith for preprints of their manuscripts in press.

REFERENCES

1. Heller, A. (1872). Ueber die Blutgefässe des Dünndarmes. Ber. Sächs. Ges. Wiss. 24:165.
2. Mall, J. P. (1888). Die Blut- und Lymphwege im Dünndarm des Hundes. Abh. Sächs. Ges. Wiss. 14:153.
3. Patzelt, V. (1936). Der Darm. Handb. Mikr. Anat. 5: 3.Teil, 1. Verlag von Julius Spring, Berlin.
4. Reynolds, D. G., Brim, J., and Sheehy, T. W. (1967). The vascular architecture of the small intestinal mucosa of the monkey. Anat. Rec. 159:211.
5. Grayson, J. (1974). The gastrointestinal circulation. *In* E. D. Jacobson and L. L. Shanbour (eds.), Gastrointestinal Physiology, pp. 105–138. University Park Press, Baltimore.
6. Mellander, S. (1960). Comparative studies on the adrenergic neuro-hormonal control of resistance and capacitance blood vessels in the cat. Acta Physiol. Scand. 50:176 (Suppl.), 1.
7. Folkow, B. (1967). Regional adjustments of intestinal blood flow. Gastroenterology 52:423.
8. Fujita, T., and Murakami, T. (1973). Microcirculation of monkey pancreas with special reference to the insulo-acinar portal system. A scanning electron microscope study of vascular casts. Arch. Histol. Jpn. 35:255.
9. Folkow, B. (1952). A critical study of some methods used in investigations on the blood circulation. Acta Physiol. Scand. 27:10.
10. Norryd, C. (1974). A study of superior mesenteric blood flow in man. Illustration Department, University Hospital, Lund.
11. Strandell, T., Erwald, R., Kulling, K. G., Lundbergh, P., Marions, O., and Wiechel, K.-L. (1973). Measurement of dual hepatic blood flow in awake patients. J. Appl. Physiol. 35:755.
12. Gore, R. W., and Bohlen, H. G. (1975). Pressure regulation in the microcirculation. Fed. Proc. 34:2031.
13. Jacobson, E. D., Linford, R. H., and Grossman, M. I. (1966). Gastric secretion in relation to mucosal blood flow studied by a clearance technic. J. Clin. Invest. 45:1.
14. Curwain, B. P., and Holton, P. (1973). The measurement of dog gastric mucosal blood flow by radioactive aniline clearance compared with amidopyrine clearance. J. Physiol. (Lond.) 229:115.
15. Aures, D., Guth, P. H., and Grossman, M. I. (1975). Use of neutral red to measure gastric mucosal blood flow. Gastroenterology 68:1057.
16. Tague, L. L., and Jacobson, E. D. (1975). Radiometric aminopyrine clearance technic for determining gastric mucosal blood flow. Gastroenterology 68:995.
17. Sapirstein, L. A. (1958). Regional blood flow by fractional distribution of indicators. Am. J. Physiol. 193:161.

18. Delaney, J. P., and Custer, J. (1965). Gastrointestinal blood flow in the dog. Circ. Res. 17:394.
19. Csernay, L., Wolf, F., and Varró, V. (1965). Der Kreislaufgradient im Dünndarm. Z. Gastroenterol. 5:261.
20. Delaney, J. P., and Grim, E. (1964). Canine gastric blood flow and its distribution. Am. J. Physiol. 207:1195.
21. Delaney, J. P., and Grim, E. (1965). Experimentally induced variations in canine gastric blood flow and its distribution. Am. J. Physiol. 208:353.
22. Ross, G. (1971). Effects of norepinephrine infusions on mesenteric arterial blood flow and its tissue distribution. Proc. Soc. Exp. Biol. Med. 137:921.
23. Grim, E., and Lindseth, E. O. (1958). Distribution of blood flow to the tissues of the small intestine of the dog. Univ. Minn. Med. Bull. 30:138.
24. Greenway, C. V., and Murthy, V. S. (1972). Effects of vasopressin and isoprenaline infusions on the distribution of blood flow in the intestine: criteria for the validity of microsphere studies. Br. J. Pharmacol. 46:177.
25. Fara, J. W. and Madden, K. S. (1975). Effect of secretin and cholecystokinin on small intestinal blood flow distribution. Am. J. Physiol. 229:1365.
26. Kety, S. S. (1951). The theory and applications of the exchange of inert gas at the lungs and tissues. Pharmacol. Rev. 3:1.
27. Zierler, K. L. (1965). Equations for measuring blood flow by external monitoring of radioisotopes. Circ. Res. 16:309.
28. Bell, P. R. F., Battersby, C., and Harper, A. M. (1967). Gastric mucosal blood-flow in the dog measured by clearance of krypton 85. The response to histamine. Br. J. Surg. 54:1003.
29. Bell, P. R. F., and Battersby, C. (1968). Effect of vagotomy on gastric mucosal blood flow. Gastroenterology 54:1032.
30. Bell, P. R. F., and Battersby, A. C. (1967). The effect of arterial pCO_2 on gastric mucosal blood flow measured by clearance of Kr^{85}. Surgery 62:468.
31. Lundgren, O. (1967). Studies on blood flow distribution and countercurrent exchange in the small intestine. Acta Physiol. Scand. 303 (suppl.):1.
32. Kampp, M., Lundgren, O., and Sjöstrand, J. (1968). On the components of the Kr^{85} wash-out curves from the small intestine of the cat. Acta Physiol. Scand. 72:257.
33. Kampp, M., and Lundgren, O. (1968). Blood flow and flow distribution in the small intestine of the cat as analysed by the Kr^{85} wash-out technique. Acta Physiol. Scand. 72:282.
34. Hultén, L., Jodal, M., Lindhagen, J., and Lundgren, O. (1976). Blood flow in the small intestine of cat and man as analyzed by an inert gas wash-out technique. Gastroenterology 70:45.
35. Norris, H. T., and Sumner, D. S. (1974). Distribution of blood flow to the layers of the small bowel in experimental cholera. Gastroenterology 66:973.
36. Hultén, L., Jodal, M., and Lundgren, O. (1969). Nervous control of blood flow in the parallel-coupled vascular sections of the colon. Acta Physiol. Scand. 335 (suppl.):65.
37. Hultén, L., Jodal, M., Lindhagen, J., and Lundgren, O. (1937). Colonic blood flow in cat and man as analyzed by an inert gas wash-out technique. Gastroenterology 70:36.
38. Billaudel, B., Morel, B., Valeyre, J., and Sutter, B. Ch. J. (1972). Mesure du débit artériel pancréatique du rat á l'aide du 133 Xénon. J. Physiol. (Paris) 65:201A.
39. Glazier, G., and Needham, T. (1974). ^{133} Xenon clearance for repeated assessment of pancreatic blood flow in the anaethetized dog. J. Physiol. (Lond.) 240:32P.
40. Ercan, T., Bor, N. M., Bekdik, C. F., and Öner, G. (1974). Measurement of pancreatic blood flow in dog by ^{133}Xe clearance technique. Pflügers Arch. 348:51.
41. Aukland, K., Bower, B. F., and Berliner, R. N. (1964). Measurement of local blood flow with hydrogen gas. Circ. Res. 14:164.
42. Aune, S., and Semb, L. S. (1969). The effect of secretin and pancreozymin on pancreatic blood flow in the conscious and anesthetized dog. Acta Physiol. Scand. 76:406.
43. Biber, B., Lundgren, O., Stage, L., and Svanvik, J. (1973). An indicator dilution method for studying intestinal hemodynamics in the cat. Acta Physiol. Scand. 87:433.

44. Wolgast, M. (1968). Studies on the regional renal blood flow with P^{32}-labelled red cells and small beta-sensitive semiconductor detectors. Acta Physiol. Scand. 313 (suppl.):1.
45. Biber, B., Lundgren, O., and Svanvik, J. (1973). Intramural blood flow and blood volume in the small intestine of the cat as analyzed by an indicator dilution technique. Acta Physiol. Scand. 87:391.
46. Lundgren, O., and Svanvik, J. (1973). Mucosal hemodynamics in the small intestine of the cat during reduced perfusion pressure. Acta Physiol. Scand. 88:551.
47. Svanvik, J. (1973). Mucosal hemodynamics in the small intestine of the cat during regional sympathetic vasoconstrictor activation. Acta Physiol. Scand. 89:19.
48. Forster, R. E. (1967). Measurement of gastrointestinal blood flow by means of gas absorption. Gastroenterology 52:381.
49. Coburn, R. F. (1968). Carbon monoxide uptake in the gut. Ann. N.Y. Acad. Sci. 150:13.
50. Levitt, M. D., and Levitt, D. G. (1973). Use of inert gases to study the interaction of blood flow and diffusion during passive absorption from the gastrointestinal tract of the rat. J. Clin. Invest. 52:1852.
51. Svanvik, J. (1973). Mucosal blood circulation and its influence on passive absorption in the small intestine. An experimental study in the cat. Acta Physiol. Scand. 385 (suppl.):1.
52. Hamilton, J. D., Dawson, A. M., and Webb, J. (1967). Limitation of the use of inert gases in the measurement of small gut mucosal blood flow. Gut 8:509.
53. Folkow, B., Lundgren, O., and Wallentin, I. (1963). Studies on the relationship between flow resistance, capillary filtration coefficient and regional blood volume in the intestine of the cat. Acta Physiol. Scand. 57:270.
54. Johnson, P. C., and Selkurt, E. E. (1958). Intestinal weight changes in hemorrhagic shock. Am. J. Physiol. 193:135.
55. Wallentin, I. (1966). Importance of tissue pressure for the fluid equilibrium between the vascular and interstitial compartments in the small intestine. Acta Physiol. Scand. 68:304.
56. Henrich, H., and Singbartl, G. (1973). Vascular adjustments in dilatory reactions. Angiologica 10:185.
57. Renkin, E. M. (1959). Transport of potassium-42 from blood to tissue in isolated mammalian skeletal muscles. Am. J. Physiol. 197:1205.
58. Dresel, P., Folkow, B., and Wallentin, I. (1966). $Rubidium^{86}$ clearance during neurogenic redistribution of intestinal blood flow. Acta Physiol. Scand. 67:173.
59. Shepherd, A. P., Mailman, D., Burks, T. F., and Granger, H. J. (1973). Effects of norepinephrine and sympathetic stimulation on extraction of oxygen and ^{86}Rb in perfused canine small bowel. Circ. Res. 33:166.
60. Alvarez, O. A., and Yudilevich, D. L. (1967). Capillary permeability and tissue exchange of water and electrolytes in the stomach. Am. J. Physiol. 213:315.
61. Jacobson, E. D. (1965). The circulation of the stomach. Gastroenterology 48:85.
62. Bruggeman, T. M. (1975). Plasma proteins in canine gastric lymph. Gastroenterology 68:1204.
63. Munro, D. R. (1974). Route of protein loss during a model protein-losing gastropathy in dogs. Gastroenterology 66:960.
64. Jansson, G., Lundgren, O., and Martinson, J. (1970). Neurohormonal control of gastric blood flow. Gastroenterology 58:425.
65. Fasth, S., and Martinson, J. (1973). On the possible role of bradykinin in functional hyperemia of cat's stomach. Acta Physiol. Scand. 89:334.
66. Haglund, U., and Lundgren, O. (1972). Reactions within consecutive vascular sections of the small intestine of the cat during prolonged hypotension. Acta Physiol. Scand. 84:151.
67. Grund, E. R., Reed, J. D., and Sanders, D. J. (1975). The effect of sympathetic nerve stimulation on acid secretion, regional blood flows and oxygen usage by stomachs of anaesthetized cats. J. Physiol. (Lond.) 248:639.
68. Reed, J. D., Sanders, D. J., and Thorpe, V. (1971). The effect of splanchnic nerve stimulation on gastric acid secretion and mucosal blood flow in the anaesthetized cat. J. Physiol. (Lond.) 214:1.

69. Reed, J. D., and Sanders, D. J. (1971). Pepsin secretion, gastric motility and mucosal blood flow in the anaesthetized cat. J. Physiol. (Lond.) 216:159.
70. Reed, J. D., and Sanders, D. J. (1971). Splanchnic nerve inhibition of gastric acid secretion and mucosal blood flow in anaesthetized cats. J. Physiol. (Lond.) 219:555.
71. Jacobson, E. D., Swan, K. G., and Grossman, M. I. (1967). Blood flow and secretion in the stomach. Gastroenterology 52:414.
72. Bynum, T. E., and Jacobson, E. D. (1971). Blood flow and gastrointestinal function. Gastroenterology 60:325.
73. Lanciault, G., Shaw, J. E., Urquhart, J., Adair, L. S., and Brooks, F. P. (1975). Response of the isolated perfused stomach of the dog to electrical vagal stimulation. Gastroenterology 68:294.
74. Lin, T-M., and Warrick, M. W. (1971). Effect of glucagon on pentagastrin-induced gastric acid secretion and mucosal blood flow in the dog. Gastroenterology 61:328.
75. Curwain, B. P., and Holton, P. (1972). The effects of isoprenaline and noradrenaline on pentagastrin-stimulated gastric acid secretion and mucosal blood flow in the dog. Br. J. Pharmacol. 46:225.
76. Wilson, D. E., Ginsberg, B., Levine, R. A., and Washington, A. (1972). Effect of glucagon on histamine- and pentagastrin-stimulated canine gastric acid secretion and mucosal blood flow. Gastroenterology 63:45.
77. Lin, T-M., and Evans, D. C. (1973). Effect of propranolol on pentagastrin-induced HCl secretion and gastric mucosal blood flow in dogs. Gastroenterology 64:1126.
78. Main, I. H. M., and Whittle, B. J. R. (1973). Gastric mucosal blood flow during pentagastrin- and histamine-stimulated acid secretion in the rat. Br. J. Pharmacol. 49:534.
79. Öztürkcan, O., de Saint Blanquat, G., and Derache, R. (1972). Le flux sanguin de la muqueuse gastrique chez le rat. Effet de l'histamine. J. Physiol. (Paris) 64:355.
80. Wilson, D. E., and Levine, R. A. (1972). The effect of prostaglandin E_1 on canine gastric acid secretion and gastric mucosal blood flow. Am. J. Dig. Dis. 17:527.
81. Semba, T., Fujii, K., and Fujii, Y. (1970). Influence of peristaltic contraction of the stomach on blood flow through the gastrosplenic vein. Hiroshima J. Med. Sci. 19:87.
82. Baez, S. (1959). Microcirculation in the intramural vessels of the small intestine in the rat. *In* S. R. M. Reynolds and B. W. Zweifach (eds.), The Microcirculation, pp. 114–128. The University of Illinois Press, Urbana.
83. Jodal, M., and Lundgren, O. (1970). Plasma skimming in the intestinal tract. Acta Physiol. Scand. 80:50.
84. Källskog, Ö., Ulfendahl, H. R., and Wolgast, M. (1972). Single glomerular blood flow as measured with carbonized 141-Ce labelled microspheres. Acta Physiol. Scand. 85:408.
85. Casley-Smith, J. R., Donoghue, P. J., and Crocker, K. W. J. (1975). The quantitative relationships between fenestrae in jejunal capillaries and connective tissue channels: Proof of "tunnel-capillaries." Microvasc. Res. 9:78.
86. Intaglietta, M., and de Plomb, E. P. (1973). Fluid exchange in tunnel and tube capillaries. Microvasc. Res. 6:153.
87. Davson, H. (1970). A Textbook of General Physiology. 4th Ed., Vol. I, p. 397. J. & A. Churchill, London.
88. Clementi, F., and Palade, G. E. (1969). Intestinal capillaries. I. Permeability to peroxidase and ferritin. J. Cell Biol. 41:33.
89. Johnson, P. C., and Richardson, D. R. (1974). The influence of venous pressure on filtration forces in the intestine. Microvasc. Res. 7:296.
90. Bohlen, H. G., and Gore, R. W. (1975). Direct measurement of microvascular pressures in innervated rat intestinal muscle and mucosa. Presented at the 1st World Congress for Microcirculation, June 15–20, Toronto.
91. Kewenter, J. (1965). The vagal control of the jejunal and ileal motility and blood flow. Acta Physiol. Scand. 65:251 (suppl.), 1.
92. Ross, G. (1973). Vascular effects of periarterial mesenteric nerve stimulation after adrenergic neurone blockade. Experientia 29:289.
93. Folkow, B., Lewis, D. H., Lundgren, O., Mellander, S., and Wallentin, I. (1964). The effect of graded vasoconstrictor fibre stimulation on the intestinal resistance and capacitance vessels. Acta Physiol. Scand. 61:445.

94. Folkow, B., Lewis, D. H., Lundgren, O., Mellander, S., and Wallentin, I. (1964). The effect of the sympathetic vasoconstrictor fibres on the distribution of capillary blood flow in the intestine. Acta Physiol. Scand. 61:458.
95. Cobbold, A., Folkow, B., Lundgren, O., and Wallentin, I. (1964). Blood flow, capillary filtration coefficients and regional blood volume responses in the intestine of the cat during stimulation of the hypothalamic defence area. Acta Physiol. Scand. 61:467.
96. Henrich, H., and Lutz, J. (1971). Das Vasculäre Escape-Phänomen am Intestinalkreislauf und seine Auslösung durch unterschiedliche vasoconstrictorische Substanzen. Pflügers Arch. 329:82.
97. Tkachenko, B. I., Medvedeva, N. Ja., and Pozdnjakov, P. K. (1974). The character of the capacitance vessel responses in the spleen and intestine under electrical stimulation of the sympathetic fibres. Experientia 30:1413.
98. Wallentin, I. (1966). Studies on intestinal circulation. Acta Physiol. Scand. 69:279 (suppl.), 1.
99. Richardson, D. R., and Johnson, P. C. (1970). Changes in mesenteric capillary flow during norepinephrine infusion. Am. J. Physiol. 219:1317.
100. Svanvik, J. (1973). The effect of reduced perfusion pressure and regional sympathetic vasoconstrictor activation on the rate of absorption of ^{85}Kr from the small intestine of the cat. Acta Physiol. Scand. 89:239.
101. Richardson, D. R., and Johnson, P. C. (1969). Comparison of autoregulatory escape and autoregulation in the intestinal vascular bed. Am. J. Physiol. 217:586.
102. Fara, J. W. (1971). Escape from tension induced by noradrenaline or electrical stimulation in isolated mesenteric arteries. Br. J. Pharmacol. 43:865.
103. Fara, J. W., and Ross, G. (1972). Escape from drug-induced constriction of isolated arterial segments from various vascular beds. Angiologica 9:27.
104. Ross, G. (1975). Norepinephrine vasoconstrictor escape in isolated mesenteric arteries. Am. J. Physiol. 228:1652.
105. Henrich, H. (1973). Adjustment behaviour of adrenergic-induced vasoconstrictions in the intestinal circulation of the cat. Angiologica 10:233.
106. Guth, P. H., and Smith, E. (1975). Escape from vasoconstriction in the gastric microcirculation. Am. J. Physiol. 228:1893.
107. Guth, P. H., Ross, G., and Smith, E. Mechanism of escape from norepinephrine in the intestinal circulation. Am. J. Physiol. In press.
108. Ross, G. (1971). Escape of mesenteric vessels from adrenergic and nonadrenergic vasoconstriction. Am. J. Physiol. 221:1217.
109. Swan, K. G., and Reynolds, D. G. (1971). Adrenergic mechanisms in canine mesenteric circulation. Am. J. Physiol. 220:1779.
110. Dresel, P., and Wallentin, I. (1966). Effects of sympathetic vasoconstrictor fibres, noradrenaline and vasopressin on the intestinal vascular resistance during constant blood flow or blood pressure. Acta Physiol. Scand. 66:427.
111. Shehadeh, Z., Price, W. E., and Jacobson, E. D. (1969). Effects of vasoactive agents on intestinal blood flow and motility in the dog. Am. J. Physiol. 216:386.
112. Shepherd, A. P., and Granger, H. J. (1973). Autoregulatory escape in the gut: a systems analysis. Gastroenterology 65:77.
113. Biber, B. (1973). Vasodilator mechanisms in the small intestine. Acta Physiol. Scand. 401 (Suppl.):1.
114. Norryd, C., Dencker, H., Lunderquist, A., Olin, T., and Tylén, U. (1975). Superior mesenteric blood flow during digestion in man. Acta Chir. Scand. 141:197.
115. Dieckhoff, D. (1973). Anteil der in Dünndarmmesenterium und Dünndarmwand gelegenen Gefässabschnitte am gesamten Strömungswiderstand der Intestinalgefässe zwischen Aorta abdominalis und Vena cava inferior. Z. Gastroenterol. 11:29.
116. Fara, J. W., Rubinstein, E. H., and Sonnenschein, R. R. (1972). Intestinal hormones in mesenteric vasodilation after intraduodenal agents. Am. J. Physiol. 223:1058.
117. Fara, J. W. (1975). Effects of gastrointestinal hormones on vascular smooth muscle. Am. J. Dig. Dis. 20:346.
118. Biber, B., Lundgren, O., and Svanvik, J. (1971). Studies on the intestinal vasodilatation observed after mechanical stimulation of the mucosa of the gut. Acta Physiol. Scand. 82:177.

119. Biber, B., Fara, J., and Lundgren, O. (1974). A pharmacological study of intestinal vasodilator mechanisms in the cat. Acta Physiol. Scand. 90:673.
120. Biber, B., Fara, J., and Lundgren, O. (1973). Intestinal vasodilatation in response to transmural electrical field stimulation. Acta Physiol. Scand. 87:277.
121. Chou, C. C., Burns, T. D., Hsieh, C. P., and Dabney, J. M. (1972). Mechanisms of local vasodilation with hypertonic glucose in the jejunum. Surgery 71:380.
122. Fara, J. W., and Madden, K. S. (1975). Effect of secretin and cholecystokinin on small intestinal blood flow distribution. Am. J. Physiol. 229:1365.
123. Biber, B. (1974). The effects of intestinal vasodilator mechanisms on the rate of ^{85}Kr absorption in the cat. Acta Physiol. Scand. 90:578.
124. Biber, B., Fara, J., and Lundgren, O. (1973). Vascular reactions in the small intestine during vasodilatation. Acta Physiol. Scand. 89:449.
125. Fasth, S., Filipsson, S., Hultén, L., and Martinson, J. (1973). The effect of the gastrointestinal hormones on small intestinal motility and blood flow. Experientia 29:982.
126. Wilson, S. E., Hiatt, J., Winston, M. and Passaro, E. (1975). Intestinal blood flow. An evaluation by clearance of xenon Xe^{133} from the canine jejunum. Arch. Surg. 110:797.
127. Jacobson, E. D., Brobmann, G. F., and Brecher, G. A. (1970). Intestinal motor activity and blood flow. Gastroenterology 58:575.
128. Sidky, M. and Bean, J. W. (1958). Influence of rhythmic and tonic contraction of intestinal muscle on blood flow and blood reservoir capacity in dog intestine. Am. J. Physiol. 193:386.
129. Semba, T., Fujii, K., and Mizonishi, T. (1973). Relation of intestinal motility to venous outflow and saturation of blood O_2 through mesenteric blood vessels. Jpn. J. Physiol. 23:541.
130. Zeigler, M. G., Barton, R. W., and Swan, K. G. (1973). Mesenteric blood flow and small intestinal motility in the dog. Surgery 73:649.
131. Geber, W. F. (1965). Intestinal blood flow pressure responses during control and induced peristalsis. Arch. Int. Pharmacodyn. Ther. 157:53.
132. Kewenter, J. (1971). Effects of graded acetylcholine infusions on intestinal motility, volume and blood flow. Scand. J. Gastroenterol. 6:435.
133. Field, M. (1974). Intestinal secretion. Gastroenterology 66:1063.
134. Winne, D. (1966). Der Einfluss einiger Pharmaka auf die Darmdurchblutung und die Resorption tritiummarkierten Wassers aus dem Dünndarm der Ratte. Naunyn-Schmiedebergs Arch. Pharmacol. 254:199.
135. Winne, D., and Ochsenfahrt, H. (1967). Die formale Kinetik der Resorption unter Berücksichtigung der Darmdurchblutung. J. Theor. Biol. 14:293.
136. Ochsenfahrt, H. und Winne, D. (1969). Der Einfluss der Durchblutung auf die Resorption von Arzneimittlen aus dem Jejunum der Ratte. Naunyn-Schmidebergs Arch. Pharmacol. 264:55.
137. Winne, D. (1971). Die Bedeutung der Blutdränage in der Pharmakokinetik der enteralen Resorption. Med. Welt. 22:632.
138. Winne, D., and Remischovsky, J. (1971). Der Einfluss der Durchblutung auf die Resorption von Harnstoff, Methanol und Athanol aus dem Jejunum der Ratte. Naunyn-Schmidebergs Arch. Pharmacol. 268:392.
139. Winne, D. (1971). Die Pharmakokinetik der Resorption bei Perfusion einer Darmschlinge mit variabler Durchblutung. Naunyn-Schmiedebergs Arch. Pharmacol. 268:417.
140. Winne, D., and Remischovsky, J. (1971). Der Einfluss der Durchblutung auf die Resorption von Polyalkoholen aus dem Jejunum der Ratte. Naunyn-Schmidebergs Arch. Pharmacol. 270:22.
141. Winne, D. (1971). Durchblutung und enterale Resorption. Z. Gastroenterol. 9:429.
142. Winne, D. (1970). Der Einfluss der Durchblutung auf die Wasser- und Salzresorption im Jejunum der Ratte. Naunyn-Schmidebergs Arch. Pharmacol. 265:425.
143. Winne, D. (1970). Formal kinetics of water and solute absorption with regard to intestinal blood flow. J. Ther. Biol. 27:1.
144. Winne, D. (1972). The influence of blood flow and water net flux on the blood-to-

lumen flux of tritiated water in the jejunum of the rat. Naunyn-Schmidebergs Arch. Pharmacol. 274:357.
145. Winne, D. (1972). The influence of blood flow and water net flux on the absorption of tritiated water from the jejunum of the rat. Naunyn-Schmidebergs Arch. Pharmacol. 272:417.
146. Winne, D. (1973). The influence of blood flow on the absorption of L- and D-phenylalanine from the jejunum of the rat. Naunyn-Schmidebergs Arch. Pharmacol. 277: 113.
147. Lichtenstein, B., and Winne, D. (1973). The influence of blood flow on the absorption of 3-*O*-methylglucose from the jejunum of the rat. Naunyn-Schmidebergs Arch. Pharmacol. 279:153.
148. Lichtenstein, B., and Winne, D. (1974). The influence of blood flow on the phlorizine-insensitive and sensitive galactose absorption in rat jejunum. Naunyn-Schmidebergs Arch. Pharmacol. 282:195.
149. Jodal, M. (1973). The Significance of the Intestinal Countercurrent Exchanger for the Absorption of Sodium and Fatty Acids. Gotab, Göteborg.
150. Haljamäe, H., Jodal, M., and Lundgren, O. (1973). Countercurrent multiplication of sodium in intestinal villi during absorption of sodium chloride. Acta Physiol. Scand. 89:580.
151. Jodal, M. (1974). An autoradiographic study of the intestinal absorption of ^{22}Na. Acta Physiol. Scand. 90:79.
152. Jodal, M., and Lundgren, O. (1975). Demonstration of tissue hyperosmolality in the tips of intestinal villi during sodium chloride absorption. Acta Physiol. Scand. 95:47A.
153. Winne, D. (1975). The influence of villous countercurrent exchange on intestinal absorption. J. Theor. Biol. 53:145.
154. Levitt, M. D., Bond, J. H., and Levitt, D. G. (1974). Does countercurrent exchange influence small bowel function? Am. J. Dig. Dis. 19:771.
155. Bond, J. H., Levitt, D. G., and Levitt, M. D. (1974). Use of inert gases and carbon monoxide to study the possible influence of countercurrent exchange on passive absorption from the small bowel. J. Clin. Invest. 54:1259.
156. Williams, J., Mager, M., and Jacobson, E. D. (1964). Relationship of mesenteric blood flow to intestinal absorption of carbohydrates. J. Lab. Clin. Med. 63:853.
157. Varró, V., Blahó, G., Csernay, L., Jung, I., and Szarvas, F. (1965). Effect of decreased local circulation on the absorptive capacity of a small intestine loop in the dog. Am. J. Dig. Dis. 10:170.
158. Pytkowski, B., and Lewartowski, B. (1972). Motility dependent absorption of amino acids in canine small intestine segment hemoperfused in vitro. Pflügers Arch. 335: 125.
159. Kampp, M., Lundgren, O., and Nilsson, N. J. (1963). Extravascular shunting of oxygen in the small intestine of the cat. Acta Physiol. Scand. 72:396.
160. Geber, W. F. (1960). Quantitative measurement of bloodflow in various areas of small and large intestine. Am. J. Physiol. 198:985.
161. Hultén, L., Jodal, M., and Lundgren, O. (1969). Extrinsic nervous control of colonic blood flow. Acta Physiol. Scand. 335 (suppl.):39.
162. Hultén, L., Jodal, M., and Lundgren, O. (1969). Local and nervous control of the consecutive vascular sections of the colon. Acta Physiol. Scand. 335 (suppl.):51.
163. Fasth, S., and Hultén, L. (1973). The effect of bradykinin of intestinal motility and blood flow. Acta Chir. Scand. 139:699.
164. Fasth, S., and Hultén, L. (1973). The effect of bradykinin on the consecutive vascular sections of the small and large intestine. Acta Chir. Scand. 139:707.
165. Semba, T., and Fujii, Y. (1970). Relationship between venous flow and colonic peristalsis. Jpn. J. Physiol. 20:408.
166. Fischer, V., Hommel, H., and Schmid, E. (1973). Continuous registration of the pancreatic blood flow after intravenous application of glucose. Experientia 29:884.
167. Eliassen, E., Folkow, B., and Hilton, S. (1973). Blood flow and capillary filtration capacities in salivary and pancreatic glands as compared with skeletal muscle. Acta Physiol. Scand. 87:11A.

168. Ohnhaus, E. E. (1973). The distribution of Rubidium[86] in the splancnic area during and after phenobarbitone administration. Arzneim. Forsch. 23:1339.
169. Lenninger, S. (1974). The autonomic innervation of the exocrine pancreas. Med. Clin. North Am. 58:1311.
170. Graham, J. D. P., Lever, J. D., and Spriggs, T. L. B. (1968). An examination of adrenergic axons around pancreatic arterioles of the cat for the presence of acetylcholinesterase by high resolution autoradiographic and histochemical methods. Br. J. Pharmacol. Chemother. 33:15.
171. Alm, P., Legrell, L., and Ehinger, B. (1967). Remarkable adrenergic nerves in the exocrine pancreas. Z. Zellf. 83:178.
172. Lever, J. D., Spriggs, T. L. B., and Graham, J. D. P. (1968). A formol-fluorescence, fine-structural and autoradiographic study of the adrenergic innervation of the vascular tree in the intact and sympathectomized pancreas of the cat. J. Anat. (Lond.) 103:15.
173. Zelander, T., Ekholm, R., and Edlund, Y. (1962). The ultrastructural organization of the rat exocrine pancreas. J. Ultrastruct. Res. 7:84.
174. Barlow, T. E., Greenwell, J. R., Harper, A. A., and Scratcherd, T. (1974). The influence of the splanchnic nerves on the external secretion, blood flow and electrical conductance of the cat pancreas. J. Physiol. (Lond.) 236:421.
175. Barlow, T. E., Greenwell, J. R., Harper, A. A., and Scratcherd, T. (1971). The effect of adrenaline and noradrenaline on the blood flow, electrical conductance and external secretion of the pancreas. J. Physiol. (Lond.) 217:665.
176. Marliss, E. B., Girardier, L., Seydoux, J., Wallheim, C. B., Kanazawa, J., Grei, L., Renold, A. E., and Porte, D., Jr. (1973). Glucagon release induced by pancreatic nerve stimulation in the dog. J. Clin. Invest. 52:1246.
177. Rappaport, A. M., Kawamura, T., Davidson, J. K., Lin, B. J., Ghira, S., Zeigler, M., Coddling, J. A., Henderson, J., and Haist, R. E. (1971). Effects of hormones and of blood flow on insulin output of isolated pancreas in situ. Am. J. Physiol. 221:343.
178. Gorczynski, R. J., Spath, J. A., Jr., and Lefer, A. M. (1974). Vascular responsiveness of the *in situ* perfused dog pancreas. Eur. J. Pharmacol. 27:68.
179. Takeuchi, O., Satoh, S., and Hashimato, K. (1974). Secretory and vascular response to various biogenic and foreign substances of the perfused canine pancreas. Jpn. J. Pharmacol. 24:57.
180. Goddhead, B., Himal, H. S., and Zanbilowicz, J. (1970). Relationship between pancreatic secretion and pancreatic blood flow. Gut 11:62.
181. Hickson, J. C. D. (1970). The secretory and vascular response to nervous and hormonal stimulation in the pancreas of the pig. J. Physiol. (Lond.) 206:299.
182. Papp, M., Varga, B., and Folly, G. (1973). Effect of secretin, pancreozymin, histamin and decholin on canine pancreatic blood flow. Pflügers Arch. 340:349.
183. Vaysse, N., Martinel, C., Lacroix, A., Pascal, J. P., and Ribet, A. (1973). Effet de la cholécystokinine-pancréozymine GIH sur la vaso-motricité du pancréas isolé du chien. Relations entre l'effet vaso-moteur et la réponse secrétoire. Biol. Gastroenterol. (Paris) 6:33.
184. Lundgren, O., and Jodal, M. (1975). Regional blood flow. Annu. Rev. Physiol. 37:395.
185. Lenninger, S. (1973). Effects of acetylcholine and papaverin on secretion and blood flow from the pancreas of the cat. Acta Physiol. Scand. 98:260.
186. Gayet, R., and Guillaumie, M. (1930). Les réactions vasomatrics du pancréas étudiées par la mesure des débits sangnins. Séanc. Soc. Biol. 103:1106.
187. Hilton, S. M., and Jones, M. (1968). The role of plasma kinin in functional vasodilatation in the pancreas. J. Physiol. (Lond.) 195:521.
188. Kaneto, A., Miki, E., and Kosaka, K. (1974). Effects of vagal stimulation on glucagon and insulin secretion. Endocrinology 95:1005.
189. Rappaport, A. M., Davidson, J. H., Kawamura, T., Lin, B. J., Zehn, S., Henderson, J., and Hint, R. E. (1968). Quantitative determination of insulin output following an intravenous glucose tolerance test in the dog. Can. J. Physiol. Pharmacol. 46:373.
190. Spanner, R. (1932). Neue Befunde über die Blutwege der Darmwand und ihre funktionelle Bedeutung. Morph. Jb. 69:394.

International Review of Physiology
Gastrointestinal Physiology II, Volume 12
Edited by Robert K. Crane

2
Gastrointestinal Motility

E. ATANASSOVA AND M. PAPASOVA

Institute of Physiology,
Bulgarian Academy of Sciences
Sofia, Bulgaria

ESOPHAGEAL MOTILITY 36
Myogenic Control 36
Nervous Control 37
Hormonal Control 38

GASTRIC MOTILITY 38
Myogenic Control 38
Nervous Control 42
Hormonal Control 45

GASTRIC EMPTYING 47

GASTRO-DUODENAL JUNCTION 48

SMALL INTESTINAL MOTILITY 49
Myogenic Control 49
Nervous Control 51
Hormonal Control 55

COLONIC MOTILITY 57
Myogenic Control 57
Nervous Control
Hormonal Control 60

MOTILITY OF THE BILIARY TRACT 60

Studies on gastrointestinal motility from 1972 to 1975 have focused on the mechanisms determining and regulating the smooth muscle contractile process and on the relationship between electrical and mechnical activities of the muscle wall. This focus has been prompted by the various specificities of the muscle layers of different organs and of different regions in the same organ (1, 2).

Detailed further information may be found in other recent reviews and monographs on gastrointestinal motility (2–12).

ESOPHAGEAL MOTILITY

Irrespective of anatomical differences in the esophagus between species, its contractile activity is almost the same. Thus, while dog esophagus is exclusively composed of striated muscle fibers, only the upper part in man is striated; the lower part is composed of smooth muscle fibers. Of the three parts of the esophagus, only the upper esophageal sphincter and the middle part or body are relatively well understood functionally. The complex mechanisms determining the opening and closing of the lower esophageal sphincter (LES) are not yet elucidated.

Myogenic Control

Slow potentials are not recorded from the wall of the esophagus; only spike potentials are found. The spike potentials have a frequency of 25 cycles per second (cps) in the striated muscle and less than 25 cps in the smooth muscle. Striated muscle contractions cause changes in the intraluminal pressure and 95–97% of the spike potential bursts are followed by such changes. Spike potentials always propagate distally with a velocity in the striated muscle region of 8.8 cm/s. The peristaltic wave propagates distally with a velocity of 2.4 cm/s (13). The smooth muscle of the esophagus does not manifest a pronounced automatism, which is known to be typical for gastric and intestinal smooth muscle (14).

The swallowing of fluids starts a peristaltic wave which, in the distal part of the esophagus, is accompanied by increases in pressure of considerably greater amplitude. The duration of relaxation in the LES is also greater than upon swallowing of a solid meal (15). Solid food elicits two peristaltic waves which have their electromyographic expression. It may be assumed that two mechanisms interact to induce the secondary wave, namely, contraction with a central mechanism and afferent information from the intraluminal bolus (16). Following esophageal transection at the thoracic and cervical levels in dogs, it was demonstrated that the presence of a bolus is necessary for the occurrence of the secondary wave. This again confirms the theory that a double mechanism participates in the generation of the secondary wave (17). The amplitude of the peristaltic wave is determined by the rate of infusion (18).

The LES comprises the most distal several centimeters of the lower esophagus. Studies of the LES have been concerned mainly with the mechanisms determining closing and relaxation. In this regard, it should be pointed out that pressure measurements in the LES by catheter actually measure the resistance to stretch of the sphincter smooth muscle (19). LES smooth muscle has rather different properties from the other parts of the esophagus. Examination of the effect of field stimulation on muscle strips showed that whereas gastric strips contract during stimulation and strips from the esophagus body contract after termination of stimulation, strips from the esophago-gastric junction relax during stimulation (20).

The LES in dog generates slow phasic potential changes with a frequency of 1.4–2.1 Hz between the periods of swallowing, when the sphincter contracts. These disappear when the sphincter relaxes (21). Regular relaxation upon swallowing is observed in human subjects in the region with maximum pressure, i.e., the region between the esophagus and the stomach with inverse pressure on both sides (22). A response in the LES is observed upon changes in the pressure in the upper part of the stomach, the change in the sphincter being related to the change in the gastric pressure in an 8:1 ratio (23). The pressure in the LES during rest is associated with reflux (24).

Nervous Control

Both central and peripheral mechanisms participate in the regulation of the esophagus. The myenteric nerve plexus is particularly well represented in the smooth muscle region which is also controlled by nonadrenergic and noncholinergic mechanisms (25). The fact that the smooth muscle activity of cat esophagus may be inhibited by mechanoreceptors through nonadrenergic vagal mechanisms of the pharynx and of the upper esophagus permits the assumption that this reflex mechanism is responsible for relaxation of the LES during swallowing. It probably functions similarly to linked reflexes (26). On the other hand, it is reported that α-adrenergic receptors are not represented in rats and that esophageal relaxation observed in rats in response to adrenergic agonists is probably attributable to the β-adrenergic receptors (27).

The character of the structural mechanisms participating in regulation of esophageal peristalsis is indicated by the great sensitivity of the circular muscle and of the LES muscle to acetylcholine and adrenaline (28). The LES is also very sensitive to caffeine. Thus, pressure in the LES sharply increases 30 min after caffeine injection in human subjects and reaches 12 ± 3.6 mm Hg. This effect is related more probably to an influence of the cyclic adenosine 3′:5′-monophosphate (cAMP) system than to the secretion of gastrin (29). After bilateral cervical vagotomy, action potentials continue to be recorded from the esophagus wall upon stimulation, but the intraluminal pressure is decreased; spike potentials with frequencies exceeding 25 cps are eliminated in the lower quarter. In the first month after vagotomy, peristalsis does not appear in 33–34% of the responses (13). Since there are no differences in the maximum amplitudes of the action potentials from the longitudinal and circular muscle layers, the fact that a change in intraluminal pressure is not always accompanied by spike potentials is assumed to be due to an incoordination between the circular and longitudinal layers. The longitudinal as well as the circular smooth muscle fibers are weakened after vagotomy. This results in incoordination, although contractile capacities are preserved (13).

Hormonal Control

Gastrin and its derivatives have no marked effects on esophageal peristalsis (30). It is considered, however, that the LES is controlled through both nervous and hormonal pathways (31). Gastrin stimulates the LES (32), this reaction being determined by the specific receptor properties of the circular muscle of the LES (33). The stimulating effect of gastrin is manifest at low doses (7). However, some authors do not agree that total serum gastrin regulates LES pressure (34, 35). Gastrin and pentagastrin raise pressure in the LES (36, 37) while glucagon inhibits this effect (37, 38). However, Jaffer et al. (38) emphasize that, like secretin, glucagon does not take part in sphincteric regulation at rest. The pressure rise caused by urecholin is explained by release of endogenous gastrin (39). However, CCK definitely reduces LES pressure, inhibiting the effect of endogenous gastrin (40). CCK also suppresses the LES response to pentagastrin (41).

GASTRIC MOTILITY

The myogenically determined pacesetter potentials (PP) from the stomach wall differ in frequency in different regions of the stomach. In an intact stomach, however, the proximal segments myogenically impose their higher intrinsic frequency on the distal segments (42). The effects of the extrinsic and intrinsic nervous systems on the electrical and motor activities of the stomach, as well as the hormonal control, are not yet completely clear.

Myogenic Control

The electrical and contractile activities of the stomach propagate aborally. Rhythmic slow potentials are always recorded from the corpus and antrum of the stomach, but not always from the fundus. However, with sufficient amplification of the recording device, slow potentials may also be led off from the fundus (1). In a fasted animal, at relative rest, the frequency of these potentials is little more than half those of other stomach regions (Figure 1*A*). Contractions of the stomach wall correspond to the frequency of the slow potentials. However, upon activation of the stomach, i.e., with the appearance of periodic motor activity, the slow potential frequency from the fundus equals that from the antrum and corpus (Figure 1*B*) (1). These differences in slow potential frequency and the contractions in the fundus and corpus are probably due to differences in the smooth muscle of the different regions; in vitro muscle strips from the antrum manifest predominantly phasic contractions, whereas the contractions of fundic strips are tonic (43). This difference may be related to the

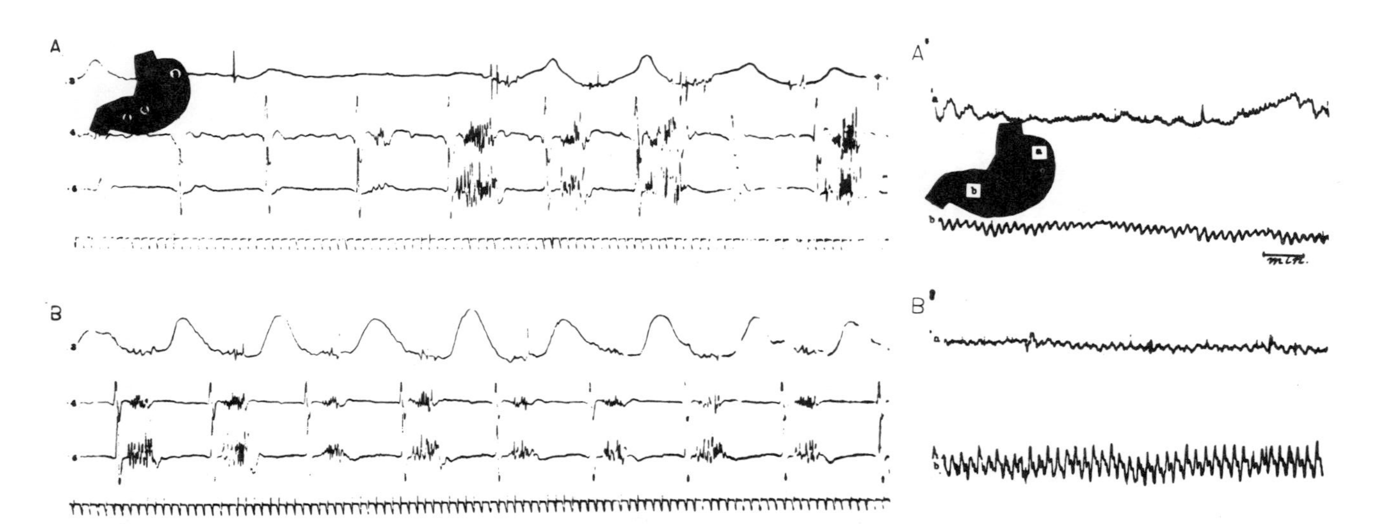

Figure 1. Switching the rhythm of the slow potentials and of the contraction waves trom the fundus region into the rhythm of the antral part of the stomach during the periods of gastric motor activity. *A*, relative rest; *B*, periodic spike activity; *A′*, contraction waves of the fundus and antrum of the stomach; *B′*, 2 min after feeding.

difference in basic function. The fundus was shown by Wilbur et al. (44) to act as a regulating reservoir and to be the main regulator of intragastric pressure after feeding.

On the other hand, differences in the type of potentials recorded from the smooth gastric muscle are observed in several animal species. In some rodents (guinea pigs, mice, and rats) the slow potential has a configuration similar to the sinusoid-type slow wave typical of cat intestinal smooth muscle. Contractions are observed only when spike potentials are superimposed on the peak of the slow wave (45). In other species (dog, cat, and man) the potential is similar to the plateau-type, typical of cardiac muscle. It is known to trigger contractions even when not accompanied by spike potentials (46). The plateau-type potential always leads to contraction; the sinusoid-type leads to contraction only if it is accompanied by spike potentials. The spike potentials of gastric smooth muscle in dog, cat, and man lead to an increased amplitude of contraction (47, 48). This does not contradict the theory of Daniel and Irwin (49) that the second PP phase is connected with the contraction. In fact, the first PP phase synchronizes the activities of a group of smooth muscle cells, while the second, slower phase results in contraction (2, 50). To a certain extent this contradicts theories that the gastric smooth muscle does not contract when there are no spike potentials (experiments on dogs) (51, 52).

Slow potentials may also be recorded upon stimulation of the gastric wall, and no matter whether the slow potential is spontaneous or evoked, it always prepares the muscle for contraction (53). Data on the connection between PP and gastric smooth muscle contraction support the hypothesis of Daniel and Irwin (54) and Code et al. (55). However, the assertion (56) that electrical stimulation of the stomach wall results in suppression of the naturally generated slow potentials (PP_n) and causes the appearance of stimulated PP (PP_s), may be disputed. Upon electrical stimulation with impulses synchronous to the slow potential (adequately amplified), a gradual change is observed in the propagation velocity of the slow potential, with gradual decrease of the period T, the distance between two potentials. This shows that stimulation results in changes in the excitability of the smooth muscle cells, which begin to generate potentials with higher frequencies (Figure 2*B*) (57, 58). In fact, upon electrical stimulation of the stomach wall the frequency of slow potential generation increases because of the increased excitability of the smooth muscle cells. In agreement, the frequency of the potential is also determined by the force of the impulse applied (59).

The intrinsic frequency of slow potential generation is different for the different regions of the stomach. Highest in the corpus, it gradually decreases aborally. The lower intrinsic PP frequency in the antral region can partly be accounted for by the smaller rheobase, faster chronaxia, and shorter refractory period compared with the corpus. These properties of the antral smooth muscle may explain the possibility of transition of the distal stomach segments to the rhythm of PP generation with the pacemaker proximal regions (60).

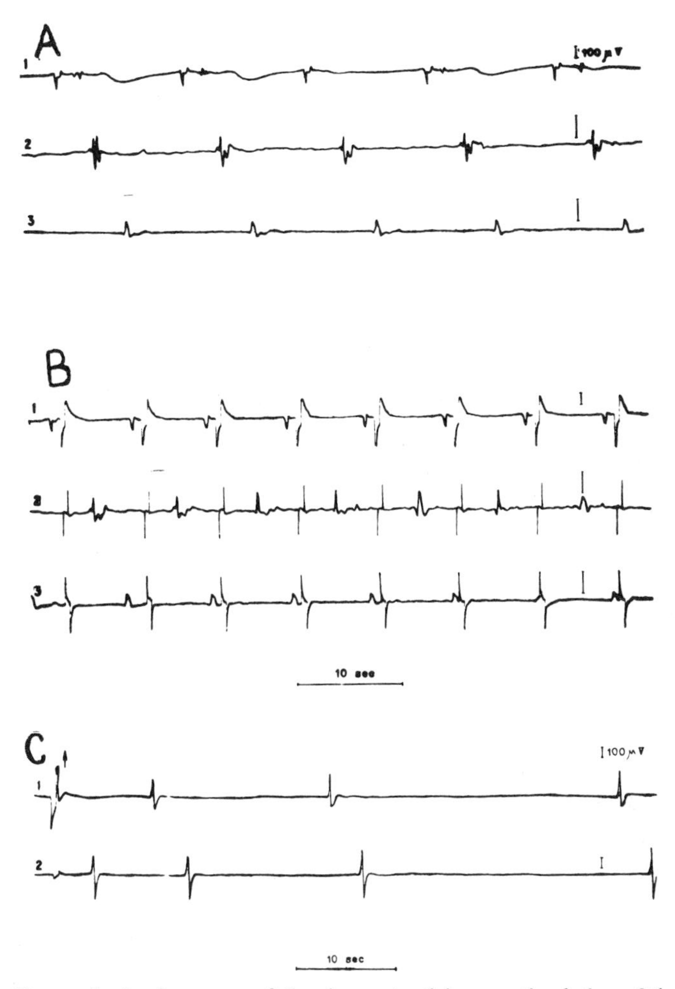

Figure 2. Changes in the frequency of the slow potential upon stimulation of the stomach wall with impulses synchronous to the slow potentials. *A*, background; *B*, after stimulation with impulse amplitude 13 V, delay 1 s, impulse duration 0.5 s; *C*, switching-off effect of stimulation.

With respect to the stomach pacemaker it is known that the pacemaker region has the property of generating higher frequency potentials (61). It has been shown recently that the value of the refractory state is smaller for the antrum than for the corpus (62). Also, both naturally generated PP and stimulated PP propagate aborally along the circumference of the gastric smooth muscle, as well as of the entire gastrointestinal tract, and are built on the principle of a system of bidirectional coupled relaxation oscillators (62–64).

In addition to the pacemaker region which imposes its rhythm of PP generation on the other regions, it is assumed that there is also a pacemaker of antiperistalsis; the greater curvature of the stomach dominating over the lesser curvature as its main course (65). This is not completely substantiated since under different functional states of the stomach, as well as under the effect of apomorphine and morphine, the beginning of the antiperistaltic wave may be observed both in the corpus and in the antral region (Figure 3). The theory of multiple pacemakers seems more feasible; under certain pathological conditions or the effect of certain drugs, these may begin antiperistaltic propagation of the slow potential (1). It is also assumed that in the intact stomach the pacemaker activity leads to strong rhythmic contractions, while asynchronous pacemaker activity maintains low amplitude mechanical activity. Synchronization of the gastric pacemaker activities is realized most frequently under the effect of activation of the stretch receptors. Fast passage of the stomach contents, however, is assumed to be influenced by the high frequency of the pacemaker activity, without neglecting the increased thickness of the muscle wall distally (66).

Nervous Control

Similarly to other organs of the gastrointestinal tract, the stomach is doubly innervated: innervation is effected by means of extrinsic vagal and sympathetic nerves and by myenteric plexuses in the stomach wall. This double innervation ensures regulation of stomach functions, including motility. Owing to double innervation, section of the extrinsic nerves results in motility disturbances that are not striking and that attenuate rapidly. Thus, after bilateral transthoracic vagotomy dysorganization is observed between the electrical activities of the corpus and antrum (67). However, this is transient and 7–10 days later the unity of generation of the PP rhythm in the two areas is restored (68), probably due to the fact that stomach regulation is undertaken entirely by the intrinsic nervous system (69). This assumption is supported by the fact that soon after vagotomy, cholinomimetics (acetylcholine or carbocholine) restore synchronization between corpus and antrum electrical activities. A biphasic effect on the basic electrical rhythm (BER) immediately after vagotomy is also reported. The BER frequency increases in the first 5–10 min after vagotomy and decreases 60–120 min later (70). Stoddard et al. (71) report that in human subjects the slow potential frequency does not change after selective and truncus vagotomy. Obviously, different experimental conditions and differences in anesthesia have different effects on the BER. Only changes observed after the passing of the effect of anesthesia should be taken into consideration.

It is believed that the branches of the vagus in different gastric regions have specific functions in regulating the evacuation of food. Thus, vagal branches in the proximal part are assumed to regulate the evacuation of fluids by controlling the transmural pressure, while branches in the antral region regulate the emptying of solid meals (72). Consequently, depending on the type of vagotomy, disturbances in food evacuation are expected depending on food consistency. Dysorganization of gastric electrical activity also depends on the degree of

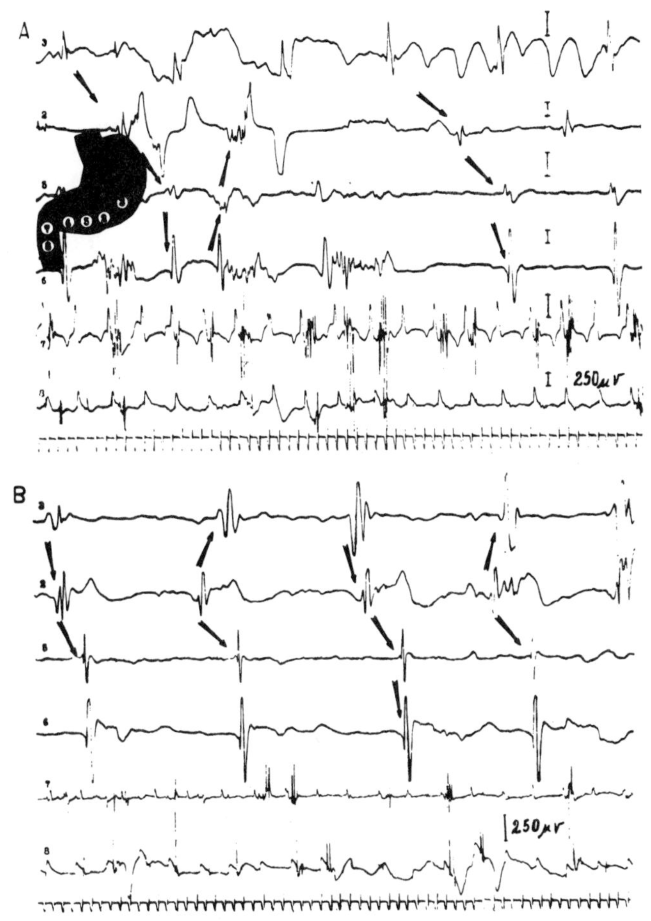

Figure 3. Antiperistaltic propagation of the slow potentials. *A*, 10 min after injection of morphine; *B*, 8 min after apomorphine.

vagotomy (73). Thus there are no changes in the rhythm and frequency of PP generation after selective vagotomy (74, 75). Marked dysorganization is observed after truncus vagotomy (73).

Changes in the character of gastric motility are also observed after vagotomy. After proximal gastric vagotomy there is a considerable increase in intragastric pressure with a decrease in the number of rhythmic contractions. These changes are related predominantly to volume adaptation (73). Usually, changes in gastric motility parallel changes in electrical activity; the disorganization between corporal and antral electrical activities shortly after vagotomy is accompanied by changes in motility. While rhythmic contractions correspond to the rhythmic slow potentials from the corpus, the bursts of low amplitude, high frequency PP groups from the antrum are not followed by contractions (Figure 4*B*). When the

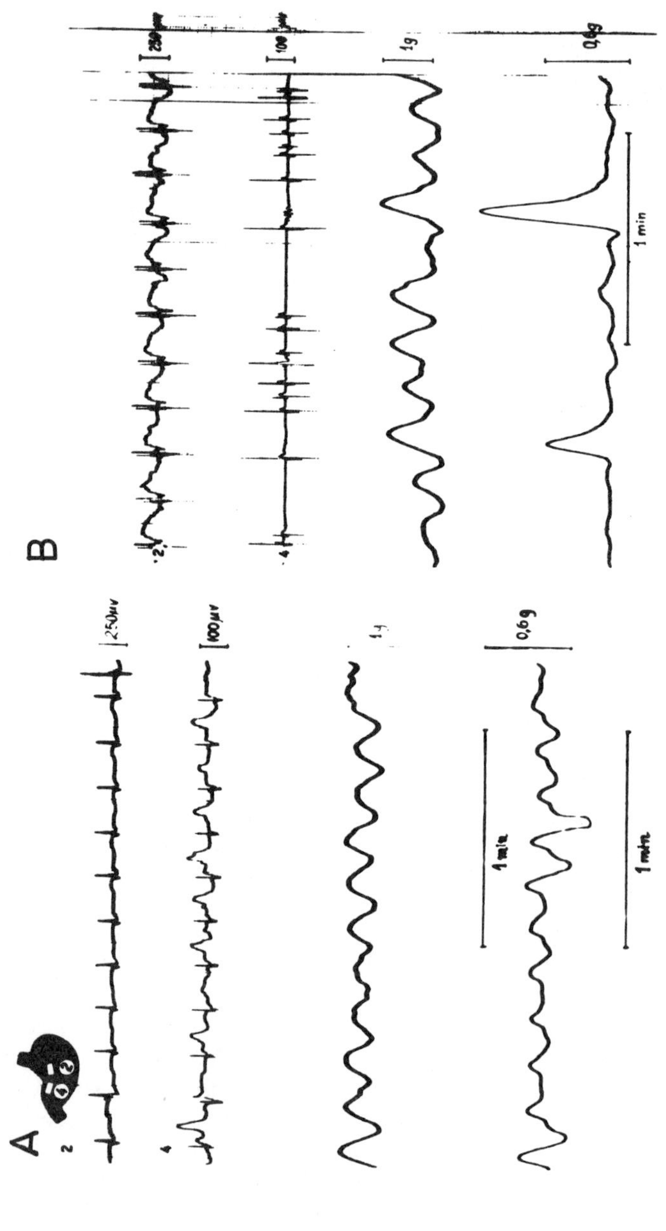

Figure 4. Incoordination of the electrical and contactile activities of the gastric corpus and antrum after vagotomy: *A*, background. Correlation between the spike activity and the changes in the stomach tone; *B*, disturbances in the slow potential propagation in the gastric antrum; after vagotomy – appearance of groups of slow potentials in faster rhythm and corresponding lowering of the antral tone.

PP amplitude reaches a definite value, it is possible to observe contractions. The incoordination between the corpus and antrum electrical and contractile activities probably also explains dilatation of the stomach after vagotomy (76), bearing in mind that truncus vagotomy results both in arrhythmia and in disturbances of the BER propagation velocity (77).

The role of the vagal nerves in regulation of gastric motility is particularly well marked following vagal stimulation. As shown by Martinson (78, 79) there are two types of nerve fibers in the vagus, one activating and the other inhibiting gastric motility. According to Miolan and Roman (80) stimulation of both types has the same effect on electrical activity, namely, an increase in BER frequency without changes in spike activity. These results are probably influenced by the anesthesia, since in chronic experiments pronounced effects on gastric electrical activity are observed, depending on the parameters of stimulation and on the functional state of the stomach (Figure 5) (81). These data agree with results indicating that the motor-evacuatory function of the stomach results in substantial changes in afferent vagal activity (82).

The vagal nerves participate in the regulation of a number of reflexes related to basic stomach functions. Thus, the vago-vagal, gastro-gastric relaxing reflex related to reservoir function and determining stomach capacity is controlled by vagal nonadrenergic fibers (83–86). This assertion is also supported by the fact that postsynaptic potentials are obtained from the gastric smooth muscle after atropinization (87). These postsynaptic potentials are excitatory (EJP) or inhibitory (IJP). IJP are regulated through transmitters released from inhibitory neurons of noncholinergic and nonadrenergic origin (88). It was also found that the reflex played an important part in regulating gastric motility (83), the activation of the reflex being effected through gastric mechanoreceptors activated by stretching (86). With respect to receptive relaxation, as established by Cannon and Lieb (89), it was found that during its occurrence impulses increased in part of the vagal fibers and decreased in other parts. Receptive relaxation is assumed to result from the decrease in activity of preganglionic synaptic fibers of the excitatory myenteric neurons, and activation of the preganglionic synaptic fibers of the inhibitory myenteric neurons (90).

Hormonal Control

Changes in gastric electrical activity under the influence of various hormones were discussed by Cooke et al. (91) and Monges and Salducci (92). The frequency of PP as well as the force of antral contractions increase, while evacuation of the stomach is delayed by pentagastrin (93). The effects of gastrin and its derivatives are dose related. Low doses intensify gastric peristaltic waves, while high doses delay emptying. Disturbances in evacuation are explained by incoordination between gastric and duodenal activities (94). On the other hand, intact innervation of the antrum is necessary for gastrin secretion when the stomach is stretched (95). Both gastrin and secretin lead to disturbances in gastric emptying. With gastrin these are the result of strong contractions of the stomach wall;

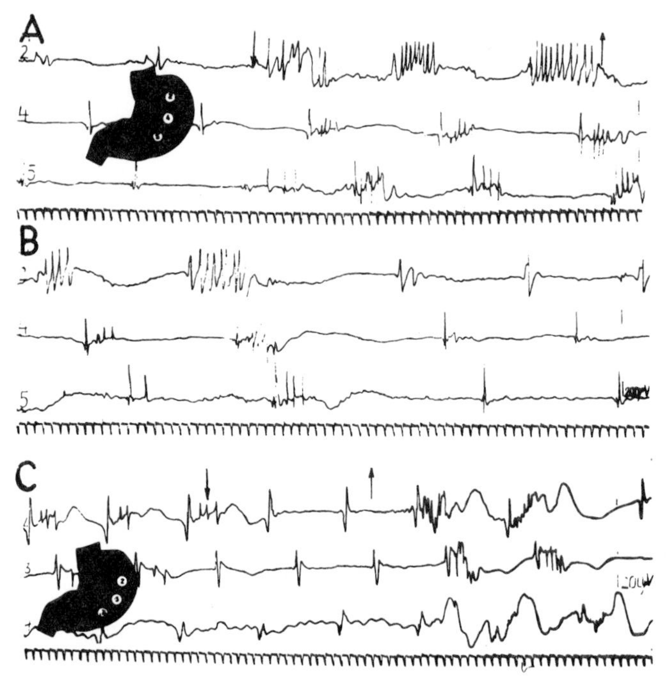

Figure 5. Increase of the electrical activity of the stomach wall upon stimulation of the nervus vagus with low frequency impulses (4 Hz, 1 ms, 3 V). *C*, inhibition of the stomach spike activity after higher frequency stimulation (30 Hz, 10 ms, for 30 s).

with secretin they are caused by inhibition of contraction (96). Parallel reductions in PP frequency and propagation velocity are observed with secretin; propagation is entirely blocked by large doses (97). These data support the assumption that secretin is a control factor for gastric electrical activity. Both caerulein and CCK have a biphasic effect on gastric motility; initially, they decrease PP frequency and spike potentials and contractions disappear; later, they cause activation (98).

The slow potentials recorded from the stomach wall of man, cat, and dog always precede contractions of the wall. In some rodents, however,(for example, mice, guineas pigs, and rats) contractions appear only when the slow potentials are accompanied by spike potentials. Electrical stimulation of the stomach wall increases the excitability of the smooth muscle cells which then generate higher frequency slow potentials. The myogenically determined electrical and contractile activities of the stomach wall are modulated and regulated by impulses originating from the extrinsic nerves and the intrinsic nervous system. Incoordination between the antrum and corpus after bilateral vagotomy is followed by disturbances in stomach contractions. Changes in gastric motility after stimulation of the nerves depend both on the parameters of the stimuli and on the functional state of the stomach. Gastric motility is also constantly modulated by hormones.

GASTRIC EMPTYING

The method of study is decisive in acquiring information on gastric evacuation. Intubation, radiological, and isotope methods all supply much information, but each has shortcomings. Certain suggestions have been aimed at filling specific gaps in various studies (99, 100), but Sheiner (12) concludes that new, better, and more precise tests are needed, especially for studying the evacuation of solid meals.

Gastric emptying depends on the functional state of the stomach. Thus, it is faster in fasted rats than in fed rats (101). All amino acids delay evacuation (102) and they are assumed to act similarly to sugars, i.e., to activate duodenal osmoreceptors which respond to osmotic pressure, pH, and the length of fatty chains. New studies confirm the dependence of emptying time on food volume (103–106). Proteins subject to hydrolysis in the stomach are retained longer than non-protein substances, as if gastric emptying "serves the needs" of the hydrolytic processes in the stomach and small intestine (107). The blood supply is very important for correct motor-evacuatory function (108); disturbances in blood supply result in disturbed nervous activity. An intact vagus is necessary for the normal functioning of receptors in the duodenum and in the proximal part of the small intestine. In chronic dog experiments the larger part of the receptors controlling stomach emptying are found in the first 50 cm (109). After truncus vagotomy, stomach evacuation of water, saline, dextrose, and fat meals is considerably extended. An additional hormonal mechanism is also assumed. According to Groisman and Begeka (110), the entero-gastric reflex is of great significance for the rate of stomach emptying in addition to the mechanical stimulation of the stomach by food. When this reflex is eliminated, the relative rate of stomach emptying progressively increases and the exponential character of emptying disappears. As a result of this the density of the duodenal chyme increases because of a smaller content of digestive juices.

Klimov et al. (111), in chronic experiments on dogs with implanted electrodes in the sympathetic branch of the stomach, in the solar plexus, and in the big splanchnic nerve, have studied the afferent impulses of the stomach during its motor-evacuatory activity. Bursts of discharges are recorded synchronously with peristaltic contractions; these bursts, the authors assume, are generated by stretch mechanoreceptors situated predominantly in the muscle wall. The first food administered leads to disappearance of the high voltage discharges and the impulsation represents a dense flux of oscillations. With the appearance of peristaltic waves, the impulse flux shows an ordered character. Longer and more powerful bursts correspond to each peristaltic contraction. During x-ray visualized propelling of food into the duodenum, a brief burst of discharges is recorded, coinciding in time with the passage of contrast matter through the pyloric sphincter. The potentials recorded may be considered as an expression of the tension developed in the stomach wall, resulting in activation of the mechanoreceptors. The rapidly adapting receptors found in the mucous membrane and in the muscle wall are probably less important for the origin of these bursts. Tension receptors and stretch receptors, which are sources of impulses in nervus vagus, are described. The biological expediency of the afferent sympathetic impulses probably consists in a connection with the motor centers, which are probably doubled along different channels.

Transection of the stomach at the level of the incisura angularis causes no change in the rate of evacuation of solid and fat food (112). Our studies (42) have shown that the structural basis for functional compensation is restoration of the stomach muscle wall through regeneration of the smooth muscle cells. However, distal antrectomy and pyloroplasty change stomach emptying (113, 114). Metoclopramide does not affect stomach evacuation in healthy subjects, but improves it considerably in patients with delayed evacuation (115). Pentagastrin increases the force and frequency of antral contractions, as well as the frequency of the duodenal slow waves (91). Gastric emptying is delayed because of incoordination in the function of the stomach and duodenum—the peristaltic waves in the duodenum begin simultaneously with the beginning of the contractions in the antrum (94). The proximal 5 cm of the duodenum are also important for inhibiting stomach emptying (116) because of the acidity of the food. This first part adjusts the acid level so that food which passes into the distal duodenum has an H^+ content of 7–8 mEq/liter.

GASTRO-DUODENAL JUNCTION

The gastro-duodenal junction is no longer considered to be an electrical insulator. Coordination has been demonstrated between the appearance of duodenal spike potentials and gastric slow potentials, though only after feeding (117). Also, electrotonic propagation of antral slow waves in the proximal part of the duodenum was pointed out (118). We have proved that there is a correlation between the gastric and duodenal spike activities, manifested in all functional states of the stomach characterized by spike activity (119). Intactness of the gastro-duodenal junction is required. Section of the pyloric sphincter results in dissociation of the gastric and duodenal spike activities, with a marked increase in the latter (120, 121). The correlation in gastric and duodenal spike activities is realized through the cholinergic part of the intrinsic nervous system (122, 123). Recent investigations with a microelectrode were made on gastro-duodenal strips including the pyloric sphincter (124). Transmural stimulation of the stomach part of the strip resulted in a postsynaptic excitatory potential, recorded from single smooth muscle cells of the duodenum (Figure 6*A*). The addition of atropine changed the character of the response. Hyperpolarization preceding relaxation of the smooth muscle strip is obtained in response to single stimulation (Figure 6*B*). This shows that nerve fibers passing from the stomach to the

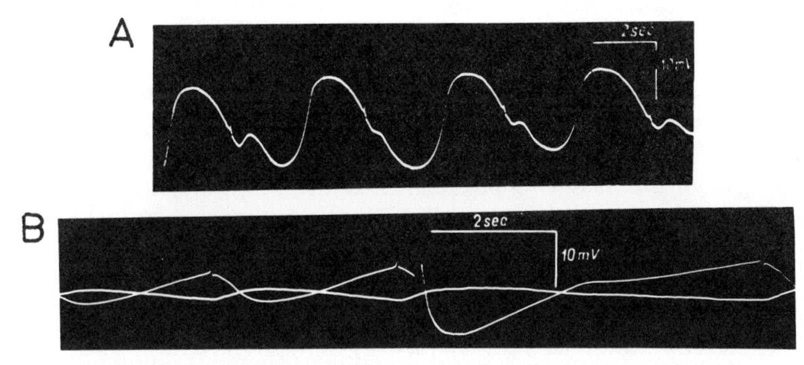

Figure 6. *A*, excitatory junction potentials evoked during the repolarization phase of the duodenal slow waves by the transmural stimulation of the stomach; *B*, inhibitory junction potential in response to single transmural stimulation after atropine. It precedes relaxation of the muscle strip.

duodenum conduct nerve impulses and are the basis for correlation in the gastric and duodenal spike activities.

Correlation between antral and duodenal contractions was also observed in human subjects (125, 126). The effect of antral contractions spreads more than 14 cm from the pyloric sphincter. Fischer and Cohen (127) characterize the gastro-duodenal junction as a gastrointestinal sphincter, that is, as a high pressure region which relaxes upon antral peristalsis and contracts upon duodenal stimulation. This theory is supported by an increase in pressure upon acidification of the duodenum and other studies (128) which demonstrate the existence of nonadrenergic inhibitory innervation.

SMALL INTESTINAL MOTILITY

Myogenic Control

The intestinal wall contracts 17–19 times per minute. The contractions are segmentary when the amplitudes are smaller and peristaltic when they are powerful and the latter propagate aborally over greater distances. There are differences in the character of the mechanical activity of the intestine of fasted and fed animals (5). The emptying of the intestine is accompanied by a rise in intraluminal pressure caused by contraction waves in the wall. This pressure is highest when the intestine is empty or just emptied (129). Comparisons have been made of the mechanical and bioelectrical activities of the intestinal wall (51, 129–131). In chronic experiments on cats, low amplitude contractions in the duodenum are preceded by slow waves with a frequency of 18 imp/min and in the ileum with 13 imp/min (130). A second type of contraction–wide, tonic, with several phasic contractions included–is preceded by prolonged spike potential bursts lasting 4–16 s, tending to cover the BER. An attempt at quantitative evaluation of the mechanical duodenal activity in human subjects according to the type of spike activity has been made (132). The duration of spike potential groups seems to correlate well with the duration and amplitude of pressure waves, as confirmed by x-ray examination, while the spike potential amplitude shows great variability and has uncertain correlations with mechanical activity. Comparison of the transmural potential difference with changes in intraluminal pressure and electrical activity suggests that changes in the transmural potential are related to the slow wave rhythm in character and frequency (133). Thus, in vivo examination of the transmural potential may provide information concerning the propagating electrical activity.

Spontaneous intestinal electrical activity is represented by constant slow waves which are rhythmic depolarizations of the smooth muscle cell membrane. By lowering the level of the membrane potential, the slow waves trigger spike potentials, i.e., fast oscillations, which cause contraction waves in the muscle wall. The electrical activity synchronizes the activity of the smooth muscle cells and thus determines the syncitial responses (6). The ion permeability of the cell membrane is of great significance for the mechanism of generation of the electrical activity and it is closely related to cell metabolism. An electrogenic pump has been demonstrated in the longitudinal muscle layer of cat duodenum. Oscillations in the level of the current result in slow waves (134). Initially, in the view of the authors, there is a sudden decrease in the velocity of the electrogenic ion transport. The inward currents remain unbalanced, depolarize the membrane, and produce the ascending phase of the slow wave. Correspondingly, the membrane voltage reaches the depolarization peak and then decreases. The cause of the sudden drop in the inward current may be either the depletion of a necessary substrate, or the release of an inhibitor. It is, of course, necessary to assume the existence of thresholds for activation and inactivation of the transport mechanism or transport delay between the sites of release and utilization.

The generated slow waves propagate aborally along the length of the intestine with propagation from one cell to the next occurring in a region with low membrane resistance, the nexus. Groups of smooth muscle fibers form muscle bundles and these are the basic effector units of smooth muscle. For details consult Bennett's monograph (3). The slow waves generated in the longitudinal muscle layer propagate electrotonically in the circular layer and the myenteric plexus is believed to participate (135).

The existence of a gradient of the frequency of the intestinal slow wave is confirmed by new studies (130, 136). The rhythm of the slow waves in the duodenum is from 18.87 ± 0.17 to 20.69 ± 0.18, in the jejunum it is from 15.57 ± 0.12 to 16.39 ± 0.10, and at the end of ileum it is from 14.46 ± 0.11 to 15.15 ± 0.20 imp/min (137). The character of this frequency gradient is reproduced by a model which depends both on the shape of the waves of the interacting oscillators and on the amplitude ratios (138). The cause for the existence of a frequency gradient may be sought in the different diameter of the smooth muscle cells, which are larger in the ileum, and in the different location of the mitochondria. The mitochondria in jejunal cells are clustered near the cell membrane, those in ileum cells being scattered in the protoplasm (139). The authors believe that the proximity of the mitochondria to the cell membrane may influence the rate at which ATP and ADP levels are restored and thus influence the frequency of the slow waves. The concept that the different rhythms depend on a different electrolyte content in duodenal and in ileal tissues has not been so successful (140).

Classification of the small intestinal spike activity is made in connection with changes in the slow wave rhythm in periods of spike activity (137). Spike activity of the first type involves a strong decrease in slow wave frequency. It is typical for the distal small intestine and it does not propagate distally. Spike activity of the second type does not always affect the slow wave frequency. Spike activity propagates aborally and corresponds to the Szurszewski myoelectrical complex. A third type of activity combines the elements of the other two, and occurs immediately before or after the electrical complexes.

Further study of the myoelectrical complex show that it starts from the stomach (141). The inflation of a balloon in the stomach, as well as feeding with milk, causes inhibition. According to Bueno et al. (142), the electrical complex controls the pressure gradient on which the flux of the contents depends. Their concept that the irregular spike activity moves the intestinal content while the regular spike activity, "though not propulsive," serves as a barrier to reflux is not very acceptable.

Nervous Control

Most studies of intestinal motor activity have examined the modulation of electrical activity by the nervous system as recently reviewed (143, 144). The cerebellum influences the motor activity of the digestive tract (145). Thus, stimulation of the nucleus fastigius increases the tone of the ileum, while the jejunum responds sometimes with an increase and sometimes with a decrease. The mechanism of this effect is believed to be an inhibition of the tonic activity of the intestinal sympathetic nerves. Disturbances of intestinal tone, as well as incoordination of the gastric, duodenal, and gall bladder motor activities, are observed after hypophysectomy (146). Extrinsic innervation is not a decisive factor for the peristaltic activity of the small intestine (147–150). Vagotomy causes almost no changes in intestinal activity (147). The myoelectrical complex is not affected, but the activity after feeding is changed, probably by the release of gastrin (148). The activity of the small intestine is restored after selective vagotomies differing in extent (149), probably by the activity of the myenteric plexus since regeneration of nerve fibers cannot be so rapid as to explain such a rapid functional restoration. Studies of intestinal adrenergic innervation show fibers ending around the ganglion cells of the myenteric plexus, and a particularly large number of fibers in the circular muscle layer, oriented along the long axis and not connected with blood vessels (150, 151). Adrenergic denervation causes the complete disappearance of all these nerve branches, but this does not affect the peristaltic activity of the small intestine. According to Burnstock and Costa (152), the widespread notion of the adrenergic nerves as the only inhibitory system should be revised. They refer to the nonadrenergic inhibitory intestinal nerves whose cell bodies are found in the intestinal wall. The effect of these nerve fibers is very powerful. Unlike those in the stomach, they are controlled only by the intrinsic cholinergic nerves. Obviously they are involved in the "descending inhibition" of peristaltic activity. In addition to these nonadrenergic inhibitory nerves, there are also noncholinergic excitatory intrinsic nerves. Consequently, four types of intrinsic nerves can be differentiated which account for the local differences in intestinal motility (147), namely, cholinergic and noncholinergic excitatory and adrenergic and nonadrenergic inhibitory fibers. There is a powerful, mainly excitatory, innervation of the small intestine (in the colon it is predominantly inhibitory) which is accounted for by regional differences in the functions of the small intestine. The cholinergic excitatory and the nonadrenergic inhibitory nerves develop simultaneously at the 17th day of gestation in the rat (153). The development of adrenergic innervation takes place later, towards the 21st day. Intracellular recordings from single neurons in the intrinsic nerve plexus demonstrate two types of nerve cells in accordance with their electrical properties (154–157). One type is S-cells characterized by an action potential and undershoot similar to those recorded in other autonomous synpathetic ganglia of mammals (155, 157, 158). In response to transmural stimulation these cells produce postsynaptic potentials though sometimes spontaneous postsynaptic potentials may also be recorded suggesting a wide synaptic input. The second type is AH-cells which demonstrate, after the action potential, a wide after-hyperpolarization related to a decrease in cellular resistance. Some of the S-cells appear to be cholinergic motoneurons (157), while the AH-cells may be nonadrenergic inhibitory neurons with recurrent collaterals to other cells (155, 157). This is compatible with the hypothesis that they are at the afferent side of the intrinsic reflex (155, 159). It appears that every neuron is affected by several pathways in the anal direction. It is possible that stretching of the intestinal wall results in excitation of the sensitive cells. The processes of these cells fall on intermediate neurons or directly on inhibitory neurons. The inhibitory nerves are directed anally and innervate the circular layer of the muscle wall predominantly, and this is probably how "descending inhibition" is effected (156). The existence of a spontaneously active inhibitory system in the intestinal nervous system is confirmed by the absence of mechanical activity in the circular muscle layer during burst-type ganglion cell activity (159, 160).

Transection of the duodenum is a particularly convenient method for elucidating the role of the intrinsic nervous system in motor function of the small intestine. The velocity of propagation of the slow waves increases after transection (161). There is a drop in the frequency of slow waves from the distal segment of the duodenum, and the frequency gradient of the slow waves disappears. The infusion of cholinergic as well as α- and β-adrenergic stimulants

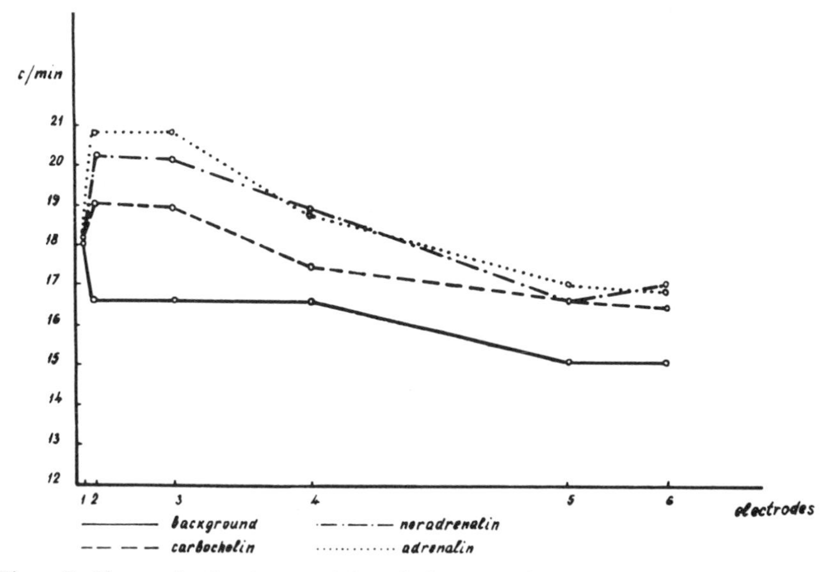

Figure 7. Changes in the transected bowel slow wave frequency gradient after repeated infusion of adreno- and cholinomimetics.

directly into the artery supplying the distal duodenal segment results in acceleration of the rhythm of the slow waves sometimes to the level of the slow waves from the proximal segment. Multiple consecutive infusions of these mimetics increases slow wave frequency both of the jejunum and the distal part of the ileum (162). Consequently, the frequency gradient of slow waves below the anastomosis is restored, the slow wave rhythm of the ileum being faster also (Figure 7). This shows that the function of the intrinsic nervous system is to maintain the frequency of the slow waves above their intrinsic, myogenically determined frequency, and to contribute to the maintenance of the frequency gradient of intestinal slow waves.

Careful follow-up of the slow wave frequency for 6 months after transection of the duodenum demonstrates partial restoration of the rhythm of the slow waves of the distal segment (163). A disturbance of the consecutive order of increase of the number of waves with higher frequency was observed. One month after the operation the histograms show a very similar distribution of the slow wave frequencies in the two duodenal segments (Figure 8*A*). Three days later the histogram of the segment under anastomosis shows a displacement to the lower frequency region (Figure 8*B*). The activity of the intramural nervous system is of great importance for the restoration of the slow wave rhythm in the segment below the anastomosis since slow wave groups with frequency similar to that of the proximal duodenum are observed during and after spike activity (163). Complete unification of the slow potential rhythm above and below the section of the stomach is observed as early as 6–8 days after the surgery (Figure

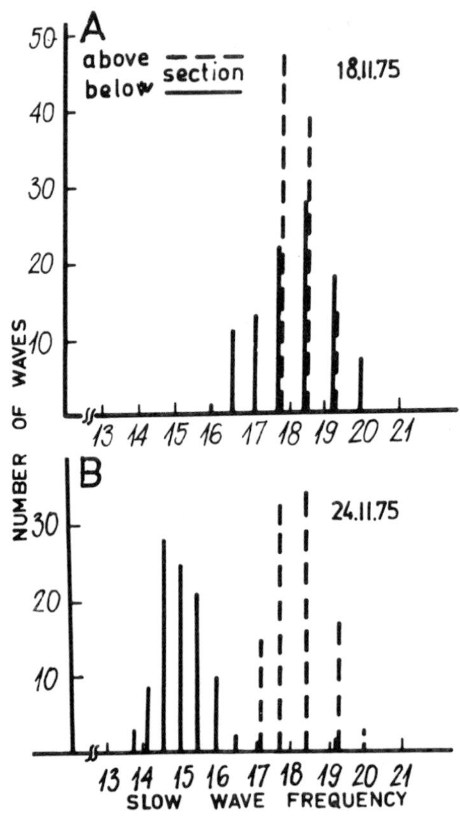

Figure 8. Distribution of the slow wave frequency of the duodenum above and below section in two different experiments 1 month after surgery.

9*C*) (42). Histological examination of the region of duodenal anastomosis shows single regenerating smooth muscle cells in the first days after operation, later in groups. At the end of the first month after the section these cells form narrow bundles which hardly ever join the two edges of the operative wound (163) (Figure 10*A*). Over the same period of time in the stomach, large bridges of regenerated smooth muscle cells cross the anastomosis to connect the two edges of the operative wound and provide a reliable basis for the unification of the rhythms of the two gastric segments (Figure 10*B*) (42). Autoradiography shows quantitatively better regeneration of the stomach, since the tritium-thymidine labeled nuclei of regenerating smooth muscle cells are found in a greater number of gastric than duodenal autoradiograms.

The propulsion of intestinal content is poorly influenced by gastric evacuation (100). Propulsion in the first three-fifths of the small intestine is faster (103, 164). In the distal part of the small intestine the content is distributed evenly and its propulsion becomes slower (164). Drugs stimulating the cholin-

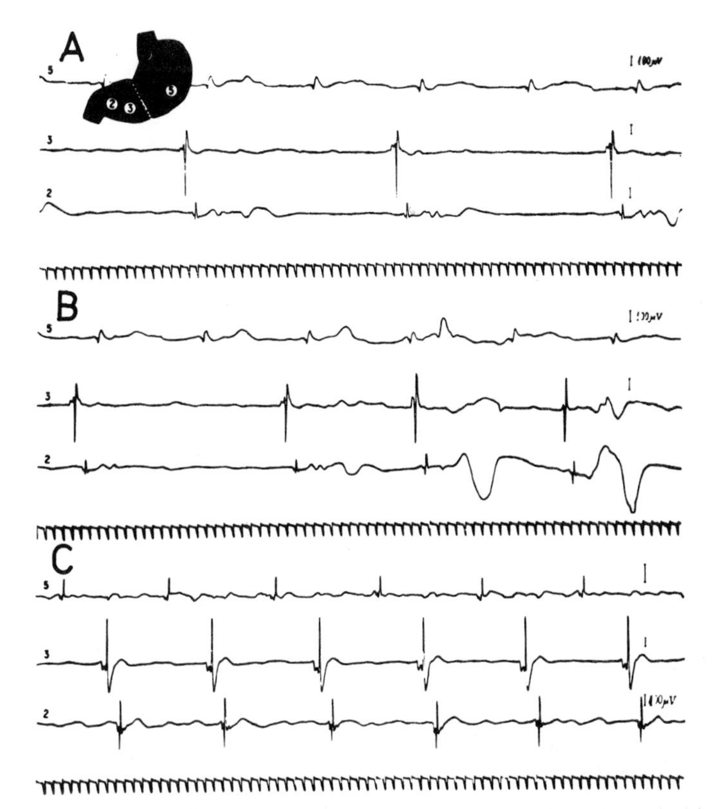

Figure 9. Unification of the slow potential rhythm of the two segments of the stomach after transection: *A*, background: dissociation of the slow potential rhythm, 2nd day after surgery; *B*, acceleration of the slow potential rhythm below the anastomosis, 5th day after section; *C*, permanently unified rhythm of the two segments, 6th day after surgery.

ergic mechanism intensify intestinal propulsion, while conversely, anticholinergic, ganglion blocking agents delay it (100).

CCK in small doses shortens transit time (165), suggesting that CCK may be involved in the normal control of intestinal motility.

Hormonal Control

In two recent reviews of the effect of hormones on the motor function of the digestive tract (7, 8) the problem of specificity of the hormones is discussed. Also discussed is the relationship of the quantity of released hormone to the area exposed to stimulation and to the concentration of the stimulus as well as the fact that the effects of certain hormones are mediated by acetylcholine. In addition to the three generally recognized hormones, secretin, gastrin, and CCK, there are another sixteen "candidate hormones" (166). One of these, gastric inhibitory peptide (GIP), is released by cells situated in the middle zone of the

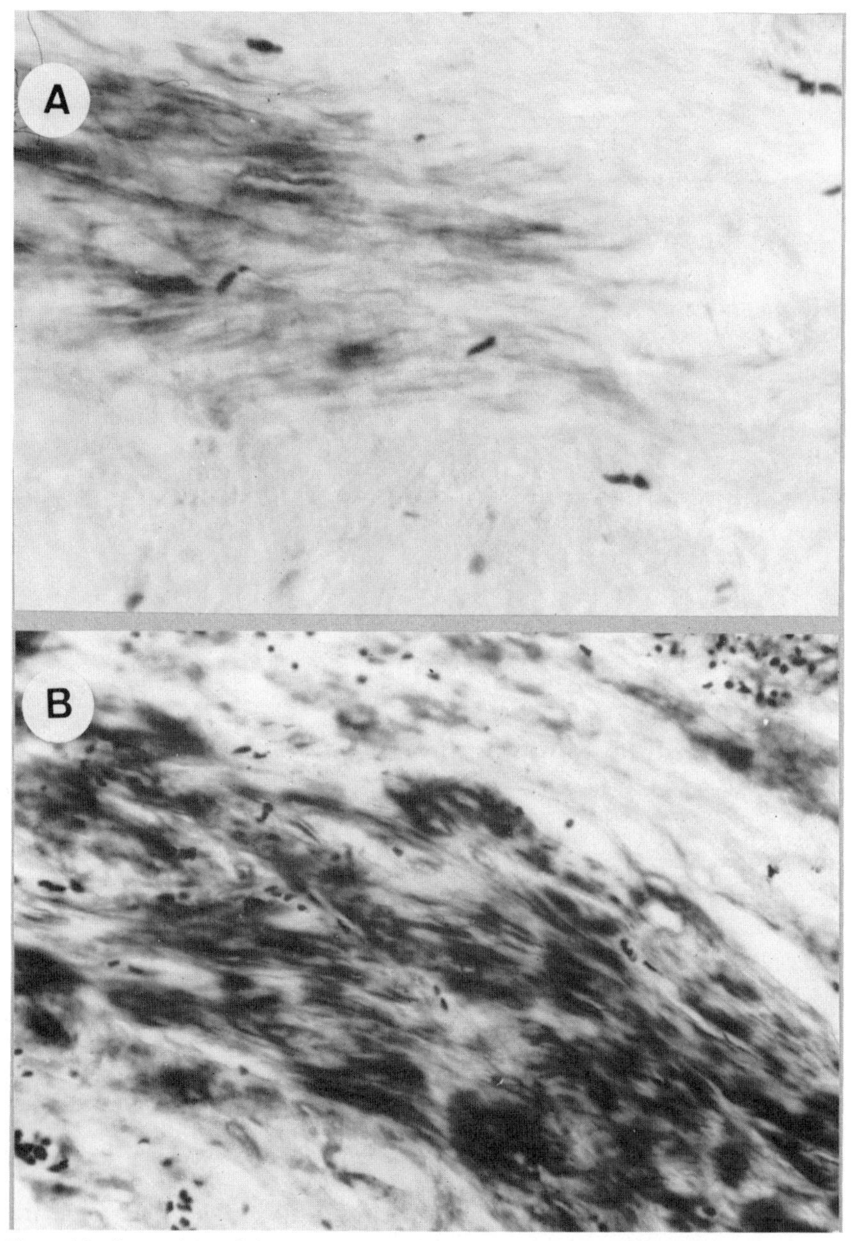

Figure 10. Comparison of the regeneration of the smooth muscle cells in the duodenum and in the stomach after transection. *A*, narrow bundles of smooth muscle cells of the duodenum infiltrate the connective tissue filling the zone of the anastomosis (33 days after surgery) (TPA after Puchtler Leblond) 700:1. *B*, bridges of smooth muscle cells of the stomach, alternating with connective tissue, which cross the anastomosis and connect the two edges of the transected gastric wall (30 days after surgery) (Mallory PTAH stain).

glands in the duodenum and to a lesser extent in the jejunum (167). Gastrin activity is now known to consist of six components with different molecular sizes (168) and it has been found that duodenal stimulation is important for its release (169).

Injection of pentagastrin in dogs transforms the fasting pattern, an aborally propagating myoelectrical complex, into a fed pattern, i.e., continuous spike activity (170). This permits the assumption that the change from a fasted to fed pattern after feeding may be due to the physiological release of gastrin. The decrease in the level of endogenous gastrin after resection of a part of antral mucosa leads to an increase in the motor and evacuatory activity of the stomach during the first 2 months after surgery (171). Pentagastrin and secretin, however, change the motility of duodenum and ileum in human subjects without an effect on the frequency of the slow waves (172, 173). CCK stimulates and secretin inhibits the motor activity of human duodenum (174). The secretin-inhibited activity is stimulated by CCK. These interrelations indicated that these two hormones are autoregulators of intestinal motility. They increase the contractile activity of the intestinal wall by releasing acetylcholine (175). Since tetrodotoxin inhibits the effect of the peptide and hexamethonium does not, it is probable that the hormones influence the non-nicotinic receptors of the ganglion cells. Caerulein increases the mechanical activity of the intestinal wall without affecting the slow wave rhythm (176, 177).

Prostaglandins are physiologically active substances, synthesized and degraded in the same tissue. Consequently, they act more as local regulators than as circulating hormones (178). Research on the prostaglandins has been reviewed (178, 179). Their effect is to decrease duodenal motor activity. Their oral and intravenous administration causes diarrhea, which is explained by a strongly increased secretion of water and electrolytes from the jejunum and ileum, with intact function of the colon (180, 181).

COLONIC MOTILITY

Anatomically the longitudinal muscle layer of the colon in man and in some animals represents an accumulation of smooth muscle cells in three muscle strips (taenia coli). Among these strips there is also an extremely thin longitudinal muscle layer. In other species, however, including dogs and cats, there are no clearly formed taenia; the longitudinal layer comprises the entire circumference of the colon. The circular layer is also well differentiated.

Myogenic Control

The diversity of contractions in the colon is probably determined by the different types of potentials which may be recorded from the colon wall. Unlike the small intestine and the stomach, the electrical potentials recorded from the colon muscle wall are neither regular nor constant. Three types of electrical activity are observed: slow waves, oscillations, and spike potentials (182).

Whereas in the small intestine the frequency gradient of the slow wave decreases aborally, an inverse gradient is observed in the colon. Thus, in cats with chronically implanted electrodes the average slow wave frequency is 26.5 c/5 min for colon ascendens, 27.4 c/5 min for colon transversum, and for colon descendens 27.8 c/5 min (183). According to Jule (184) the slow waves in the proximal colon are regular, while in the distal colon they are irregular in frequency and in amplitude. Moreover, after colon section in in vitro experiments, there is an inhibition of the frequency of slow wave generation in the proximal segment (down from 4.5 c/min to 1.9 c/min) and an increase of the frequency in the most distal part (up from 6 c/min to 7.3 c/min) (185). In other in vitro experiments the slow wave frequency ranges from 2.2 to 2.6 c/min (186). Intracellular studies have shown that the membrane potential from the smooth muscle cells of the circular muscle layer is −52 ± 4.5 mV (187) approaching the values found by Christensen et al. (188). Pointing to the linear dependence between the resting potential and the slow wave amplitude, the same authors emphasize that, using adequate amplification, each contraction is preceded by a slow wave with a duration of 3–19 s.

Velocity propagation in the transverse direction is 7.5–9.6 mm/s and 2.3–2.6 mm/s in the longitudinal direction (10). These values approach those accepted by Christensen and Rasmus (189) for a single colon smooth muscle cell. The great velocity propagation of the slow wave in a transverse direction explains haustral contractions (186). There are differences in propagation velocity in different parts of the colon. Propagation velocity in the longitudinal axis in the middle of the colon is the same as the velocity in colon ascendens; in the transverse axis it is higher. According to Christensen and Rasmus (189) this indicates that the oscillators are either larger, or that the coupling nature is different. Differences from the small intestine are also observed with respect to slow wave generation. In the colon the slow wave is generated in the circular muscle layer and propagated electrotonically into the longitudinal layer (190). The slow wave is Ca- and Na-dependent. Ca is the basis for the inward transport current and Na is involved in the genesis of the slow wave, probably in the repolarization phase (190). Electrical activity in the rectum is usually recorded from the longitudinal and not from the circular layer (191).

Spike potentials can be recorded from cat colon. Typically, spike potentials appear from time to time on the slow wave and migrate distally. Upon section the migrating spike potentials are inhibited (185). Moreover, there are oscillatory potentials with a frequency of 35–40 cpm which often appear in combination with spike potentials (183). In rabbit the frequency of the oscillator is 9–13 oscillations/min (192). Whereas the slow wave is generated in the circular muscle layer, the oscillations are assumed to represent activity of the longitudinal muscle (193).

The three types of electrical activity may be observed in dog colon (chronic experiments) and sometimes slow waves with a frequency exceeding that found in terminal ileum can be observed in colon ascendens (Papasova and Mizhorkova, unpublished data).

There are not enough data to connect the observed electrical phenomena of the colon muscle wall to its various movements. The division of colon movements into propulsive and nonpropulsive is a convenience only (11). One theory holds that these colon movements have different functions, that they are related to activities of different muscle layers and that they are regulated along different channels. It is known, however, that the two muscle layers contract simultaneously, though this unit may be disturbed under the influence of certain drugs. Thus, the longitudinal muscle layer of rabbit terminal colon and rectum is more sensitive to acetylcholine, while prostaglandins inhibit the activity of the circular layer and excite the longitudinal (194).

Mechanisms controlling ileo-caecal activity are practically unstudied. Migration of ileal spike potential groups into the colon is reported to cause the appearance of oscillatory potentials in the colon (195). On the other hand, spike potentials and oscillations do not usually pass from the colon to the ileum, though they pass into the caecum. Spikes of this nature are observed 20 min after feeding (196). About 80% of ileal myoelectrical complexes pass into the caecum (6). Migration of the slow wave from the terminal ileum into the caecum can be observed in a dog with chronically implanted electrodes under a definitive functional state (Papasova and Mizhorkova, unpublished data). It is assumed that the muscle of the ileo-caecal junction has a high resistance to stretch compared with the muscle of the adjoining regions of the colon and ileum. The muscle of the ileo-caecal junction resembles the muscle of the gastro-duodenal and eso-phago-gastric junctions (197).

Nervous Control

Both extrinsic and intrinsic nerves play an important role in the control of colon motility. The ileo-caecal valve, colon ascendens, and colon transversum are innervated by the vagal nerves, while colon decendens and the sigmoid colon areas are innervated by nn pelvici. Sympathetic innervation of the proximal colon is by plexus mesentericus superior. Both inhibitory and excitatory impulses are transmitted along the vagus as is demonstrated by the fact that vagal stimulation results in both inhibitory (IJP) and excitatory (EJP) junction potentials. The inhibitory potentials are caused by stimulation of the nonadrenergic inhibitory fibers of the vagus. Velocity propagation is 1.01 m/s for the excitatory fibers and 0.50 m/s for the inhibitory. Innervation of the proximal and distal colon parts is assumed to be double inhibitory, one performed by the adrenergic fibers of the sympathetic nerves and the other by the nonadrenergic fibers of the parasympathetic nerves (198, 199). IJP and EJP may also be obtained upon electrical stimulation of nn pelvici. The excitatory and inhibitory potentials are neither adrenergic nor serotonin-like (200). The activating role for coordination of the longitudinal and circular muscle layer belongs to the nervous system (201, 202).

Excitation of the sympathetic nerves has two effects on colon myoelectrical activity. The first effect is characterized by hyperpolarization of the smooth muscle cells and disappearance of spontaneous activity. The second is characterized by disappearance of EJP due to the stimulation of nn pelvici (200). It may be assumed that the dependence established between the contractions of guinea pig colon under the effect of 5-hydroxytryptamine is also valid for the colon of other animal species.

Colon excitatory effects caused by the stimulation of the hypothalamic sympatho-inhibitory area are regulated by the sympathetic lumbar colonic nerves, while the excitatory responses, accompanied also by a rise in the blood pressure, take place by means of the parasympathetic nn pelvici (203). It is also established that the excitatory effect obtained upon stimulation of the anterior ectosylvian gyrus is controlled by the vagus. However, the inhibitory effects obtained by stimulation of the same area are effected by means of lumbo-colonic nerves. The stimulating effects on the colon of stimulation of nucleus amygdalis are also realized by the vagal nerves (204).

Hormonal Control

There are data to indicate that colon motility is influenced by a number of extrinsic factors. Gastrin is believed to activate colon motility which would explain the intensification of colon contractions after feeding, as well as the acceleration of its evacuation. According to Bennett (205), hormones released from the upper part of the gastrointestinal tract have no effect on colon motility. It is reported, however, that under the effect of pentagastrin a slow rhythm appears in the recording made from human recto-sigma. Parallel with this there is an increase in the amplitude of the remaining faster rhythm (206). It can also be demonstrated that CCK has a stimulating effect on the colon and that the CCK-induced activity is inhibited by secretin. Neither secretin nor CCK, however, affect the activity of the rectum (207, 208).

MOTILITY OF THE BILIARY TRACT

Studies on bile duct motility in recent years have included investigations of electrical activity of the bile duct and gall bladder walls, changes of intraluminal pressure, and radiological examination under different functional states or under the action of various drugs, in human subjects as well as various animal species.

In chronic experiments with dogs in a period of relative rest, oscillations in gall bladder pressure range from 5 to 10 mm H_2O (209). The slow waves of the gall bladder and common bile duct have a duration of 3.32 s, similar to the duodenum (210, 211). The slow waves explain the constant fluctuations in the contour of the gall bladder as established through series of x-ray cholegrams; they cause changes in its transverse and longitudinal dimensions (211). Mechanical activity is preceded by spike potentials which occur in bursts on the rhythm of the slow waves of the sphincter of Oddi, and the sphincter of Lütkens. In fasted animals bile flux may be observed when there are no spike potentials in all recordings, or when there are spikes in the gall bladder and in the common bile duct, with an absence of spikes in the sphincter of Oddi (212).

During feeding a relaxation of the gall bladder for 15–30 s is observed. Gastric peristaltic waves are followed by bursts of spike potentials on the slow waves of the gall bladder, the common bile duct, and Oddi's sphincter. In the first 90 min after feeding, rhythmic pressure changes are observed in the gall bladder (150–200 mm H_2O), synchronous to the rhythm of the slow gastric potentials (209, 210, 212). X-ray cineradiography of human subjects shows again the existence of fluctuations in the gall bladder related to the mixing of different bile fractions (213).

Electrical stimulation of the big splanchnic nerve results in contraction of Oddi's sphincter, but the gall bladder shows variable responses (215). A brief increase of bile flux suggests its mechanical expulsion owing to an increase in the tone of the gall bladder wall (216).

A more detailed discussion of biliary tract motility is to be found in chapter 7.

REFERENCES

1. Papasova, M. (1970). Electrophysiological Study of the Motor Activity of the Stomach, p. 189. Publishing House of Bulgarian Academy of Sciences, Sofia.
2. Prosser, L. C. (1974). Diversity of electrical activity in gastro-intestinal muscles. *In* E. E. Daniel (ed.), Proceedings of the Fourth International Symposium on Gastrointestinal Motility, p. 21. Mitchell Press, Vancouver.
3. Bennett, M. R. (1972). Autonomoic Neuromuscular Transmission, p. 274. University Press, Cambridge.
4. Bortoff, A. (1972). Digestion: Motility. Annu. Rev. Physiol. 34:261.
5. Christensen, J. (1974). The physiology of the gastrointestinal tract. Med. Clin. North Am. 58:1165.
6. Duthie, H. L. (1974). Electrical activity of gastrointestinal smooth muscle. Gut 15:669.
7. Gregory, R. A. (1974). The gastrointestinal hormones: a review of recent advances. J. Physiol. (Lond.) 241:1.
8. Makhlouf, G. M. (1974). The neuroendocrine design of the gut. Gastroenterology 67:159.
9. Desoucher, G. (1974). Physiologie de l'oesophage. Rev. Med. 15:2217.
10. Daniel, E. E. (1975). Electrophysiology of the colon. Gut 16:298.
11. Misiewicz, J. J. (1975). Colonic motility. Gut 16:311.
12. Sheiner, H. J. (1975). Gastric emptying tests in man. Gut 16:235.
13. Ueda, M., Schlegel, F., and Code C. F. (1972). Electric and motor activity of innervated and denervated feline oesophagus. Am. J. Dig. Dis. 17:1075.
14. Roman, C., and Tieffenbach, L. (1972). Enregistrement de l'activité unitaire des fibres motrices vagales destinées à l'oesophage du babouin. J. Physiol. (Paris) 64:479.
15. Dodds, W. J., Hogan, W. J., Roid, D. P., Stewart, E. T., and Andorfer, R. C. (1973). A comparison between primary oesophageal peristalsis following wet and dry swallows. J. Appl. Physiol. 35:851.
16. Janssens, J., Valembois, P., Hellemans, J., Vantrappen, G., and Pellemans W. (1974). Studies on the necessity of a bolus for the progression of secondary peristalsis in the canine oesophagus. Gastroenterology 67:245.
17. Christensen, J., Anuras, J., and Hauser, R. L. (1974). Migrating spike bursts and electrical slow waves in the cat colon: effect of sectioning. Gastroenterology 66:240.
18. Hollis, J. B., and Castell, D. O. (1972). Amplitude of oesophageal peristalsis as determined by rapid infusion. Gastroenterology 63:417.
19. Kaye, M. D., and Showalter, J. P. (1974). Measurement of pressure in the lower oesophageal sphincter. Am. J. Dig. Dis. 19:860.

20. Christensen, J., Conklin, J. L., and Freeman, B. W. (1973). Physiologic specialisation at the oesophagogastric junction. Am. J. Physiol. 225:1265.
21. Arimori, M., Code, C. F., Schlegel, J. E., and Sturm, R. E. (1970). Electrical activity of the canine oesophagus and gastroesophageal sphincter. Am. J. Dig. Dis. 15:191.
22. Kaye, M. D., and Philip, J. (1972). Normal deglutive responses of the human lower oesophageal sphincter. Gut 13:352.
23. Diamant, N. E., and Akin, A. N. (1972). Effect of gastric contractions on the lower oesophageal sphincter. Gastroenterology 63:38.
24. Thurer, L. R., Demeester, T. H., and Johnson, L. F. (1974). The distal oesophageal sphincter and its relationship to gastrooesophageal reflux. J. Surg. Res. 16:418.
25. Tuch, A., and Cohen, S. (1973). Lower oesophageal sphincter relaxation studies on the neurogenic inhibitory mechanism. J. Clin. Invest. 52:14.
26. Abrahamsson, H., and Jansson, G. (1973). Reflex vagal inhibition of oesophageal motility. Acta Physiol. Scand. 89:600.
27. Buckner, C. K., and Christopherson, R. C. (1974). Adrenergic receptors of rat oesophageal smooth muscle. J. Pharmacol. Exp. Ther. 189:467.
28. Cohen, S., and Green, F. (1973). Force-velocity characteristics of oesophageal muscle; effect of acetylcholine and norepinephrine. Am. J. Physiol. 226:1250.
29. George, L. T., Dennish, W., and Castell, D. O. (1972). Caffeine and the lower oesophageal sphincter. Am. J. Dig. Dis. 17:993.
30. Hollis, J. B., Levine, S. M., and Castell, D. O. (1972). Differential sensitivity of the human oesophagus to pentagastrin. Am. J. Physiol. 222:870.
31. Affolter, H. (1972). The gastro-oesophageal closing mechanism. Digestion 5:311.
32. Lipshutz, W., and Cohen, S. (1972). Interaction of gastrin I and secretin on gastrointestinal circular muscle. Am. J. Physiol. 222:775.
33. Cohen, S. (1972). The hormonal regulation of lower oesophageal sphincter competence. Digestion 6:231.
34. Sturdevant, R. A. L. (1974). Is gastrin the major regulator of lower oesophageal sphincter pressure? Gastroenterology 67:551.
35. Dodds, W. J., Hogan, W. N., Miller, R. F., Borreras, R. F., Ardorfer, R. C., and Stef, J. J. (1975). Relationship between serum gastrin concentration and lower oesophageal sphincter pressure. Am. J. Dig. Dis. 20:201.
36. Nebel, O. T., and Castell, D. O. (1973). Inhibition of the lower oesophageal sphincter by fat–a mechanism for fatty food intolerance. Gut 14:270.
37. Jennewein, H. M., Waldeck, F., Siewert, R., Weiser, F., and Thimm, R. (1973). The interaction of glucagon and pentagastrin on the lower oesophageal sphincter in man and dog. Gut 14:861.
38. Jaffer, S. S., Makhlouf, G. M., Schorr, B. A., and Zfass, A. M. (1974). Nature and kinetics of inhibition of lower oesophageal sphincter pressure by glucagon. Gastroenterology 67:42.
39. Roling, G. T., Farrel, R. L., and Castell, D. O. (1972). Cholinergic response of the lower oesophageal sphincter. Am. J. Physiol. 222:967.
40. Fisher, R. S., Dimarino, A. J., and Cohen, S. (1975). Mechanism of cholecystokinin inhibition of lower oesophageal sphincter pressure. Am. J. Physiol. 228:1469.
41. Studervant, R. A. L., and Kun, T. (1974). Interaction of pentagastrin and the octapeptide of cholecystokinin on the human lower oesophageal sphincter. Gut 15:700.
42. Atanassova, E., Jurukova, Z., and Zacheva, I. (1974). Regeneration of the smooth muscle cells–a structural basis for restoration of the slow-potential rhythm in the distal part of the stomach after transection. J. Physiol. (Paris) 68:291.
43. Boev, K., Golenhofen, K., and Lukanov, J. (1973). Selective suppression of phasic and tonic activation mechanisms in stomach smooth muscle. Pflugers Arch. 343:R56.
44. Wilbur, B. G., Kelly, K. A., and Code, C. F. (1974). Effect of gastric fundectomy on canine gastric electrical and motor activity. Am. J. Physiol. 226:1445.
45. Milenov, K., and Boev, K. (1972). Bioelectric activity of the stomach in guinea-pig. Bull. Inst. Physiol. Bulg. Acad. Sci. 14:151.
46. Papasova, M., and Boev, K. (1976). The slow potential and its relation to the gastric smooth muscle contraction. *In* E. Bülbring and M. Shuba (eds.), Physiology of Smooth Muscle, p. 209. Raven Press, New York.

47. Papasova, M., Boev, K., Milenov, K., and Atanassova, E. (1966). Mechanical and bioelectrical activity of the stomach wall. Bull. Inst. Physiol. Bulg. Acad. Sci. 10:15.
48. Papasova, M., Altuparmarov, I., and Boev, K. (1972). On the character of the bioelectrical activity of the smooth muscle in human stomach. C. R. Bulg. Acad. Sci. 25:545.
49. Daniel, E. E., and Irwin, J. (1968). Electrical activity of gastric musculature. *In* C. F. Code (ed.), Handbook of Physiology, Sect. 6, The Alimentary Canal, Vol. 4, p. 1969. Williams & Wilkins Co., Baltimore.
50. Papasova, M., Nagai, T., and Prosser, L. (1969). Two-component slow waves in smooth muscle of cat stomach. Am. J. Physiol. 214:695.
51. Klimov, P. K., and Ustinov, V. N. (1973). Bioelectrical activity of the smooth muscles of the gastrointestinal tract and its relation to the contractile activity. Usp. Physiol. Nauk (USSR) 4:3.
52. Ustinov, V. N. (1974). The biopotential pattern of the stomach and duodenum smooth muscles. Sechenov Physiol. J. (USSR) 60:961.
53. Kelly, K. A., and La Force, R. C. (1972). Role of the gastric pace setter potential defined by electrical pacing. Can. J. Physiol. Pharmacol. 50:1017.
54. Daniel, E. E., and Irwin, J. (1971). Electrical activity of the stomach and upper intestine. Am. J. Dig. Dis. 16:602.
55. Code, C. F., Szurszewski, J. H., Kelly, K. A., and Smith, I. B. (1968). A concept of control of gastrointestinal motility. *In* C. F. Code (ed.), Handbook of Physiology, Sect. 6, The Alimentary Canal, Vol. 4, p. 2881. Williams & Wilkins Co., Baltimore.
56. Kelly, K. A., and La Force, R. C. (1972). Pacing the canine stomach with electrical stimulation. Am. J. Physiol. 222:588.
57. Ustinov, V. N., and Papasova, M. (1974). The method of bioelectric stimulation of the stomach smooth muscle. Sechenov Physiol. J. (USSR) 60:831.
58. Papasova, M., Ustinov, V. N., Mizhorkova, Z., and Atanassova, E. (1974). On the mechanism of changes in the frequency of generation of the slow potential from the stomach upon electric stimulation. Acta Physiol. Pharmacol. Bulg. 2:7.
59. Sarna, S. K., and Daniel, E. E. (1973). Electrical stimulation of gastric electrical control activity. Am. J. Physiol. 225:125.
60. Kelly, K. A. (1974). Differential responses of the canine gastric corpus and antrum to electric stimulation. Am. J. Physiol. 226:230.
61. Sarna, S. K., Daniel, E. E., and Kingma, Y. J. (1972). Simulation of the electric control activity of the stomach by an array of relaxation oscillators. Am. J. Dig. Dis. 17:299.
62. Sarna, S. K., and Daniel, E. E. (1974). Threshold curves and refractoriness properties of gastric relaxation oscillators. Am. J. Physiol. 226:749.
63. Sarna, S. K., Daniel, E. E., and Kingma, V. J. (1972). Effects of partial cuts on gastric electrical control activity and its computer model. Am. J. Physiol. 223:332.
64. Sarna, S. K., Daniel, E. E., and Kingma, V. J. (1972). Premature control potentials in the dog stomach and in the gastric computer model. Am. J. Physiol. 222:1518.
65. Okada, R. (1972). Electromyographical studies on a location of origin (pace maker) and characteristic of the action potential of canine stomach. Jpn. J. Smooth Muscle 8:99.
66. Dahl, G. P., and Berger, W. K. (1974). Limited excitation spread and regional contractions in spontaneously active isolated bundles of smooth muscle of frog stomach. Pflugers Arch. 351:147.
67. Kelly, K. A., and Code, C. F. (1969). Effect of transthoracic vagotomy on canine gastric electrical activity. Gastroenterology 57:51.
68. Kahn, I. H., and Bedi, B. S. (1972). Effect of vagotomy and pyloroplasty on the interdigestive myoelectrical complex of the stomach. Gut 13:841.
69. Papasova, M., and Atanassova, E. (1972). Changes in the bioelectric activity of the stomach after bilateral transthoracal vagotomy. Bull. Inst. Physiol. Bulg. Acad. Sci. 14:121.
70. Bogach, P. G., Groisman, S. D., Lukazkyi, R. A., Novosselova, A. I., and Plotkin, E. L. (1974). The effect of vagotomy on the stomach electrical responses in dogs. Sechenov Physiol. J. (USSR) 60:251.
71. Stoddard, C. J., Smallwood, R., Brown, B. H., and Duthie, H. L. (1975). The

immediate and delayed effects of different types of vagotomy on human gastric myoelectrical activity. Gut 16:165.
72. Wilbur, B. G., and Kelly, K. A. (1973). Effect of proximal gastric and truncal vagotomy on canine gastric electric activity, motility and emptying. Ann. Surg. 178:295.
73. Stoddard, C. J., Brown, B. H., Whittaker, G. E., Waterfall, W. E., and Duthie, H. L. (1973). Effect of varying the extent of vagotomy on the myoelectrical and motor activity of the stomach in man. Br. J. Surg. 60:307.
74. Stadaas, O. J. (1975). Intragastric pressure, volume relationship before and after proximal gastric vagotomy. Scand. J. Gastroenterol. 10:129.
75. Stoddard, C. J., Waterfall, W. E., Brown, B. H., and Duthie, H. L. (1973). The effect of varying the extent of vagotomy on the myoelectrical and motor activity of the stomach. Gut 14:657.
76. Papasova, M., Atanassova, E., and Boev, K. (1976). Disturbances in the electrical and contractile gastric activities after bilateral transthoracic vagotomy. Acta Physiol. Pharmacol. Bulg. 2:15.
77. Kuwashina, T. (1974). Effects of various types of vagotomy on electrical and contractile activities of the canine stomach. Gastroenterol. Jpn. 9:407.
78. Martinson, J. (1964). The effect of graded stimulation of efferent vagal nerve fibres on gastric motility. Acta Physiol. Scand. 62:256.
79. Martinson, J. (1965). Vagal relaxation of the stomach. Experimental re-investigation of the concept of the transmission mechanism. Acta Physiol. Scand. 64:453.
80. Miolan, J. P., and C. Roman. (1971). Modification de l'électromyogramme gastrique du chien par stimulation des nerfs extrinsques. J. Physiol. (Paris) 63:561.
81. Papasova, M., and Atanassova, E. (1973). Excitatory and inhibitory effects of the stimulation of the vagus on the myoelectric activity of the stomach. Sechenov Physiol. J. (USSR) 59:1074.
82. Tchernigovsky, V. N., Klimov, P. K., and Nozdratchev, A. D. (1972). The vagal afferents and the stomach motor activity. Sechenov Physiol. J. (USSR) 58:297.
83. Abrahamsson, H. (1974). Reflex adrenergic inhibition of gastric motility elicited from gastric antrum. Acta Physiol. Scand. 90:14.
84. Abrahamsson, H. (1973). Vagal relaxation of the stomach induced from the gastric antrum. Acta Physiol. Scand. 89:406.
85. Abrahamsson, H. (1973). Studies on the inhibitory nervous control of gastric motility. Acta Physiol. Scand. 390 (suppl.):1.
86. Abrahamsson, H., and Jansson, J. (1973). Vago-vagal gastro-gastric relaxation in the cat. Acta Physiol. Scand. 88:289.
87. Atanassova, E., Vladimirova, I. A., and Shuba, M. F. (1972). Non-adrenergic inhibitory postsynaptic potentials of stomach muscle cells. Neurophysiology (Kiev) 4:216.
88. Kinoshita, S. (1973). Electrophysiological analysis of the gastric smooth muscle fibres of guinea-pig and rabbit. Tohoku J. Exp. Med. 110:119.
89. Cannon, W. B., and Lieb, C. W. (1911–1912). The receptive relaxation of the stomach. Am. J. Physiol. 29:267.
90. Miolan, J. P., and Roman, C. (1974). Décharge unitaire des fibres vagales efférentes lors de la relaxation receptive de l'estomach du chien. J. Physiol. (Paris) 68:693.
91. Cooke, A. R., Chvasta, T. E., and Weisbrodt, N. W. (1972). Effect of pentagastrin on emptying and electrical and motor activity of the dog stomach. Am. J. Physiol. 223:934.
92. Monges, H., and Salducci, J. (1972). Variations of the gastric electrical activity in man produced by administration of pentagastrin and by introduction of water or liquid nutritive substance into the stomach. Am. J. Dig. Dis. 17:333.
93. Kowalewski, K., Zajac, S., and Kolody, A. (1975). The effect of drugs on the electrical and mechanical activity of the isolated porcine stomach. Pharmacology 13:86.
94. Klimov, P. K., Barashkova, G. M., Braginsky, V. I., Kotelnikova, V. I., Linar, E. Y., Pavlova, N. S., Rozova, E. I., Troitskaya, V. B., Ustinov, V. N., Fokina, A. A., and Tchipen, G. I. (1972). Changes in function of the stomach, duodenum and the biliary

system, following the syntehtic pentagastrin administration. Sechenov Physiol. J. (USSR) 58:579.

95. Stadaas, E. J., Schrumpf, E., and Haffner, J. F. W. (1974). The effect of gastric distention on gastric motility and serum gastrin in acutely vagotomized pigs. Scand. J. Gastroenterol. 9:127.
96. Barashkova, G. M. (1975). Gastrin, cholecystokinin-pancreozymin, secretin and interaction between motor functions of the stomach, duodenum and bile-excreting apparatus. Sechenov Physiol. J. (USSR) 61:763.
97. Rozé, C., Couturier, D., Langneau, P., Gilles, M. R., and Debray, C. (1972). Action de la sécretine sur l'activité électrique de l'estomach chez le porc. C. R. Acad. Sci. Paris 275:1783.
98. Bartolotti, M., and Labo, C. (1975). Comparaison des effects de la céruleine et de la cholecystokinine-pancreozymine sur l'activité électrique et manométrique de l'estomac humain. Acta Gastrolenterol. Belgica 38:111.
99. Hunt, J. N. (1974). A modification to the method of George for studying gastric emptying. Gut 15:812.
100. Nilsson, F., and Johansson, H. (1973). A double isotope technique for the evaluation of drug action on gastric evacuation and small bowel propulsion studied in rat. Gut 14:475.
101. Poulakos, L., and Kent, T. H. (1973). Gastric emptying and small intestinal propulsion in fed and fasted rats. Gastroenterology 64:968.
102. Cooke, A. R., and Moulang, J. (1972). Control of gastric emptying by amino acids. Gastroenterology 62:528.
103. Groissman, S. D., and Begeka, A. D. (1972). Gastric emptying of solid meals. Sechenov Physiol. J. (USSR) 58:1596.
104. Purdon, R. A., and Bass, P. (1973). Gastric and intestinal transit in rats measured by a radioactive test meal. Gastroenterology 64:968.
105. Ochia, B. A. (1973). Gastric emptying in young pigs. J. Physiol. (Lond.) 233:467.
106. Payler, D. K., Pomare, E. W., Heaton, K. W., and Harvey, R. F. (1975). The effect of wheat bran on intestinal transit. Gut 16:209.
107. Korotko, G. F., and Abliasov, A. A. (1973). Some aspects of the regulation of gastric emptying activity. Sechenov Physiol. J. (USSR) 59:337.
108. Korotko, G. F., Aripov, A. N., and Ustinov, V. N. (1973). Motor-evacuatory function of the stomach in disturbed blood-supply conditions. Sechenov Physiol. J. (USSR) 59:1058.
109. Shahidullah, M., Kennedy, T. L., and Parks, T. G. (1975). The vagus, the duodenal brake and gastric emptying. Gut 16:331.
110. Groisman, S. D., and Begeka, A. D. (1974). Analysis of mechanisms of gastric emptying. Sechenov Physiol. J. (USSR) 60:805.
111. Klimov, P. K., Nosdrachev, A. D., and Chernigovsky, V. N. (1973). On the stomach sympathetic pathways in relation to its motor-evacuatory activity. Sechenov Physiol. J. (USSR) 59:1268.
112. Groisman, S. D., and Khartchenko, N. M. (1973). The stomach evacuatory function after its transection and anastomosing. Sechenov Physiol. J. (USSR) 59:1257.
113. Dozois, R. R., Kelly, K. A., and Code, C. F. (1971). Effect of distal antrectomy on gastric emptying of liquids and solids. Gastrenterology 61:675.
114. Clarke, R. J., and Alexander-Williams, J. (1973). The effect of preserving antral innervation and of a pyloroplasty on gastric emptying after vagotomy in man. Gut 14:300.
115. Hancock, B. D., Bowen-Jones, E., Dixon, R., Dymolk, I. W., and Cowley, D. J. (1974). The effect of metoclopramide on gastric emptying of solid meals. Gut 15:462.
116. Cooke, A. R. (1974). Duodenal acidification. Role of the first part of the duodenum in gastric emptying and secretion in dogs. Gastroenterology 67:85.
117. Allen, G. L., Poole, E. W., and Code, C. F. (1962). A search for relationships between antral and duodenal electrical activity. Fed. Proc. 21:261.
118. Bortoff, A., and Davis, R. S. (1968). Myogenic transmission of antral slow waves across the gastroduodenal junction *in situ*. Am. J. Physiol. 215:889.

119. Atanassova, E. (1969). The role of the gastroduodenal junction in correlating the spike activities of the gastric and duodenal walls. C. R. Bulg. Acad. Sci. 22:947.
120. Atanassova, E. (1970). Bioelectrical activity of the stomach and duodenum after cutting the gastroduodenal junction. Bull. Inst. Physiol. Bulg. Acad. Sci. 13:211.
121. Bedi, B. S., and Code, C. F. (1972). Pathway of co-ordination of postprandial antral and duodenal action potentials. Am. J. Physiol. 222:1295.
122. Atanassova, E. (1969). The role of the intramural nervous system in correlating the spike activity between the stomach and duodenum. C. R. Bulg. Acad. Sci. 22:1337.
123. Atanassova, E. (1970). On the mechanism of correlation between the spike activities of the stomach and duodenum. Bull. Inst. Physiol. Bulg. Acad. Sci. 13:229.
124. Atanassova, E. (1976). The role of the intrinsic nervous system in the correlation between the spike activities of stomach and duodenum. *In* E. Bülbring and M. Shuba (eds.), Physiology of Smooth Muscle, p. 127.
125. Rosé, C., Couturier, D., and Debray, C. (1972). Relation between antral and duodenal pressure waves in man–a semi-automated method of analysis. Rendic. Gastroenterol. 4:134.
126. Couturier, D., Rosé, C., and Debray, C. (1972). Motor activity of the duodenum in man: correlation with antral contractions. Digestion 6:1.
127. Fischer, R., and Cohen, S. (1973). Physiological characteristics of the human pyloric sphincter. Gastroenterology 64:67.
128. Anuras, S., Cooke, A. B., and Christensen, J. (1974). An inhibitory innervation at the gastroduodenal junction. Gastroenterology 66:660.
129. Drewes, W. M. (1972). Small intestinal cinematography correlated to pressure waves. Differential pressure and action potentials. Am. J. Dig. Dis. 17:352.
130. Weisbrodt, N. W. (1974). Electrical and contractile activities of the small intestine of the cat. Am. J. Dig. Dis. 19:93.
131. Oigaard, A., and Dorph, S. (1974). The relative significance of electrical spike potentials and intraluminal pressure waves as quantitative indicators of motility. Am. J. Dig. Dis. 19:797.
132. Oigaard, A., and Dorph, S. (1974). Quantitative analysis of motility recordings in the human small intestine. Am. J. Dig. Dis. 19:804.
133. Wingate, D., Green, R., Symes, J., and Pilot, M. (1974). Interpretation of fluctuation of transmural potential difference across the proximal small intestine. Gut 15:515.
134. Connor, J. A., Prosser, C. L., and Weems, W. A. (1974). A study of pace maker activity in intestinal smooth muscle. J. Physiol. (Lond.) 240:671.
135. Gonella, J., and Vienot, J. (1972). Action des gangloiplégiques sur la propagation du péristaltisme duodénal. J. Physiol. (Paris) 64:623.
136. Kortezova, N., and Papasova, M. (1972). Change in slow wave frequency gradient of small intestine in different functional states. C. R. Bulg. Acad. Sci. 25:1289.
137. Kortezova, N. (1974). On the slow wave frequency gradient in the small intestine. M. D. thesis, Sofia.
138. Reshodko, L. V., Ilin, V. N., and Bogach, P. G. (1973). Analogous model of the gradient of small intestinal motility. Sechenov Physiol. J. (USSR) 59:808.
139. Job, D. D., Griffing, W. J., and Rodda, B. E. (1974). A possible origin of intestinal gradients and their relation to motility. Am. J. Physiol. 226:1510.
140. Job, D. D., Bloomquist, W. E., and Briggeforth, J. (1974). Correlation between the electrolyte content and a spontaneous electrical activity in intestinal muscle. Am. J. Physiol. 226:1502.
141. Code, C. F., and Marlett, J. A. (1975). The interdigestive myoelectric complex of the stomach and small bowel of dogs. J. Physiol. (Lond.) 246:289.
142. Bueno, L., Fioramonti, J., and Ruckebusch, Y. (1975). Rate of flow of digestive and electrical activity of the small intestine in dogs and sheep. J. Physiol. (Lond.) 249:69.
143. Bennett, A., and Stockley, H. L. (1975). The intrinsic innervation of the human alimentary tract and its relation to function. Gut 16:443.
144. Furness, J. B., and Costa, M. (1974). The adrenergic innervation of the gastrointestinal tract. Erg. Physiol. 69:1.
145. Lissander, B., and Martner, J. (1974). Influences on gastrointestinal and bladder motility by the fastigial nucleus. Acta Physiol. Scand. 90:792.

146. Klimov, P. K., Fokina, A. K., Tchubarova, N. I., Kotelnikova, V. I., and Barashkova, G. M. (1972). Effect of the hypophyseal-adrenal system on the motor function of the digestive tract. Sechenov Physiol. J. (USSR) 58:265.
147. Bennett, A. (1974). Relation between gut motility and innervation in man. Digestion 11:392.
148. Weisbrodt, N. W., Copeland, E. M., Moore, E. P., Kearley, K. W., and Johnson, L. R. (1975). Effect of vagotomy on electrical activity of the small intestine of the dog. Am. J. Physiol. 228:650.
149. Bogach, P. G., Groisman, S. D., and Tchan Zui Nga. (1974). The effects of different kinds of parasympathetic denervation of the stomach, duodenum and jejunum on their motor responses to insulin hypoglycemia. Sechenov Physiol. J. (USSR) 60:1446.
150. Ahlman, H., Enerbäck, L., Keventer, J., and Storm, B. (1973). Effects of extrinsic denervation on the fluorescence of monoamines in the small intestine of the cat. Acta Physiol. Scand. 89:429.
151. Silva, D. G., Ross, G., and Osborne, L. W. (1971). Adrenergic innervation of the ileum of the cat. Am. J. Physiol. 220:347.
152. Burnstock, G., and Costa, M. (1973). Inhibitory innervation of the gut. Gastroenterology 64:141.
153. Gershon, M. D., and Thompson, E. B. (1973). The maturation of neuromuscular function in a multiple innervated structure: development of the longitudinal smooth muscle of the foetal mammalian gut and its cholinergic excitatory, adrenergic inhibitory, and non-adrenergic inhibitory innervation. J. Physiol. (Lond.) 234:257.
154. Hirst, G. D. S., Holman, M. E., Prosser, C. L., and Spence, I. (1972). Some properties of the neurones of Auerbach's plexus. J. Physiol. (Lond.) 225:60P.
155. Nishi, S., and North, A. R. (1973). Intracellular recording from the myenteric plexus of the guinea-pig ileum. J. Physiol. (Lond.) 231:471.
156. Hirst, G. D. S., and McKirdy, H. C. (1974). A nervous mechanism for descending inhibition in guinea-pig small intestine. J. Physiol. (Lond.) 238:129.
157. Hirst, G. D. S., Holman, M., and Spence, I. (1974). Two types of neurones in the myenteric plexus of duodenum in the guinea-pig. J. Physiol. (Lond.) 236:303.
158. Nosdrachev, A. D., Katchalov, I. P., and Gnetov, A. V. (1975). Spontaneous activity of neurones in myenteric plexus of intestine intact segments in rabbits. Sechenov Physiol. J. (USSR) 61:725.
159. Wood, J. D. (1973). Electrical discharge of single enteric neurones of guinea-pig small intestine. Am. J. Physiol. 225:1107.
160. Mayer, C. J., and Wood, J. D. (1973). Properties of mechanosensitive neurones within Auerbach's plexus of the small intestine of the cat. Pflugers Arch. 357:35.
161. Diamant, N. E., Wong, J., and Chen, L. (1973). Effects of transection on small intestinal slow waves propagation velocity. Am. J. Physiol. 225:1497.
162. Attanassova, E., and Kortezova, N. (1974). On the role of the intrinsic nervous system in the small bowel slow bioelectrical activity. Bull. Inst. Physiol. Bulg. Acad. Sci. 16:135.
163. Atanassova, E., Jurukova, Z., and Kortesova, N. Changes in the frequency of the duodenal slow waves below transection resulting from intrinsic nervous system activity and smooth muscle cell regeneration. *In* Proceedings of Fifth International Symposium on Gastrointestinal Motility, September 3–6, 1975. In press.
164. Bogach, P. G., Gubkin, V. A., and Groisman, S. D. (1973). On the propulsion rate of the content in the proximal small intestine of dogs. Sechenov Physiol. J. (USSR) 59:1081.
165. Levant, J. A., Kun, T. L., Jachna, J., Sturdevant, R. A. L., and Isenberg, J. I. (1974). The effect of graded doses of the C-terminal octapeptide of cholecystokinin on small intestinal transit time in man. Am. J. Dig. Dis. 19:207.
166. Grossman, M. I. (1974). Candidate hormones of the gut. Gastroenterology 67:730.
167. Polak, J. M., Bloom, S. R., Kuzio. M., Brown, J. C., and Pearse, A. G. E. (1973). Cellular localisation of gastric inhibitory polypeptide in the duodenum and jejunum. Gut 14:284.
168. Rehfeld, J. F. (1974). What is gastrin? A progress report on the heterogeneity of gastrin in serum and tissue. Digestion 11:397.

169. Hayes, J. R., Ardill, J., Kennedy, T. L., and Buchanan, K. D. (1974). A duodenal role in gastrin release. Gut 15:626.
170. Weisbrodt, N. W., Copeland, E. M., Kearley, R. W., Moore, E. P., and Johnson, L. R. (1974). Effects of pentagastrin on electrical activity of small intestine of the dog. Am. J. Physiol. 227:425.
171. Klimov, P. K., Rozova, E. I. and Barashkova, G. M. (1972). Influence of the lowering of the level of the endogenous gastrin on the motor function of the gastrointestinal tract. Sechenov Physiol. J. (USSR) 58:1586.
172. Waterfall, W. E., Brown, B. H., Duthie, H. L., and Whittaker, W. (1972). The effects of humoral agents on the myoelectrical activity of the terminal ileum. Gut 13:528.
173. Waterfall, W. E., Duthie, H. L., and Brown, B. H. (1973). The electrical and motor actions of gastrointestinal hormones on the duodenum in man. Gut 14:689.
174. Gutiérrez, J. G., Chey, W. Y., and Dinoso, V. P. (1974). Actions of cholecystokinin and secretin on the motor activity of the small intestine in man. Gastroenterology 67:35.
175. Vizi, S. E., Bertaccini, G., Impicciatore, M., and Knoll, J. (1973). Evidence that acetylcholine released by gastrin and related polypeptides contributes to their effect on gastrointestinal motility. Gastroenterology 64:268.
176. Labo, G., Barbara, L., Lanfranchi, G. A., Bartolotti, M., and Miglioli, M. (1972). Modification of the electrical activity of the human intestine after serotonin and caerulein. Am. J. Dig. Dis. 17:363.
177. Leccini, S., and Gonella, J. (1973). Modification by caerulein of action potential activity in circular smooth muscle of isolated small intestine. J. Pharm. Pharmacol. 25:261.
178. Waller, S. (1973). Prostaglandins and the gastrointestinal tract. Gut 14:402.
179. Markov, H. M. (1970). Prostaglandins. Usp. Fiziol. Nauk 1:98.
180. Cummings, J. H., Newmann, A., Misiewicz, J. J., Milton-Thompson, G. J., and Billings, J. A. (1973). Effects of intravenous prostaglandin $F_{2\alpha}$ on small intestinal functions in man. Nature 243:169.
181. Milton-Thompson, G. J., Cummings, J. H., Newmann, N., Billings, J. A., and Misiewicz, J. J. (1975). Colonic small intestinal response to intravenous prostaglandin $F_{2\alpha}$ and E_2 in man. Gut 16:42.
182. Wienbeck, M. (1972). The electrical activity of the cat colon in vivo. I. The normal electrical activity and its relationship to contractile activity. Res. Exp. Med. 158:268.
183. Wienbeck, M., Christensen, J., and Weisbrodt, N. W. (1972). Electromyography of the colon in the unanesthetized cat. Am. J. Dig. Dis. 17:356.
184. Julé, J. (1974). Etude in vitro de l'activité électromyographique du côlon proximal et distal de lapin. J. Physiol. (Paris) 68:305.
185. Christensen, J., Anuras, S., and Hauser, L. (1974). Migrating spike bursts and electrical slow waves in the cat colon: effect of sectioning. Gastroenterology 66:240.
186. Christensen, J., and Hauser, L. (1971). Circumferential coupling of electric slow waves in circular muscle of cat colon. Am. J. Physiol. 221:1033.
187. Ito, J., and Kuriyama, H. (1973). Membrane properties and inhibitory innervation of the circular muscle cells of guinea-pig caecum. J. Physiol. (Lond.) 231:455.
188. Christensen, J., Caprilli, R., and Lund, G. F. (1969). Electric slow waves in circular muscle of cat colon. Am. J. Physiol. 217:771.
189. Christensen, J., and Rasmus, S. C. (1972). Colon slow waves: size of oscillators and rates of spread. Am. J. Physiol. 223:1330.
190. Caprilli, R., and Onori, L. (1972). Origin, transmission and ionic dependence of colonic electrical slow waves. Scand. J. Gastroenterol. 7:65.
191. Vanasin, B., Bass, D. D., Mendeloff, A. I., and Schuster, M. M. (1973). Alteration of electrical and motor activity of human and dog rectum. Am. J. Dig. Dis. 18:403.
192. Gardette, B. (1973). Etude de l'activité électrique du colon de chat et de lapin in vivo. Effects de la stimulation des nerfs extrinséques parasympathiques. Thesis, Marseilles.
193. Gonella, J., and Gardette, B. (1974). Etude électromyographiques in vitro de la commande nerveuse extrinséque parasympathique du colon. J. Physiol. (Paris) 68: 395.
194. McKirdy, H. C. (1972). Functional relationship of longitudinal and circular layers of the muscularis externa of the rabbit large intestine. J. Physiol. (Lond.) 227:839.

195. Wienbeck, M., and Janssen, H. (1973). Der Einfluss von Nahrungsaufnahme auf die electrische Activität des Ileo-kolons. Gastroenterol. 11:717.
196. Wienbeck, M., and Janssen, H. (1973). Electrical control mechanisms at the ileo-colic junction. *In* E. E. Daniel (ed.), Proceedings of Fourth International Symposium of Gastrointestinal Motility. Vancouver.
197. Conklin, J. L., and Christensen, J. (1974). Neuromuscular properties of ileo-caecal junction. Fed. Proc. 33:1021.
198. Julé, J. (1975). Modification de l'activité électrique du colon proximal de lapin in vivo par stimulation des nerfs vagues et splanchniques. J. Physiol. (Paris) 70:5.
199. Stockley, H. L., and Bennett, A. (1974). The intrinsic innervation of the sigmoid colonic muscle in man. *In* E. E. Daniel (ed.), Proceedings of Fourth International Symposium on Gastrointestinal Motility, p. 163. Vancouver.
200. Julé, J., and Gonella, J. (1972). Modification de l'activité électrique du colon terminal de lapin par stimulation des fibres nerveuses pelviennes et sympathiques. J. Physiol. (Paris) 64:599.
201. Pahomov, A. I. (1974). On the mechanism of the inhibitory responses of the cat colonic longitudinal and circular muscle layers to transmural stimulation. Sechenov Physiol. J. (USSR) 60:628.
202. Pahomov, A. I. (1975). On the role of the intrinsic nervous system of the cat colon in its rhythmic contractile activity. Sechenov Physiol. J. (USSR) 56:731.
203. Rostad, H. (1973). Colonic motility in the cat. IV. Peripheral pathways mediating the effects induced by hypothalamic and mesencephalic stimulation. Acta Physiol. Scand. 89:154.
204. Rostad, H. (1973). Colonic motility in the cat. V. Influence of telecephalic stimulation in the peripheral pathways mediating the effects. Acta Physiol. Scand. 89:169.
205. Bennett, A. (1975). Pharmacology of colonic muscle. Gut 16:307.
206. Taylor, I., Duthie, H. L., Smallwood, R., Brown, B. H., and Linkens, D. (1974). The effect of stimulation on the myoelectrical activity of the rectosigmoid in man. Gut 15:599.
207. Dinoso, V. P., Meshkinpour, J. H., Lorber, S. H., Guiterrez, J. G., and Chey, W. P. (1973). Motor responses of the sigmoid colon and rectum to exogenous cholecystokinin and secretin. Gastroenterology 65:438.
208. Harvey, R. F., and Read, A. E. (1973). Effect of cholecystokinin on colonic motility and symptoms in patients with the irritable bowel syndroms. Lancet i:1.
209. Ustinov, V. N., and Kotelnikova, V. I. (1971). Interrelation of bioelectric and motor activity of the stomach and duodenum, and the gall-bladder. Sechenov Physiol. J. (USSR) 57:284.
210. Klimov, P. K., Fokina, A. A., and Antonova, G. M. (1971). Recherches sur l'interdependance des activités motrices gastriques, duodénales et biliares. J. Med. Chir. Prat. 142:51.
211. Klimov, P. K., and Kotelnikova, V. I. (1973). The bioelectrical activity of the gall-bladder, the Oddi and Lutkens sphincters and the stomach. Sechenov Physiol. J. (USSR) 59:140.
212. Klimov, P. K., Kotelnikova, V. I., and Ustinov, V. N. (1974). Bile excretion and the bioelectric activity of the digestive tract organs. Sechenov Physiol. J. (USSR) 60: 1249.
213. Krugliakov, I. O., and Tcherkai, A. D. (1974). The cinerhoentgenographic study of the fluctuating vibrations of the gall bladder. Sechenov Physiol. J. (USSR) 60:310.
214. Sarles, J.-C., Midjean, A., and Gayne, F. (1974). Etude électromyographique du sphincter d'Oddi. Biol. Gastroenterol. (Paris) 7:19.
215. Persson, C. G. A. (1973). Dual effect on the sphincter of Oddi and the gallbladder induced by stimulation of the right great splanchnic nerve. Acta Physiol. Scand. 87:334.
216. Essipenko, B. E., Kostromina, A. P., and Syromyatnikov, A. V. (1974). On the role of sympathetic nerves in regulation of the hepato-biliary system functions. Sechenov Physiol. J. (USSR) 60:179.

International Review of Physiology
Gastrointestinal Physiology II, Volume 12
Edited by Robert K. Crane
Copyright 1977 University Park Press Baltimore

3
Gastrointestinal Hormones

S. R. BLOOM

Department of Medicine, Hammersmith Hospital
Royal Postgraduate Medical School
London, England

SCOPE OF REVIEW 72

IMPORTANCE AND COMPLEXITY OF GUT HORMONES 73

ADVANCES IN METHOD OF STUDY 73
Hormone Purification 73
Immunocytochemistry 74
Radioimmunoassay 74

GASTRIN 77
Chemistry 77
Pharmacology 77
Localization 79
Pattern of Release 79
Physiological Actions 80
Pathology 81

SECRETIN 81
Chemistry 81
Localization 82
Pattern of Release 82
Actions 82
Pathology 83

CHOLECYSTOKINEN-PANCREOZYMIN 83
Actions and Localization 84
Assay 84

PANCREATIC GLUCAGON 84
Role in the Gut 85

ENTEROGLUCAGON 85
Physiology 85

MOTILIN 87
Pharmacology 87
Physiology 88

GASTRIC INHIBITORY PEPTIDE 89
Actions 89
Physiology 90

VASOACTIVE INTESTINAL PEPTIDE 90
Actions 90
Physiology and Pathology 91

SOMATOSTATIN 92
Actions 93

PANCREATIC POLYPEPTIDE 93
Physiology and Pathology 94

THE FUTURE 94

SCOPE OF REVIEW

Hormones were first discovered in the gut over 70 years ago (1, 2), but understanding of the endocrinology of the gut has progressed very slowly. A massive literature was built up in the early part of the century with the existence of many hormones being postulated in order to account for many observations. However, gut endocrinology could not be put on a firm footing until proposed hormones were purified and methods of measuring their plasma levels worked out. This has recently happened and there is now literally an explosion of information on the gut hormones and what they do. There is still need for caution because most so-called hormones have not been provided with rigorous proof of this status; for these Grossman (3) has coined the term "candidate hormones." Rigorous proof requires a clear demonstration that administration of a totally pure peptide produces specific physiological actions at the same circulating concentration of the peptide found to occur after the normal physiological stimulus for that action. Rigorous proof is still rare; most gut hormones rest on "adequate proof" which consists of suggestive information such as the fact that the peptide is found in an endocrine type cell and has a spectrum of actions similar to those known to occur in a given physiological situation, or has been found to be released into the circulation after a known physiological stimulus. This review deals with both rigorously proved hormones, of which there is only one, gastrin, and candidate hormones, for which the evidence is adequate though not rigorous.

IMPORTANCE AND COMPLEXITY OF GUT HORMONES

Gut hormones probably play an important role in normal physiology elsewhere than in the gut. There are at least 13 different types of endocrine cell in the gut as defined by electron microscopy granule type (4), and the total mass of endocrine tissue in the gut greatly exceeds that of any other endocrine gland in the body. Traditionally, gut hormones control gut functions, but these interactions are undoubtedly only part of the total physiological role of gastrointestinal hormones. For example, when gut factors are absent, as in animals with intestines removed, gross disturbance of hepatic metabolism results (5). Also, hepatic extraction of glucose given intravenously is of a much lower order than that which is absorbed from the alimentary tract (6).

Gut hormone abnormalities as the major pathology in diseases states has also a high probability. For example, common human obesity may reflect a relative deficiency of normal metabolic control and hormonal influences at the satiety center. Maturity-onset diabetes may be the result of deficient gut hormone control of carbohydrate metabolism. Peptic ulcer and spastic colon may be a result of hormonal imbalance.

The chief reason why it has been difficult to unravel gut endocrinology is that the hormones have complex interreactions (7). In the gut, the simple situation of a single controlling "master" hormone such as insulin, the lack of which is obvious, does not occur. Gut hormones overlap in their actions; most agonists are balanced by antagonists and the absence of one is easily compensated. Assessment of the role of an individual hormone is thus very difficult. Current knowledge is structured in terms of individual hormones and this review will be the same. However, future reviews might do better to regard hormones from the viewpoint of physiological situations such as control of the pancreas, control of gastric acid secretion, etc.

ADVANCES IN METHOD OF STUDY

Hormone Purification

It took altogether 60 years from the discovery of secretin in 1902 (1) to its purification in 1961 (8) owing to the delay in the development of such preparative techniques as gel filtration, ion exchange chromatography, counter current distribution, and, most recently, the highly specific method of affinity chromatography. In the classical situation, the existence of a hormone was postulated from physiological observation and attempts were made to purify it. Today, peptides are generally found in the course of other work and purified before a biological action is tested for. Such is the case with pancreatic polypeptide found during insulin purification (9), and two new hormones found in the semi-pure peptide fractions from which secretin and cholecystokinin-pancreozymin are prepared.

Gastric inhibitory peptide (GIP) was found as an impurity in cholecystokinin-pancreozymin and was isolated (10) and sequenced (11). As indicated in Figure 1, GIP has sequence similarities with secretin and glucagon. A vasoactive intestinal peptide (VIP) was found in the semi-pure secretin fraction, isolated (12), sequenced (13), and found also to have sequence similarities with these hormones (Figure 1). It is apparent that altogether these substances constitute a structural family; they have a very similar spectrum of hormonal actions.

Immunocytochemistry

When a hormone has been purified, specific antibodies can be raised in animals and these can be used to localize the cell producing the hormone. Visualization is provided by antibody coupled with the fluorescent dye, fluorescein (14), or an enzymic peroxidase label (15) which can be used to produce a brown color reaction. Thus the specific hormone-producing cell can be identified and its specific chemical staining reactions studied. A finding that these staining reactions are typical of endocrine tissue and that the cell is a member of the APUD series (amine content and/or amine precursor uptake and decarboxylation) (16) is useful confirmation of the hormonal nature of the purified peptide.

The serial thin, semi-thin technique recently developed by Polak and colleagues (17) also allows identification of the cells by electron microscopy characteristics. Using this technique, it has been possible to attribute hormone production to the majority of the cells in the amended Weisbaden classification based on EM granule appearance (4).

Such techniques allow investigation of the normal development of the cells and their endocrine product, their response to physiological changes, and also pathological processes. They are essential to understanding increased or decreased hormone production.

Radioimmunoassay

After a hormone has been purified and its cellular site of origin defined, the next step in understanding its role in physiology is to measure plasma levels. Previously this was done by means of bioassays which appear now to be of value only as reference and backup systems. They suffer from three serious faults: 1) they are not specific; as already mentioned many of the gut hormones have overlapping actions; 2) they are extremely insensitive, usually being incapable of measuring the rather low circulating concentrations of gut hormones in plasma; and

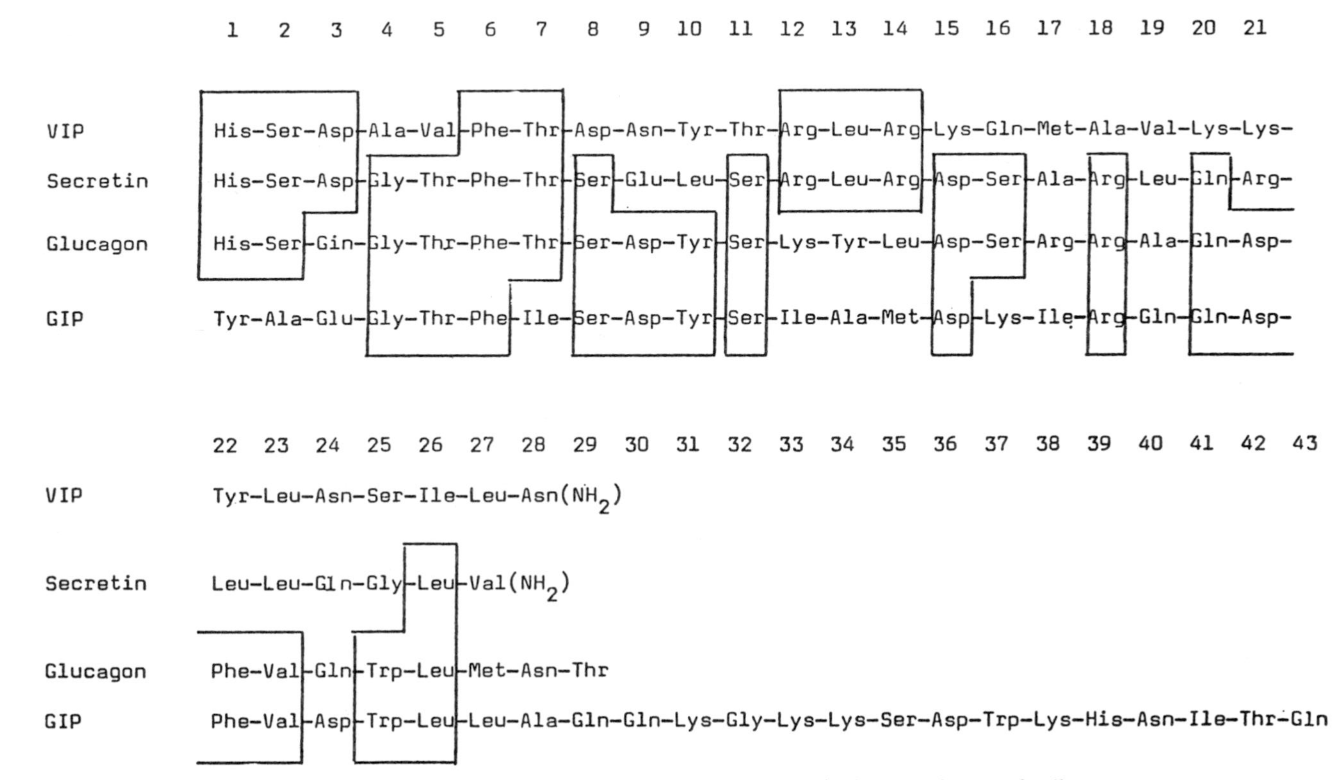

Figure 1. The amino acid sequence of the hormones from the secretin-glucagon family.

3) they are extremely expensive and laborious. By contrast the technique of radioimmunoassay is very cheap, sensitive, and specific.

The principles of radioimmunoassay are well known; the measurement depends on a highly reproducible quantitative combination of antibody and antigen. In general terms the technique has been a great success, but it has certain flaws. For example, large discrepancies in results are reported from different laboratories and results agreeing within 50% are extremely rare. The three major reasons for this are 1) the use of antibodies of different specificity, 2) the uncontrolled intereference of "nonspecific" effects in the antibody-antigen reaction, and 3) bad practical technique. The physiologist expects a measured hormone level to reflect the degree of hormone action. A radioimmunoassay, however, depends on the reaction of an antibody with a specific amino acid sequence present in the hormone which does not necessarily reflect biological effectiveness. Different laboratories use different antibodies which measure slightly different amino acid sequences. The presence of hormone fragments and different forms of the hormone will, therefore, result in different numerical values.

An ideal solution to the problem might, in theory, be to use isolated receptor sites instead of antibodies. In practice, however, this has not succeeded because (*a*) the affinity of isolated receptors tends to be low, leading to an insensitive assay system, and (*b*) receptor specificity is altered by isolation and further changes are produced by the abnormal environment of the test tube. In addition, synthetic fragments with no biological action bind to receptors (18, 19).

In order to avoid some of these problems assays have been devised in which isolated cells or small pieces of tissue are exposed to the hormone and the early, very sensitive stages of activation rather than a physiological effect are measured (20). These assays have proved far more sensitive than conventional bioassays but they still have a number of drawbacks. They do not measure one hormone specifically but measure the hormone milieu; in gastric acid production, for example, secretin will measure as negative gastrin. Also, tissues tend to be used from one species, e.g., the guinea pig, where expertise has been developed, but the assay is done on plasma from another species, e.g., man, in which sensitivity to the active hormones differs. Again, tissue isolation and denervation will alter relative sensitivity to the hormone environment.

A number of recent improvements have been made in radioimmunoassay. First, it has become apparent that iodination techniques which damage the hormone, for example, at the methionine or tryptophan residues (21) must be avoided. This is done by trace iodination of less than 5% of the hormone followed by high resolution ion exchange chromatography to purify out the pure monoiodinated undamaged hormone (22). Second, synthetic chemistry technology has been adapted to the preparation of highly specific immunogens. For example, to produce an antibody reacting with the biologically active COOH-terminal end of the hormone, animals are immunized with the hormone coupled to an immunogenic carrier molecule by its NH_2-terminal leaving the COOH-terminal end free for antibody induction. Third, problems of "nonspecific assay interference," due to the presence in plasma of factors which interfere with the antibody-antigen reaction and masquerade in the assay as hormone, have been solved by removing the hormone from part of an unknown plasma sample by means of a specific immunoabsorbent or affinity chromatograph (23). The assay of the plasma containing no hormone is then compared with the assay of untreated plasma to give a highly accurate result. Finally, progress is being made in international exchange of plasma samples, standards, and antisera to allow comparison and standardization.

GASTRIN

The idea that "gastrin" was responsible for the stimulation of acid after food was proposed in 1905 (2). Gastrin was isolated, however, only in 1964 (24) but it was almost immediately synthesized (25). The free availability of 17 amino acid synthetic gastrin has greatly stimulated research and is responsible for the easy access to gastrin radioimmunoassay. In 1972 Yalow and Berson described the presence of a second major form of gastrin in the circulation—big gastrin (26). This has been subsequently isolated and sequenced and found to consist of 34 amino acids (27). Still other forms have now been described, including big big gastrin (28), intermediate gastrin (29), and little little gastrin (30, 31), but at the present time it is doubtful whether these latter forms have any physiological import.

Chemistry

The minimal fragment of gastrin with reasonable activity is the last three amino acids (32) at the COOH-terminal end (Figure 2). This fragment shows the full range of actions of the larger molecule. A derivative of the tetrapeptide (pentagastrin) is widely used clinically to maximally stimulate gastric acid secretion (33). The presence or absence of sulfate on the typrosine (position 29, Figure 2) makes no difference to the biological potency (24). Big gastrin is composed of the same 17 amino acids as gastrin with an additional 17 amino acid NH_2-terminal sequence (27). Gastrin can therefore be produced by enzymic splitting of big gastrin.

Pharmacology

Gastrin has a wide range of actions in the gastrointestinal tract (34). In the stomach it stimulates acid secretion and increases muscle contractility. In the pancreas it is a strong stimulant of enzyme secretion and a weak stimulant of alkaline juice and insulin. It stimulates the output of watery bicarbonate into bile by the liver and of mucus from Brunner's glands into the duodenum. It increases contractility of the intestines and gall bladder. It inhibits the absorption of water and salts from the intestine. It increases the rate of mucosal growth

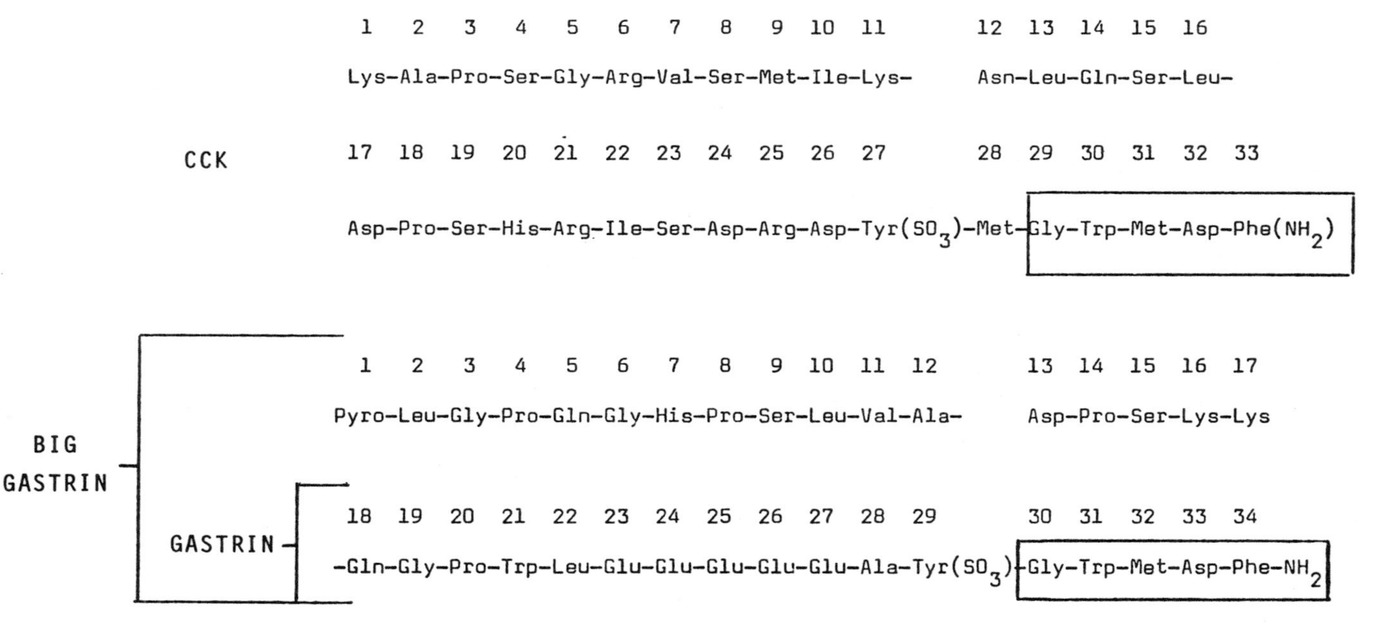

Figure 2. The amino acid sequence of human big gastrin (*lower*) and porcine cholecystokinin (*CCK*) (*upper*).

in the stomach and upper small intestine and increases pancreatic growth. Chronic administration of large amounts of gastrin results in hypertrophic mucosa of the stomach and hyperacidity. There are no differences in action between big gastrin (G34) and gastrin (G17) (35).

Localization

Extracts of human gastrointestinal tissue show that most gastrin occurs in the antrum of the stomach but that about one-third of the total is in the upper small intestine. There is none in the normal pancreas (36). In abnormal situations such as anacidity (pernicious anemia) gastrin in the antrum greatly increases and gastrin is also found in the fundus of the stomach (37). In starvation extractable gastrin greatly decreases (38). Indirect immunofluorescence localizes the gastrin-producing cell to the midpoint of the mucosa, where the cell of origin is designated the G cell (39).

Pattern of Release

The pattern of gastrin release in the fasting state is not yet fully worked out. Several reliable radioimmunoassays have shown relatively high fasting levels of gastrin when compared with the increase seen after a meal stimulus (40–42). The fasting gastrin level is thus disproportionately high compared with the low fasting output of gastric acid. Thus, much of the fasting gastrin in the circulation, as measured by radioimmunoassay, is biologically inactive. Chromatographic studies show a considerable proportion of a high molecular weight substance, "big gastrin," which remains constant in amount even after a meal, when the level of the smaller forms of gastrin rise considerably (28). Some assays do not seem to detect big big gastrin as readily as others and thus give rise to very much lower readings of fasting gastrin (43, 44). In the fasting state serum gastrin is not a good index of the drive to the stomach to produce acid.

The main physiological stimulus to gastrin release in the ingestion of a meal, and gastrin levels remain elevated for some hours postprandially. Analysis of this response shows the major stimulatory agent of food to be the protein component, particularly peptides and amino acids (34, 44). On the other hand, acid actively inhibits the release of gastrin (45) and a much greater rise is seen if acid production is neutralized by addition of an antacid. It has been demonstrated that the release of gastrin is greater after a normally appetizing meal and diminished if the meal is homogenized prior to administration. The presence of a vagal component in gastrin release has long been postulated and indeed well demonstrated in the dog (34). Insulin hypoglycemia which stimulates vagal activity gives rise to a considerable release of gastrin and in the dog this is abolished by vagal section. Similarly, direct electrical stimulation of the canine vagus also releases gastrin (46). In man, however, the release of gastrin during insulin hypoglycemia is actually enhanced after vagal section and gastrin release is also increased after atropine (47). Subjects who have undergone vagotomy have higher levels of fasting gastrin and greater increases of gastrin after a meal (48). This is partly due to the removal of the inhibitory vagal action and partly to the decreased inhibition of gastrin release secondary to diminution of gastric acid. Infusions of gastrin in man show a half-life of about 5 min and of big gastrin, 42 min (49). Because of the slower clearance of big gastrin it circulates at about four times the concentration of gastrin although a greater amount of gastrin is normally secreted by the antrum. Gastrin is more biologically active, however, and thus in spite of its lower concentration in the circulation gastrin is still the major stimulus to acid production (34). Big gastrin and gastrin are probably both released by a single cell and there is no evidence at present that there is any circumstance where differential release occurs. The elevation of gastrin levels seen in patients with nephrectomy suggests that the kidney is the important site of gastrin destruction.

Physiological Actions

Numerous actions of gastrin have been mentioned under "Localization"; how many of these are physiological?

There seems little doubt that gastrin is the major stimulant of gastric acid production after a meal. Infusions of gastrin adjusted to give levels in the postprandial range result in a considerable increase in gastric acid output. Two other major physiological roles are postulated for gastrin about which there is less certainty.

1. The importance of gastrin in the control of the motor functions of the upper small intestine, in particular the lower esophageal sphincter (LES), is not clear. Bolus injections of gastrin cause an increase in pressure in the LES and after introduction of food into the stomach the rise of endogenous gastrin correlates very well with the increase in LES pressure (50, 51). On the other hand, gastrin infusion at a concentration which produces physiological blood levels does not appear to affect LES pressure at all (52, 53). Furthermore, it can be shown that under some conditions the endogenous release of gastrin is associated with no change of LES pressure. Thus, there are two opposing views on the role of gastrin in LES control. A possible compromise is that gastrin is of importance under some conditions, but not under others though it is probably not the only or even the major controlling influence.
2. A postulated physiological role of gastrin is the control of growth in the gastric mucosa, the pancreas, and the mucosa of the upper small intestine. The elegant experiments of Johnson et al. in the rat have demonstrated that the atrophy in these areas induced by parenteral feeding can be reversed by concomitant administration of pentagastrin or gastrin (54). It has also been suggested that the amount of gastrin needed is within the physiological range. This latter point is extremely important and needs further study. Furthermore, what is true of the rat may very well not apply to other mammalian species.

Pathology

Circulating gastrin levels are abnormally high in two circumstances: in achlorhydria (37, 55) where the normal feedback suppression of gastrin by acid is absent and in tumorous production of gastrin (Zollinger-Ellison syndrome) (56). An important clinical use of gastrin radioimmunoassay is to diagnose the latter condition, and the estimation of plasma gastrin has become almost a routine screening test. As a result most laboratories are now plagued with borderline results–gastrin levels higher than normal but not sufficiently high to be definitively diagnostic of the Zollinger-Ellison syndrome. Some conditions are known to produce slightly elevated gastrin, e.g., vagotomy (48), uremia (57), and high plasma calcium (58). Furthermore, a patient with a low acid output, though not actually achlorhydric, can be expected to have elevated gastrin levels. Diagnosis of the Zollinger-Ellison syndrome is best made by finding a high gastrin in the face of a high gastric acid output. In case of doubt, use can be made of the different response of tumors and normal G cells to the administration of secretin (59) although this test is not totally reliable in early cases of tumor. A few patients have very high gastrin levels and high gastric acid output but no discoverable pancreatic tumor. In some of these cases the source of the gastrin is found to be the G cells of the pyloric antrum itself (60). This condition is known as antral G cell hyperplasia. An important open question is gastrin's exact role in the etiology of common duodenal ulcer. While it is clearly the responsible agent for ulcer in the Zollinger-Ellison syndrome, the fasting gastrin concentration in duodenal ulcer patients is not significantly raised. It has been shown that the release of gastrin after a meal is increased in duodenal ulcer patients (61) but there is no information on the influence of the rate of gastric emptying and previous medication. In view of the difficulties in interpreting fasting plasma gastrin levels and the incomplete data in duodenal ulcer patients, it is not possible to say whether the excess acid secretion seen in these patients is the result, in part, of excess gastrin production.

SECRETIN

Secretin was discovered in 1902 (1) by Bayliss and Starling, who were able to demonstrate that acid placed in the duodenum elicited the secretion of a pancreatic watery bicarbonate juice even when all nervous connections were severed with the duodenum. They further demonstrated that extracts of the duodenum injected intravenously would produce the same effect on the pancreas. A colleague, W. B. Hardy, coined the term hormone from the Greek word meaning "I arouse to activity." Pure secretin was finally obtained in 1961 (8).

Chemistry

Secretin is composed of 27 amino acids. It was synthesized in 1966 (62) and the synthetic material had all the properties of the natural product. Although secretin in dry form appears to be quite stable, it very rapidly loses its biological activity when dissolved in water. Three other hormonal peptides have been found to have considerable sequence similarities to secretin (Figure 1) and altogether these hormones are referred to as the secretin-glucagon group. This group has considerable similarities in biological actions and are alike in that the whole molecule appears to be necessary for the full range of biological activity, contrary to the gastrin-cholecystokinin group where the COOH-terminal peptides carry full activity.

Localization

Radioimmunoassay of tissue extracts has shown that most of the secretin is found in the mucosa of the duodenum and upper jejunum (63). The total amount of secretin in the jejunum exceeds that of the duodenum but the amount per gram of tissue is somewhat lower. This distribution agrees with the distribution of secretin cells (S cells) as demonstrated by indirect immunofluorescence and electron microscopy (64). No secretin cells are found in the antrum of the stomach.

Pattern of Release

Reliable radioimmunoassays for plasma secretin levels have been developed only recently and less is known of the pattern of secretin release than that of gastrin. It is generally agreed that secretin is rapidly released by intraduodenal acidification (65, 66), as had been predicted by work with bioassays where the threshold for secretin release had been set at pH 4.5 (67). Bioassays, however, are unable to distinguish secretin from VIP, which has identical actions on pancreatic bicarbonate production (68), while radioimmunoassays are completely specific, showing no cross-reaction whatsoever with VIP. In contradiction of an early report, several different workers have now found that secretin release does not follow introduction of glucose into the duodenum (22, 69). Similarly, neither protein nor fat releases secretin (69). It is therefore perhaps not surprising that no significant secretin release has been found after a normal meal (70, 71) since it was already established that after a meal very little of the duodenum becomes significantly acidified (72, 73). The author of this review has been unable to confirm recent reports that secretin is released either during prolonged starvation (74) or after administration of ethanol (75). Thus there is no known "physiological" stimulus producing a release of secretin and this molecule's current status as a hormone is less secure than is the case with gastrin. Accordingly, the nature of the stimulus for postprandial pancreatic bicarbonate production is now somewhat uncertain.

Actions

Secretin increases the watery bicarbonate juice flow from the pancreas and gives rise to an initial release of enzymes, the so-called wash-out effect (77). In isolated pancreatic acinar cell preparations secretin very potently increases the output of cyclic adenosine 3′:5′-monophosphate (cyclic AMP) (76). Secretin also increases the flow of alkaline bile and is a stimulant of gastric pepsin but inhibits the output of gastric acid. It decreases mucosal protein turnover and markedly inhibits gastric and small intestinal muscle tone (77). High doses of secretin give rise to a very rapid release of insulin. Measurements of plasma levels of secretin after exogenous administration suggest that the only action which may be physiological in man is on pancreatic juice production because this effect occurs with blood levels similar to those produced endogenously after duodenal acidification (78). On the other hand, administration of enough exogenous secretin to inhibit gastric acid yields blood levels 10 times greater than detected after endogenous stimuli (79).

Pathology

No pure secretin-secreting tumor has yet been convincingly described, nor has any other condition in which secretin release is excessive. A failure of secretin release has been reported in patients suffering from duodenal ulcer (65). It is known that these patients have a relatively lower duodenal pH than normal subjects and some workers have reported a failure of neutralization of exogenous duodenal acid (80). Duodenal ulceration is thought to be caused by excess acid in the duodenum, and maneuvers which reduce duodenal acidification cure the ulcer. As pancreatic bicarbonate juice is the major neutralizing influence in the duodenum and control is thought to be by secretin, failure of secretin release would be regarded as significant in patients with duodenal ulcer. Studies on duodenal ulcer patients with a healed ulcer after vagotomy and pyloroplasty showed secretin release within the normal range (81). The vagus is not thought to affect release of secretin (82); therefore the return to normal in these treated patients might indicate that the failure of secretin release mentioned above was a secondary phenomenon.

CHOLECYSTOKININ-PANCREOZYMIN

Ivy and Oldberg in 1928 postulated the existence of a hormone which caused gall bladder contraction (83) and named it cholecystokinin. Harper and Raper in 1943 postulated a hormone stimulating the pancreas to secrete enzyme-rich juice and named it pancreozymin (84). By 1964 these two hormonal substances had been purified and shown to co-exist in the same material (85). The hybrid name cholecystokinin-pancreozymin is therefore commonly in use. Some prefer cholecystokinin because this action was the first to be found but recently pancreozymin was put forward as the preferred name (86). The hormone is composed of 33 amino acids (87) of which the entire activity is found in the last 8 amino acids (88). As the last 5 amino acids are identical with those found in gastrin (Figure 2), and many of the actions of these two hormones are the same, they are clearly part of a single hormonal family group. More recently a slightly larger form of CCK has been isolated (89), called CCK Variant with 39 amino acids.

The proportion of natural cholecystokinin that exists as CCK Variant is not known.

Actions and Localization

Apart from its strong stimulation of gall bladder contraction and enzyme secretion, CCK is also a stimulant of gastric pepsin secretion and Brunner's glands secretion. Furthermore, it stimulates the output of hepatic biliary water and bicarbonate, and also enhances the actions of secretin on pancreatic bicarbonate output (90). CCK stimulates motor activity in both the stomach (91) and the intestine (92). It stimulates basal gastric acid output but can competitively inhibit stimulated acid production (93). It has been shown to enhance pancreatic growth (94) and may possibly, in very large doses, affect the appetite. Of this list the most likely true physiological actions are those of gall bladder contraction and uniform secretion of all pancreatic enzymes (95). The COOH-terminal octapeptide of cholecystokinin is rather more powerful in stimulating these actions than the whole molecule (88) but is less selective and is thus a much stronger stimulant of gastric acid output. A similar peptide, caeruelin, is found naturally in the skin of the frog (96). Perhaps this protects the frog from predators by producing severe belly ache on ingestion.

CCK is found in highest concentrations in the jejunum but also occurs throughout the ileum (22, 70). It is localized to a specific endocrine cell type of the APUD series (97), the base of which, containing hormone granules, is applied to the basement membrane and a long thin apical process reaches into the lumen of the gut. On electromicroscopy the cell type is classified "I."

Assay

CCK is thought to circulate in very low concentrations in plasma (98) and there has been considerable difficulty in setting up a sufficiently sensitive radioimmunoassay to detect it. This is because the amount of CCK purified each year needed to raise high affinity antibodies is extremely small. CCK also appears to be rather a poor antigen; it has a sulfated tyrosine which is difficult to label with ^{125}I, and it may well be rather unstable in solution. Furthermore, if CCK is similar to gastrin and perhaps circulates in several different forms, it will be necessary for assays specific to each form to be developed because of the difference in relative biological potencies that would be expected. Several CCK assays have indeed been reported (99–103) and some very exciting findings have been described (104). For example, extremely elevated CCK levels have been reported in patients with pancreatic exocrine deficiency (102), so that this condition could be readily diagnosed by taking a single fasting plasma sample. Confirmation of this report is eagerly awaited.

PANCREATIC GLUCAGON

Glucagon was first found as a contaminant of insulin preparations and was clearly of pancreatic origin. When immunoassays for glucagon were developed it was found that glucagon-like immunoreactivity was also present in the gastrointestinal tract. It was later found that this intestinal material was of a different nature (105). Recently, experiments with pancreatectomized dogs led to the discovery that glucagon indistinguishable from pancreatic glucagon was being produced by the canine stomach (106). The amount of this "gastric pancreatic glucagon" was approximately one-third of the total extractable canine pancreatic glucagon. This finding does not apply generally to other species as in man less than 0.3% of pancreatic glucagon is found outside the pancreas (70). In man plasma pancreatic glucagon levels fall to an effective zero after total pancreatectomy (23).

Role in the Gut

Pancreatic glucagon has a number of effects on the gastrointestinal tract (107). It inhibits gastric acid and gastric motility but stimulates bile flow and Brunner's glands secretions. It is a strong inhibitor of pancreatic enzyme output and small intestinal and large intestinal motor function but greatly increases celiac axis blood flow. These actions are only seen with administration of large doses of glucagon and are probably pharmacological. Any special role for canine "gastric pancreatic glucagon" as opposed to pancreatic glucagon is unknown.

ENTEROGLUCAGON

The glucagon-like immunoreactivity which early immunoassays demonstrated in the gut was soon found to be different from pancreatic glucagon. Extracts containing glucagon-like immunoreactivity (enteroglucagon) when purified by gel chromatography yielded two components, one of high molecular weight (in the 7,000 position) and one of medium molecular weight (in the 3,500 position) (108). Enteroglucagon has not yet been purified completely. Its amino acid sequence is unknown, though presumably there are similarities to pancreatic glucagon to account for cross-reacting antibodies. It is probably a member of the glucagon-secretin group (Figure 1). Enteroglucagon is found in highest concentration in the ileum and, in the primate at least, high concentrations are also found in the colon (70). The enteroglucagon-producing cell (EG cell) is found in the basal part of the mucosal glands and demonstrates such specific features as darkfield luminescence and fluorescence with *O*-phthalaldehyde (109). Only one type of EG cell has been found, suggesting that the multiple forms of enteroglucagon, like those of gastrin, are variants of a single hormone released simultaneously from a single cell.

Physiology

Enteroglucagon is released both by long chain triglycerides and by glucose (110, 111), two substances which inhibit the release of pancreatic glucagon. Thus after a normal hospital lunch, mostly composed of carbohydrate and fat, there occurs a good rise of enteroglucagon but a suppression of pancreatic glucagon (70) (Figure 3). As enteroglucagon is not yet available for pharmacological studies,

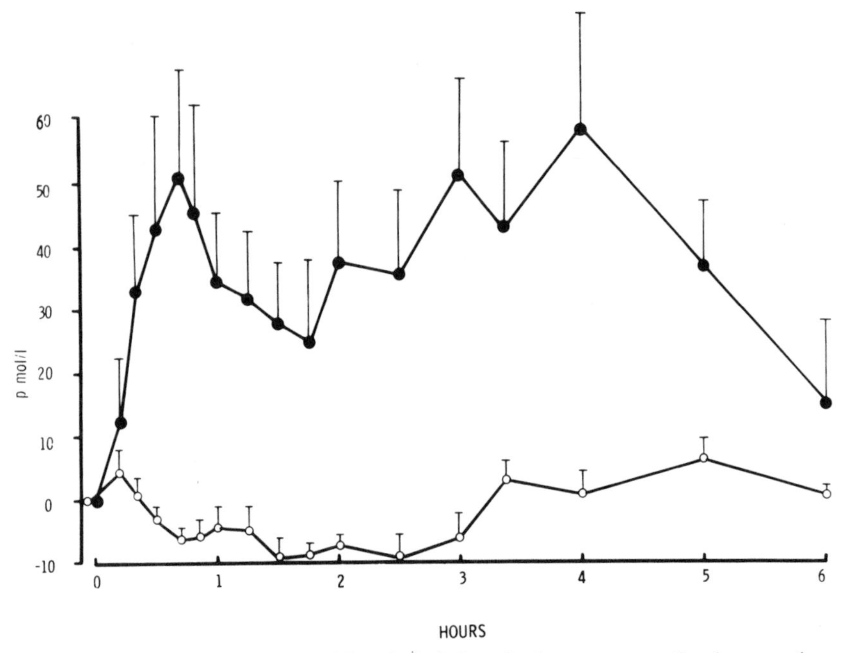

Figure 3. Plasma enteroglucagon (*closed circles*) and plasma pancreatic glucagon (*open circles*) in 10 normal volunteers after a hospital lunch.

the consequences of this plasma rise are unknown. In situations of gastrointestinal hurry when foodstuff passes unduly rapidly down the small intestine, a rather greater than normal rise in enteroglucagon occurs, possibly because a larger number of EG cells are stimulated; e.g., in the dumping syndrome the rise in plasma enteroglucagon is extremely great (112). A good correlation also exists between the rate of gastric emptying (113), the fall in plasma volume (114), and the rate of rise of plasma enteroglucagon after an oral glucose load. This implies that when a larger area of the small intestine is exposed to hypertonic glucose, a larger number of EG cells are stimulated. Coincident with the peak plasma concentration of enteroglucagon in patients with dumping syndrome a change occurs in bowel motility; hypermotility gives way to a period of stasis. One patient with an enteroglucagon-producing tumor (115) showed gross intestinal stasis and an increase in mucosal growth (116). Studies have since been carried out on gut-resected and hyperphagic animals and it is apparent that under these conditions the EG cells hypertrophy and the enteroglucagon levels rise (117). Thus one physiological role of enteroglucagon may be to prevent unabsorbed carbohydrate or fat passing too rapidly down the small intestine by decreasing motility and another may be to improve absorption by increasing mucosal growth.

MOTILIN

Motilin was discovered when alkali was accidentally placed in the duodenum of the dog and strong gastric contractions were noted (118). A stimulant of gastric motor activity was subsequently purified from the semi-pure secretin fractions prepared from porcine duodenal extracts (119). Motilin is a 22-amino acid peptide quite dissimilar from other gut hormones (120, 121) and it is found in highest concentration in the duodenum and jejunum (70) (Figure 4). Immunofluorescence localization shows the motilin-producing cell to be identical with the well known enterochromaffin (EC) cell (122). The motilin-producing cells are confined to the upper small intestine, however, and thus form only a small portion of the total alimentary enterochromaffin cell system.

Pharmacology

Natural motilin has been shown to stimulate motor activity in both antral and fundic gland area pouches of the stomach of dogs (119) and to coordinated emptying of the stomach. In 1973 the synthesis of motilin was reported with norleucine substituted for methionine at position 13 to increase stability (123). Synthetic motilin is not thought to differ in its actions from natural motilin (124). In the guinea pig and the rat motilin potency is low but in the rabbit it has a powerful contracting activity in the duodenum and jejunum (125). In man activity is greater higher in the alimentary tract, with peak contraction in the antrum and duodenum (126). These effects appear to be direct and not mediated through the local innervation (125). Cyclic AMP is not involved (127). Perfusion studies in man demonstrated that the main effect is a decrease in the speed of gastric emptying, possibly due to increased sphincter activity at the stomach outlet. The minimum effect on gastric emptying was seen at an infusion level of 0.1 μg/kg/hr, the maximum at an infusion rate of 0.4 μg/kg/hr (128).

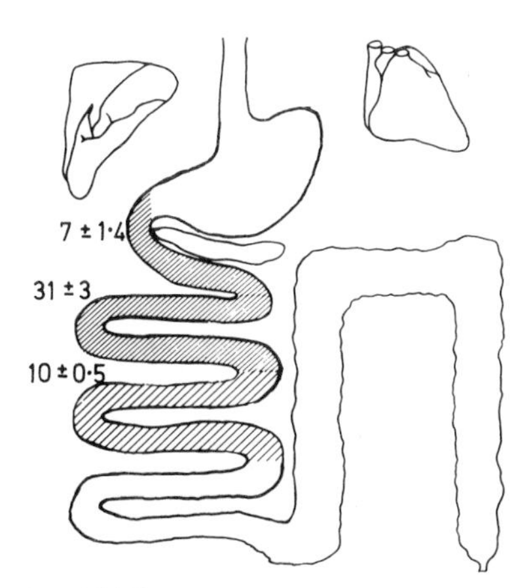

Figure 4. The concentration (*shading*) and total amount (*numbers*) of motilin in extracts of whole bowel taken from 4 baboons.

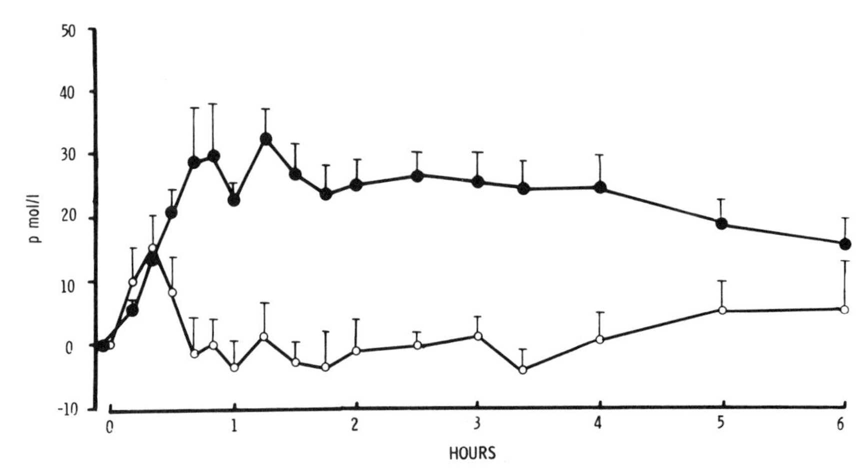

Figure 5. Plasma GIP (*closed circles*) and motilin (*open circles*) in 10 normal volunteers after ingestion of a hospital lunch (as in Figure 3).

Motilin, like secretin, stimulates pepsin output (128) and decreases gastric mucosal growth (129).

Physiology

Radioimmunoassay of motilin is straightforward as it is highly antigenic (130, 131). Gel chromatography of human gut extracts shows only a single component of motilin immunoreactivity eluting in the identical position to natural porcine motilin (132). Fasting plasma levels of motilin are easily measured but considerable individual variations are seen which remain constant over many months. The change in motilin after food is seen in Figure 5. There is a small, rather

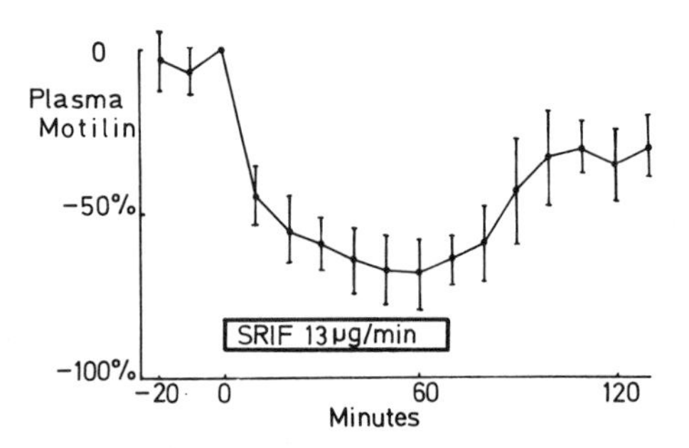

Figure 6. Change in plasma motilin concentrations during infusion of somatostatin (*SRIF*) in 4 healthy volunteers.

variable rise in the first half-hour after which motilin levels are not significantly different from those seen in the fasting state (131). It is reported that the only agent to release motilin is alkali (130) but the author of this review has been unable to confirm this at pH values up to 8.5. A significant suppression of fasting motilin levels occurs during infusion of somatostatin (Figure 6) (131). Plasma levels of motilin measured during motilin infusion (132) which produce a maximal inhibition of gastric emptying (128) can be considered physiological as a few healthy subjects have continuous fasting motilin levels that are actually higher. It seems likely that motilin has a physiological role in the control of gastric emptying in man. The situation in the dog is probably different because motilin is reported to enhance the emptying rate in this species (119). Motilin has also been reported to stimulate interdigestive myoelectric activity in dogs (134).

GASTRIC INHIBITORY PEPTIDE

In early work impure CCK preparations were found to be potent inhibitors of gastric acid secretion whereas pure preparations were not. These observations led to the isolation of a gastric inhibitory peptide (GIP) in 1969 (135). GIP has been purified and sequenced (136). It belongs to the secretin glucagon group of hormones (Figure 1). GIP has been recently synthesized (137) but proof of identity between the natural and synthetic compounds has not yet been published. GIP is found in highest concentration in the jejunum (138). The gut distribution is very similar to that of motilin (Figure 4), the total amounts down the gut being duodenum 7 ± 0.9, jejunum 29 ± 4, and upper ileum 18 ± 2. The GIP-producing cell resides in the middle zone of the jejunal mucosal glands (139).

Actions

GIP is a powerful inhibitor of gastric acid secretion. Pederson and Brown using Bickel type denervated pouches of the body of the stomach of conscious dogs were able to show that an infusion of 4 μg/kg/hr produced a 40% inhibition of acid secretion stimulated by histamine and a 45% inhibition of acid secretion stimulated by insulin hypoglycemia. An infusion of 0.5 μg/kg/hr, on the other hand, produced a 55% inhibition during pentagastrin stimulation. Pepsin output was similarly inhibited (140). An inhibition of fundic motor activity has also been described (141). Like glucagon and VIP, GIP has been shown to stimulate secretion of intestinal juice (142).

GIP stimulates insulin release in the rat (143), baboon (144), dog (145), and man (146). It shares this action with the other members of the secretin-glucagon group of hormones but GIP appears to be the most potent insulintropic hormone. The potent insulin release activity attributed to CCK has now been shown to be caused by GIP in the impure preparations.

Physiology

Radioimmunoassay of circulating GIP shows that it is released after a meal and that the main stimulating components are fat (147) and carbohydrate (148). GIP does not appear to be significantly released by acid. It has been known since 1886 that fat suppressed gastric acid secretion and in 1930 the putative hormone was named enterogastrone (149). Experiments to ascertain whether GIP is the same as enterogastrone have so far proved inconclusive (148). GIP infused at a rate to produce detectable inhibition of gastric acid secretion appears to result in plasma levels somewhat above those produced by physiological stimuli. This may, however, be attributable to the fact that the GIP infusions were not tested in an optimal physiological setting. On the other hand, the insulin-releasing action of GIP on the pancreatic β cell may well occur at physiological blood levels (146). Thus GIP may be the hormonal mediator of the enteroinsular axis, which is the mechanism put forward to explain the observation that glucose taken orally produces a very much larger insulin release than the same amount of glucose given intravenously. As maturity-onset diabetes is first detected by the inability to assimilate an oral glucose load normally, it was of interest to see if a failure of GIP release could be a possible cause. In fact GIP release was found to be either normal (150) or actually increased (151) in maturity-onset diabetics studied after oral glucose. Further investigation is clearly required, however, to ascertain if a separate subgroup of diabetics exists whose disease is caused by a failure of the enteroinsular axis through defective GIP release.

VASOACTIVE INTESTINAL PEPTIDE

Vasoactive intestinal peptide (VIP) was originally isolated, as its name implies, by purification of a peripheral vasodilator substance from porcine intestine (12). When sequenced, it was found to belong to the secretin-glucagon family (13) but it was notable for its strong positive charge at neutral pH. Synthetic VIP was prepared in 1973 (152) and shown to have the full activity of the natural substance (13). VIP in the gut is present in quantities 40 times greater than those of secretin (138). It is also unusual among the gut hormones in that it is very evenly distributed, being present in the esophagus, pancreas, and rectum (Figure 7) (138). It has also recently been demonstrated to be present in the adrenal gland and brain (153). The VIP-producing cell has been localized by immunofluorescence and shown to be pyramidal in shape. A long apical process extending towards the lumen of a gland can sometimes be demonstrated (154).

Actions

VIP has a wide spectrum of activity. Like the others of the glucagon-secretin group it stimulates insulin release and inhibits gastric acid production (142). This latter is an extremely powerful effect. VIP can reverse histamine stimulation at doses lower than equivalently effective amounts of GIP. Like glucagon, VIP

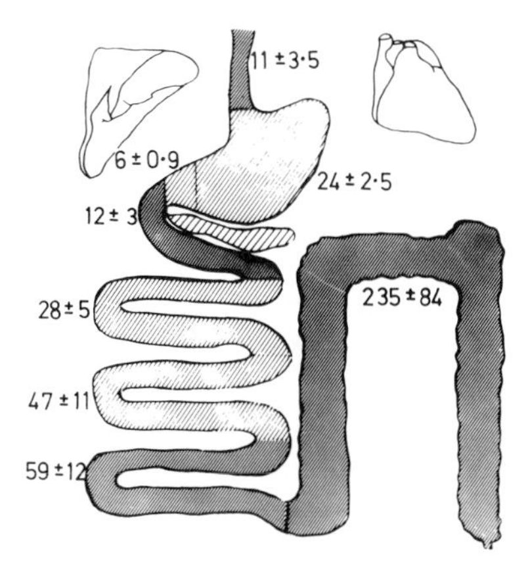

Figure 7. The concentration (*shading*) and total amount (*numbers*) of VIP in extracts of whole bowel taken from 4 baboons (as in Figure 4).

causes hepatic glycogenolysis, though it is relatively less potent (155). The action of VIP on the exocrine pancreas is indistinguishable from that of secretin (68) and, although it is rather less potent (156), the greater amounts of VIP found in the gastrointestinal tract make VIP as likely a candidate for the physiological control of pancreatic juice as secretin. Indeed it is possible that the active principle in the extract made by Bayliss and Sterling in 1902 was, in fact, VIP and not the peptide known today as secretin. VIP stands out from the other members of the secretin-glucagon group in powerfully stimulating small intestinal juice production and increasing mucosal cyclic AMP levels (157). Other actions of VIP which have been described include relaxation of smooth muscle, vasodilatation, and lowering of blood pressure, and an inotropic action on the heart, similar to that of glucagon (12).

Physiology and Pathology

No release of VIP is seen after a meal and fasting blood levels in man are extremely low (70), probably beyond the detection limit of current radioimmunoassays. At the present time there are no reports of any physiological stimulus to VIP release. It is therefore very difficult to speculate on any possible normal physiological role for VIP. It may be a locally acting hormone or even a neurotransmitter substance and therefore not normally released into the circulation.

Very elevated plasma VIP levels have been reported in one particular circumstance. This is the syndrome of severe watery diarrhea associated with an endocrine tumor, usually of pancreatic origin, but tumors of the sympathetic chain have also been described (ganglioneuroblastomas) (158). The syndrome was classically described by Verner and Morrison (159) and is also known as pancreatic cholera or the WDHA syndrome (water diarrhea, hypokalaemia, and achlorhydria) from its main features. Removal of the causative tumor gives immediate cure but during an average time from onset of symptoms to diagnosis of 3 years, 50% of the tumors have metastasized and patients then inevitably succumb to the diarrhea (160). Early diagnosis is thus clearly important. Estimation of plasma VIP levels appears very satisfactory in this respect as there are very few other conditions which give similar gross elevations (161) (Figure 8). In cases where tumor removal proves impossible, VIP levels are also useful in the monitoring of therapy (streptocytocin) (162). The existence of these "VIPomas" allows observation of the effect of long term elevations of VIP levels as a clue to the possible physiological role of VIP. Thus, although named after its vasoactivity, VIP would appear to be more important in controlling small intestinal juice production and its physiological role is most likely to lie in this direction.

SOMATOSTATIN

Somatostatin, or growth hormone release-inhibiting hormone, was first purified from the ovine hypothalamus as a substance inhibiting growth hormone release (163). It is a 14-amino acid polypeptide with a cystine bridge. It has been synthesized in large quantities and is now freely available for experimental purposes. Although originally extracted from the hypothalamus it has now been shown to occur more widely in the brain and high concentrations are also found in the gastrointestinal mucosa and in the pancreas (164). Indirect immunofluorescence has localized the somatostatin-producing cell of the gut to the classic D-cell which exists in both the gastrointestinal mucosa and the islets of Langerhans (165). In the brain somatostatin is found in nerve cell bodies in the periventricular region of the anterior hypothalamus and also more widely in nerve terminals (166).

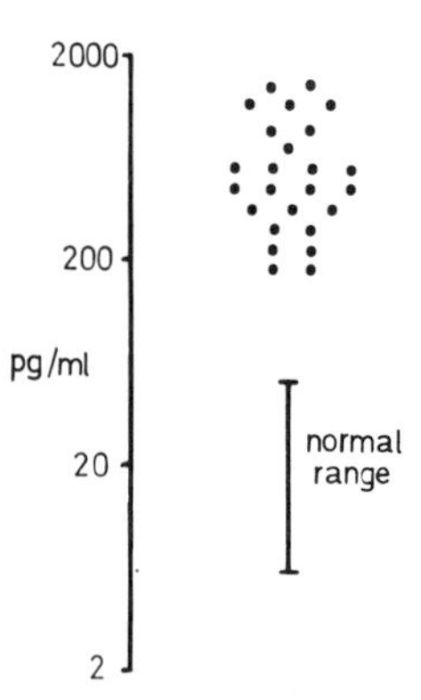

Figure 8. VIP concentrations in the initial plasma sample received from 25 patients suffering from the Verner-Morrison syndrome (*solid dots*) compared with the normal range seen in 35 hospitalized patients suffering from other conditions (*solid line*).

Actions

Somatostatin is a potent inhibitor not only of growth hormone release but also of thyroid-stimulating hormone (TSH) release during TRH stimulation (167). It also completely inhibits the release of insulin and glucagon from the islets of Langerhans (168, 169). It has extreme potency in this direction because all known stimuli of insulin and glucagon release can be totally blocked by somatostatin (170). It has been postulated to act beyond the stage of cyclic AMP control and may interfere with membrane release of hormone (163). It is specific for certain endocrine cells because no effect at all is seen on the release of adrenocorticotropic hormone (ACTH), luteinizing hormone (LH), or follicle-stimulating hormone (FSH) for example (163).

Recently somatostatin has been found to have powerful gastrointestinal effects. It inhibits the release of gastrin (171) and also, independently of its effect on gastrin, directly inhibits the release of gastric acid (172, 173). Consequently it is one of the most effective inhibitors of gastric acid, as acting both on the G cell and the parietal cell, it prevents feedback escape due to increased gastrin release. During somatostatin infusion gastric pH rapidly rises to greater than 7 even in patients with gastrin-producing tumors (174). Somatostatin also inhibits pepsin secretion (173) and delays gastric emptying (131). As can be seen in Figure 6 it inhibits the release of motilin (131) and it has also been claimed by some to inhibit the release of secretin (175, 176), though this is denied by others (177). Somatostatin totally blocks the effect of CCK in stimulating gall bladder contraction or enzyme secretion from the pancreas (175, 177).

The wide spectrum of actions of somatostatin make it unlikely that it acts by general release into the circulation. It is more probable that it exerts local tissue control. It is probably a member of the locally active paracrine system and is one of a growing number of important local hormonal substances, such as nerve growth factor, prostaglandins, 5-hydroxytryptamine, etc. The physiological importance of somatostatin is difficult to estimate. Its potential importance is, of course, enormous as somatostatin could be responsible for such common conditions as diabetes if released in excess, or duodenal ulcer if deficient.

PANCREATIC POLYPEPTIDE

In 1968 Kimmel in the course of purifying chicken insulin observed another peak of peptide material which he isolated and termed avian pancreatic polypeptide (9). This peptide was shown to be derived from endocrine cells located in the exocrine pancreas (178). Subsequently, a bovine pancreatic polypeptide (BPP) was purified (179) and found to have 15 of its amino acids homologous with the avian material (180, 181). Similar pancreatic polypeptides from pig, sheep, and human differ from BPP in only 1 or 2 residues at positions 2, 6, or 23 (182). Immunocytochemical localization of pancreatic polypeptide-producing (PP-producing) cells in mammals showed the majority to be present in the exocrine pancreas with a few cells seen in the islets of Langerhans and also a few cells in the gastrointestinal mucosa (183). In the dog PP cells were concentrated in the head of the pancreas where numerous F cells had been previously described in EM studies (183).

BPP administered to dogs was found at the lowest doses to inhibit gall bladder contraction, increase bile duct tone, and inhibit pancreatic enzyme output. It also had a biphasic effect on pancreatic juice production, first stimulating then inhibiting. At slightly higher doses BPP was shown to stimulate basal gastric acid secretion but to inhibit pentagastrin-stimulated acid secretion. At still higher doses it had a general inhibitory action on gastrointestinal motility (181, 184).

Physiology and Pathology

A considerable elevation of plasma PP is seen after meal ingestion in man, the rise being only slightly smaller than that of insulin (185, 186). As most of the PP cells are located in the pancreas this postprandial release is presumably not caused by direct contact with food but is the result of some enteric signal to the pancreas. The most sensitive pharmacological effects of pancreatic polypeptide oppose the actions of CCK and this is, therefore, the most reasonable surmise as to its physiological role.

It has recently been reported that PP is produced in considerable amounts by endocrine pancreatic tumors (gastrinomas, VIPomas, glucagonomas, and insulinomas (186)). PP production by these tumors appears to be an intrinsic property and is seen equally in the tumor metastases. In some cases the PP content and plasma PP elevations are considerably greater than those of the hormone from which the tumor gets its name. As PP production is seen equally in all four types of endocrine pancreatic tumors, its clinical effects are presumably less strong than those of the more classic hormones which have given rise to the recognized clinical syndromes. A tumor-producing PP alone has not yet been described and the clinical features to be expected are unknown. PP cell involvement is of interest both from the viewpoint of understanding why PP cells should so closely associate themselves with endocrine pancreatic tumor cells and also because of the potential for easy tumor detection by PP measurement. At the present time considerable difficulty is encountered in the diagnosis of early, and therefore curable, cases of pancreatic endocrine tumor and the presence of another tumor marker would be of great clinical benefit.

THE FUTURE

Testimony to the presence in the gut of more hormones yet to be identified is the large number of endocrine cells in the gut mucosa which have been described and classified but whose product is not yet known (4). A number of possible hormones have been proposed (3). Anderson has coined the term bulbogastrone to explain the phenomenon of marked inhibition of gastric acid when the duodenal bulb has been acidified (187). Adelson and Rothman are working with a chymotrypsin-stimulating peptide, chymodenin (188). Harper et al. have proposed a pancreas inhibiting factor present in the colonic mucosa (189). Urogastrone, a material extracted from human urine which inhibits gastric acid, has proved to be identical with epidermal growth factor (190), holding out the exciting possibility that a single substance would both inhibit gastric acid and enhance the growth and repair of the duodenal mucosa after duodenal ulcer. Various other phenomena may indicate other hormones, for example, acid stimulation by substances introduced into the small intestine and acid inhibition by other substances placed in the stomach. In any case, it is clear that gut endocrinology is still an infant science.

REFERENCES

1. Bayliss, W. M., and Starling, E. H. (1902). The mechanism of pancreatic secretion. J. Physiol. 28:325.
2. Edkins, J. S. (1905). On the chemical mechanism of gastric secretion. Proc. R. Soc. Lond. (Biol.) 76:376.
3. Grossman, M. I., et al. (1974). Candidate hormones of the gut. Gastroenterology 67:730.
4. Solcia, E., Pearse, A. G. E., Grube, O., Kobayashi, S., Bussola, G., Creutzfeldt, W., and Gepts, W. (1973). Revised Wiesbaden classification of gut cells. Rendic. Gastroenterol. 5:13.
5. Price, J. B., Takeshige, K., Parsa, M., and Voorhees, A. B. (1971). Characteristics of animals maintained without splanchnic portal organs. Surgery 70:768.
6. Lund, B., Schmidt, A., and Deckert, T. (1975). Portal and cubital serum insulin during oral, portal and cubital glucose tolerance tests. Acta Med. Scand. 197:275.
7. Bloom, S. R. (1975). Gastrointestinal hormones. Proc. R. Soc. Med. 68:710.
8. Jorpes, E., and Mutt, V. (1961). On the biological activity and amino acid composition of secretin. Acta Chem. Scand. 15:1790.
9. Kimmel, J. R., Pollock, H. G., and Hazelwood, R. L. (1968). Isolation and characterization of chicken insulin. Endocrinology 83:1323.
10. Brown, J. C., Mutt, V., and Pederson, R. A. (1970). Further purification of a polypeptide demonstrating enterogastrone activity. J. Physiol. (Lond.) 309:56.
11. Brown, J. C., and Dryburgh, J. R. (1971). A gastric inhibitory polypeptide II: The complete amino acid sequence. Can. J. Biochem. 49:867.
12. Said, S. I., and Mutt, V. (1970). Polypeptide with broad biological activity: Isolation from small intestine. Science 169:1217.
13. Bodanszky, M., Klausner, Y. S., and Said, S. I. (1973). Biological activities of synthetic peptides corresponding to fragments of and to the entire sequence of the vasoactive intestinal peptide. Proc. Natl. Acad. Sci. U.S.A. 70:382.
14. Polak, J. M., Pearse, A. G. E., Bloom, S. R., and Joffe, S. N. (1975). Quantification of physiological changes in gut endocrine cells: Immunocytochemical and radioimmunoassays of secretin cell function. Acta Hepatogastroenterol. 22:137.
15. Robinson, G., and Dawson, I. (1975). Immunochemical studies of the endocrine cells of the gastrointestinal tract. I. The use and value of peroxidase-conjugated antibody techniques for the localization of gastrin-containing cells in the human pyloric antrum. Histochem. J. 7:321.
16. Pearse, A. G. E. (1968). Common cytochemical and ultrastructural characteristics of cells producing polypeptide hormones (the APUD series) and their relevance to thyroid and ultimobranchial C cells and calcitonin. Proc. R. Soc. Lond. (Biol.) 170:71.
17. Polak, J. M., Pearse, A. G. E., and Heath, C. M. (1975). Complete identification of

endocrine cells in the gastrointestinal tract using semithin-thin sections to identify motilin cells in human and animal intestine. Gut 16:225.
18. Lin, M. C., Wright, D. E., Hruby, V. J., and Rodbell, M. (1975). Structure-function relationships in glucagon: Properties of highly pruified des-his, monoiodo-, and (des asn^{28}, thr^{29}) (homoserine lactone^{27})-glucagon. Biochemistry 14:1559.
19. Davis, J. O., Freeman, R. H., Johnson, J. A., and Spielman, W. S. (1974). Agents which block the action of the renin-angiotensin system. Circ. Res. 34:279.
20. Loveridge, N., Bloom, S. R., Welbourn, R. B., and Chayen, J. (1974). Quantitative cytochemical estimation of the effect of pentagastrin (0.005–5 pg/ml) and of plasma gastrin on the guinea pig fundus in vitro. Endocrinology 3:389.
21. Shima, K., Sawazaki, N., Tanaka, R., Tarui, S., and Nishikawa, M. (1975). Effect of an exposure to chloramine-t on the immunoreactivity of glucagon. Endocrinology 96:1254.
22. Bloom, S. R. (1974). Hormones of the gastrointestinal tract. Brit. Med. Bull. 30:62.
23. Barnes, A. J., and Bloom, S. R. (1976). Pancreatectomised man: A model for diabetes without glucagon. Lancet 11:734.
24. Gregory, R. A., and Tracy, H. J. (1964). The constitution and properties of two gastrins extracted from hog antral mucosa. Gut 5:103.
25. Anderson, J. D., Barton, M. A., Gregory, R. A., Hardy, P. M., Kenner, G. W., MacLeod, J. K., Preston, J., Sheppard, R. D., and Morley, J. S. (1964). The antral hormone gastrin II. Synthesis of gastrin. Nature 204:933.
26. Yalow, R. S., and Berson, S. A. (1970). Size and charge distinctions between endogenous human plasma gastrin in peripheral blood and heptadecapeptide gastrins. Gastroenterology 58:609.
27. Gregory, R. A., and Tracy, H. J. (1975). The chemistry of the gastrins: Some recent advances. *In* J. C. Thompson (ed.), International Symposium on Gastrointestinal Hormones, Oct. 9–12 1974, Galveston Texas, pp. 13–24.
28. Yalow, R. S., and Wu, N. (1973). Additional studies on the nature of big big gastrin. Gastroenterology 65:19.
29. Rehfeld, J. F., Stadil, F., and Vikelsoe, J. (1974). Immunoreactive gastrin components in human serum. Gut 15:102.
30. Gregory, R. A., and Tracy, H. J. (1974). Isolation of two minigastrins from Zollinger-Ellison tumour tissue. Gut 15:683.
31. Hailet, D., Walsh, J. H., and Grossman, M. I. (1974). Pure human minigastrin: secretory potency and disappearance rate. Gut 15:686.
32. Lin, T. M. (1972). Gastrointestinal actions of the C-terminal tripeptide of gastrin. Gastroenterology 63:922.
33. Petersen, H., and Myren, J. (1975). Pentagastrin dose-response in peptic ulcer disease. Scand. J. Gastroenterol. 10:705.
34. Walsh, J. H., and Grossman, M. I. (1975). Gastrin. N. Engl. J. Med. 292:1324.
35. Walsh, J. H., Debas, H. T., and Grossman, M. I. (1974). Pure human big gastrin: Immunochemical properties, disappearance half time, and acid stimulating action in dogs. J. Clin. Invest. 54:477.
36. Bloom, S. R., Polak, J. M., and Pearse, A. G. E. (1973). Distribution of gut hormones by radioimmunoassay and immunofluorescence. *In* Endocrinology, pp. 91–99. Heinemann, London.
37. Polak, J. M., Hoffbrand, A. V., Reed, P. I., Bloom, S. R., and Pearse, A. G. E. (1973). Qualitative and quantitative studies of antral and fundic G cells in pernicious anaemia. Scand. J. Gastroenterol. 8:361.
38. Lichtenberger, L. M., Lechago, J., and Johnson, L. R. (1975). Depression of antral and serum gastrin concentration by food deprivation in the rat. Gastroenterology 68:1473.
39. McGuigan, J. E., and Greider, M. H. (1971). Correlative immunochemical and light microscopic studies of the gastrin cell of the antral mucosa. Gastroenterology 60:223.
40. Brandsborg, D., Brandsborg, M., Rokkjaer, M., Bone, J., Lovgreen, N. A., and Amdrup, E. (1975). Variations in serum gastrin concentration after feeding in man, dog and pig. J. Surg. Res. 19:1.
41. Stremple, J. F., and Elliott, D. W. (1975). Gastrin determinations in sympotomatic patients before and after standard ulcer operations. Arch. Surg. 110:875.

42. Bieberdorf, F. A., Walsh, J. H., and Fordtran, J. S. (1975). Effect of optimum therapeutic dose of poldine on acid secretion, gastric acidity, gastric emptying, and serum gastrin concentration after a protein meal. Gastroenterology 68:50.
43. Bloom, S. R., Mortimer, C. H., Thorner, M. O., Besser, G. M., Hall, R., Gomez-Pan, A., Roy, V. M., Russell, R. C. G., Coy, D. H., Kastin, A. J., and Schally, A. V. (1974). Growth hormone release inhibiting hormone: Inhibition of gastrin and gastric acid secretion. Lancet ii:1106.
44. Blair, E. L., Greenwell, J. R., Grund, E. R., Reed, J. D., and Sanders, D. J. (1975). Gastrin response to meals of different composition in normal subjects. Gut 16:766.
45. Walsh, J. H., Richardson, C. T., and Fordtran, J. S. (1975). pH dependence of acid secretion and gastrin release in normal and ulcer subjects. J. Clin. Invest 55:462.
46. Smith, C. L., Kewenter, J., Connell, A. M., Ardill, J., Hayes, R., and Buchanan, K. (1975). Control factors in the release of gastrin by direct electrical stimulation of the vagus. Am. J. Dig. Dis. 20:13.
47. Farooq, D., and Walsh, J. H. (1975). Atropine enhances serum gastrin response to insulin in man. Gastroenterology 68:662.
48. Kronborg, O., Stadil, F., Rehfeld, J., and Christiansen, P. M. (1973). Relationship between serum gastrin concentrations and gastric acid secretion in duodenal ulcer patients before and after selective and highly selective vagotomy. Scand. J. Gastroenterol. 8:491.
49. Walsh, J. H., Maxwell, V., and Isenberg, J. I. (1975). Biological activity and clearance of human big gastrin in man. Clin. Res. 23:259A.
50. Giles, G. R., Mason, M. C., Humphries, C., and Clark, C. G. (1969). Action of gastrin on the lower oesophageal sphincter in man. Gut 10:730.
51. Castell, D. O., and Harris, L. D. (1970). Hormonal control of gastroesophageal -sphincter strength. N. Engl. J. Med. 282:886.
52. Grossman, M. I. (1974). What is physiological?. Gastroenterology 65:994.
53. Grossman, M. I. (1974). What is physiological? Round 2. Gastroenterology 67:766.
54. Johnson, L. R., Lichtenberger, L. M., Copeland, E. M., Dudrick, S. J., and Castro, G. A. (1975). Action of gastrin on gastrointestinal structure and function. Gastroenterology 68:1184.
55. Ganguli, P. C., Cullen, D. R., and Irvine, W. J. (1971). Radioimmunoassay of plasma gastrin in pernicious anaemia, achlorhydria without pernicious anaemia, hypochlorhydria and controls. Lancet i:155.
56. Isenberg, J. I., Walsh, J. H., and Grossman, M. I. (1973). Zollinger-Ellison syndrome. Gastroenterology 65:140.
57. Gedde-Dahl, D. (1975). Serum gastrin response to food stimulation in male azotemic patients. Scand. J. Gastroenterol. 10:683.
58. McGuigan, J. E., Colwell, J. A., and Franklin, J. (1974). Effect of parathyroidectomy on hypercalcaemic hypersecretory peptic ulcer disease. Gastroenterology 66:269.
59. Schrumpf, E., Petersen, H., Berstad, A., et al. (1973). The effect of secretin on plasma gastrin in the Zollinger-Ellison syndrome. Scand. J. Gastroenterol. 8:145.
60. Polak, J. M., Stagg, B., and Pearse, A. G. E. (1972). Two types of Zollinger-Ellison syndrome: immunofluorescent, cytochemical and ultrastructural studies of the antral and pancreatic gastrin cells in different clinical states. Gut 13:501.
61. McGuigan, J. E., and Trudeau, W. L. (1973). Differences in rates of gastrin release in normal persons and patients with duodenal ulcer disease. N. Engl. J. Med. 288:64.
62. Bodanszky, M., Ondetti, M. A., and Levine, S. D. (1966). Synthesis of a heptacosa peptide amide with the hormonal activity of secretin. Chem. Industr. 42:1757.
63. Bloom, S. R., and Bryant, M. G. (1973). Distribution of radioimmunoassayable gastrin, secretin, pancreozymin and enteroglucagon in rat, dog and baboon gut. J. Endocrinol. 59:44.
64. Polak, J. M., Bloom, S. R., Coulling, I., and Pearse, A. G. E. (1971). Immunofluorescent localisation of secretin in the canine duodenum. Gut 12:605.
65. Bloom, S. R., and Ward, A. S. (1975). Failure of secretin release in patients with duodenal ulcer. Br. Med. J. 1:126.
66. Boden, G., Essa, N., Owen, O. E., and Reichle, F. A. (1974). Effects of intraduodenal administration of HCl and glucose on circulating immunoreactive secretin and insulin concentrations. J. Clin. Invest 53:1185.

67. Meyer, J. H., Way, L. W., and Grossman, M. I. (1970). Pancreatic response to acidification of various lengths of proximal intestine in the dog. Am. J. Physiol. 219:971.
68. Said, S. I., and Mutt, V. (1972). Isolation from porcine intestinal wall of a vasoactive octacosapeptide related to secretin and to glucagon. Eur. J. Biochem. 28:199.
69. Boden, G., Essa, N., and Owen, O. E. (1975). Effects of intraduodenal amino acids, fatty acids, and sugars on secretin concentrations. Gastroenterology 68:722.
70. Bloom, S. R., Bryant, M. G., and Cochrane, J. P. S. (1975). Normal distribution and post-prandial release of gut hormones. Clin. Sci. Mol. Med. 49:3P.
71. Chey, W. Y., Rhodes, R. A., Lee, K. Y., and Hendricks, J. (1975). Radioimmunoassay of secretin: Further studies. *In* J. C. Thompson (ed.), Gastrointestinal Hormones, pp. 269–281. University of Texas Press, Austin.
72. Rhodes, J., and Prestwich, C. J. (1966). Acidity at different sites in the proximal duodenum of normal subjects and patients with duodenal ulcer. Gut 7:509.
73. Meldrum, S. J., Watson, B. W., Riddle, H. C., Bown, R. L., and Sladen, G. E. (1972). pH profile of gut as measured by radiotelemetry capsule. Br. Med. J. 1:104.
74. Henry, R. W., Flanagan, R. W. J., and Buchanan, K. D. (1975). Secretin: a new role for an old hormone. Lancet ii:202.
75. Straus, E., Urbach, H., and Yalow, R. S. (1975). Alcohol stimulated secretion of immunoreactive secretin. N. Engl. J. Med. 293:1031.
76. Deschodt-Lanckman, M., Robberecht, P., Pector, J. C., and Christophe, J. (1975). Effects of somatostatin on pancreatic exocrine function. Interaction with secretin. European Pancreatic Club VIII Symposium, p. 123. Saint George, Paris.
77. Hubel, K. A. (1972). Secretin: A long progress note. Gastroenterology 62:318.
78. Bloom, S. R. (1975). The development of a radioimmunoassay for secretin. *In* J. C. Thompson (ed.), Gastrointestinal Hormones, pp. 257–268. University of Texas Press, Austin.
79. Ward, A. S., and Bloom, S. R. (1974). The role of secretin in the inhibition of gastric secretion by intraduodenal acid. Gut 15:889.
80. Wormsley, K. G. (1972). Reactions to acid in the intestine in health and disease. Gut 13:40.
81. Ward, A. S., and Bloom, S. R. (1976). Effect of vagotomy on secretin release in man. Gut 16:951.
82. Konturek, S. J., Popiela, T., and Thor, P. (1971). Effect of vagotomy and pyloroplasty on pancreatic dose-response curves to secretin in man. Am. J. Dig. Dis. 16:1087.
83. Ivy, A. C., and Oldberg, E. (1928). A hormone mechanism for gallbladder contraction and evacuation. Am. J. Physiol. 86:599.
84. Harper, A. A., and Raper, H. S. (1943). Pancreozymin, a stimulant of the secretion of pancreatic enzymes in extracts of the small intestine. J. Physiol. (Lond.) 102:115.
85. Jorpes, E., Mutt, V., and Toczko, K. (1964). Further purification of cholecystokinin and pancreozymin. Acta Chem. Scand. 18:2408.
86. IUPAC-IUB (1974). The nomenclature of peptide hormones. IUPAC-IUB commission on biochemical nomenclature recommendations. J. Biol. Chem. 250:3215.
87. Jorpes, J. E. (1968). The isolation and chemistry of secretin and cholecystokinin. Gastroenterology 55:157.
88. Debas, H. T., and Grossman, M. I. (1973). Pure cholecystokinin: Pancreatic protein and bicarbonate response. Digestion 9:469.
89. Mutt, V., and Jorpes, J. E. (1968). Structure of porcine cholecystokinin-pancreozymin I cleavage with thrombin and with trypsin. Eur. J. Biochem. 6:156.
90. Vaysse, N., Laval, J., Duffaut, M., and Ribet, A. (1974). Effect of secretin and graded doses of CCK-PZ on pancreatic secretion in man. Am. J. Dig. Dis. 19:887.
91. Debas, H. T., Farooq, O., and Grossman, M. I. (1975). Inhibition of gastric emptying is a physiological action of cholecystokinin. Gastroenterology 68:1211.
92. Dollinger, H. C., Berz, R., Raptis, S., von Uexkull, T., and Goebell, H. (1975). Effects of secretin and cholecystokinin on motor activity of human jejunum. Digestion 12:9.
93. Odori, Y., and Magee, D. F. (1970). Cholecystokinin-pancreozymin as a physiological mediator of gastric acid inhibition. Pflügers Arch. 318:287.

94. Barrowman, J. A. (1975). The trophic action of gastrointestinal hormones. Digestion 12:92.
95. Robberecht, P., Cremer, M., Vandermeers, A., Vandermeers-Piret, M., Cotton, P., De Neef, P., and Christophe, J. (1975). Pancreatic secretion of total protein and of three hydrolases collected in healthy subjects via duodenoscopic cannulation. Gastroenterology 69:374.
96. Dockray, G. J., and Hopkins. C. R. (1975). Caerulein secretion by dermal glands in *Xenopus laevis*. J. Cell Biol. 64:724.
97. Polak, J. M., Pearse, A. G. E., Bloom, S. R., Buchan, A. M. J., Rayford, P. L., and Thompson, J. C. (1975). Identification of cholecystokinin-secreting cells. Lancet ii:1016.
98. Bloom, S. R. (1974). Progress report: radioimmunoassay of intestinal hormones. Gut 15:502.
99. Young, J. D., Lazarus, L., and Chisholm, D. J. (1969). Radioimmunoassay of pancreozymin cholecystokinin in human serum. J. Nucl. Med. 10:743.
100. Go, V. L. W., Ryan, R. J., and Summerskill, W. H. J. (1971). Radioimmunoassay of porcine cholecystokinin-pancreozymin. J. Lab. Clin. Med. 77:684.
101. Johnson, A. G., and McDermott, S. J. (1973). Sensitive bioassay of cholecystokinin in human serum. Lancet ii:589.
102. Harvey, R. F., Dowsett, L., Hartog, M., and Read, A. E. (1974). Radioimmunoassay of cholecystokinin-pancreozymin. Gut 15:690.
103. Reeder, D. D., Becker, H. D., Smith, N. J., Rayford, P. L., and Thompson, J. C. (1973). Measurement of endogenous release of cholecystokinin by radioimmunoassay. Ann. Surg. 178:304.
104. Low-Beer, T. S., Harvey, R. F., Davies, E. R., and Read, A. E. (1975). Abnormalities of serum cholecystokinin and gallbladder emptying in coeliac disease. N. Engl. J. Med. 292:961.
105. Unger, R. H., and Eisentraut, A. M. (1967). Glucagon. *In* Hormones in Blood, Ch. 5, p. 83. Academic Press, New York.
106. Sasaki, H., Rubalcava, B., Baetens, D., Blazquez, E., Srikant, C. B., Orci, L., and Unger, R. H. (1975). Identification of glucagon in the gastrointestinal tract. J. Clin. Invest. 56:135.
107. Bloom, S. R. (1975). Glucagon. Br. J. Hosp. Med. 13:150.
108. Valverde, I., Rigopoulou, D., Exton, J., Ohneda, A., Eisentraut, A., and Unger, R. H. (1968). Demonstration and characterization of a second fraction of glucagon-like immunoreactivity in jejunal extracts. Am. J. Med. Sci. 255:415.
109. Polak, J. M., Bloom, S. R., Coulling, I., and Pearse, A. G. E. (1971). Immunofluorescent localisation of enterglucagon cells in the gastrointestinal tract of the dog. Gut 12:311.
110. Holst, J. J., Christiansen, J., and Kuhl, C. The enteroglucagon response to intrajejunal infusion of glucose, triglycerides and sodium chloride and its relation to jejunal inhibition of gastric acid secretion in man. Scand. J. Gastroenterol. In press.
111. Bottger, I., Dobbs, R., Faloona, G. R., and Unger, R. H. (1973). The effects of triglyceride absorption upon glucagon, insulin and gut glucagon-like immunoreactivity. J. Clin. Invest. 52:2532.
112. Bloom, S. R., Royston, C. M. S., and Thomson, J. P. S. (1972). Enteroglucagon release in the dumping syndrome. Lancet ii:789.
113. Ralphs, D. N. L., Bloom, S. R., Lawson-Smith, C., and Thompson, J. P. S. (1975). The relationship between gastric emptying rate and plasma enteroglucagon concentration. Gut 16:406.
114. Thomson, J. P. S., Bloom, S. R., Haynes, S., and Ogawa, O. (1973). Plasma enteroglucagon and plasma volume changes after oral hypertonic glucose: their relationship to the dumping syndrome. Br. J. Surg. 60:308.
115. Bloom, S. R. (1972). An enteroglucagon tumour. Gut 13:520.
116. Gleeson, M. H., Bloom, S. R., Polak, J. M., Henry, K. and Dowling, R. M. (1971). An endocrine tumour in kidney affecting small bowel structure, motility and absorptive function. Gut 12:733.
117. Jacobs, L. R., Polak, J. M., Bloom, S. R., and Dowling, H. R. (1976). Does

enteroglucagon play a trophic role in intestinal adaptation. Clin. Sci. Mol. Med. 50:14P.
118. Brown, J. C., Johnson, L. P., and Magee, D. F. (1976). Effect of duodenal alkalinization on gastric motility. Gastroenterology 50:333.
119. Brown, J. C., Mutt, V., and Dryburgh, J. R. (1971). The further purification of motilin, a gastric motor activity stimulating polypeptide from the mucosa of the small intestine of hogs. Can. J. Physiol. Pharmacol. 49:399.
120. Brown, J. C., Cook, M. A., and Dryburgh, J. R. (1973). Motilin, a gastric motor activity stimulating polypeptide: The complete amino acid sequence. Can. J. Biochem. 51:533.
121. Schubert, H., and Brown, J. C. (1973). Correction to the amino acid sequence of porcine motilin. Can. J. Biochem. 52:7.
122. Pearse, A. G. E., Polak, J. M., Bloom, S. R., Adams, C., Dryburgh, J. R., and Brown, J. C. (1974). Enterochromaffin cells of the mammalian small intestine as the source of motilin. Virchows Arch. (Zellpathol.) 16:111.
123. Wunsch, E., Brown, J. C., Deimer, K. H., Drees, F., Jaeger, E., Musiol. J., Scharf, R., Stocker, H., Thamm, P., and Wendleberger, G. (1973). Zur Synthese von norleucin-LW-motilin. The total synthesis of norleucine-LW-motilin (preliminary communication). Z. Naturforsch. 28c:235.
124. Strunz, U., Domschke, W., Domschke, S., Mitznegg, P., Wunsch, E., Jaeger, E., and Demling, L. (1976). Gastroduodenal motor response to natural motilin and synthetic position 13 substituted motilin analogues. A comparative in vitro study. Scand. J. Gastroenterol. 11:199.
125. Strunz, U., Domschke, W., Mitznegg, P., Domschke, S., Schubert, E., Wunsch, E., Jaeger, E., and Demling, L. (1975). Analysis of the motor effects of 13-norleucine motilin on the rabbit, guinea pig, rat and human alimentary tract in vitro. Gastroenterology 68:1485.
126. Mitznegg, P., Strunz, U., Domschke, W., Wunach, E., and Demling, L. (1975). Analysis of the motor effect of synthetic motilin on animal and human intestinal smooth muscle in vitro. Arch. Pharmacol. Suppl. 287:R45.
127. Schubert, E., Mitznegg, P., Strunz, U., Domschke, W., Domschke, S., Wunsch, E., Jaeger, E., Demling, L., and Heim, F. (1975). Influence of the hormone analogue 13-NLE-motilin and of 1-methyl-3-isobutylxanthine on tone and cyclic 3′,5′-AMP content of antral and duodenal muscle in the rabbit. Life Sci. 16:263.
128. Ruppin, H., Domschke, S., Domschke, W., Wunsch, E., Jaeger, E., and Demling, L. (1975). Effects of 13-NLE-motilin in man–inhibition of gastric evacuation and stimulation of pepsin secretion. Scand. J. Gastroenterol. 10:199.
129. Mitznegg, P., Domsche, W., Domsche, S., Belohlavek, D., Sprugel, W., Strunz, U., Wunsch, E., Jaeger, E., and Demling, L. (1975). Protein synthesis in human gastric mucosa: effects of pentagastrin, secretin and 13-NLE-motilin. Acta Hepatogastroenterol. 22:333.
130. Dryburgh, J. R., and Brown, J. C. (1975). Radioimmunoassay for motilin. Gastroenterology 68:1169.
131. Bloom, S. R., Ralphs, D. N., Besser, G. M., Hall, R., Coy, D. H., Kastin, A. J., and Schally, A. V. (1975). Effect of somatostatin on motilin levels and gastric emptying. Gut 16:834.
132. Bryant, M. G., and Bloom, S. R. (1975). Characterisation of the new gastrointestinal hormones. Gut 16:840.
133. Mitznegg, P., Bloom, S. R., Domsche, W., Domsche, S., Wunsche, E., and Demling, L. Disappearance half-time of exogenous and endogenous motilin in man. Arch. Pharmacol. Suppl. 288 In press.
134 Itoh, Z., Hizhwh, I., and Takeuchi, S. Hunger contraction and motilin. Fifth International Symposium on Gastrointestinal Motility. In press.
135. Brown, J. C., Pederson, R. A., Jorpes, E., and Mutt, V. (1969). Preparation of a highly active enterogastrone. Can. J. Physiol. Pharmacol. 47:113.
136. Brown, J. C., and Dryburgh, J. R. (1971). A gastric inhibitory polypeptide II: The complete amino acid sequence. Can. J. Biochem. 49:867.
137. Yajima, H., Ogawa, H., Kubota, M., Tobe, T., Fujimura, M., Henmi, K., Torizuka, K., Adachi, H., Imura, H., and Taminato, T. (1975). Synthesis of the trietetracontapep-

tide corresponding to the entire amino acid sequence of gastric inhibitory polypeptide. J. Am. Chem. Soc. 97:5593.
138. Bloom, S. R., Bryant, M. G., and Polak, J. M. (1975). Distribution of gut hormones. Gut 16:821.
139. Polak, J. M., Bloom, S. R., Kuzio, M., Brown, J. C., and Pearse, A. G. E. (1973). Cellular localisation of gastric inhibitory polypeptide in the duodenum and jejunum. Gut 14:284.
140. Pederson, R. A., and Brown, J. C. (1972). Inhibition of histamine, pentagastrin and insulin stimulated canine gastric secretion by pure "gastric inhibitory polypeptide." Gastroenterology 62:393.
141. Pederson, R. A. (1971). The isolation and physiological actions of gastric inhibitory peptide. Thesis, University of British Columbia.
142. Barbezat, G. O., and Grossman, M. I. (1971). Intestinal secretion: Stimulation by peptides. Science 174:422.
143. Shabaan, A. A., Turner, D. S., and Marks, V. (1974). Sustained insulin release in response to intravenous infusion of insulin releasing polypeptide (IRPP) in the rat. Diabetes 23:902.
144. Turner, D. S., Etheridge, L., Jones, J., Marks, V., Meldrum, B., Bloom, S. R., and Brown, J. C. (1974). The effect of the intestinal polypeptides, IRP and GIP on insulin release and glucose tolerance in the baboon. Clin. Endocrinol. 3:489.
145. Pederson, R. A., Schubert, H. E., and Brown, J. C. (1975). The insulintropic action of gastric inhibitory polypeptide. Can. J. Physiol. 53:217.
146. Dupre, J., Ross, S. A., Watson, D., and Brown, J. C. (1973). Stimulation of insulin secretion by gastric inhibitory peptide in man. J. Clin. Endocrinol. 37:826.
147. Catland, S., Crockett, S. E., Brown, J. C., and Mazzaferri, E. L. (1974). Gastric inhibitory polypeptide (GIP) stimulation by oral glucose in man. J. Clin. Endocrinol. 39:223.
148. Cleator, I. G. M., and Gourlay, R. H. (1975). Release of immunoreactive gastric inhibitory polypeptide (IR-GIP) by oral ingestion of food substances. Am. J. Surg. 130:128.
149. Kosaka, T., and Lim, R. K. S. (1930). Demonstration of the humoral agent in fat inhibition of gastric secretion. Proc. Soc. Exp. Biol. 27:890.
150. Bloom, S. R. (1975). GIP in diabetes. Diabetologia 11:334.
151. Ross, S. A., Brown, J. C., Dryburgh, J., and Dupre, J. (1973). Hypersecretion of gastric inhibitory polypeptide in diabetes mellitus. Clin. Res. 21:1029.
152. Bodanszky, M., Klausner, Y. S., Yang Lin, C., Mutt, V., and Said, S. I. (1974). Synthesis of the vasoactive intestinal peptide (VIP). J. Am. Chem. Soc. 96:4973.
153. Bryant, M. G., Bloom, S. R., Polak, J. M., Albuquerque, R. H., Modlin, I., and Pearse, A. G. E. (1976). Possible dual role for vasoactive intestinal peptide as gastrointestinal hormone and neurotransmitter substance. Lancet i:991.
154. Polak, J. M., Pearse, A. G. E., Garaud, J. C., and Bloom, S. R. (1974). Cellular localization of a vasoactive intestinal peptide in the mammalian and avian gastrointestinal tract. Gut 15:720.
155. Kerins, C., and Said, S. I. (1973). Hyperglycaemic and glycogenolytic effects of vasoactive intestinal polypeptide. Proc. Soc. Exp. Biol. 142:1014.
156. Konturek, S. J., Thor, P., Dembinski, A., and Krol, R. (1975). Comparison of secretin and vasoactive intestinal peptide on pancreatic secretion in dogs. Gastroenterology 68:1527.
157. Schwartz, C. J., Kimberg, D. V., Sheerin, H. E., Field, M., and Said, S. I. (1974). Vasoactive intestinal peptide stimulation of adenylate cyclase and active electrolyte secretion in intestinal mucosa. J. Clin. Invest. 54:536.
158. Polak, J. M., and Pearse, A. G. E. (1973). Vasoactive intestinal peptide and watery-diarrhoea syndrome. Lancet ii:14.
159. Verner, J. V., and Morrison, A. B. (1974). Endocrine pancreatic islet disease with diarrhoea. Arch Intern. Med. 133:492.
160. Kraft, A. R., Tompkins, R. K., and Zollinger, R. M. (1970). Recognition and management of the diarrhoeal syndrome caused by non-beta islet cell tumours of the pancreas. Am. J. Surg. 119:163.
161. Bloom, S. R., and Polak, J. M. (1975). The role of VIP in pancreatic cholera. *In* J. C.

Thompson (ed.), Gastrointestinal Hormones, pp. 635–642. University of Texas Press, Austin.
162. Kahn, C. R., Levy, A. G., Gardner, J. D., Miller, J. V., Gordon, P., and Schein, P. S. (1975). Pancreatic cholera: beneficial effects of treatment with streptozotocin. N. Engl. J. Med. 292:941.
163. Brazeau, P., and Guillemin, R. (1974). Somatostatin: newcomer from the hypothalamus. N. Engl. J. Med. 290:963.
164. Arimura, A., Sato, H., Nishi, N., and Schally, A. V. (1975). Somatostatin: abundance of immunoreactive hormone in rat stomach and pancreas. Science 189:1007.
165. Polak, J. M., Pearse, A. G. E., Grimelius, L., Bloom, S. R., and Arimura, A. (1975). Growth-hormone release-inhibiting hormone in gastrointestinal and pancreatic D cells. Lancet i:1220.
166. Hokfelt, T., Efendic, S., Hellerstrom, C., Johansson, O., Luft, R., and Arimura, A. (1975). Cellular localization of somatostatin in endocrine-like cells and neurons of the rat with special references to the A_1-cells of the pancreatic islets and to the hypothalamus. Acta Endocrinol. (Kbh) Suppl. 80:5.
167. Siler, T. M., Yen, S. S. C., Vale, W., and Guillemin, R. (1974). Inhibition by somatostatin on the release of TSH induced in man by thyrotropin-releasing factor. Clin. Endocrinol. 38:742.
168. Mortimer, C. H., Carr, D., Lind, T., Bloom, S. R., Mallinson, C. N., Schally, A. V., Tunbridge, W. M. G., Yeomans, L., Coy, D. H., Kastin, A., and Besser, G. M. (1974). Growth hormone release inhibiting hormone: effects on circulating glucagon, insulin and growth hormone in normal, diabetic, acromegalic and hypopituitary patients. Lancet i:697.
169. Alberti, K. G. M. M., Christensen, N. J., Christenson, S. E., Hansen, A. A. P., Iversen, J., Lundbaek, K., Seyer-Hansea, K., and Orskov, H. (1973). Inhibition of insulin secretion by somatostatin. Lancet ii:1299.
170. Johnson, D. G., Ensinck, J. W., Koerker, D., Palmer, J., and Goodner, C. J. (1975). Inhibition of glucagon and insulin secretion by somatostatin in the rat pancreas perfused in situ. Endocrinology 96:370.
171. Bloom, S. R., Mortimer, C. H., Thorner, M. O., Besser, G. M., Hall, R., Gomez Pan, A., Roy, V. M., Russell, R. C. G., Coy, D. H., Kastin, A. J., and Schally, A. V. (1974). Inhibition of gastrin and gastric acid secretion by growth hormone release inhibiting hormone. Lancet ii:1106.
172. Barros D'Sa, A. A. J., Bloom, S. R., and Baron, J. H. (1975). Direct inhibition of gastric acid by growth hormone release inhibiting hormone in dogs. Lancet i:886.
173. Gomez-Pan, A., Reed, J. D., Albinus, M., Shaw, B., Hall, R., Besser, G. M., Coy, D. H., Kastin, A. J., and Schally, A. V. (1975). Direct inhibition of gastric acid and pepsin secretion by growth-hormone release-inhibiting hormone in cats. Lancet i:888.
174. Bloom, S. R., Russell, R. C. G., Barros D'Sa, A. A. J., Baron, J. H., Besser, G. M., Hall, R., Coy, D. H., Kastin, A. J., and Schally, A. V. (1975). Inhibition of gastrin and gastric acid by growth hormone release inhibiting hormone. Gut 16:396.
175. Creutzfeldt, W., Lankisch, P. G., and Folsch, U. R. (1975). Hemmung der sekretin- und cholezystokinin-Pankreozymin-induzierten saft-und-enzymsekretion des pankreas und der gallenblasen-kontraktion beim menschen durch somatostatin. Dtsch. Med. Wochenschr. 100:1135.
176. Boden, G., Sivitz, M. C., Owen, O. E., Essa-Koumar, N., and Landor, J. H. (1975). Somatostatin suppresses secretin and pancreatic exocrine secretion. Science 190:163.
177. Bloom, S. R., Joffe, S. N., and Polak, J. M. (1975). Effect of somatostatin on pancreatic and biliary function. Gut 16:836.
178. Larsson, L. I., Sundler, F., Hakanson, R., Pollock, H. G., and Kimmel, J. R. (1974). Localization of APP, a postulated new hormone, to a pancreatic endocrine cell type. Histochemistry 42:377.
179. Lin, T. M., and Chance, R. E. (1972). Spectrum gastrointestinal actions of a new bovine pancreas polypeptide (BPP). Gastroenterology 62:852.
180. Kimmel, J. R., Pollock, H. G., and Hazelwood, R. L. (1971). A new pancreatic polypeptide hormone. Fed. Proc. 30:1318.
181. Lin, T. M., and Chance, R. E. (1974). Gastrointestinal actions of a new bovine

pancreatic peptide (BPP). *In* W. Y. Chey and F. P. Brooks (eds.), Endocrinology of the Gut, pp. 143–145, Charles B. Slack, Inc., New Jersey.
182. Lin, T. M., and Chance, R. E. (1974). Bovine pancreatic polypeptide (BPP) and avian pancreatic polypeptide (APP). Gastroenterology 67:737.
183. Larsson, L-I., Sundler, F., and Hakanson, R. (1976). Pancreatic polypeptide–a postulated new hormone: Identification of its cellular storage site by light and electron microscopic immunocytochemistry. Diabetologia 12:211.
184. Lin, T., Chance, R., and Evans, D. (1973). Stimulatory and inhibitory actions of a bovine pancreatic peptide on gastric and pancreatic secretion of dogs. Gastroenterology 64:179, 865.
185. Floyd, J. C., Jr., Chance, R. E., Hayashi, M., Moon, N. E., and Fajans, S. S. (1973). Concentrations of a newly recognized pancreatic islet polypeptide in plasma of healthy subjects and in plasma and tumors of patients with insulin-secreting islet cell tumors. Cent. Soc. Metab. 12:535A.
186. Polak, J. M., Bloom, S. R., Adrian, T. E., Heitz, P., Bryant, M. G. and Pearse, A. G. E. (1976). Pancreatic polypeptide in insulinomas, gastrinomas, vipomas and glucagonomas. Lancet i:328.
187. Anderson, S., Nilsson, G., and Uunas, B. (1967). Effect of acid in proximal and distal duodenal pouches on gastric secretory responses to gastrin and histamine. Acta Physiol. Scand. 71:368.
188. Adelson, J. W., and Rothman, S. S. (1974). Selective pancreatic enzyme secretion due to a new peptide called chymodenin. Science 183:1087.
189. Harper, A. A., Hood, A. J. C., Muskens, J., and Smy, T. R. (1974). Pancreotone: an inhibitor of pancreatic secretion in extracts of ileal and colonic mucosa. J. Physiol. (Lond.) 253:32P.
190. Gregory, H. (1975). The isolation and structure of urogastrone and its relationship to epidermal growth factor. Nature 257:325.

International Review of Physiology
Gastrointestinal Physiology II, Volume 12
Edited by Robert K. Crane

4
Morphology and Physiology of Salivary Myoepithelial Cells

J. A. YOUNG AND E. W. VAN LENNEP
Departments of Physiology and of Histology and Embryology
University of Sydney, N.S.W. 2006, Australia

HISTORICAL BACKGROUND 106

MORPHOLOGY 107
Identification 107
General Configuration 109
Ultrastructure 110
Pre- and Postnatal Development 112
Innervation 113

CELL BIOLOGY 115

PHYSIOLOGY AND PHARMACOLOGY 116

FUNCTIONS OF MYOEPITHELIAL CELLS 119

SUMMARY 120

The morphology of salivary glands and the physiology and pharmacology of the secretory process were all topics of great interest to research workers in the last century and the early part of the present one, but they fell somewhat into neglect during the 1940s and 1950s. The development of new investigative techniques, such as micropuncture and microperfusion of ductal elements, and the perfection of electron microscopic, histochemical, and cytochemical procedures, has caused a reawakening of interest in salivary glands, as can be attested by the appearance of a number of reviews on various aspects of the subject (1–6) including an extensive one on the physiology of salivary secretion in the predecessor to the present volume (7) and one on the morphology of salivary glands (8). For this reason it does not seem appropriate at present to attempt another review on such a general scale—rather, particular facets of the secretory process could, with profit, be developed, more or less in isolation. The present review on myoepithelial cells represents such an attempt; considerable stress is placed on morphological studies since these have provided much of our insight into the function of these fascinating cells.

HISTORICAL BACKGROUND

Harvey (9) and Malpighi (10) realized that many glands developed as branching ductular organs in which the secretory elements were attached to the extremities like grapes on the stalks of a bunch. However, even when Johannes Müller (11) published his monograph *De Glandularum* in 1830, the smallest glandular subdivisions (i.e., the *acini* as they were called) that had been identified corresponded only to what we call a lobule. By the 1850s, with the availability of improved microscopical techniques, it was realized that the "acinus" could be subdivided further into a number of berry-like or tubular endpieces, each consisting of several cells (12). (These endpieces have variously been called acini (13), alveoli (14), *Adenomeren* (15), etc., but, in order to overcome the confusion that has arisen, we prefer the general term secretory endpiece (8) (cf. *sekretorisches Hauptstück,* as used by Zimmermann (16)), which may be applied to all salivary glands regardless of the shape or histology of the structure being considered). Even before the secretory function of the cells composing the endpiece was appreciated, physiologists were aware that each endpiece was enveloped in a "basement membrane" to which many were inclined to ascribe a secretory role. When it was realized that this role belonged to the epithelial cells, it seemed natural to ascribe a supportive function to the basement membrane and to suggest that it might constitute a permeability barrier that could modify the composition of the secretion by a process of sieving.

In 1865 Krause (17) first recognized cellular structures associated with the basement lamina of the secretory endpieces of the cat parotid gland which he termed *sternförmige Zellen* on account of their star-shaped (Figure 1) or multipolar morphology. Subsequently they were described surrounding the endpieces of salivary glands in many other species (14, 18–20) and, although opinion was divided initially as to whether the cells lay inside or outside or formed part of the basement membrane, there was general agreement that they formed a supportive trellis for the endpiece. Boll (21) wrote that the cells *bilden um die Alveolen eine korbartige Umhüllung*, using an analogy with wickerwork; in time, his term *Korbzellen* (basket cells) came into general use. However, studies in sweat glands performed at about this time suggested that cells similar to *Korbzellen* actually had a contractile role to play in the process of sweat expulsion and, in 1881, Unna (22) put forward the idea that salivary basket cells

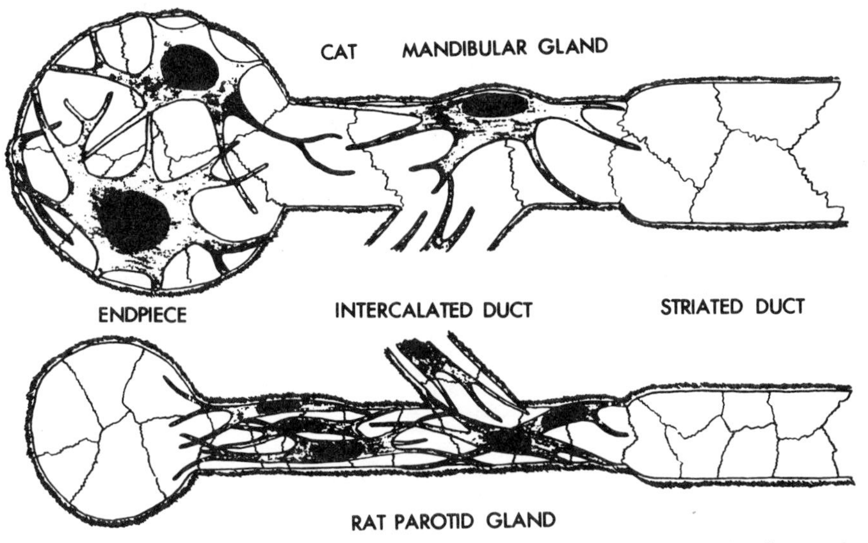

Figure 1. Diagrammatic representation of the arrangement of myoepithelial cells seen in typical salivary glands such as the cat mandibular gland (*above*) and the atypical arrangement (*below*) reported to exist in the rat parotid (37). In typical glands there are two types of myoepithelial cell, multipolar cells investing the endpieces and spindle-shaped cells applied to the intercalated ducts. In the rat parotid, the former (multipolar) type is absent and the latter (spindle-shaped) type extends from the intercalated ducts onto the base of the endpiece. (Modified from Garrett (114).)

also were contractile. In 1897 Renaut (23) first coined the name myoepithelial cell, the term now favored by all authors.

Since the time of their first discovery, the cells have been described not only in salivary glands (24–39) and sweat glands (40, 41), but also in other ectodermal glands such as mammary glands (29, 42), lacrimal glands (43, 44), Harderian glands (44), and the prostate (45) and even in the endodermal bronchial glands (46).

MORPHOLOGY

Identification

Originally, myoepithelial cells were identified from gland preparations in which fresh tissue was macerated in solutions such as acetic acid, potassium dichromate, potassium hydroxide, and chromic acid (14, 17, 20). Although such procedures can establish the existence of the cells and reveal their general configuration, clearly, they can tell us nothing of their structure. Unfortunately the cells are not visible in ordinary histological sections, although use of special histochemical procedures, particularly the Gomori procedure for demonstration of nonspecific alkaline phosphomonoesterase (47), can frequently, but not always, be used to indicate the presence of the cells (Figures 2 and 3).

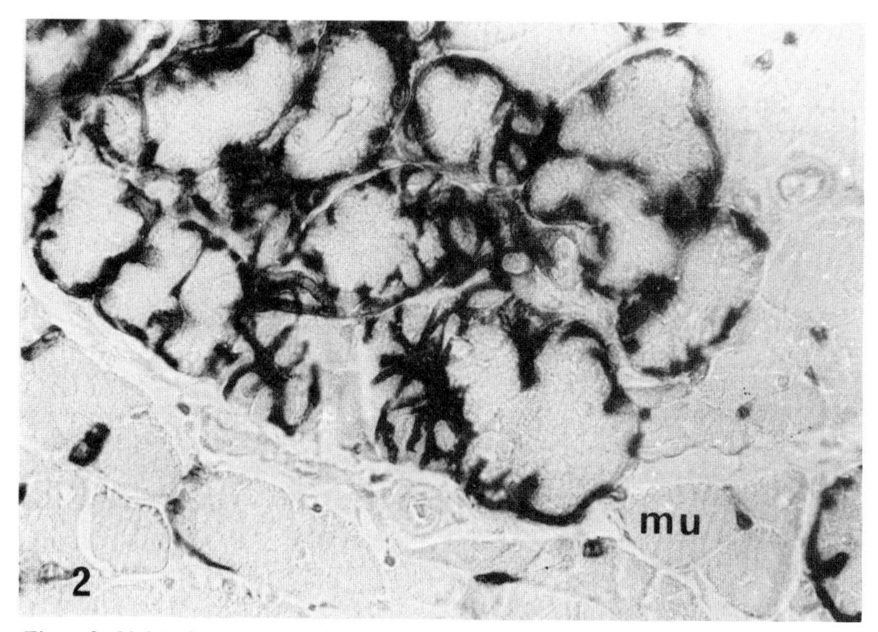

Figure 2. Multipolar myoepithelial cells investing the endpieces of von Ebner's gland in the rat, demonstrated with Gomori's cobalt sulfide method for alkaline phosphatase without use of a counterstain. The glandular tissue is surrounded by the striated muscle fibers (*mu*) of the tongue. × 410.

Although many authors have investigated the activity of nonspecific alkaline phosphatase in salivary glands (24, 25, 37, 48–56), the results reported have proven rather contradictory. This is due largely to false localization of precipitated phosphate resulting from diffusion away from the site of enzyme activity, the consequence of using excessively long incubation periods, although, even when experimental conditions are good, interpretation of the stained microscopical sections is difficult. Following the demonstration that, under well controlled conditions, it was the myoepithelial cells that gave the strongest reaction for alkaline phosphatase in sheep parotid (24) and rat mandibular (57) glands (Figure 3), the notion grew up that this histochemical reaction would provide a reliable means of demonstrating myoepithelial cells in all glands. Since this is not, in fact, the case (e.g., myoepithelial cells in salivary glands of man, dog, and the Australian marsupial, *Trichosurus vulpecula* do not stain positively for alkaline phosphatase) considerable confusion has grown up in the literature. This confusion was not reduced by the demonstration that myoepithelial cells in the salivary glands of man stain strongly for another enzyme, Mg^{2+}-adenosine triphosphatase (58), since this reaction is negative for myoepithelial cells in cats and dogs (25, 37, 38). Histochemical demonstration of phosphorylase has been used to visualize the myoepithelial cells in rat salivary glands (59), but it remains

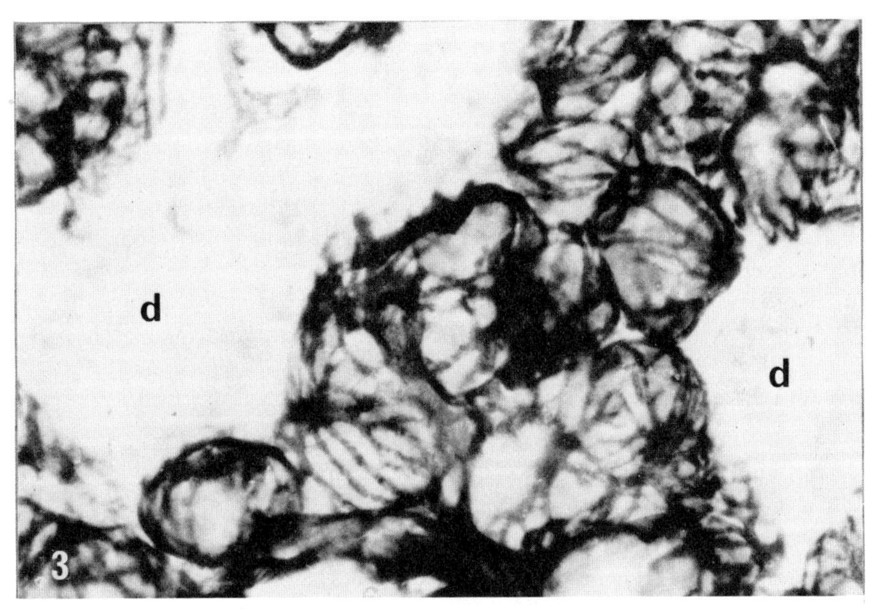

Figure 3. Multipolar myoepithelial cells surrounding the endpieces of the rat mandibular gland, demonstrated by the technique referred to in the legend to Figure 2. The striated and granular ducts, not being invested with myoepithelial cells, show only as blank spaces (*d*) in the micrograph. × 480.

to be seen if this method is universally applicable. Specific identification may eventually be achieved by use of immunohistochemical methods or myosin-specific staining reactions (see below) but, at the present time, the only certain method of identification requires the use of the electron microscope when the cells can be seen to have a striking and characteristic morphology.

General Configuration

In salivary glands, the configuration of a myoepithelial cell depends on whether it lies in association with an endpiece or an intercalated duct. Those cells that are associated with endpieces usually lie, in Tamarin's (32) words, "like an octopus sitting on a rock" (Figure 1). It is usual for there to be only one myoepithelial cell per endpiece, but two or three are not uncommon. Each cell consists of a central body with four to eight processes radiating from it (Figure 1). Each process subdivides to give rise to two or more generations of branches so that, in all, there may be 20 or more processes tending always to diverge from one another. In general, the primary processes follow the long axis of the endpiece and the secondary ones radiate laterally in the manner of a rib cage; they often give the appearance of molding the shape of the endpiece. Although this is the general pattern, it is by no means always the case. Thus, at one extreme, the cells may be very sparse and, at least in the case of the endpieces of the rat parotid, they may be absent altogether.[1] At the opposite extreme, they may be so conspicuous as to form a continuous enveloping "muscular" coat, as in the case of the sublingual gland of the monotreme, *Tachyglossus aculeatus* (60).

The myoepithelial cells investing intercalated ducts differ strikingly from those investing endpieces, being spindle-shaped, without processes. They lie longitudinally along the ducts and do not give the impression of molding their shape. At the ends of these ducts the myoepithelial cells tend to overlap onto the cells, comprising, respectively, the base of the endpiece and the start of the intralobular (striated or granular) ducts (Figure 1).

Myoepithelial cells are not normally described in association with other parts of the salivary duct system but an exception exists in the case of the sheep parotid gland where the cells are very conspicuously applied to the striated ducts (115).

Ultrastructure

The most detailed descriptions of the ultrastructure of salivary myoepithelial cells come from studies on the rat mandibular gland (32) and the human submandibular (33) and labial (34) salivary glands. An excellent ultrastructural study has also been published on the pre- and postnatal development of these cells in the rat mandibular gland (39). The following description is based largely on these papers supplemented by our own observations in a variety of species including some marsupials and monotremes.

The cells possess all the usual intracellular organelles (Figure 4). The nuclei are ellipsoidal with only occasional ribosomes on the outer envelope. Similarly, the small cisternae of the endoplasmic reticulum, found in the perinuclear cytoplasm, carry relatively few ribosomes. Free ribosomes are quite numerous in the area around the nucleus, but some may also be found in the cell processes, scattered between the bundles of filaments. A Golgi complex, rather elongated and associated with a few small vesicles, can also be seen. Mitochondria are scattered throughout the cell body but are seen less frequently within the cell processes (Figures 4 and 5).

The most striking feature of the cells, of course, is the presence of bundles of cytoplasmic filaments that traverse the cell cytoplasm (Figure 5). The filaments comprising a bundle are fairly uniform in thickness, but the density of the bundle varies greatly, depending on how tightly packed the filaments are at a particular point. Bundles may be so tightly packed as to obscure their fila-

---

[1] Much confusion exists in the literature concerning the occurrence of myoepithelial cells in the rat parotid gland. Scott and Pease (28) state definitely that they do not occur while Parks (62) and some others (36, 63) are equally positive that they envelop the endpieces and intercalated ducts. Recently, Garrett and Parsons (37) reinvestigated the rat parotid using a combined histochemical and ultrastructural approach and concluded that, while myoepithelial cells normally invest the intercalated ducts, only processes of ductal myoepithelial cells overlap onto the endpieces and there are no distinct multipolar cells investing them (see Figure 1).

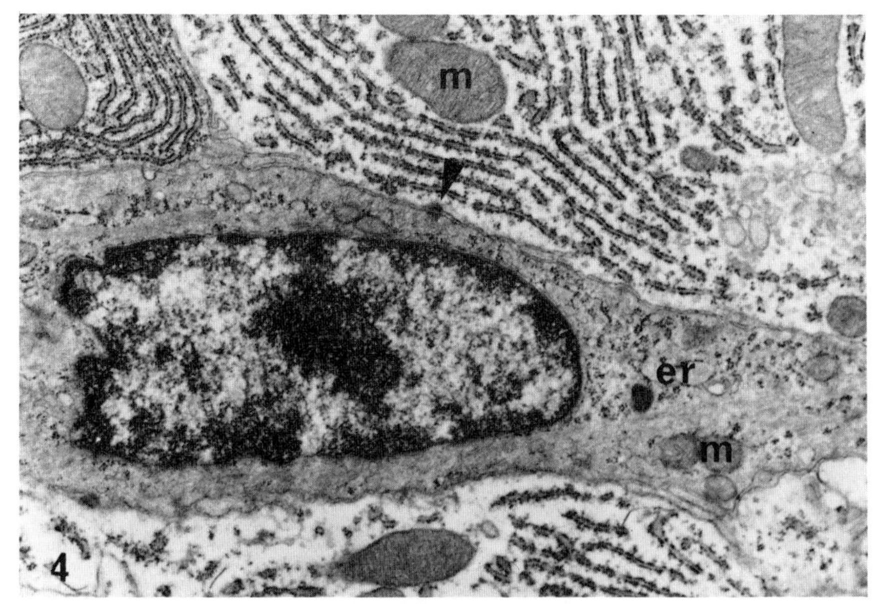

Figure 4. Electron micrograph of a myoepithelial cell from the mandibular gland of the guinea pig showing the perinuclear region of the cell between two endpiece secretory cells. The perinuclear cytoplasm of the myoepithelial cell contains some small mitochondria (*m*), a few small cisternae of rough endoplasmic reticulum (*er*), scattered polysomes, and some filaments. The *arrowhead* points to a desmosomal connection between the myoepithelial cell and an endpiece secretory cell. × 19,400.

mentous structure, thereby coming to resemble the "dense elongated bodies" seen in smooth muscle (62). Generally, bundles of filaments may fuse, but do not intersect, with other bundles. The primary bundles tend to run in the longitudinal axis of the cell and, in multipolar cells, they diverge and follow the ramifications of the cell processes. In addition to the bundles of fine uniform filaments resembling actin filaments, some scattered filaments with a diameter of 10 nm have been observed (see Figure 7). Microfilament attachment plaques, resembling hemidesmosomes, are seen frequently on the plasma membrane at the surface of the cell facing the basement lamina.

Since the ultrastructure of the cell near its two surfaces is not the same, it is convenient to denote these surfaces with the terms *stromal* (facing the interstitium) and *visceral* (facing the secretory cells of the endpiece or the epithelial cells lining the intercalated duct). On the stromal side of a myoepithelial cell there are numerous flask-shaped structures that most probably may be interpreted as caveolar invaginations of the plasma membrane (Figure 5).

The visceral surface of the cell is rather smooth and does not follow the plications of the plasma membrane of the underlying epithelial cell so that a trellis-like space can usually be discerned between the two cell types. (However, in the sublingual gland of *Tachyglossus aculeatus* this space is obliterated and the

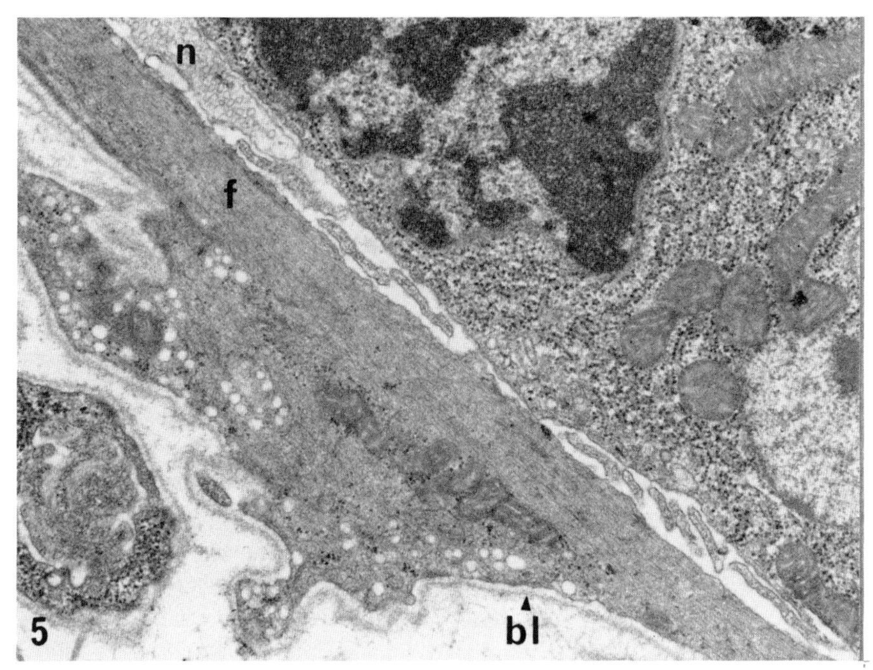

Figure 5. Electron micrograph of a myoepithelial cell from the mandibular gland of the guinea pig showing a myoepithelial cell process between the basement lamina (*bl*) and the base of an endpiece secretory cell. It is separated from the secretory cell by a narrow intercellular space containing basal plications of the secretory cell as well as a hypolemmal nerve terminal (*n*). The myoepithelial cell process contains bundles of filaments (*f*) and small mitochondria with a dense matrix. The numerous vesicles in the basal region of the myoepithelial cell are caveolae for which the communications with the exterior are not well shown owing to the tangential plane of the section. × 31,800.

plasma membranes of the two cell types are very closely applied (60).) In rare cases, a few ciliary processes of the myoepithelial cell are observed penetrating deeply into the cytoplasm of neighboring epithelial cells (see Figure 5 in the paper by Tandler et al. (34)). It has been proposed that these might constitute points of information exchange between the two cell types (41) but, even if this were so, it seems unlikely that they subserve any exclusive role of importance since they are excessively rare. However, desmosomal attachments between myoepithelial and epithelial cells are seen quite frequently, although not so often as they can be found between two neighboring epithelial cells (see Figure 4).

Pre- and Postnatal Development

It is not uncommon in salivary glands to observe, in association with the endpieces, scattered electron-lucent cells that are similar in shape to myoepithelial cells but lack filamentous bundles and have relatively few organelles and inclusions. These cells are thought to be myoepithelial cell precursors since all stages of transition between the clear cells and unmistakable myoepithelial cells can be encountered (33) (Figure 6). Recent studies on developing mandibular glands in pre- and postnatal rats confirm this impression (39). On the 18th day of prenatal development, flattened clear cells, the presumptive myoepithelial cells, can be seen wedged between the outer surface of the developing secretory cells of the endpiece and the enveloping basement lamina. They develop by cytodifferentiation of epithelial stem cells in the end-buds of the terminal tubules, and they remain capable of undergoing mitotic division until they develop into mature, myosin-containing cells, a process that begins on the first day post partum and takes about 7 days to reach completion (39); of course, additional myoepithelial cells continue to develop as new endpieces form, up until about 6–8 weeks post partum (8, 39).

There is a possibility that some elements of the salivary gland tissues could be of neural crest origin (64). If this should prove to be the case, then, of all salivary parenchymal cells, the myoepithelial cells would be the most likely candidates since at no stage of their development do they show any evidence of secretory activity. There would be opportunity for them to enter the developing terminal buds on day 15 or 16 of prenatal developmental when the basement lamina is discontinuous (39, 65).

Innervation

In most salivary glands, the endpieces are the structures that are most densely innervated, with cholinergic fibers and also, when they are present, with adrenergic fibers. Neuroeffector junctions consist merely of short, varicose segments of the nerve terminal, bare of Schwann cell covering, at which there is an accumulation of cytoplasmic vesicles containing a neurotransmitter substance. The pattern of innervation is variable, among species, among glands within the one species, and at different stages of development within a single gland (8). Endpieces in all glands receive a cholinergic innervation, but endpieces in at least a few species lack appreciable adrenergic supply. Furthermore, in some glands the nerve fibers are of hypolemmal type (i.e., they penetrate the basement lamina and come to within 20 nm of the endpiece cell membrane; see Figure 5) while in others they are only of epilemmal type (i.e., they lie outside the basement membrane at a distance of 100 nm or more from the endpiece cell membrane) (8).

Since myoepithelial cells form part of the secretory endpiece, it is inevitable that they lie in close proximity to whatever neuroeffector junctions are present, be they sympathetic or parasympathetic (66, 69) and, indeed, it is not uncommon to observe both types of autonomic neuroeffector junction in close proximity to a single myoepithelial cell. When hypolemmal fibers are observed, they often lie between a secretory cell and a myoepithelial cell, thereby coming into close contact with the plasma membranes of both cell types (33, 70–72); from a morphological point of view it seems that such fibers can be either adrenergic or cholinergic (36, 66, 72–75).

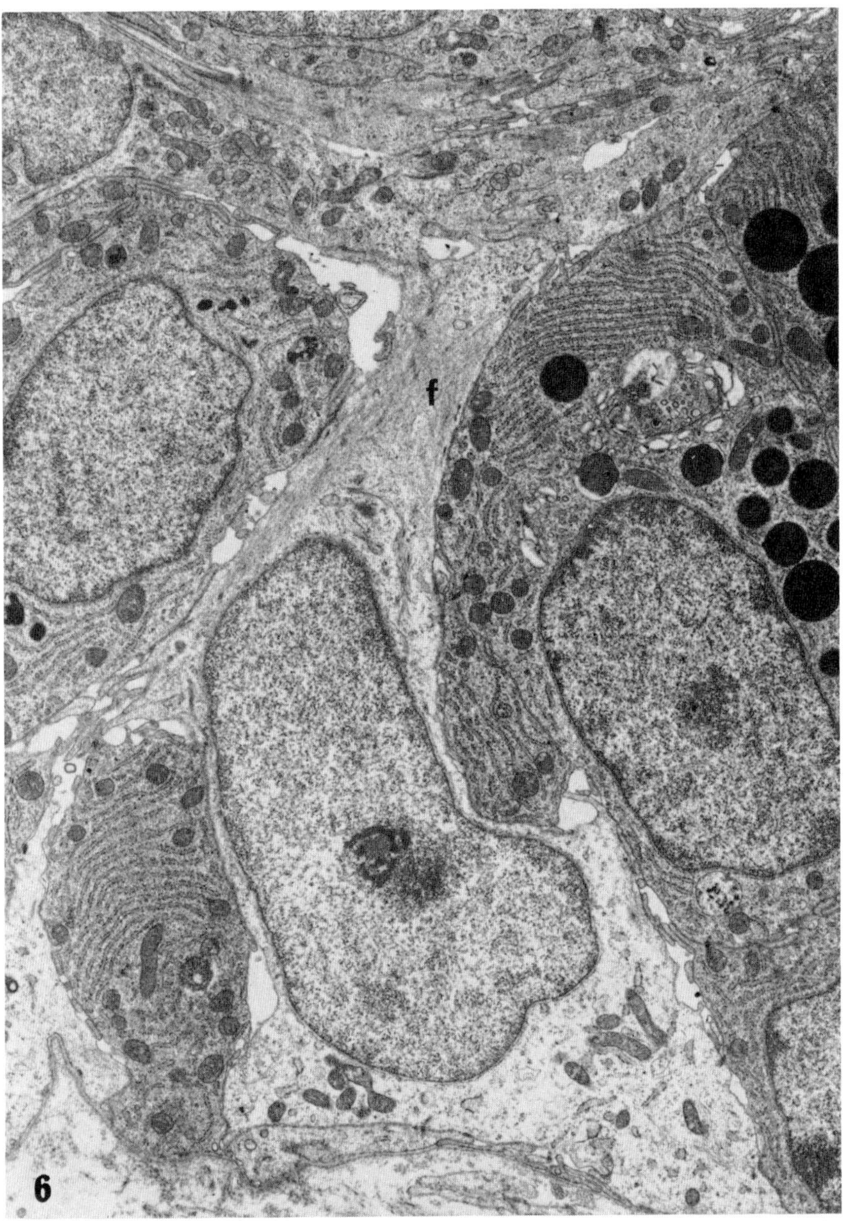

Figure 6. Differentiating myoepithelial cells in the terminal tubules of the parotid gland of a pouch young of an Australian marsupial, the brush-tail possum (*Trichosurus vulpecula*). The basal part of the myoepithelial cell still has the appearance of an undifferentiated "clear" cell, but the apical part contains bundles of cytoplasmic filaments (*f*). The cells surrounding the myoepithelial cell are immature secretory cells. × 8,100.

CELL BIOLOGY

The morphological similarities between smooth muscle cells and myoepithelial cells are striking and suggest strongly that the myoepithelial cells subserve a contractile function. The physiological and pharmacological evidence supporting this conclusion is discussed in the following section, but it seems appropriate first to make some more detailed comments on the morphological evidence, which is quite convincing, albeit indirect.

As mentioned above, the myoepithelial cell filaments resemble the actin filaments seen in smooth and skeletal muscle but, although electron microscopic identification of actin, using heavy meromyosin (76), and immunohistochemical identification, using specific antibodies to actin (77), have both been attempted for muscle cells, the results of any such attempt in relation to myoepithelial cells have yet to be published. Myosin, which can be identified histochemically (with myosin-specific dyes such as TP-Levanol fast cyanine 5RN) and immunohistochemically, has been demonstrated convincingly in myoepithelial cells (78–81), and its postnatal development in presumptive myoepithelial cells of rat mandibular gland has been followed (80). Such methods do not, however, permit a precise intracellular localization of the myosin protein. Evidently there are some differences between myoepithelial myosin and that seen in smooth muscle since thick filaments, approximately 15 nm in diameter, in the shape of rods (82) or ribbons (83, 84), the forms thought to be assumed by smooth muscle myosin, have not been observed in salivary myoepithelial cells.

The fibers of intermediate thickness (10 nm) that we have observed in myoepithelial cells of some salivary glands (Figure 7) occur also in related cells of the rat exorbital lacrimal gland (85) and of the anterior medial gland in the nasal septum of the mouse (86). They are seen quite commonly in smooth muscle, particularly in tissue culture preparations and in embryonic tissue (87). The functional significance of these fibers, however, is only a matter of speculation.

Two other structural features of myoepithelial cells that suggest a similarity to smooth muscle are the filament attachment plaques (see above), which are also quite common in smooth muscle (88), and the numerous caveolae. Caveolae should not be confused with pinocytotic vesicles since they are always open to the exterior (a fact that can easily be demonstrated by use of tracers such as lanthanum hydroxide) and they have necks of fairly constant diameter. In the case of smooth muscle, these caveolae have been shown to have a close topographical relation to the sarcoplasmic reticulum (89) from which the idea arises that they may represent areas of plasma membrane specialized for the purpose of ion exchange; the relation is reminiscent of the triad in striated muscle fibers formed by the transverse tubules and the sarcoplasmic reticulum. In myoepithelial cells, to our knowledge, there has not been any report of a close topographical relation between caveolae and smooth-surfaced endoplasmic reticulum, although occasionally we have observed smooth vesicles interposed between

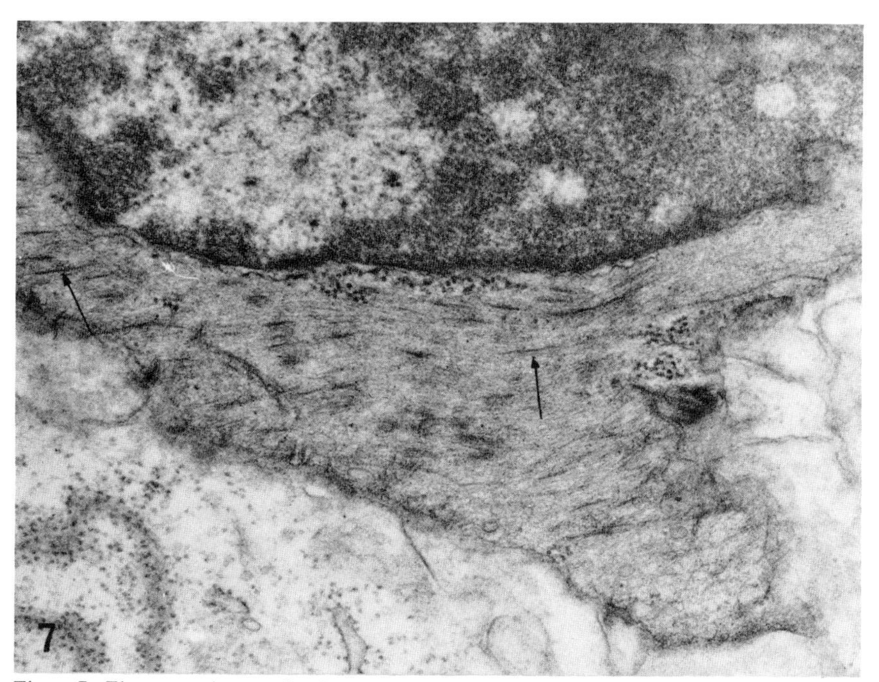

Figure 7. Electron micrograph of a myoepithelial cell from the submandibular gland of a human subject suffering from cystic fibrosis. Scattered among the conspicuous bundles of fine actin filaments are some of intermediate (10 nm) diameter, the function of which is not yet defined. × 35,900.

caveolae and mitochondria in the myoepithelial cell processes of the parotid of *Tachyglossus aculeatus* (60). In addition, a careful examination of the myoepithelial cell pictured in Figure 6*a* of Kerjaschki's paper (86) on the nasal septum glands of the mouse shows cisternae and vesicles of smooth endoplasmic reticulum lying directly over caveolae.

PHYSIOLOGY AND PHARMACOLOGY

In the case of sweat glands (90) and mammary glands (42), the contractility of the myoepithelial cells has been established by direct observation. Indeed, in the mammary gland (42) a functional distinction has been made between myoepithelial cells investing the endpiece, where contraction causes obliteration of the lumen, and those applied to the ducts, where contraction causes shortening of the duct and widening of its lumen. It hardly seems possible that such direct evidence will ever be obtained for salivary glands but, nevertheless, there is abundant indirect experimental evidence pointing to the conclusion that sympathetic nerve stimulation causes contraction of salivary myoepithelial cells (30, 35, 91–96). In addition, there is a smaller, less generally accepted body of evidence to implicate parasympathetic nerves in the contraction process (97, 98).

In recent times, several workers have attempted to obtain more direct evidence by cannulating the main ducts of the salivary glands of dogs and cats and then measuring either the salivary flow rate, in an open system with a variable outflow resistance, or the intraductal pressure, in a closed system.[2] The first of such studies was carried out on cat salivary glands by Emmelin et al. (99), who had been prompted by Garrett's electron microscope observations that the myoepithelial cells of these animals seemed to receive a dual autonomic innervation (67, 69). From their studies Emmelin and his collaborators concluded that myoepithelial cells contracted readily in response to single shock parasympathetic nerve stimulation and gave only equivocal responses to sympathetic nerve stimulation (99). Subsequently, Darke and Smaje (100) criticized these experiments and repeated them with a less compliant measuring system. They found no evidence to show that single shock parasympathetic nerve stimulation caused myoepithelial cell contraction although, since they found that infused bradykinin did (following studies on the dog of Emmelin et al. (101)), they felt that sustained parasympathetic nerve stimulation might produce such an effect by promoting intraglandular bradykinin production. Single shock parasympathetic nerve stimulation has also been claimed to cause myoepithelial cell contraction without initiating secretion in dog salivary glands (102), but the objection of Darke and Smaje (100) would seem likely to apply here also.

Like Emmelin et al. (99), Darke and Smaje (100) obtained only equivocal evidence to suggest that sympathetic nerve stimulation caused myoepithelial cell contraction in the cat mandibular gland, although they were inclined to believe that sustained stimulation definitely had such an effect. Fortunately, studies in the dog seem much less equivocal. Arising from studies on the effects of the β-adrenergic stimulant isoproterenol on dog mandibular secretion (103), Emmelin and his co-workers have been able to demonstrate that there is a complete separation between α-adrenergic activation, which causes myoepithelial cell contraction, and β-adrenergic activation, which causes secretion (102, 104–106). Thus secretion can be elicited from the dog mandibular by sympathetic nerve stimulation or by administration of isoproterenol or adrenalin, and these effects can be blocked by prior administration of the specific β-blocker propranolol. On the other hand, increases in intraductal pressure, indicative either of myoepithelial cell contraction or stimulated secretion, produced by sympathetic nerve stimulation or administration of phenylephrine or adrenalin, could not be blocked by prior administration of propranolol (which blocks secretion) but could be blocked by prior administration of α blockers such as dihydroergotamine, phenoxybenzamine, and phentolamine. In the pres-

---

[2] It would be instructive to perform such experiments on the rat parotid gland, which lacks endpiece myoepithelial cells (37), since one might thereby gain additional insight into the role of these cells in the expulsion of the salivary secretion.

ence of α-blockers, initiation of secretion by stimulation of beta receptors is only apparent after a short delay, and the intraductal pressure never rises to the levels observed when both α and β receptors are stimulated. These results lead unequivocally to the conclusion that β sympathetic stimulation results in an expulsion of saliva only after the lumina of the endpieces and intercalated ducts have become distended, but that α sympathetic stimulation, by activating myoepithelial cells, provides support for these structures and normally prevents or minimizes this distension. This support might be of special importance in the dog mandibular gland since sympathetically produced saliva is relatively slight in volume but extremely viscous (13, 18, 107, 108) and could be expected to offer great resistance to flow. Emmelin and Gjörstrup (105) also made another important observation in their studies of α and β activation of dog mandibular gland. They noticed that during combined α and β stimulation, intraductal pressure in a closed system was not only higher than during β stimulation alone but could also be sustained, whereas the lower pressures observed during β stimulation declined slowly toward zero. They inferred from this that, during β stimulation, secreted fluid slowly leaked back across the gland epithelium into the interstitium but that this was prevented by α activation of myoepithelial cells.

Although myoepithelial cell contraction is thus unequivocally established as occurring during sympathetic stimulation, the real physiological importance of such an event remains obscure since the physiological importance of sympathetically evoked salivation itself is obscure (see however Harrop and Garrett (109)). The extent of myoepithelial cell contraction during parasympathetic nerve stimulation, as mentioned above, is a matter of some dispute since both myoepithelial and secretory cells possess only muscarinic receptors, so it is not possible to block secretion easily without also affecting myoepithelial cell activation. Nevertheless, the balance of evidence would seem to suggest that myoepithelial cells do contract during parasympathetic nerve activation, if not by direct action of acetylcholine, then certainly, indirectly, via the action of bradykinin (101, 102, 104). Recent pharmacological studies point more strongly to a direct action of acetylcholine (110). Thus decentralization of the dog mandibular, either of the parasympathetic or of the sympathetic innervation, causes a classic postjunctional supersensitivity of myoepithelial cells to adrenergic and cholinergic agents; in addition, sympathetic ganglionectomy causes a prejunctional sensitization. The authors concluded that not only are the myoepithelial cells, considered collectively, innervated by both divisions of the autonomic nervous system, but that most individual cells must possess a dual innervation. Recently Emmelin et al. (111) have reinvestigated the effect of parasympathetic and sympathetic stimulation and of infused kinins on myoepithelial cells in cat salivary glands in which atrophy had been induced by ligation of the main ducts. This atrophy results in a profound reduction (but not abolition) of the secretory flow produced by nerve stimulation, which is associated morphologically with atrophy of endpiece cells (a nonspecialized epithelial cell remains) and structural changes in, but not atrophy of, myoepithelial cells. Under such circumstances the authors could obtain pressure responses (albeit reduced) in ligated glands that they attributed to myoepithelial cell contraction and heightened (supersensitive) myoepithelial cell responses to kinin infusion. Although strongly suggestive that cat myoepithelial cells are indeed innervated with cholinergic fibers, the results were not unequivocal, and the question must still be considered an open one.

FUNCTIONS OF MYOEPITHELIAL CELLS

If, as would seem reasonable from the foregoing, we may accept that the myoepithelial cells are contractile, it still remains to define what role their contraction plays in the secretion process. Since the cells may be sparse and confined only to the intercalated ducts, as in the rat parotid (37), or may be absent altogether, as in the pancreas (112), it is clear that they are not absolutely essential for secretion by an exocrine gland to take place. Nevertheless, their widespread occurrence suggests that they ought to subserve a function of more than incidental importance.

Studies on the mammary gland (42) reveal that myoepithelial cell contraction causes obliteration of the lumen of the secretory endpiece, with simultaneous expulsion of its contents, and a widening and shortening of the small ducts, which would thereby facilitate passage of the secretion to the exterior. Clearly, myoepithelial cell contraction might be expected to produce a similar result in salivary glands, and it would be likely to be of importance whenever the salivary secretion was especially viscous. Good evidence that myoepithelial contraction does produce this effect comes from studies on the sheep parotid (35, 94, 95, 113) since sympathetic stimulation in this animal is without parotid secreto-motor effect, but it can cause a temporary increase in secretion rate merely by initiating myoepithelial contraction. The possibility that the cells of the intercalated duct epithelium of the sheep parotid may themselves contract cannot entirely be excluded since actin filaments are exceptionally abundant in these cells and hypolemmal nerve terminals are commonly seen beside them (115).

Given that, initially, myoepithelial contraction does speed up entry of saliva into the mouth, it is difficult to see what advantage this would confer, except in animals using their saliva for offensive or defensive purposes, since the augmentation can only be transitory, if myoepithelial cell contraction is sustained, or intermittent, if contraction is intermittent. Tandler (34) suggests that the sudden contraction might play a role in the rupture and discharge of mucous granules from endpiece cells. Perhaps the most plausible explanation of myoepithelial cell function comes from Emmelin and his colleagues (99) who point out that, in addition to expelling pre-formed saliva, by contraction, myoepithelial cells "act as support for the underlying structures against distension which the secretion might otherwise induce." Their experiments (99, 102, 104–106) show beyond doubt that, when secretion occurs in the absence of myoepithelial cell contraction, there is distension of the glandular lumen with leakage of secretion back into the interstitium; this distension and leakage are prevented when myoepithelial cell contraction occurs. In our own experiments, we find that retrograde injection along the duct system of the rat parotid can be carried out at much lower pressures than are possible for the rat mandibular gland, presumably because of the greater ease with which fluid can be driven across the unsupported endpiece epithelium of the rat parotid. Reference has already been made to the sublingual gland of the monotreme anteater *Tachyglossus aculeatus* (60) in which the myoepithelial cells show extraordinary hypertrophy—perhaps not surprisingly, we note also that the secretion of this gland is exceptionally viscous.

From the above we may list the following possible functions for salivary myoepithelial cells:

A. The cells act as a support for the endpieces, preventing their distension when secretion is taking place.
B. By contracting, the cells widen and shorten the intercalated ducts, thereby lowering the outflow resistance that might otherwise retard the formation of secretion or cause distension of the endpieces.
C. Contraction may speed entry of pre-formed saliva into the mouth at the onset of stimulation, thereby providing saliva when needed urgently in some species for offensive or defensive purposes.
D. Contraction may aid in the rupture of cells packed with mucous secretion granules.

SUMMARY

Myoepithelial cells, which occur in many but not all exocrine glands, are usually quite conspicuous in salivary glands. Their contractile nature seems established beyond doubt, and the morphological similarities to smooth muscle cells are close enough to suggest that they contract by a similar, if not the same, mechanism. Morphologically speaking, they seem to possess a dual autonomic innervation, but, functionally speaking, the sympathetic control seems usually to be more important that the parasympathetic. While their contraction undoubtedly causes a transitory increase in the rate of delivery of saliva to the oral cavity, the principal function of the cells seems to be to prevent distension of the endpieces during secretion. This is achieved partly by providing support for the endpiece and partly by a widening and shortening of the intercalated ducts, which has the effect of lowering the outflow resistance.

ACKNOWLEDGMENTS

We are grateful to Mr. Alan Kennerson for his assistance and to Mrs. Eva Vasak for her expert section cutting. The research support of the National Health and

Medical Research Council of Australia and of the University of Sydney is gratefully acknowledged.

REFERENCES

1. Burgen, A. S. V. (1967). Secretory processes in salivary glands. *In* C. F. Code (ed.), Alimentary Canal, Handbook of Physiology, Sect. 6, Vol. II, pp. 561–579. American Physiological Society, Washington.
2. Schneyer, L. H., and Schneyer, C. A. (1967). Inorganic composition of saliva. *In* C. F. Code (ed.), Alimentary Canal, Handbook of Physiology, Sect. 6, Vol. II, pp. 497–530. American Physiological Society, Washington.
3. Schneyer, L. H., Young, J. A., and Schneyer, C. A. (1972). Salivary secretion of electrolytes. Physiol. Rev. 52:720.
4. Young, J. A. (1973). Electrolyte transport by salivary epithelia. Proc. Aust. Physiol. Pharmacol. Soc. 4:101.
5. Emmelin, N., Schneyer, C. A., and Schneyer, L. H. (1973). The pharmacology of salivary secretion. *In* P. Holton (ed.), Pharmacology of Gastrointestinal Motility and Secretion, Vol. I, pp. 1–39. Pergamon, Oxford.
6. Emmelin, N., and Trendelenburg, U. (1972). Degeneration activity after parasympathetic or sympathetic denervation. Ergeb. Physiol. 66:147.
7. Schneyer, L. H., and Emmelin, N. (1974). Salivary secretion. *In* E. D. Jacobson and L. L. Shanbour (eds.), MTP International Review of Science. Gastro Intestinal Physiology, pp. 183–226. MTP Press, Lancaster.
8. Young, J. A., and van Lennep, E. W. The morphology of salivary glands. Springer, Berlin-Heidelberg–New York. In press.
9. Harvey, W. (1651). On animal generation. *In* R. Willis (trans.), The Works of William Harvey, M.D. Sydenham Society, London.
10. Malpighi, M. (1687). Epistolae Anatomicae: de Viscerum Structura Exercitatio Anatomica, pp. 51–144. Robert Littlebury, London.
11. Müller, J. (1830). De Glandularum Secernentium Structura Penitiori Earumque Prima Formatione in Homine atque Animalbus. Leopold Voss, Leipzig.
12. Kölliker, A. (1854). Manual of Human Histology, Vol 2, p. 26. Sydenham Society, London.
13. Heidenhain, R. (1878). Ueber secretorische und trophische Drüsennerven. Pflügers Arch. 17:1.
14. Pflüger, E. (1871). Die Speicheldrüsen. *In* S. Stricker (ed.), Handbuch der Lehre von den Geweben des Menschen und der Thiere, Vol. 1, pp. 306–332. Engelmann, Leipzig.
15. Schaffer, J. (1927). Das Epithelgewebe. *In* W. von Möllendorff (ed.), Handbuch der mikroskopischen Anatomie des Menschen, Vol 2, Part 1, pp. 1–231. Springer, Berlin.
16. Zimmermann, K. W. (1927). Die Speicheldrüsen der Mundhöhle und die Bauchspeicheldrüse. *In* W. von Möllendorff (ed.), Handbuch der mikroskopischen Anatomie des Menschen, Vol. 5, Part 1, pp. 61–244. Springer, Berlin.
17. Krause, W. (1865). Ueber die Drüsennerven. II. Die Nervenendigung in den Drüsen. Z. Rat. Med. 23:46.
18. Heidenhain, R. (1883). Die Speicheldrüsen und die verwandten Drüsen der Schleimhäute. *In* L. Hermann (ed.), Handbuch der Physiologie, Vol. 5, Part 1, pp. 14–90. Vogel, Leipzig.
19. Lavdowsky, M. (1877). Zur feineren Anatomie und Physiologie der Speicheldrüsen, insbesondere der Orbitaldrüse. Arch. Mikr. Anat. 13:281.
20. Boll, F. (1868). Ueber den Bau der Thränendrüse. Arch. Mikr. Anat. 4:146.
21. Boll, F. (1869). Die Bindesubstance der Drüsen. Arch. Mikr. Anat. 5:334.
22. Unna, P. G. (1881). Zur Theorie der Drüsensecretion, insbesondere des Speichels. Zbl. Med. Wiss. 19:257.
23. Renaut, J. L. (1897). Traité d'Histologie Pratique, Vol. 2. Lecrosnier and Babé, Paris. (Cited by Zimmermann (16).)

24. Silver, I. A. (1954). Myoepithelial cells in the mammary and parotid glands. J. Physiol. (Lond.) 125:8P.
25. Leeson, C. R. (1956). Localization of alkaline phosphatase in the submaxillary gland of the rat. Nature 178:858.
26. Shear, M. (1966). The structure and function of myoepithelial cells in salivary glands. Arch. Oral Biol. 11:769.
27. Jacoby, F., and Leeson, C. R. (1959). The post-natal development of the rat submaxillary gland. J. Anat. 93:201.
28. Scott, B. L., and Pease, D. C. (1959). Electron microscopy of the salivary and lacrimal glands of the rat. Am. J. Anat. 104:115.
29. Takahashi, N. (1958). Electron microscopic studies on the ectodermal secretory glands in man. II. The fine structures of the myoepithelium in the human mammary and salivary glands. Bull. Tokyo Med. Dent. Univ. 5:177.
30. Travill, A. A., and Hill, M. F. (1963). Histochemical demonstration of myoepithelial cell activity. Q. J. Exp. Physiol. 48:423.
31. Tamarin, A., and Screebny, L. M. (1965). The rat submaxillary salivary gland: a correlative study by light and electron microscopy. J. Morphol. 117:295.
32. Tamarin, A. (1966). Myoepithelium of the rat submaxillary gland. J. Ultrastruct. Res. 16:320.
33. Tandler, B. (1965). Ultrastructure of the human submaxillary gland. III. Myoepithelium. Z. Zellforsch. 68:852.
34. Tandler, B., Denning, C. R., Mandel, I. D., and Kutscher, A. H. (1970). Ultrastructure of human labial salivary glands. III. Myoepithelium and ducts. J. Morphol. 130:227.
35. Blair-West, J. R., Coghlan, J. P., Denton, D. A., Nelson, J., Wright, R. D., and Yamauchi, A. (1969). Ionic, histological and vascular factors in the reaction of the sheep's parotid to high and low mineralocorticoid status. J. Physiol. (Lond.) 205:563.
36. Harrop, T. J., and Mackay, B. (1968). Electron microscopic observations on myoepithelial cells and secretory nerves in rat salivary glands. J. Can. Dent. Assoc. 34:481.
37. Garrett, J. R., and Parsons, P. A. (1973). Alkaline phosphatase and myoepithelial cells in the parotid gland of the rat. Histochem. J. 5:463.
38. Garrett, J. R., and Harrison, J. D. (1970). Alkaline phosphatase and adenosine triphosphatase histochemical reactions in the salivary glands of cat, dog and man, with particular reference to the myoepithelial cells. Histochemie 24:214.
39. Cutler, L. S., and Chaudhry, A. P. (1973). Differentiation of the myoepithelial cells of the rat submandibular gland *in vivo* and *in vitro*: an ultrastructural study. J. Morphol. 140:343.
40. Bunting, H., Wislocki, G. B., and Dempsey, E. W. (1948). The chemical histology of human eccrine and apocrine sweat glands. Anat. Rec. 100:61.
41. Ellis, R. A. (1965). Fine structure of the myoepithelium of the eccrine sweat glands of man. J. Cell Biol. 27:551.
42. Linzell, J. L. (1955). Some observations on the contractile tissue of the mammary glands. J. Physiol. (Lond.) 130:257.
43. Leeson, C. R. (1960). The electron microscopy of the myoepithelium in the rat exorbital lacrimal gland. Anat. Rec. 137:45.
44. Leeson, C. R. (1960). The histochemical identification of myoepithelium, with particular reference to the Harderian and exorbital lacrimal glands. Acta Anat. (Basel) 40:87.
45. Rowlatt, C., and Franks, L. M. (1964). Myoepithelium in mouse prostate. Nature 202:707.
46. Sorokin, S. P. (1965). On the cytology and cytochemistry of the opossum's bronchial glands. Am. J. Anat. 117:311.
47. Pearse, A. G. (1968). Histochemistry, Theoretical and Applied, Vol. 1, 3rd. Ed. Churchill, Livingstone, London, Edinburgh.
48. Bourne, G. (1944). The distribution of alkaline phosphatase in various tissues. Q. J. Exp. Physiol. 32:1.
49. Noback, C. R., and Montagna, W. (1947). Histochemical studies of the basophilia, lipase and phosphatases in the mammalian pancreas and salivary glands. Am. J. Anat. 81:343.

50. Deane, H. W., (1947). A cytochemical survey of phosphatases in mammalian liver, pancreas and salivary glands. Am. J. Anat. 80:321.
51. Feyel-Cabanes, T. (1949). Présence et rôle d'une phosphatase (monophosphoesterase I) dans la sous-maxillaire de souris. C. R. Soc. Biol. (Paris) 143:230.
52. Junqueira, L. C., Fajer, A., Rabinovitch, M., and Frankenthal, L. (1949). Biochemical and histochemical observations on the sexual dimorphism of mice submaxillary glands. J. Cell. Comp. Physiol. 34:129.
53. Hill, C. R., and Bourne, G. H. (1954). The histochemistry and cytology of the salivary gland duct cells. Acta Anat. (Basel) 20:116.
54. Dewey, M. M. (1958). A histochemical and biochemical study of the parotid gland in normal and hypophysectomized rats. Am. J. Anat. 102:243.
55. Rauch, S. (1959). Die Speicheldrüsen des Menschen. Thieme, Stuttgart.
56. Quintarelli, G., and Chauncey, H. H. (1959). Istochimia delle ghiandole salivari. Rass. Trim. Odontoiat. 40:541.
57. Bogart, B. I. (1968). The fine structural localization of alkaline and acid phosphatase activity in the rat submandibular gland. J. Histochem. Cytochem. 16:572.
58. Shear, M. (1964). Histochemical localization of alkaline phosphatase and adenosine triphosphatase in the myoepithelial cells of rat salivary glands. Nature 203:770.
59. Ohanian, C. (1973). Histochemical studies on phosphorylase activity in the tissues of the albino rat under normal and experimental conditions. III. Myoepithelial cells. Acta Anat. (Basel) 86:15.
60. van Lennep, E. W., and Kennerson, A. Light and electron microscopical studies on the salivary glands of the echidna (*Tachyglossus aculeatus*, Monotremata). In preparation.
61. Rhodin, J. A. G. (1962). Fine structure of vascular walls in mammals with special reference to smooth muscle component. Physiol. Rev. Suppl. 5:48.
62. Parks, H. F. (1961). On the fine structure of the parotid gland of mouse and rat. Am. J. Anat. 108:303.
63. Leeson, C. R., and Leeson, T. S. (1968). Fine structure and possible secretory mechanism of rat palatine glands. J. Dent. Res. 47:653.
64. Johnston, M. C., and Listgarten, M. A. (1972). Observations on the migration, interaction and early differentiation of orofacial tissues. *In* H. C. Slavkin and L. A. Bavetta (eds.), Developmental Aspects of Oral Biology, pp. 53–80. Academic Press, New York.
65. Cutler, L. S., and Chaudhry, A. P. (1973). Intercellular contacts at the epithelial–mesenchymal interface during the prenatal development of the rat submandibular gland. Dev. Biol. 33:229.
66. Garrett, J. R. (1966). The innervation of salivary glands. I. Cholinesterase-positive nerves in normal glands of the cat. J. R. Micro. Soc. 85:135.
67. Garrett, J. R. (1966). The innervation of salivary glands. II. The ultrastructure of nerves in normal glands of the cat. J. R. Micro. Soc. 85:149.
68. Garrett, J. R. (1967). The innervation of normal human submandibular and parotid salivary glands. Arch. Oral Biol. 12:1417.
69. Garrett, J. R. (1972). Neuro-effector sites in salivary glands. *In* N. Emmelin and Y. Zotterman (eds.), Oral Physiology, pp. 83–97. Pergamon, Oxford.
70. Bolande, R. P., and Towler, W. F. (1973). Terminal autonomic nervous system in cystic fibrosis. Arch. Pathol. 95:172.
71. Kahn, N., Mandel, I., Licking, J., Wassermann, A., and Morea, D. (1969). Comparison of the effects of parasympathetic and sympathetic nervous stimulation on cat submaxillary gland saliva. Proc. Soc. Exp. Biol. Med. 130:314.
72. Tandler, B., and Ross, L. L. (1969). Observations of nerve terminals in human labial salivary glands. J. Cell Biol. 42:339.
73. Bogart, B. I. (1970). Fine structural localization of cholinesterase activity in the rat submandibular gland. J. Histochem. Cytochem. 18:730.
74. Bogart, B. I. (1971). The fine structural localization of acetylcholinesterase activity in the rat parotid and sublingual glands. Am. J. Anat. 132:259.
75. Kagayama, M., and Nishiyama, A. (1972). Comparative aspect on the innervation of submandibular glands in cat and rabbit: an electron microscopic study. Tohoku J. Exp. Med. 108:179.

76. Huxley, H. E. (1963). Electron microscope studies on the structure of natural and synthetic protein filaments from striated muscle. J. Mol. Biol. 7:281.
77. Bray, D. (1974). Anti-actin. Nature 251:187.
78. Archer, F. L., Beck, J. S., and Melvin, J. M. O. (1971). Localization of smooth muscle protein in myoepithelium by immunofluorescence. Am. J. Pathol. 63:109.
79. Archer, F. L., and Kao, V. C. Y. (1968). Immunohistochemical identification of actinomyosin in myoepithelium of human tissues. Lab. Invest. 18:669.
80. Line, S. E., and Archer, F. L. (1972). The postnatal development of myoepithelial cells in the rat submandibular gland. An immunohistochemical study. Virchows Arch. (Zellpathol.) 10:253.
81. Puchtler, H., Waldrop, F. S., Carter, M. G., and Valentine, L. S. (1974). Investigation of staining, polarization and fluorescence microscopic properties of myoepithelial cells. Histochemistry 40:281.
82. Somlyo, A. P., Devine, C. E., Somlyo, A. V., and Rice, R. V. (1973). Filament organization in vertebrate smooth muscle. Philos. Trans. R. Soc. Lond. (Biol.) 265:223.
83. Lowy, J., and Small, J. V. (1970). The organization of myosin and actin in vertebrate smooth muscle. Nature 227:46.
84. Sobieszek, A., and Small, J. V. (1973). The assembly of ribbon-shaped structures in low ionic strength extracts obtained from vertebrate smooth muscle. Philos. Trans. R. Soc. Lond. (Biol.) 265:203.
85. Alexander, J. H., Young, J. A., and van Lennep, E. W. (1973). The ultrastructure of the duct system in the rat extraorbital lacrimal gland. Z. Zellforsch. 144:453.
86. Kerjaschki, D. (1974). The anterior medial gland in the mouse nasal septum: an uncommon type of epithelium with abundant innervation. J. Ultrastruct. Res. 46:466.
87. Uehara, Y., Campbell, G. R., and Burnstock, G. (1971). Cytoplasmic filaments in developing and adult vertebrate smooth muscle. J. Cell Biol. 50:484.
88. Gabella, G. (1973). Fine structure of smooth muscle. Philos. Trans. R. Soc. Lond (Biol.) 265:7.
89. Devine, C. E., Somlyo, A. V., and Somlyo, A. P. (1973). Sarcoplasmic reticulum and mitochondria as cation accumulation sites in smooth muscle. Philos. Trans. R. Soc. Lond. (Biol.) 265:17.
90. Hurley, H. J., and Shelley, W. B. (1954). The role of the myoepithelium of the human apocrine sweat gland. J. Invest. Dermatol. 22:143.
91. Mathews, A. (1898). The physiology of secretion. Ann. N.Y. Acad. Sci. 11:293.
92. Anrep, G. V. (1922). Observations on augmented salivary secretion. J. Physiol. (Lond.) 56:263.
93. Babkin, B. P. (1950). Secretory Mechanism of the Digestive Glands, 2nd Ed. Hoeber, New York.
94. Kay, R. N. B. (1958). The effects of stimulation of the sympathetic nerve and of adrenaline on the flow of parotid saliva in sheep. J. Physiol. (Lond.) 144:476.
95. Coats, D. A., Denton, D. A., Goding, J. R., and Wright, R. D. (1956). Secretion by the parotid gland of the sheep. J. Physiol. (Lond.) 131:13.
96. Langley, L. L., and Smith, J. A. (1959). Effects of vagosympathetic stimulation on parotid saliva flow in the dog. Am. J. Physiol. 197:821.
97. Holzlöhner, E. (1931). Die Drüsentätigkeit bei Nervenreizung. 1. Mitteilung. Sekretionssachogramme der Glandula submaxillaris bei Reizung der Chorda tympani. Z. Biol. 91:531.
98. Holzlöhner, E., and Cammann, O. (1934). Die Drüsentätigkeit bei Nervenreizung. 4. Mitteilung. Sekretion gegen Überdruck–Versuche an der Glandula submaxillaris. Z. Biol. 95:235.
99. Emmelin, N., Garrett, J. R., and Ohlin, P. (1968). Neural control of salivary myoepithelial cells. J. Physiol. (Lond.) 196:381.
100. Darke, A. C., and Smaje, L. H. (1971). Myoepithelial cell activation in the submaxillary salivary gland. J. Physiol. (Lond.) 219:89.
101. Emmelin, N., Garrett, J. R., and Ohlin, P. (1970). Action of kinins on salivary myoepithelial cells. J. Physiol. (Lond.) 207:539.

102. Emmelin, N., Ohlin, P., and Thulin, A. (1969). The pharmacology of salivary myoepithelial cells in dogs. Br. J. Pharmacol. Chemotherap. 37:666.
103. Emmelin, N., and Holmberg, J. (1967). The presence of beta-receptors in the submaxillary gland of the dog. Br. J. Pharmacol. Chemotherap. 30:371.
104. Emmelin, N., Garrett, J. R., and Ohlin, P. (1969). Motor nerves of salivary myoepithelial cells in dogs. J. Physiol. (Lond.) 200:539.
105. Emmelin, N., and Gjörstrup, P. (1973). On the function of myoepithelial cells in salivary glands. J. Physiol. (Lond.) 230:185.
106. Emmelin, N., and Gjörstrup, P. (1974). The physiology of salivary myoepithelial cells. *In* N. A. Thorn and O. H. Petersen (eds.), Secretory Mechanisms of Exocrine Glands, pp. 29–41. Munksgaard, Copenhagen.
107. Heidenhain, R. (1868). Beiträge zur Lehre von der Speichelabsonderung. Stud. Physiol. Inst. Breslau. 4:1.
108. Dische, Z., Kahn, N., Rothschild, C., Danilchenko, A., Licking, J., and Wang, S. C. (1970). Glycoproteins of submaxillary saliva of the cat: Differences in composition produced by sympathetic and parasympathetic nerve stimulation. J. Neurochem. 17:649.
109. Harrop, T. J., and Garrett, J. R. (1974). Effects of preganglionic sympathectomy on secretory changes in parotid acinar cells of rats on eating. Cell Tissue Res. 154:135.
110. Emmelin, N., and Thulin, A. (1973). Action of drugs on denervated myoepithelial cells of salivary glands. Br. J. Pharmacol. Chemotherap. 48:73.
111. Emmelin, N., Garrett, J. R., and Ohlin, P. (1974). Secretory activity and the myoepithelial cells of salivary glands after duct ligation in cats. Arch. Oral Biol. 19:275.
112. Garrett, J. R., Lenninger, S., and Ohlin, P. (1970). Concerning possible contractile mechanisms in the pancreas–myoepithelial cells. Experientia 26:741.
113. Kay, R. N. B. (1958). Continuous and reflex secretion by the parotid gland in ruminants. J. Physiol. (Lond.) 144:463.
114. Garrett, J. R. (1976). Structure and innervation of salivary glands. *In* B. Cohen and I. Kramer (eds.), Scientific Foundations of Dentistry, Section X, Ch. 44. Heinemann, London.
115. van Lennep, E. W., Kennerson, A. R., and Compton, J. C. Preliminary investigations on the ultrastructure of the sheep parotid. In preparation.

International Review of Physiology
Gastrointestinal Physiology II, Volume 12
Edited by Robert K. Crane

5
Gastric Secretion

G. SACHS, J. G. SPENNEY, and W. S. REHM

Laboratory of Membrane Biology, Department of Medicine and Department of Physiology
University of Alabama in Birmingham,
Birmingham, Alabama

HORMONE-RECEPTOR INTERACTION 128

TRANSDUCTION OF STIMULUS 132
- **Amphibia** 132
- **Mammals** 132
- **Phosphodiesterases** 134
- **cAMP-activated Phosphokinase** 134
- **Guanylate Cyclase** 134
- **Role of Intracellular Ca^{2+}** 134

MORPHOLOGY OF SECRETORY MEMBRANE 135
- **Membrane Turnover** 137

SECRETORY PROCESS 137
- **Proton Pump Mechanisms** 137
- **Membrane Isolation** 141
- **Membrane-bound ATPase** 144
  - ***Mg^{2+}ATPase*** 144
  - ***Transport ATPase*** 144

ENERGY SOURCE FOR ACID SECRETION 148
- **Substrates** 148
- **Inhibitors** 148
- **Metabolite Measurements** 149

TISSUE PROPERTIES 152

This work was supported by National Institutes of Health Grants AM 15878, National Science Foundation Grant GB 31075, the Humboldt Foundation, and AM 17315, and Veterans Administration Project 8059-01 and 02, Veterans Administration Hospital, Birmingham, Alabama.

SUBCELLULAR TRANSPORT 157

SUMMARY 163

There are currently several concentrated areas of research on transport across the gastric mucosa which may be conveniently divided into (*a*) hormone-receptor interaction, (*b*) stimulus transduction, (*c*) morphological correlates of secretion, (*d*) secretory membrane properties, (*e*) energy source for secretion, (*f*) transport by the intact tissue, and (*g*) subcellular transport. These fields have not received uniform emphasis over a period of years nor is there equal sophistication in the approaches used.

Apart from the usual medical importance, there are certain features of the stomach which make it unique. Paramount among these is the generation of the largest ion gradient in biological systems, with more than a million-fold difference of $[H^+]$ between blood and lumen being commonplace in mammals. The biochemical and electrical correlates of the production of this gradient are among the most controversial in epithelial transport, and an understanding of H^+ transport by this tissue could well be expected to have far-reaching consequences in many systems which generate or utilize a proton gradient.

Several reviews of gastric acid secretion have appeared recently (1–5), and in an attempt to emphasize different areas from these we have relied heavily on those areas in which we have had direct experience and have not attempted to do justice to all the recent excellent work.

HORMONE-RECEPTOR INTERACTION

The nature of the primary stimulus of acid secretion remains in doubt. The problem may be defined by asking whether a terminal receptor exists with a single histamine site, whether a receptor exists with multiple sites, or whether three independent receptors are present.

Histamine is the best known stimulant of acid secretion, but it has proved difficult to establish its physiological role in the stomach. For example, following a meal, the level of circulating gastrin alone is adequate to account for the quantity of acid secreted (6). Measurable quantities of histamine appear to efflux from the mucosa during secretion (7), but there is no apparent quantitative relationship with the amount of gastric secretion.

Compelling evidence for a role of histamine has, however, come from the action of histamine antagonists on induced acid secretion. With the development of antihistamines capable of antagonizing the vascular effects of histamine (8), it was found that they did not block histamine-induced gastric acid secretion (9). To explain this, two types of histamine receptor were postulated, H1 and H2, the H1 receptor being the situs of known antihistamine action, the H2 receptor being responsible for acid secretion.

In correlation with two postulated receptors, two forms of histamine were postulated (10), corresponding, as a first suggestion, to open and closed structures due to an extended or folded ethylamine side chain. More recently, the structure of a selective H2 agonist, 4-methylhistamine, has led to the suggestion that the H1 form of histamine is the fully extended transconformation with maximum separation between the ammonium group and the ring N_1 nitrogen atom, which 4-methylhistamine cannot achieve (11). As known H1 antagonists differ from histamine in the ring structure but retain the essential features of the side chain, usually in the form of *N,N*-dimethylethylamine, H1 and H2 receptors were assumed to differ in the conformational parameters required of their respective agonists.

Accordingly, when the imidazole, or 4-methylimidazole ring structure was retained, it was found possible to produce H2 antagonists by side chain modification. For example, when butyl thiourea was substituted for the ethylamine, a compound, burimamide, was produced which selectively blocked the action of histamine on both the myometrium and the stomach (12). CPK models of burimamide and other H2 antagonists show that, as predicted, the side chain can achieve a folded conformation with minimum distance between N_1 and the ammonium group. One of these, cimetidine, shows distinct promise as a therapeutic antisecretory agent (13). Figure 1 shows the structural formulas of some of the compounds discussed.

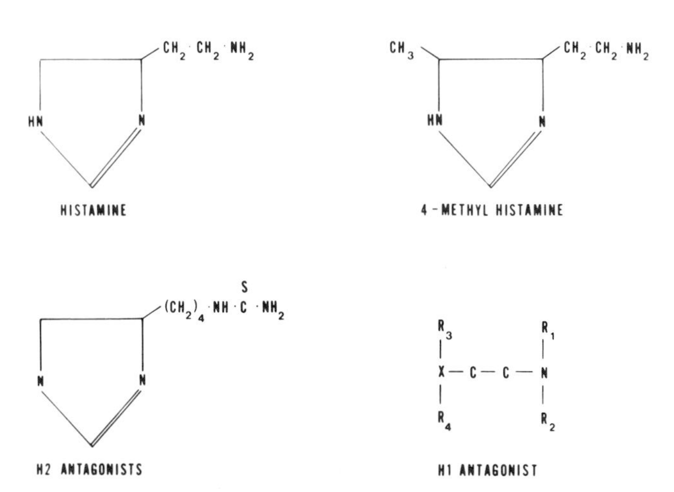

Figure 1. The structure of histamine, an H2 agonist, burimamide, and the general formula for H1 antagonists were $R_{1,2}$ are usually $< CH_3$.

H2 antagonists block acid secretion in vivo in several species, regardless of whether the secretion is induced by gastrin, histamine, or cholinergic stimulus (12), although the last is rather refractory to H2 blockade, and response varies as a function of the H2 antagonist used. Arguments continue as to whether there are separate receptors for each gastric stimulus or whether the receptors are sequential with the histamine receptor as final mediator, but the fact remains that H2 anatagonists block all secretory stimuli in vivo.

Inasmuch as hormone receptors are currently believed to be located at the cell surface or at one specific intracellular site, a concept of multiple receptors linked to a common transducer is generally favored as an explanation of hormone action. For example, different hormones stimulate adenylate cyclase (14) in the same cell, e.g., the fat cell, and the hormone effects are, in general, not additive. In contrast, a concept of sequential receptors, as for gastric stimuli, would appear to require that the different receptors be localized in different cell types or in different regions of the same cell. A multisite receptor concept is even more elaborate as the same receptor would have to be capable of reacting with such dissimilar compounds as gastrin, histamine, and acetylcholine. However, there is indeed evidence that interaction occurs among these stimuli.

The fact that H2 antagonists, in vivo, block secretion irrespective of stimulus revives not only the sequential receptor concept but also the idea that histamine release is a final common pathway (15).

Effective acid secretion in vivo requires two distinct sites of stimulation; one site is the blood supply (see Chapter 1, section on gastric circulation), the other is the acid-secreting cell. In the non-secreting stomach, the mucosal cell layer receives little blood. However, immediately prior to secretory onset arteriolar sphincters dilate (16). Secretion inhibitors like vasopressin act by inhibition of blood flow (17) and inhibit all secretory stimuli regardless of the quantity of stimulus. This does not happen in vitro because blood supply plays no role in in vitro acid secretion. H2 antagonists are active in vitro and must act on the acid-secreting cell, where they may block some critical process or receptor activation (Figure 2). However, it is as yet difficult to be confident

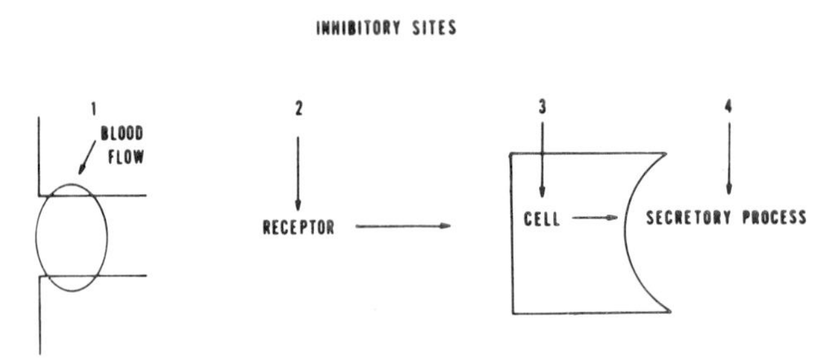

Figure 2. The possible modes of inhibition of gastric secretion by inhibition of blood flow, receptor activation, metabolism, and the secretory process.

about the specificity or selectivity of H2 antagonists, although some useful information has been obtained.

Frog mucosa mounted in an Ussing chamber usually shows spontaneous secretion, which the addition of burimamide blocks. Associated with inhibition is a relative oxidation of the respiratory chain and a decrease in oxygen consumption (18), neither of which occurs when burimamide is added to a resting mucosa. The changes are reversed if histamine is added to the bathing solution in the continued presence of burimamide (18). Thus, the action of burimamide depends on acid secretion, it involves a histamine-sensitive locus, and it is not a simple metabolic inhibition. The simplest interpretation is that burimamide and other H2 antagonists compete with histamine at a receptor site (19) which, it is logical to assume, is the final hormone-linked step in acid secretion (Figure 3). Other evidence supports the same concept. For example, when resting frog mucosa is repeatedly subjected to cholinergic stimulus, it becomes refractory to this stimulus; that is, it exhibits tachyphylaxis, though it remains sensitive to gastrin. A gastrin-stimulated mucosa also displays tachyphylaxis, and such a mucosa will respond to histamine but not to acetylcholine. A mucosa which is refractory to histamine is responsive only to cyclic adenosine 3′:5′-monophosphate (cAMP) (20).

From these data combined it is possible to justify a sequantial model of stimulation as shown in Figure 3, though the postulated histamine receptor has never been established by direct experimentation.

It may be emphasized again that the H2 receptor differs from the H1 receptor in recognizing a different conformation of histamine; it is the receptor

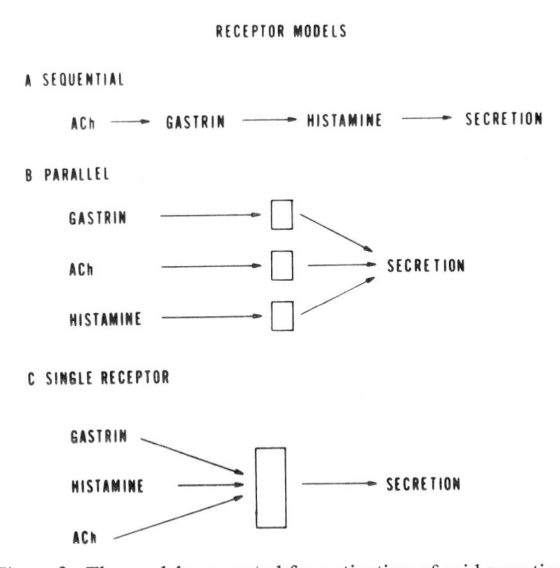

Figure 3. The models suggested for activation of acid secretion.

structure that determines the binding conformation. Specific agonists are structures that cannot achieve the alternate conformation, and specific antagonists bind only to one type of receptor. At this stage the nature and location of the H2 receptor are the critical questions awaiting direct experimentation.

TRANSDUCTION OF STIMULUS

It has now become dogma that the binding of a hormone at its high affinity site on the cell membrane results in a concetration change of a second messenger inside the cell. Accordingly, most research on stimulus transduction in the gastric mucosa has focused on adenylate cyclase and cAMP (21). To establish a role for cAMP in acid secretion, four criteria must be met: 1) inhibitors of cAMP phosphodiesterase should augment or mimic hormone effects; 2) cAMP should reproduce the effects of hormone; 3) the level of cAMP should change in accord with changes in secretion; and 4) adenylate cyclase should be present and be stimulated by hormones at physiological levels. In general, these criteria have been met; nevertheless, there are enough negative findings to provide a basis for reasonable doubt as to the involvement of cAMP in acid secretion, especially in the mammal.

Amphibia

Studies in these species have been confined to Ussing chamber studies. Early experiments with phosphodiesterase inhibitors showed that theophylline and other xanthine derivatives increased the acid rate in spontaneously secreting frog mucosa (22), as did cAMP (23). These findings were suggestive but were not proof of cAMP involvement in the secretory stimulation process since, for example, certain substrates have a similar action (3). The situation became a little clearer with the demonstration that either theophylline or cAMP could initiate acid secretion in the nonsecreting *Necturus* gastric mucosa (24). Finally, it has been shown that resting frog mucosa, obtained by exhaustive washing (25) or by anti-H2 treatment (19) or rendered refractory to histamine (20), can be stimulated by cAMP. The fact that H2 blockade does not affect cAMP-induced secretion is clear evidence that H2 antagonists do not act by metabolic blockade and is highly suggestive that cAMP is the final mediator of acid secretion. The cAMP level of frog mucosa also increases with increases in acid secretion (23). However, the exact time course of these changes has not been well defined. Recent investigations on adenylate cyclase have shown that *Necturus* (26) and frog (27) mucosa contain an adenylate cyclase consistently sensitive to histamine and occasionally to gastrin, though effective doses are pharmacological. In sum, it appears likely that cAMP is involved in stimulation of amphibian acid secretion in vitro.

Mammals

In mammals only some of the above criteria have been met, and these only in part. In fact, a recent review has concluded that cAMP is not the second messenger in acid secretion (28). However, the point is still regarded as controversial.

Phosphodiesterase inhibitors have a variable effect on acid secretion in mammals. Caffeine increases acid secretion in man (29), but theophylline has no effect in dog, either alone or in combination with other stimuli such as histamine (30). cAMP or its dibutyryl derivative is without effect on in vivo secretion, and some reports have emphasized its inhibitory action (31). However, in dog, using pooled samples, a rise in cAMP has been reported in gastric juice and in tissue prior to changes in secretory rate (32), as would be compatible with the second messenger role of cAMP.

A membrane-bound adenylate cyclase has been found in all species examined. The question then is whether there is correlation between hormone action and cyclase activity in the appropriate range of sensitivity. The most detailed study of mammalian gastric cyclase activity has been performed on rabbit tissue in which adenylate cyclase is sensitive to histamine and responds to histamine analogs in the same way as gastric secretion (33). Table 1 shows that

Table 1. Effect of histamine analogs on adenylate cyclase[a]

Compound	Gastric secretion	Adenylate cyclase (% control)
Control	None	100
Histamine	+++	342
N-Methylhistamine	+++	288
N,N-Dimethylhistamine	+++	253
4-Methylhistamine	+++	301
2-Methylhistamine	++	230
3-(2′-Aminoethyl)-1,2,4-triazole	++	234
3-(2′-Aminoethyl)-pyrazole	++	142
2-Amino-4-aminoethyl thiazole	++	192
2-aminoethylpyridine	+	137
N-tert-butylhistamine	+	115
3-Methylhistamine	+	85
N-Acetylhistamine	+	123
4-(2-Methyl-2-amino propionyl)-imidazole		89
Histidine		103
Imidazole 4-acetic acid		93
2-Methyl-4-aminomethyl imidazole		105
4-chloroethyl imidazole		102
4-hydroxyethyl imidazole		102
4-thioamidoethyl imidazole		74
4-(2′-aminoethyl)-1,2,3-triazole		102

[a]The compounds were tested at a final concentration of 10^{-3} M in the standard adenylate cyclase system. The adenylate cyclase data are means of two experiments expressed as percent of control. Gastric secretion data were obtained in the Shay rat or the conscious dog. +++, 10–100% of histamine output (on a molar basis); ++, 1–10% of histamine output; +, 0.1–1% of histamine output; – 0.1% of histamine output.

the secreting response and the cyclase response correlate well. These data confirm studies with guinea pig tissue (34) and rat tissue. However, H1 as well as H2 antagonists block the adenylate cyclase response in contrast to the gastric secretory response. Also, the histamine-sensitive adenylate cyclase has not been localized to the parietal cell. Hence, a final judgment as to the significance of these data cannot be made. It is interesting that brain cerebellar adenylate cyclase shows similar activation characteristics (35). The response of gastric cyclase in the rabbit to gastrin was more variable (33) but often present. However, little if any response to cholinergic compounds was detected in the crude gastric fraction. Some difficulty has been found in obtaining a hormone-responsive cyclase in dog mucosa (36) but recently a histamine-sensitive cyclase has been described (37).

In vitro studies of mammalian acid secretion have been limited because of the difficulty in obtaining a suitable preparation (38), though a recent piglet preparation seems encouraging (39). However, to our knowledge no report of a cAMP effect on such a preparation has appeared.

Phosphodiesterases

There appear to be two types of cAMP diesterase in gastric mucosa (40, 41), as in other tissues, as well as a cyclic guanosine 3′:5′-monophosphate (cGMP) diesterase. These enzymes appear to be soluble, and surface mucosal cells appear to contain a higher activity of cGMP diesterase than do deeper cells (40). There is no evidence as yet for a bound form of either diesterase.

cAMP-activated Phosphokinase

One prevalent hypothesis holds that functional change is induced by cAMP-dependent phosphorylation of specific proteins at various sites in the cell (42). Of particular interest in this regard is the demonstration of adenylate cyclase activity on the basal surface of renal tubular cells and of a cAMP-activated phosphokinase on the apical surface (43). Gastric mucosa contains a soluble protein phosphokinase (44). In a preliminary note, a membrane-bound phosphokinase has been reported (45).

Guanylate Cyclase

A soluble guanylate cyclase has been described in gastric mucosa (33) and, like cGMP diesterase, it is more prevalent in the superficial cells. Currently the literature is devoid of data relating cGMP to acid secretion, although there has been recent interest in a role of cGMP in hormone effects on many tissues including the pancreas (46).

Role of Intracellular Ca^{2+}

In many hormone responsive tissues, changes in the level of intracellular free Ca^{2+} have been postulated as an alternative or a parallel second messenger. Removal of Ca^{2+} from both sides of the in vitro mucosa produces a biphasic change in tissue parameters. An initial rise in resistance, with decrease in secretion, is followed by a fall in resistance as the intercellular junctions disintegrate (47, 48). Removal of Ca^{2+} from the serosal surface only, abolishes secretion with maintenance of the electrical parameters. Secretion is restored by the readmission of Ca^{2+} (49). Measurements of $^{45}Ca^{2+}$ flux in isolated oxyntic cells show an increased efflux in the presence of cAMP or theophylline. These data suggest that Ca^{2+} may play a critical role in the tissue response to hormonal stimulation. A summarizing hypothesis is presented in Figure 4.

MORPHOLOGY OF SECRETORY MEMBRANE

The amphibian oxyntic or mammalian parietal cell undergoes a series of remarkable morphological changes during the transition between rest and secretion (50). Cells in the gastric mucosa at rest are characterized by 1) a large number of mitochondria and 2) at the apical surface of the oxyntic cell or the region of the intracellular canaliculus of the parietal cell a larger number of mitochondria and 3) at the apical surface of the oxyntic cell or the region of the intracellular canaliculus of the parietal cell a larger number of cytoplasmic membrane-bound inclusions or tubulo-vesicles (51). Studies using La^{3+} show that there is no morphologically defined communication between these tubulo-vesicles and the mucosal surface. When the cells are secreting, the surfaces become folded into microtubular form and the cytoplamic tubulo-vesicles decrease in number as the microtubules increase. Morphometry has established a correlation between the surface area of the tubulo-vesicles disappearing and the microtubule surface area appearing (52). However, it has recently been reported that the appearance of acid secretion is not quantitatively correlated with the appearance of microtubules (53). Several other cell types are similar in morphology to the acid-secreting cell; the KCl-secreting cell of the blowfly salivary gland has a regular microtubular arrangement (54), and the KCl-secreting cell of the *Cecropia* midgut also contains microtubules, though in *Crecopia* each microtubule contains a mitochondrion (55). In the acid-secreting cell there is no evidence for a direct anatomical path between the apical surface and mitochondria, just close association.

The transition between tubulo-vesicles and microtubules is a major morphological problem (56). For example, if the transition from rest to secretion

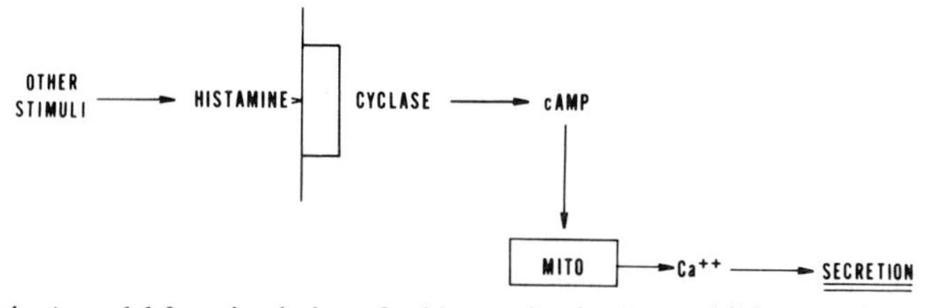

Figure 4. A model for stimulation of acid secretion in the amphibian gastric mucosa.

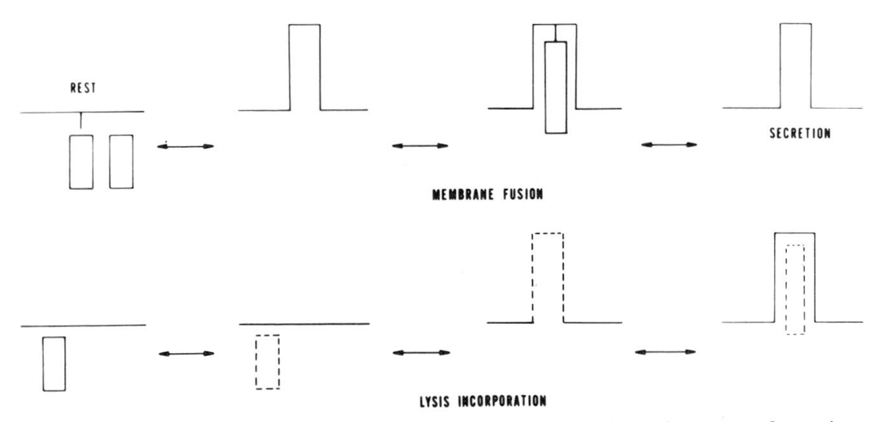

Figure 5. A fusion or lysis incorporation model for the observed membrane transformation in stimulation of acid secretion.

involves eversion of the tubulo-vesicles and the converse involves inversion, a mechanical process must be involved. During this process, one membrane would slide over another, producing double layered membrane forms (Figure 5), which might explain the large number of double membrane forms in the membrane fraction isolated from secreting hog stomach (77) (Figure 6). There is close

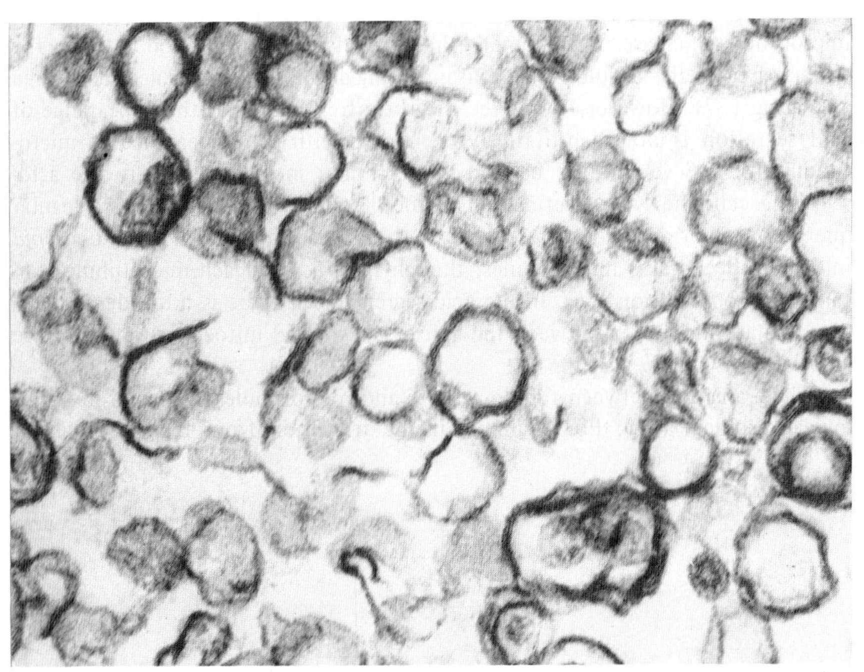

Figure 6. The double membrane and pentalaminar form membranes found in the first electrophoretic peak of the denser membrane fragments from hog gastric mucosa.

contact between adjacent surfaces of oxyntic tubules which may restrict access of luminal fluid to the base of the inter-microtubular space and create a compartment for ions that is extracellular. It may be noted that such large changes in the surface area of an acid secreting cell should be corrected for when analyzing the electrical events surrounding changes in secretion (39). Recently it has been shown that microvilli contain tubular forms internally and that multi-vesicular bodies are present (56). None of these morphological studies establish membrane fusion, but they do confirm membrane transition.

Membrane Turnover

The few extant studies on membrane turnover in gastric mucosa are probably outdated. Phospholipid turnover has been measured in amphibian mucosa (57), and alterations in $^{32}P_i$ incorporation were noted in several other hormone-sensitive epithelia (58). However, in these studies changes in protein components were not looked for and relatively impure subcellular fractions were used. These studies need to be repeated with more current techniques.

SECRETORY PROCESS

The major point of interest in gastric mucosa is its capacity to generate a 10^6-fold proton gradient and to provide a closely associated active Cl^- transport. In the last decade, H^+ transport and proton gradients have assumed a central role in energy transduction by biomembranes for both net flow of material and chemical transformations such as ATP synthesis. Current views of the original chemiosmotic hypothesis tend to agree that in chloroplasts, mitochondria, and bacteria (59) energy-yielding reactions are vectorially oriented across the membrane. These reactions generate proton and/or potential gradients across the membrane, and dissipation of a gradient is coupled to solute flow or ATP synthesis.

As a model system in which to investigate the molecular details of the production or utilization of these gradients, the simplest is the purple membrane of halophile bacteria (60). This system generates a proton gradient upon illumination of a membrane containing a retinal type of pigment and a 46,000 molecular weight protein (61). The gradient can then be used to drive ATP synthesis provided mitochondrial F_1 ATPase with F_0 coupling factor are also present in the membrane (62).

Proton Pump Mechanisms

In all ion pumps, the critical question is the nature of the pump site(s) and the mechanism of coupling of ion flow to a chemical reaction. The type of pump in the stomach is at present unknown, but it may be hopefully anticipated that it resembles one that is already known. Figure 7 details some simple possible electrical models.

Classification of Pumps

Uniport

(1) Monoionic

$$PD = \frac{E_A \cdot g_A}{g'_A \cdot g_A} = V_{AB}$$

$$I_A = (E_A - V_{AB})\, g_A$$

(2) Polyionic

$$PD = \frac{E_A g_A}{g'_A + g_A + g_B}$$

$$I_A = (E_A - V_{AB}) g_A$$

$$I_B = V_{AB} g_B$$

Uniport pump mechanisms in biological membranes would be invariably electrogenic, and for the monoionic uniport pump the total conductance of the membrane would be due to the actively transported ion. Any chemical reaction coupled to the pump would then be increased by the addition of any conductance g_Y across the membrane as a function of the reduction of V_{AB}. In polyionic uniport mechanisms, the transference number of the active ion is less than unity, and there is redistribution of B as a function of the ratio of g'_A and g'_B. Addition of a conductance g_Y would increase the reaction rate of the pump, but if g_Y is sufficiently large, the net flux of B would approach zero.

Antiport/Symport These systems differ from one another only in whether like charges exchange or unlike charges cotransport, but both react with the pump mechanism.

(1) Bi-ionic

$$PD = \frac{E_A g_A - E_B g_B}{g_A + g_B} = V_{AB}$$

$$I_A = (E_A - V_{AB}) g_A$$

$$I_B = (E_A + V_{AB}) g_B$$

(2) Polyionic

$$PD = \frac{E_A g_A - E_B g_B}{g_A + g_B + g_C} = V_{AB}$$

I_H, I_B equations are the same and $I_C = V_{AB} g_C$

In antiport/symport at least two ions are actively transported, and there must be distinguishing features in ionic interaction with the pump. In the case of the bi-ionic uniport or symport mechanism the potential developed by the pump will be a function of the difference between $E_A g_A$ and $E_B g_B$. If these terms are

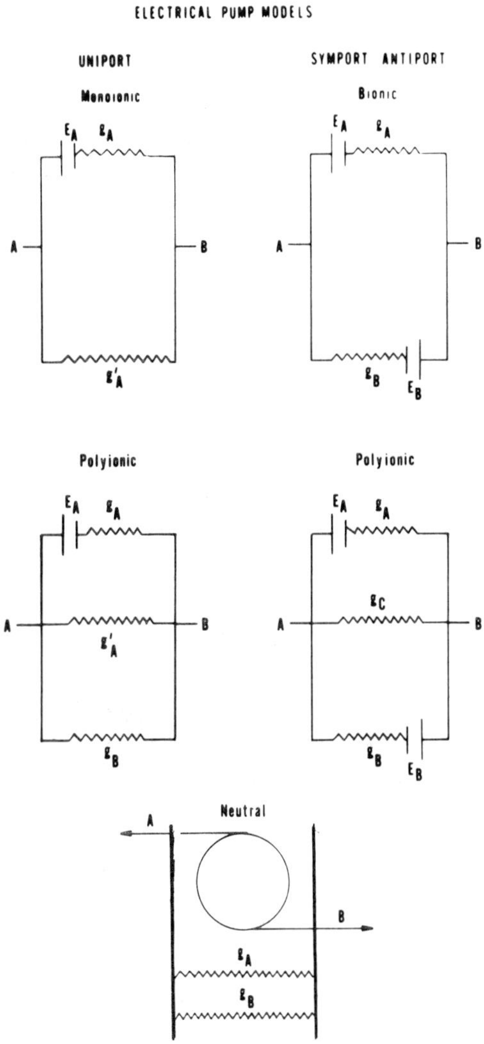

ELECTRICAL PUMP MODELS

UNIPORT

Monoionic

E_A g_A A B g'_A

SYMPORT ANTIPORT

Bionic

E_A g_A A B g_B E_B

Polyionic

E_A g_A A B g'_A g_B

Polyionic

E_A g_A g_C A B g_B E_B

Neutral

A B g_A g_B

Figure 7. Different electrical representation of pump models as a function of the number of ionic pumps and the membrane conductances.

equal, the pump is nonelectrogenic. If they are not equal, i.e., $t_A \neq t_B$, application of a shunt conductance g_X will result in a decrease of V_{AB} with an increase in I_A and a decrease in I_B. However, when g_X is infinite and I_B becomes zero, this pump becomes a uniport, bi-ionic model.

For the polyionic situation, an energy-dependent flux of *A, B,* and *C* is observed. Addition of a conductance, *X,* reduces flux of *C* to zero and operates on flux of *A* and *B* in a manner similar to the bi-ionic case. Also in the polyionic model, electrogenicity depends on the relationship of $E_A g_A$ and $E_B g_B$ as for the bi-ionic case.

In all these models, it is assumed that equilibrium conditions can be attained and that the initial flux rates with energization can be studied; hence, passive limbs have no diffusion potential of significance. At long time intervals, however, diffusion potentials must be taken into account. With appropriate measurements of pump rate and fluxes in vesicles and the addition of appropriate conductances, these different mechanisms are distinguishable. This type of experiment is possible in defined vesicles but difficult in intact epithelia.

Neutral The obligatory neutral or nonelectrogenic pump has been discussed recently in relation to violation of the Ussing flux ratio equation (63). This pump differs from antiport/symport, discussed above, since in antiport/symport electrogenicity is a function of the EMF and conductance of each battery or pump. Hence, by altering the conductance of one of the limbs, a nonelectrogenic situation can convert into an electrogenic one. The assumption is made that ions move across membranes by conducting mechanisms.

It is possible, however, that there are obligatory 1:1 antiport or symport mechanisms which are nonconductive. These obligatory neutral mechanisms would be unaffected by voltage or conductance changes. Also, the membrane may contain conductance pathways physically distinct from the pump so that when concentration gradients are developed, diffusion potentials would be observed. This is a major problem with epithelia, but in vesicles it is possible to set conditions of zero concentration gradient and, hence, of zero diffusion potential at the outset. The key question is the distinction between the situation in this neutral mechanism and the situation in the antiport/sumport where $E_A g_A = E_B g_B$.

A simple model for the neutral mechanism is the ionophore, nigericin, which moves across hydrophobic regions in the neutral form and functions as a cation exchange carrier. Nigericin may dissipate a diffusion potential due to cation gradients but not a pump potential.

With a neutral pump, the rate of ion flux is a function of the ion concentration and is not affected by the electrical properties of the membrane. With the antiport/symport mechanism, even if $E_A g_A$ were the same as $E_B g_B$, a reduction in A and an increase in B would generally result in equality no longer holding. Hence, the potential would transiently differ from zero resulting in flux rate alteration and changes in ion conductances. This is not the case with the neutral mechanism.

It is also probable that these differences in electrical characteristics reflect differences in chemical structure of the pump site. For example, assuming a mobile site, as appears to be the case for the K^+-ATPase discussed below, the electrogenic characteristics of this site can be predicted.

If the site is mobile only when it is electrically neutral, e.g., a phosphate binding two cations at pH 7.4, then the pump may exhibit only neutrality characteristics. This would imply strong charge restrictions for the singly charged negative form in terms of mobility. However, if the site is mobile in the charged state, e.g., one cation bound per phosphate group, then the pump is electrogenic, or potentially so as for any of the antiport or symport mechanisms discussed.

Membrane Isolation

Membrane isolation from the gastric mucosa involves three points of difficulty: separation of the parietal cell plasma membrane from other plasma membranes, separation of the plasma membrane from other cellular membranes, and separation of the apical and basal membranes of the polar parietal cell.

Isolation of viable parietal or oxyntic cells has been achieved by a variety of enzyme procedures (64, 65), the most commonly used being a mixture of fungal proteases known as pronase. This enzyme mixture provides the best yield and viability for both amphibia and mammals. Other procedures such as scraping or EDTA treatment have not produced viable cells.

Cell separation has been achieved in several ways: 1) Incubating tissue as a flat sheet and removing successive layers of cells during the incubation has proved successful for preparing relatively pure oxyntic cells from amphibia such as frog or *Necturus* (64) or preparing isolated intact tubules (66). 2) The surface of the mucosa may be treated with 3-M salt solution to destroy the surface cells. (67); the underlying layers of cells may then be isolated. 3) In the most generally useful procedure, a mixed cell suspension is prepared and the cells are separated by centrifugation on either isokinetic or isopycnic gradients (68, 69). The parietal cell mass in the mammal comprises at most 20% of the total and such centrifugation techniques result in a 2- to 3-fold enrichment.

Isolated cells have thus far proved useful for studies of metabolism (70), secretagogue response (71), and membrane isolation. Because of the low yield of purified parietal cells most membrane studies have been performed on whole tissue homogenates or on tissue after the surface cells have been removed. Species investigated include frog (72), *Necturus* (73), dog (74), rabbit (75), and hog (76, 77) using a combination of differential and density gradient centrifugation and, more recently, free flow electrophoresis (77).

Of the several marker enzymes which have been considered for identification of gastric cell plasma membranes, the current candidate as a marker specific for microtubular or tubulo-vesicular structures is K^+-activated ATPase (78). Fractionation of hog gastric mucosa using differential centrifugation, zonal density gradient centrifugation, and free flow electrophoresis results in the separation of a membrane fraction free of 5′-nucleotidase and 40-fold enriched in K^+-ATPase. Some HCO_3^--stimulated ATPase is retained in this fraction, although the major amount is located in the mitochondrial component of the homogenate (Figure 8). The K^+-ATPase fraction has a remarkably simplified peptide pattern on polyacrylamide gel elctrophoresis (Figure 9).

Direct proof of the parietal cell origin of this fraction is lacking, since little is known about the histochemistry of gastric mucosa. It has been suggested that the HCO_3^--ATPase is localized in the apical surface (79) along with carbonic

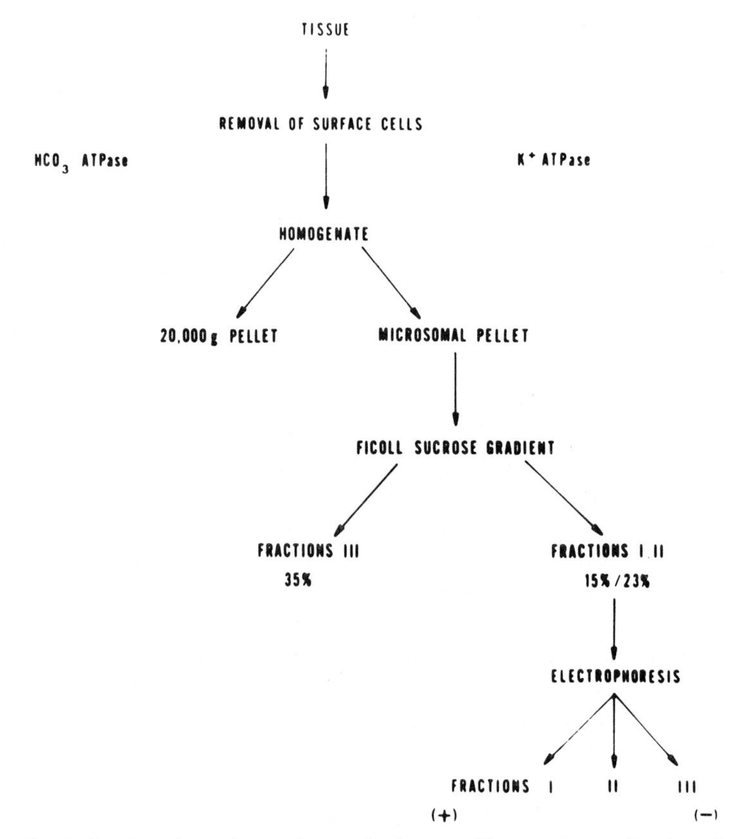

Figure 8. A fractionation scheme for producing specific membrane fractions from hog gastric mucosa.

anhydrase (80). Preliminary evidence indicates that the K^+-ATPase is localized in the microtubules (81).

The peptides of the purified hog mucosa fraction consist of two higher molecular weight protein types, 100,000 and 84,000 M_r and a group of low molecular weight peptides localized at the buffer front in gel electrophoresis. The amino acid composition of a less pure membrane fraction from the dog seems unremarkable (74) as do the carbohydrate or lipid components of fractions from rabbit (75) or hog (76).

Preliminary topology of dog gastric membrane peptides using proteolysis, crosslinking, and iodination has been described (82). The major fraction of the 100,000 M_r peptide is located internally and is particularly subject to lateral crosslinking. An 82,000 M_r peptide apparently spans the membrane and is presumably equivalent to the 84,000 M_r peptide of hog mucosa. As will be discussed later, K^+-activated *p*-nitrophenylphosphatase is reactive on the external surface of the cell (83) during acid secretion and the 100,000 M_r peptide is

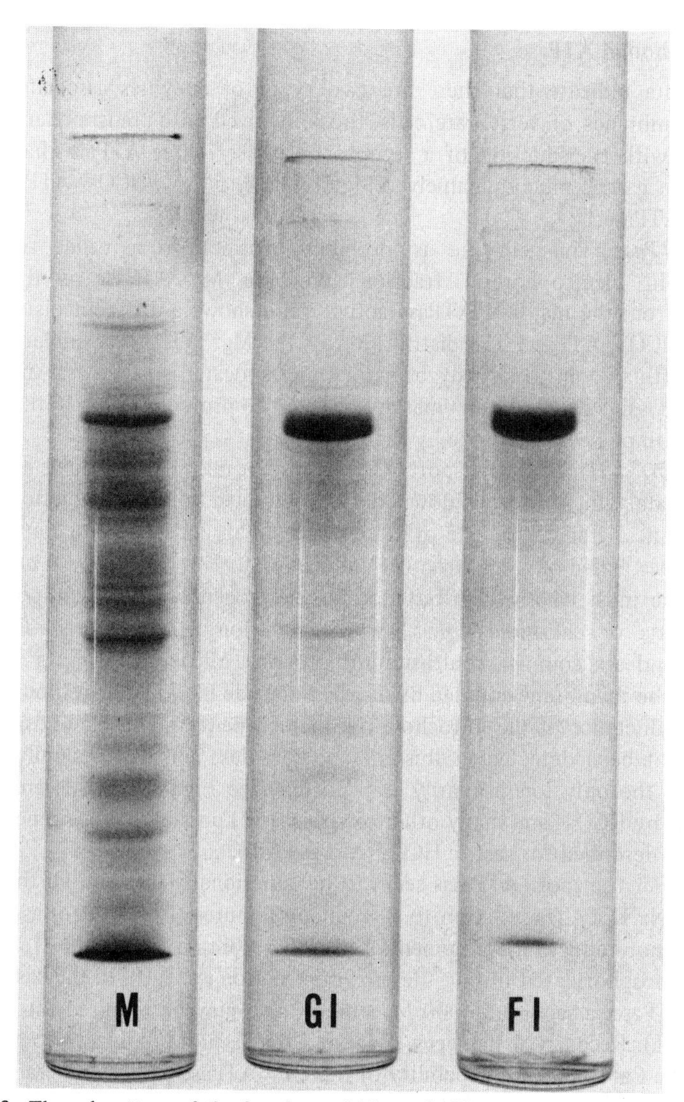

Figure 9. The gel pattern of the fractions of Figure 8, *M* corresponding to the microsomal pellet, *GI* to the lower density zonal peak, and *FI* to the anodic component on free flow electrophoresis.

phosphorylated by [γ-^{32}P] ATP, suggesting that this peptide spans the bilayer and must be closely associated with another peptide to allow cross-linking.

The most purified electrophoretic fraction contains several double membrane forms, some separate and some fused (cf. Refs. 53 and 56). There is also evidence that this membrane fraction contains vesicles, randomly oriented with respect to ATP binding and hydrolysis.

Membrane-bound ATPase

Current data indicate that there are two types of enzymes associated with plasma membranes of vertebrate cells, those involved with transport and those associated with transduction of a receptor response. Three ATPases have been described in gastric mucosa, namely, a Mg^{2+}ATPase, a Mg^{2+}, HCO_3^--ATPase, and a $Mg^{2+}K^+$-ATPase.

Mg^{2+}ATPase This enzyme is probably distinct from other transport ATPases. In electrophoretic fractionation, the Mg^{2+}ATPase is distinctly separated from the major K^+ATPase activity and shows a different distribution from the HCO_3^--ATPase. The distribution of the Mg^{2+}ATPase is similar to that of 5-nucleotidase, and both may be ectoenzymes located on the basal surface of gastric mucosal cells. Both enzymes are resistant to proteolysis and both may be involved with nucleoside uptake by the tissue (84).

Transport ATPase A transport ATPase system may be defined as a protein complex catalyzing the breakdown of ATP coupled to the production of an electrochemical gradient of one or more solutes. This class seems to contain two distinct types, based on both structure and function.

The one most widely distributed is the electrogenic ATPase component of mitochondria or chloroplasts (85) which function as generators of proton gradients and are complex multi-subunit proteins. Mitochrondrial Mg^{2+}ATPase appears to be membrane-coupled by another peptide F_0 which is responsible for the H^+ conductance of the mitochondrial membrane (86). Peptide phosphorylation has not been detected in this enzyme complex, and it is generally agreed that H^+ is the only ion primarily transported (see Figure 7). This enzyme is stimulated by $HCO_3^-$and many other oxybases (87) possibly due to general base catalysis as described for gastric HCO_3^--ATPase (88).

The other transport ATPases seem to be functionally distinct. Of these, the Ca^{2+} and (Na^+K^+)-ATPases seem to share many features. They are not associated with oxidation-reduction components but rather contain a 100,000 M_r subunit which is phosphorylated during the enzyme reaction (89, 90). In the case of the (Na^+K^+)-ATPase, a second 55,000 M_r subunit also may be required for enzyme activity (91). Recently it has been suggested that a proteolipid may be responsible for the Ca^{2+} transport capability of the Ca^{2+}-ATPase (92). A general scheme for this type of enzyme is as follows:

$$\text{ATP} + \text{ENZ} \xrightarrow{Mg^{+7},\ Na^+} \text{ATP-ENZ}$$

$$\text{ATP–ENZ} \longrightarrow \text{P}\cdot\text{E}_1\text{-ADP}$$

$$\text{PE}_1\text{-ADP} \longrightarrow \text{E}_1\text{P} + \text{ADP}$$

In the case of Na^+K^+-ATPase, Mg^{2+} and Na^+ are required for the formation of E_1P, the successive sequence being

$$\text{E}_1\text{P} \xrightarrow{Mg^{2+}} \text{E}_2\text{P}$$

$$E_2P \xrightarrow{K^+} E_2P + P_i$$

$$E_2 \longrightarrow E_1$$

K^+ is required for the dephosphorylation of E_2P, and E_1P reacts with ADP and not K^+, whereas E_2P reacts with K^+ and not ADP (93). There is also evidence for an EP form sensitive to ADP and a form insensitive to ADP (94) which could be explained by a dimer form. Ca^{2+}-ATPase shows similar complex teaction patterns. It is clear that uniport electrogenicity is not a primary property of the enzyme and that only specific ion gradients can be vectorially coupled to ATP synthesis or breakdown (see Figure 7). The phosphoenzyme in this type of ATPase is a β-aspartylphosphate, sensitive to NH_2OH (96, 97). The mitochondrial ATPase in contrast can be satisfied by various conductance paths such as those induced by protonophores (95) and no phosphoenzyme has been detected. For both types of enzyme, the major problem is to understand the precise coupling between the chemical reaction and the ion gradients.

Cardiotonic steroids specifically inhibit (Na^+K^+)-ATPase, and dicyclohexyl carbodiimide inhibits both the Ca^{2+}- and (Na^+K^+)-ATPase, but F_1 ATPase is inhibited only in the presence of F_0 in an oriented membrane system. Acetyl, carbamyl, and *p*-nitrophenylphosphate are hydrolyzed by the (Na^+K^+)- and Ca^{2+}-ATPases (98–100) but not by the F_1 ATPase.

Since gastric mucosa transports both H^+ and Cl^-, the question is whether the known ATPases are responsible for this transport and, if so, to what class they belong. Two ATPases may be considered: a HCO_3^--sensitive ATPase (101) or a K^+-activated ATPase (76, 78).

HCO_3^--ATPase The first enzyme to achieve prominence as perhaps relating to acid secretion was a SCN^--inhibited ATPase (101); SCN^- had long been known as an inhibitor of H^+ secretion (102) (see review (4)). However, SCN^- inhibited not only gastric microsomal ATPase but also rat liver mitochondrial ATPase (103). SCH; was later shown to inhibit K^+ movement in mitochondria (104) and in gastric mucosa (105). Subsequently HCO_3^- and other oxyganions were shown to stimulate the SCN^--sensitive ATPase (88). This was also a property of mitochondrial ATPase (87) and it was pointed out that there were too many analogies between mitochondrial ATPase and this gastric HCO_3^--ATPase to feel comfortable about the latter's suggested role in H^+ secretion (106). However, detailed measurements of the distribution of this enzyme and of mitochondrial makrers in the mucosal homogenate showed that, although HCO_3^-ATPase was mainly present in the mitochondrial fraction, its distribution extended into membrane fractions free of measurable mitochonrial contamination (74, 77). What has been said is true of dog and hog; the rat shows no HCO_3^- sensitivity of the microsomal ATPase (107). Thus, it seems reasonable to conclude that the SCN^--sensitive, HCO_3^--activated ATPase in gastric mucosa is a mitochondrial enzyme.

However, there are two points of possible significance. First, it has been suggested that the enzyme might be responsible for HCO_3^--Cl^- exchange at the basal surface of the oxyntic cell (108), though only one publication has reported evidence for Cl^- sensitivity (109) like the Cl^--sensitive ATPase in plant tissue (110). Second, HCO_3^- transport is mechanistically easier to visualize as a direct enzymic process, rather than as the transport of H^+/OH^-, particularly at the 160 mM level of gastric secretion. This may be true of other proton gradient generating systems as well. The following scheme illustrates this point:

$$H_2O \leftrightarrow H^+ + OH^-$$

$$OH^- + CO_2 \leftrightarrow HCO_3^- \rightarrow \text{transport}$$

These reactions occur at one face of the membrane. The ATPase, be it F_1 ATPase or another type of membrane-bound ATPase, then reacts with HCO_3^- or OH^- or another base and removes HCO_3^- (hence OH^-), generating a pH gradient. This mechanism avoids direct interaction of the enzyme protein with the highly reactive OH^- or H^+, as is illustrated in Figure 10. The difficulty of this scheme is in the localization of the HCO_3^--ATPase in the mucosa. However, for mitochondrial pH gradient generation by the ATPase neither the pH value realized nor the localization presents a problem.

K^+-ATPase A role of K^+ in acid secretion was established through finding that K^+ removal resulted in inhibition of secretion (111); restoration of K^+ to the mucosal bathing solution led to an especially rapid restoration of acid secretion (112), and morphological changes due to K^+ were similar to those found for other inhibitors (113). However, analysis showed that less than 10% of tissue K^+ had been lost at a time when K^+ removal had reduced H^+ secretion to zero. As pointed out above, SCN^- was shown to block K^+ transport in tissue and mitochondria (104, 105) along with H^+ secretion.

Yeast cells have long been known to secrete H^+, and much evidence suggests that H^+ secretion involves symport of a permeant anion such as succinate if K^+ is absent in the external medium (114). If K^+ is present externally, H^+ secretion

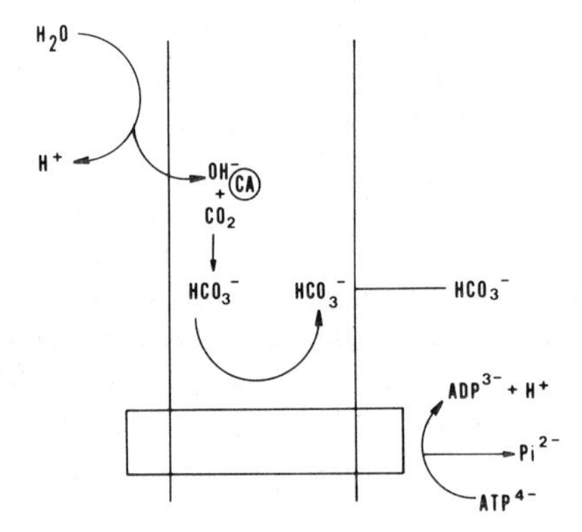

Figure 10. A model using HCO_3^--ATPase to generate an $H+/HCO_3^-$ gradient which may be applicable in various organs other than the stomach.

occurs via a K^+:H^+ antiport (115). More recently H^+ gradients have been shown to be involved in active solute transport by yeast (116) by a symport mechanism. In mitochondria, as an alternative to the exclusive proton gradient system, a K^+:H^+ exchange has also been suggested as the ion gradient coupling mechanism for ATP synthesis (117).

Altogether there is much suggestive evidence, both in gastric secretion and in other proton gradient systems, that K^+ may play a critical role.

Suggestions that a K^+-activated enzyme having unique properties might be present in the stomach came not long after the initial description of the HCO_3^--ATPase. A K^+-activated *p*-nitrophenylphosphatase (*p*-NPPase) (118, 119) or K^+-activated acetylphosphatase (120) was described which had the unique property of being ouabain insensitive. Other properties of this K^+-phosphatase were not at the time directly related to H^+ secretion. Recently, however, it has been shown that the K^+-activated *p*-NPPase behaves as an ectoenzyme in secreting mucosa (83) and that the rate of hydrolysis of *p*-nitrophenylphosphate added in the mucosal solution is a function of the prevailing acid secretion rate and a function of the presence of oxyntic cells (83).

It was a logical step to show the presence of a K^+-activated ATPase in gastric mucosa. Although it was first described in hog mucosa, the significance of this enzyme became clear in studies on rabbit mucosa (78). K^+-ATPase has been described in rabbit (75), frog (72), dog (82), and hog (76, 77) but so far has not been shown in rat or *Necturus.* Indeed in the latter species it has not been shown that a K^+-activated phosphatase is present (121). It is also apparently absent in the pancreas (121) and has not been demonstrated in yeast (123).

There are many similarities between K^+-ATPase and the transport ATPases, (Na^+ + K^+)- and Ca^{2+}-ATPase (122). K^+-ATPase requires a monovalent cation for activity, and the activation sequence is $Tl > K > Rb > Cs > Na$, Li as found for (Na^+K^+)-ATPase (124). This activation is for the dephosphorylation step, and there seems to be no cation requirement for the phosphorylation step (125). Formation of the NH_2OH-sensitive phosphorylated intermediate is blocked by a variety of inhibitors such as Zn^{2+}, F^-, DCCD, and *p*-CMBS; whereas *N*-ethylmaleimide reduces dephosphorylation. Interestingly, HCO_3^- also accelerates dephosphorylation, but the effect is not additive with K^+ (82).

K^+-ATPase is localized exclusively on the apical surface of the oxyntic or parietal cell (53) and shows a distribution pattern characteristic of a plasma membrane marker (76) but different from that of Mg^{2+}ATPase or 5′-nucleotidase (77). It has also been shown to be present in ion-tight vesicles derived either from hog (18) or dog gastric mucosa (127) and also to require K^+-active ionophores for maximum activity (18, 127, 128) in fresh preparations.

As will be discussed later, localization of this enzyme in a vesicular fraction of the gastric homogenate allows a determination of its ion transport activity, though the major problem is to determine its physiological significance. Several pieces of information make it hard to accept K^+ movement across the apical membrane of the cell as a necessary accompaniment of H^+ secretion. The measured chemical flux of K^+ is no more than 10% of the $J_H{}^+$ in amphibia (104). Voltage clamping in K_2SO_4 solutions does not bypass the SO_4^{2-} restriction on acid secretion. Under these conditions a K^+:H^+ exchange mechanisms might be deficient if K^+ did not enter the mucosal solution due to the presence of SO_4^{2-} and the absence of an electrogenic Cl^- pump. In high K^+ mucosal solutions this restriction should disappear, even in the presence of SO_4^{2-}. Although $J_H{}^+$ does increase with addition of K^+ to the mucosal SO_4^{2-} solution, this could well be accounted for by the change in tissue potential difference (129).

ENERGY SOURCE FOR ACID SECRETION

An alternative approach is provided by studies of the energy source for H^+ secretion in the hope that evidence will be provided for ATP as the direct energy source. However, despite increasing technical refinement, the answer as to the energy source of acid secretion has become increasingly elusive.

Measurements of changes in oxygen consumption of the resting and secreting dog mucosa show that a 16-fold change occurs (130). There is also a change in blood perfusion of the mucosa. Thus, it could be hoped that significant metabolite changes would be found due both to shift in utilization and shifts in access of substrate to the cell.

Substrates

In vitro amphibian mucosa shows remarkably little dependence on added substrate. Stimulation of acid secretion by addition of substrates such as fatty acids and by cofactors such as lipoate have been interpreted as being due to a block at the pyruvate dehydrogenase step (131). Some direct evidence has been obtained for this by measuring pyruvate dehydrogenase enzyme levels in frog mucosa, but similar measurements have given negative data in rabbit mucosa (132). Measurement of [^{14}C]glucose utilization has led to the postulation of activation of the hexose monophosphate shunt (133) in the rat as being necessary for secretion, but the data are indirect and lack critical information. In fact, frog data directly contradict this view (134).

Assessment of the dependence of acid secretion on fatty acid oxidation has led to the challenging finding that short chain fatty acid analogs (e.g., bromobutyrate) can stimulate a resting mucosa (135). This has been shown to be blocked by H2 antagonists; hence, it is due to release of secretagogue rather than specific metabolic stimulation (136).

Inhibitors

Few inhibitor studies have been carried out with full appreciation of the nonspecific as well as specific effects to be expected. In general, anoxia and oxidation-reduction inhibitors such as amytal clearly inhibit acid secretion with a corresponding decrease in ATP levels (137). Phosphorylation inhibitors such as atractylate or aurovertin block acid secretion and oxygen consumption (138), whereas uncouplers only block secretion (139). With amytal inhibition, menadione + ascorbate are successful in restoring oxygen and ATP levels, but there is no bypass of the secretion inhibition (140) emphasizing the combined requirement of adequate ATP levels and mitochondrial redox flow.

Metabolite Measurements

To serve as a framework for the ensuing discussion two extreme situations may be assumed: 1) gastric secretion is ATP dependent, and stimulation simply activates ATP utilization, or 2) gastric secretion depends on a membranal oxidation-reducation reaction and is independent of ATP (Figure 11). Assumption 1) would predict a decrease in the phosphorylation potential, i.e., ATP/ADP$\cdot P_i$ and a fall in the PCr + ATP/Cr + ADP ratio, at least during the initial phase of stimulation, and an associated change in the mitochondrial oxidation-reduction system. Assumption 2) would predict changes in a spectroscopically observable oxidation-reduction component dependent on the relative rates of

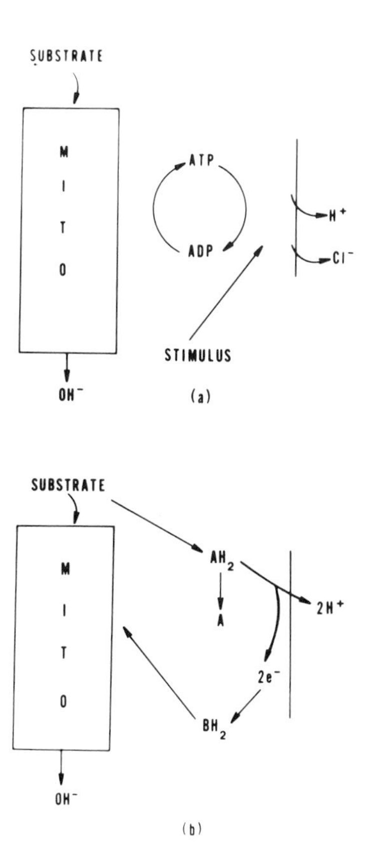

Figure 11. A model representation of a simple ATP system with interaction of stimulus at the secretory site and a redox model using a cytoplasmic electron acceptor as a reducing equivalent shuttle.

formation and consumption by the transport reaction and on the subsequent path for the electrons produced by the transport reaction. Neither assumption takes into account changes in metabolic substrate flux or cofactor levels.

Spectroscopic observations have been made on amphibian mucosa and allow measurement of the oxidation-reduction state of (NAD, $NADP^+$) FAD and the cytochromes (141–143). Chemical measurements of NAD^+, $NADP^+$, and their reduced counterparts have been made on dog mucosa and allow distinctions to be drawn between the components, as well as assessment of their compartmental concentrations.

The spectroscopic data, whether obtained by stimulating a resting mucosa induced by H2 antagonists (18) or by repeated washing (144), show a marked relative reduction in all the oxidation-reduction components measured, i.e., an increase in AH:A ratio which is difficult to reconcile with a simple ATP

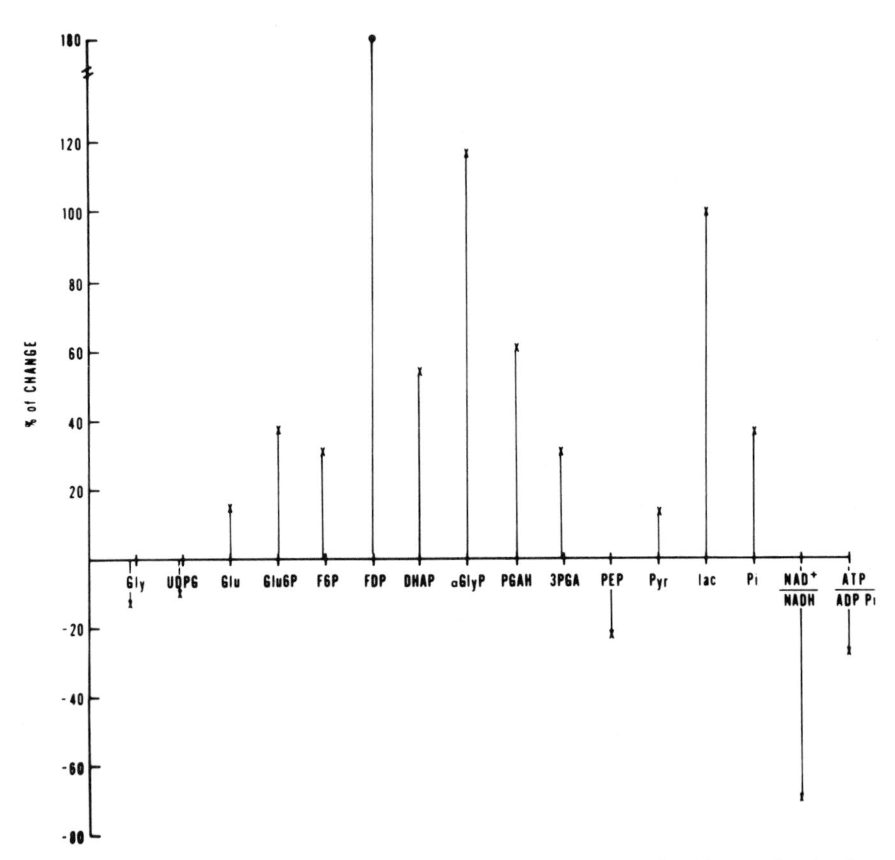

Figure 12. The alteration in metabolite levels found with onset of acid secretion in dog gastric mucosa with respect to glycolytic intermediates.

depletion mechanism. The data may reflect a shift in intracellular pH inasmuch as the reaction

$$NADH + H^{+} + A \leftrightarrow NAD^{+} + AH_2$$

is obviously sensitive to changes of $[H^{+}]$, though direct measurements of $NADP^{+}$ in the dog argue for a constant pH in the parietal cell in vivo.

Chemical measurement of pyridine nucleotide shows that, with secretory onset, cytosol NAD^{+} is reduced along with a large increase in lactate/pyruvate ratio and mitochondrial nucleotide is oxidized, as would be more compatible with an ATP mechanism (145).

Several measurements have been made of adenine nucleotide levels in gastric mucosa. It was shown earlier that the prevailing secretory rate of amphibian mucosa was related to the tissue ATP level. More detailed measurements in the same species have shown a slight fall in the ATP:ADP ratio with a rise in phosphocreatine (146). Measurements in the dog, where there is a 60-fold greater change in secretory rate, showed slight change in adenine nucleotide levels or in phosphocreatine but did show a fall in the phosphorylation potential

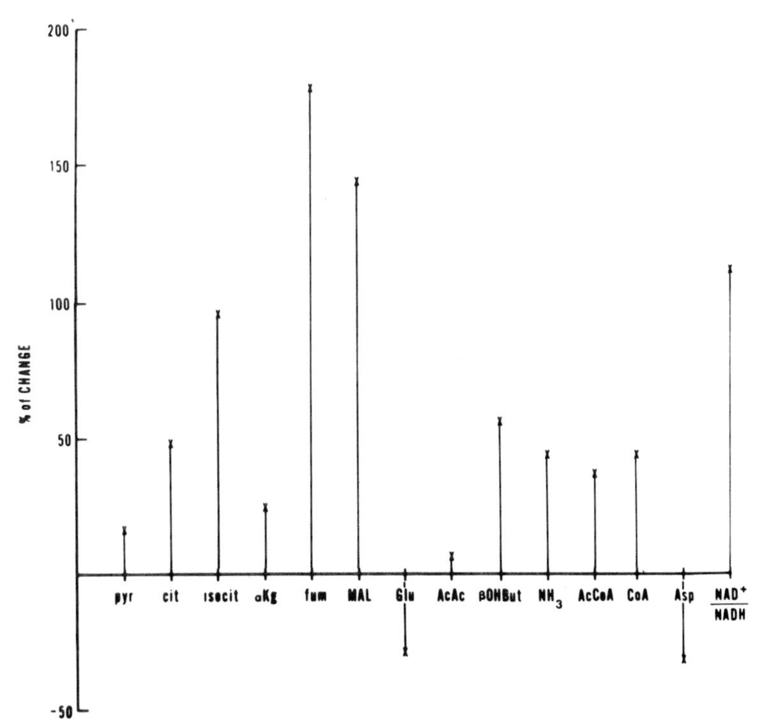

Figure 13. The alteration in metabolite levels of Krebs' cycle and other metabolites found with onset of acid secretion in dog gastric mucosa.

due to an unexplained increase in phosphate. These data are summarized in references 18, 145, and 146.

Measurement of several other metabolites in dog showed changes in levels of glycolytic and Krebs' cycle intermediates (Figures 12 and 13) which would result from activation of glycolysis at the phosphofructokinase step, as with aldosterone action in renal tissue (147), and a large increase in Krebs' cycle activity. No evidence for changes in fatty acid oxidation was seen and no hormone-sensitive lipase was present, although such has been suggested in the frog (3).

Overall, the spectroscopic and chemical data generally suggest that metabolic transitions are considerably more complex than assumption 1) would predict. The data do illustrate the remarkable capacity of this tissue to sustain a large change in metabolic rate with little shift in energy intermediates. It is tempting to speculate that there is a tight feedback between secretion and metabolism as would be predicted from a combined ATP/oxidation-reduction scheme similar in concept to the chemiosomotic theory. Perhaps the most puzzling finding is the constancy of phosphocreatine in contrast to what is found in other tissues subjected to a work load increase (148).

Altogether, approaches used so far have not provided an unequivocal distinction between an ATP-based or oxidation-reduction for H^+ secretion. (For further discussions, references 145, 146, and 149 may be consulted.)

TISSUE PROPERTIES

This section deals with the epithelial properties of stomach and attempts to emphasize those points which will ultimately have to be explained with a simpler experimental model.

Studies of intact epithelia have emphasized the mechanism of integration of function of the different surface membranes of the different cell types with respect to the movement of solute from one side of the tissue to the other. In the case of gastric fundus the solutes in question are the ions H^+, Na^+, K^+, and Cl^-, and the cells in question are surface, mucous neck, and oxyntic (or parietal and peptic).

Only the fundic and antral stomach regions have been studied in vitro. These regions have been studied in vitro. These regions exhibit major differences in their properties. The amphibian fundus, which secretes HCl, has very little in the way of paracellular or shunt conductance, as demonstrated by a variety of techniques (150, 151). Hence, its electrical properties reflect the properties of the cellular conductance pathway. In contrast, the antrum which does not secrete H^+, and indeed may secrete HCO_3^- (or absorb H^+) (152), has 80% of its conductance via the paracellular route (153), which places it halfway between fundus and small intestine (184) (Table 2).

In tissue of both regions, the conductance of the basal surface is largely due to K^+ and Cl^- (155, 156). In *Necturus* antrum the conductance of the luminal

Table 2.

	Toad bladder[a] (190) Ω cm²	Necturus gall bladder (191) Ω cm²	Rabbit ileum (154) Ω cm²	Necturus antrum[b] (153) Ω cm²	Necturus fundus[a] (151) Ω cm²
Tissue	3,735	307	100	1,730	2,229
Cellular	7,260	7,350	720	7,697	2,825
Shunt	13,245	324	120	2,241	10,573
R_t/R_s	0.28	0.95	0.83	0.77	0.21

[a]Mean of two methods used.
[b]Mean of three methods used.

surface is due largely to an amiloride-sensitive Na^+ path (152), whereas in bullfrog HCO_3^- and Cl-ion are also involved in this surface. The paracellular path in antrum is cation selective at neutral pH in the sequence Rb > Ca > K > Na > Li and is dependent on the presence of free carboyxl groups with a pK between 3 and 4. Relative anion selectivity can be induced by blocking carboxyl groups with carbodiimide reagents (157). Similar data have been reported for piglet fundus (158).

In the fundus it is generally agreed that the transport of H^+ and Cl^- is active. The basal surface of the tissue is selectively permeable to K^+ and Cl^-, which accounts for essentially all the conductance of that surface as measured on intact tissue (159) or by use of microelectrodes (151, 155). Problems arise in attempting to account for the conductance of the luminal surface. Whether directly or with the use of microelectrodes, it is possible to infer that the conductance of the tubular gland region exceeds that of the surface cell component (159). Hence, in dealing with the mucosa mounted in an Ussing chamber, one is largely concerned with the conductance of the luminal surface of the acid-secreting cell, although it may be that both cell types are electrically integrated by means of electronic coupling (66, 151, 160). Such coupling occurs in the atrum (153).

The movement of protons across a membrane may occur in two fundamentally different ways: 1) Transport of H^+ may be primarily electrogenic and the circuit completed by the transport of Cl^- (161) or the absorption of $Na^+ + K^+$ in processes which may occur at the same site in the membrane or at separate locations (162). 2) Transport of H^+ may be nonelectrogenic as HCl and involve a forced exchange such as H^+:Na^+ or H^+:K^+. Gastric mucosa was the first tissue in which electrogenic H^+ transport was suggested (163), and this early finding has had considerable influence on the development of transport concepts generally.

A membrane electrogenic process has several components, namely, the primary electrogenic battery for movement of H^+, E_{H^+}, an internal resistance, R_{H^+}, and either another ionic conductance pathway, R_{Cl^-}, or a Cl^- battery of opposite polarity (Figure 7). A K^+ counterion circuit may also be substituted. From such a simple equivalent circuit one may make a variety of predictions, remembering that the H^+ limb may be switched on and off as a function of the presence of stimuli. These predictions are described below.

First, with onset of secretion there should be an increase in membrane conductance. Second, during secretion changes of H^+, Cl^-, or K^+ activity in the luminal solution should occasion a change in potential. Third, the potential difference (PD) across the luminal surface should be oriented positive with respect to the interior. Fourth, clamping the potential across the membrane should alter secretory rate as a function of the orientation and magnitude of the potential. Fifth, alteration of the Cl^- (or K^+) conductance should result in predictable changes in potential and current so that when g_{Cl^-} approaches zero the current across the membrane is equal to the H^+ current, and hence the potential is a linear function of the secretory rate under these conditions. Sixth, other things being equal, reduction of g_{Cl^-} should reduce the H^+ secretion rate, but this should be reversed by insertion of another ionic conductance, such as ionophores, lipid permeant anions, or cations.

In general, stimulation of secretion is associated with a fall in resistance. Under certain conditions, this is not the case. For example, with Ba^{2} present in the nutrient solution there is an increase in the resistance of frog mucosa, and stimulation of secretion further increases the resistance (164). This has been interpreted as being due to reduction of intracellular Cl^- with consequent reduction of Cl^- current across the secretory membrane. The fall in resistance which occurs under other conditions may also be due to the large increase in surface area of the apical membrane which occurs with stimulation. The corollary also holds that with inhibition of secretion there is an increase of resistance. This also may have an anatomic mechanism by reduction in surface area. To our knowledge, there has not yet been a correlative study of electrical and morphological parameters. Analysis of electrical transients has been carried out, and in part the changes in capacitance could also be explained by surface area changes.

Alterations of pH, Cl^-, or cation concentration in the luminal solution have at best only a transient effect on electrical parameters, contrary to what could be predicted from the simple equivalent circuit (Figure 7). Only in SO_4^{2-} solutions, where there is low anionic conductance, do changes of $[K^+]$ in the lumen alter the potential across the tissue (165). With other anions, such as acetate or glucuronate, K^+ changes in the luminal solution are without effect (166). Accordingly, the luminal surface behaves as if it were electrically silent or non-permselective, quite contrary to the predictions of a simple electrogenic hypothesis. Measurements of the orientation of the potential across the luminal surface of the oxyntic cell are technically difficult, the major problem being visualization of the site of puncture. Nevertheless, frog data have been published supporting the presence of a potential due to an outward electrogenic H^+ pump (167) (i.e., basal membrane PD $>$ luminal membrane PD). Although no evidence for this orientation was obtained in *Necturus,* changing the bathing solution from Cl^- to SO_4^{2-} resulted in increase of potential across the luminal membrane of the tissue with little change across the serosal surface. The net result is that the potential inverts across the tissue. These findings would be quite compatible with the simple elctrogenic circuit illustrated.

The earliest evidence suggesting the possibility that H^+ transport by gastric mucosa was electrogenic came from voltage clamp studies, in which increasing the positivity of the luminal surface decreased acid secretion and vice versa (163). Such studies indicate that the primary process is electrogenic H^+ transport, since opposite data would obtain if Cl^- were the primary electrogenic process. Indeed it proved possible to reversibly abolish secretion by voltage clamping. This may be regarded as strong evidence for H^+ secretion being via a conductive pathway. However, evidence has also been published showing that applied voltage may influence H^+ secretion opposite to what would be predicted (168).

Perhaps the strongest evidence for electrogenic H^+ transport comes from studies in which the Cl^- of the bathing medium is replaced by $SO_4{}^{2-}$. Under these conditions, the potential across the tissue is inverted and the acid rate is a linear function of the potential (Figure 14). This would be predicted from the simple equivalent circuit illustrated, assuming that $SO_4{}^{2-}$ only affects the Cl^- limb and that the inhibitor added (e.g., dinitrophenol, 2-deoxyglucose) does not affect the $R_{SO_4^{2-}}$. This can be seen from the following equations (169): Considering a simple equivalent circuit with a Cl^- e.m.f., E_{Cl^-}, and a Cl^- resistance, $R_{Cl_X^-}$; an H^+ e.m.f., $E_H{}^+$, and $R_H{}^+$, with SO_4^{2-}; substitution, another ion or Cl^- is used to complete the circuit with an e.m.f., E_X, and a resistance, R_X. the PD across one limb is

$$PD_X = E_X - I_H \cdot R_X$$

and across the other

$$PD_H = E_H + I_H \cdot R_H$$

Assuming that the inhibitor does not affect the parameters of the return limb, then clearly there is a linear relationship between the PE and acid rate.

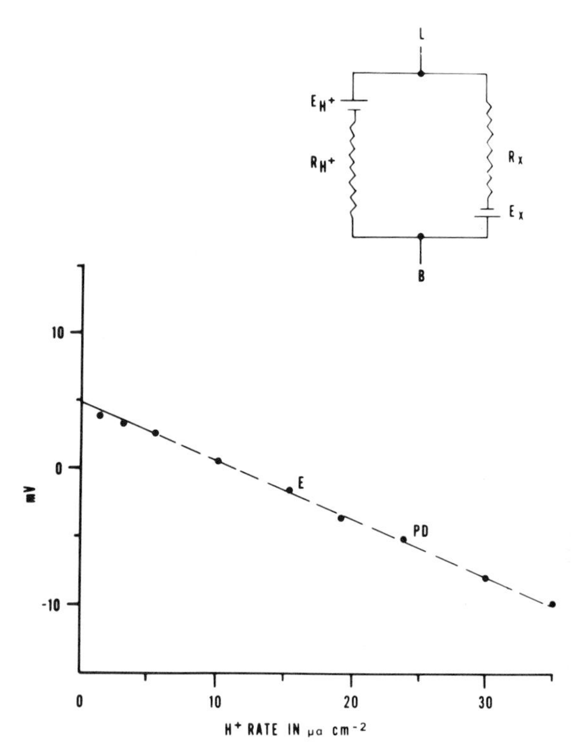

Figure 14. The variation of PD with secretory rate of frog mucosa bathed in $SO_4{}^{2-}$ solutions (redrawn from Dr. W. Rehm).

Under the above conditions, it is reasonable to assume that the effect of SO_4^{2-} is to reduce the anionic conductance of the secretory membrane so that the electrogenicity of the H^+ pump can be seen. Hence ionophores such as valinomycin in the K^+ or amphotericin, or lipid permeable anions such as tetraphenylborate, or cations such as triphenylmethylphosphonium should increase the H^+ rate in presence of SO_4^{2-} This does not occur, and no effect of any of these ionophores has been detected in the in vitro frog mucosa (170). This may be due to poor accessibility of the oxyntic cell apical surface. The endogenous K^+ conductance of the oxyntic cell cannot bypas the SO_4^{2-} block. Thus, a 10-fold increase in K^+ concentration in the luminal solution produces, in SO_4^{2-}, a change in tissue PD equal to that obtained with a serosal change, though only a small increase in acid rate is observed. Hence the effect of SO_4^{2-} is more complex, or, put another way, the Cl^-requirement for maximal secretion is not simply due to the need for an anionic conductance to allow H^+ extrusion.

SUBCELLULAR TRANSPORT

Transport at the subcellular level has been studied in several systems, such as mitochondria (171), chloroplasts (172), and bacterial vesicles (173). Here, generation of H^+ gradients by electrogenic pumps occurs. In sarcoplasmic vesicles (174) or (Na^+ + K^+)ATPase containing liposomes (175) ion gradients apparently are produced by nonelectrogenic means. The most promising approach to transport in vesicles has been a combination of reconstitution of vectorial systems in spherical liposomes (176) and detailed topology of the natural membrane (177). These approaches have succeeded in establishing the asymmetric orientation of, for example, mitochondrial oxidation-reduction components (178) and the sufficiency of H^+ or other ion gradients to explain ATP synthetase reaction. In the case of the Ca^{2+}-ATPase, in particular, initial success has been obtained in dissociating the hydrolytic peptide from the transport proteolipid (179). Another approach has been the insertion of fragments of Ca_2-ATPase into a black lipid membrane with the demonstration of conductance changes (180).

In studies relating to ion movements across vesicular membranes, various tools and approaches are available. The initial problem, which is often very difficult to solve for all conditions, is whether binding or uptake into vesicular water has occurred or, if a mixture of both is present, what proportion can be assigned. This can be at least partially resolved in one of three ways. First, uptake but not binding would be osmotically sensitive; second, uptake but not binding would be sensitive to loss of vesicular form; and third, membrane active complexones are thought to dissipate trans-membrane gradients, hence detect uptake rather than binding. Of these, the first approach can be interpreted directly and gives a direct measure of the ratio of bound to transported ions, since at infinite osmolarity (i.e., zero volume) the fraction remaining is due to binding.

Active transport of an ion can occur across a membrane in several ways. The classification used in Figure 7 is as follows: I. *Uniport:* (*a*) a primary electrogenic pump with an active uniport mechanism (i.e., only one ionic species is actively transported). All the conductance of the membrane can be accounted for by the conductance of the active ion. An example of this is the H^+ chemiosmotic pump of mitochondria (59), (*b*) a primary electrogenic pump with an active uniport mechanism but passive conductance of another ion of compensating charge. II. *Symport/antiport:* (*a*) where only two ions move across the membrane and a potential results if $t_A \neq t_B$ and (*b*) where under physiological conditions more than two ionic conductances are present. III. *A neutral pump:* A major question then following the demonstration of ion uptake is the electrical signal associated with its transport. To this end, membrane active complexones and lipid permeable ions have provided useful tools in attempting to answer this question.

If the movement of an ion is a primary electrogenic process, then for transport to occur the circuit has to be completed by the presence of another conductance. For example, if H^+ uptake into vesicles occurs by such an electrogenic process either 1) cation conductance must be present so that cation efflux can occur coupled to the H^+ potential (it should be noted that the electrogenic H^+ e.m.f. is oriented opposite in sign to the diffusion e.m.f. of H^+), hence internal cation must be present, or 2) anion conductance is present so that anion uptake can occur in response to the potential. In intact tissue, short circuiting a preparation removes the electrorestriction. Since electrical short circuiting is impossible in vesicles, alterations of membrane membrane permeability are used. Hence, adding lipid permeable ions or providing specific ion conductance paths (not nonelectrogenic antiport pathways as with nigericin) will increase the transport rate, provided a sufficient change in conductance is induced by the addition of the ion or ionophore. Alternatively, the change in potential can be measured by analyzing the distribution of lipid permeable ions by direct radioactive techniques or by use of the black lipid membrane as a lipid permeable ion electrode (181). If all these imposed conditions do not affect transport rate, or no change in potential can be measured, it is generally a safe conclusion that transport is nonelectrogenic. Ionophores can also be used to assess the inherent conductance of the vesicle membrane. Again, given the formation of a H^+ gradient, with a K^+ gradient present of opposite orientation, the addition of an H^+ selective ionophore will not dissipate the gradient if no other conductance is present. From this, one can conclude that little or no K^+ conductance is present, nor is there any anion conductance. Equally, if valinomycin, a K^+-seletive ionophore, dissipates the gradient, the membrane possesses an inherent H^+ conductance.

It is also important to realize that the passive properties of the vesicle membrane may be different from the active properties and all these tests should be performed both in the presence and absence of pump activity.

If the process studied is due to electrogenic or nonelectrogenic sym(anti)port of two ions, then the flux of the ion participating in the symport or antiport process has to be measured, e.g., cation exchange or Cl^- transport with H^+ uptake in a KCl medium. In either case it would be necessary to show such ion movement is obligatory and not nonspecific. Again this is usually done by short circuiting a membrane by passage of electric current, the flux ratio equation predicting unequal movement of only the active ion. This is not possible in vesicles, but, as we have discussed, other approaches can be taken.

In this discussion, a distinction has been made between various types of pumps ranging from a primary electrogenic system whose function is dependent on the specific conductance of the passive area of the membrane to the pumped ion to a nonelectrogenic system whose operation depends on equal specific ion pathways associated with the pump. Biological systems probably fall between the two extremes. Thus, although the mitochondrial ATPase system can be thought of as primarily electrogenic, (Na^+K^+)-ATPase largely fulfills the specific ion conductance criteria.

Since gastric mucosa transports H^+ and, on the other hand, contains a K^+-ATPase with properties reminiscent of the $(Na^+ + K^+)$-ATPase, it is of particular importance to study the ion transport associated with gastric vesicles. The surprising conclusion, contrary to much of the evidence from the intact amphibian mucosa, is that H^+ uptake by gastric vesicles is largely nonelectrogenic.

The ultimate goal of vesicle studies is to provide in a subcellular system a simplified model of acid secretion by the intact tissue. The key parameters that one would like to establish are 1) the nature of the activation of the apical surface of the cell; 2) the energy source for H^+ secretion; 3) the ionic fluxes associated with H^+ secretion, i.e., the flux of Cl^-, K^+, and Na^+; 4) the electrogenicity of the similar H^+ process; 5) the structure of the secretory vesicle; 6) the peptide arrangement of the pump; 7) the structure of the hydrolytic site of the pump, i.e., the reaction steps; 8) the nature of the transport site, if different from the hydrolytic site; 9) the transport steps; and 10) the lipid specificity for the transport as opposed to the chemical reaction. Many more questions could be added, but many or most of these could more simply be defined in a vesicle with transport properties.

An exciting development in the field of gastric secretion came with the finding that dog microsomes were capable of active H^+ uptake with the addition of ATP at pH 6.1. A combination of ionophores, valinomycin, and a protonophore, was necessary to abolish H^+ uptake. This led to the surprising conclusion that the H^+ uptake was nonelectrogenic (127). It should be pointed out that the use of pH 6.1 was determined by the pH change resulting from the hydrolysis of ATP at pH values different from 6.1 at the Mg^{2+} concentration used. These data have been considerably amplified using hog preparations (18, 126, 182), and the relationship of the work in the system to the parameters discussed above will now be considered.

In the original work, the microsomal fraction was composed of a mixture of membranes from various cell locations and mitochondria (127). This fraction shows not only an alkalinization of the medium with the addition of ATP (i.e., uptake or binding of H^+) but a similar alkalinization with NADH. The oligomycin insensitivity of the H^+ uptake, however, argued against mitochondrial ATPase as being responsible.

Subfractionation of the hog microsomal fraction, as discussed above, using density gradients and free flow electrophoresis, produced a fraction containing only two Coomassie blue staining regions on sodium dodecyl sulfate (SDS) gel electrophoresis at 100,000 M_r and at 84,000 M_r and much enriched in K^+-ATPase and K^+-*p*-NPPase activity but devoid of other membrane markers such as 5′-AMPase and also free of NADH oxidase and mitochondrial markers. Of various nucleotides and redox substrates, only ATP addition resulted in the formation of a gradient. The existence of a gradient rather than binding was established by showing that H^+ uptake was a function of the vesicular volume.

No correlation was established between the secretory status of the stomach at the time of slaughter and the efficiency of uptake by the vesicles. However, it was found that usually about 50% of the K^+-ATPase activity was latent, i.e., activity doubled with hypotonic treatment of the vesicles. This might be taken to mean that there was a random orientation of the vesicles, i.e., that half were inside out with respect to their orientation in the cell. If the original cell population were at maximum secretion, this would imply perhaps equally distributed microtubules and tubulo-vesicular orientation of the acid secreting system of the parietal cell. Activation may then simply be a morphological reorientation of the cell membrane dependent perhaps on activation of a mechanical system, or on membrane phosphorylation. However, it is known that mitochondrial ATPase contains an endogenous inhibitor (183), and the above data may result from removal of the inhibitor during fractionation procedures. If a gastric ATPase inhibitor were found in the stomach, this then might suggest that secretory activation might result from deinhibition of the ATPase.

In the purified vesicular system, the only effective energy source for transport found was ATP. However, the method of purification may result in the loss of essential factors for the utilization of other substrates such as NADH. It could also be that, in the intact tissue, there is coupling between the ATPase used as a marker in this purification and a redox system, which coupling is abolished during purification. Be that as it may, the only known energy source for generation of subcellular H^+ gradients is ATP. Strangely, also, in view of the strong analogy between this ATPase and the Ca^{2+}- and Na^+ + K^+-ATPases, *p*-NPP is unable to generate ion gradients in this vesicle preparation.

In considering the ionic fluxes that occur across the vesicle membrane, this sort of study can define the passive and active permeation processes as well as the probable conductances of the energized and nonenergized state of the membrane.

Intact vesicles when placed in KCl solution do not produce any change of medium pH. This could be due to several factors. For example, low conductance to K^+ or H^+ would result in this observation. The addition of an H^+ conductor, such as tetrachlorsalicylanilide, should then result in H^+ extrusion, provided a K^+ conductance were present. This does not occur, but the addition of a K^+ conductance in the form of valinomycin results in H^+ extrusion. This is interpreted as a potential-coupled extrusion of H^+, the potential arising from the inward K^+ gradient. Hence, these vesicles, in the nonenergized state, have a higher H^+ conductance than K^+ conductance. There is, however, passive uptake of alkali metal cations such as Rb^+, but the selectivity of uptake corresponds to free solution mobility, i.e., Cs > Rb > K > Na > Li, suggesting that this pathway is electrically inactive.

With addition of ATP the uptake of H^+, which occurs at a ratio of 3.5 or more moles H^+ per mole ATP hydrolyzed, is associated with a simultaneous efflux of K^+ or Rb^+ ($^{86}Rb^+$ was used in these studies), and the ratio H^+/Rb^+ would appear to be about 1. The counterion for H^+ uptake seems more effective when initially in the interior of the vesicle, and with an outward K^+ gradient there is a further enhancement of H^+ uptake most simply explained as electrical coupling between K^+ outward and H^+ inward movement during pump activity. Hence, the energized vesicles contain both K^+ and H^+ conductance. SO_4^{2-} substitution for Cl^- reduces both H^+ and Rb^+ movement with energization, indicating a role for anion in the observed cation movements.

The action of ionophores or lipid permeable ions argues against any significant potential being developed across the vesicles during the above phenomena. For example, with an electrogenic, inward H^+ pump and cation effluxing passively as a result of the PD, protonophores should increase pump activity (i.e., ATPase activity) and lipid permeable cations should reduce counterflow of cation. Neither occurs. Similarly, with H^+ uptake occurring in response to an outward electrogenic H^+ movement, H^+ uptake should be stimulated by protonophores or shunted by lipid permeable ions. Again, neither occurs. Hence the process under the conditions of study (i.e., zero cation gradient, pH 6.1) appears nonelectrogenic. The SO_4^{2-} restriction, however, implies that under different conditions electrogenic transport could be observed. Hence the most likely model at pH 6.1 is a polyionic conductance K^+:H^+ antiport pump (Figure 15).

As pointed out above, this transport is a property of a vesicle containing only a few peptides. Of major interest is the 100,000 M_r peptide, which, based on its phosphorylation by [γ-^{32}P] ATP is a subunit of the K^+-ATPase. Since invariably K^+-ATPase inhibitors inhibit both H^+ and cation efflux, the role of the ATPase in transport by this vesicle is clear.

Although the ATPase activity in the transporting vesicle is latent, the *p*-NPPase activity is not, i.e., no change in *p*-NPPase activity occurs, although ATPase activity doubles in hypotonic conditions of assay. Hence, as found for the intact stomach (83), *p*-NPP hydrolysis can occur on either surface of the cell membrane. Thus, the hydrolytic site of the ATPase (presumably the *p*-NPPase site) is present on both sides of the membrane, whereas the ATP binding size is assymmetric again in confirmation of the intact tissue data. This suggests 1) that the 100,000 M_r peptide spaons the bilayer and 2) that the hydrolytic site is

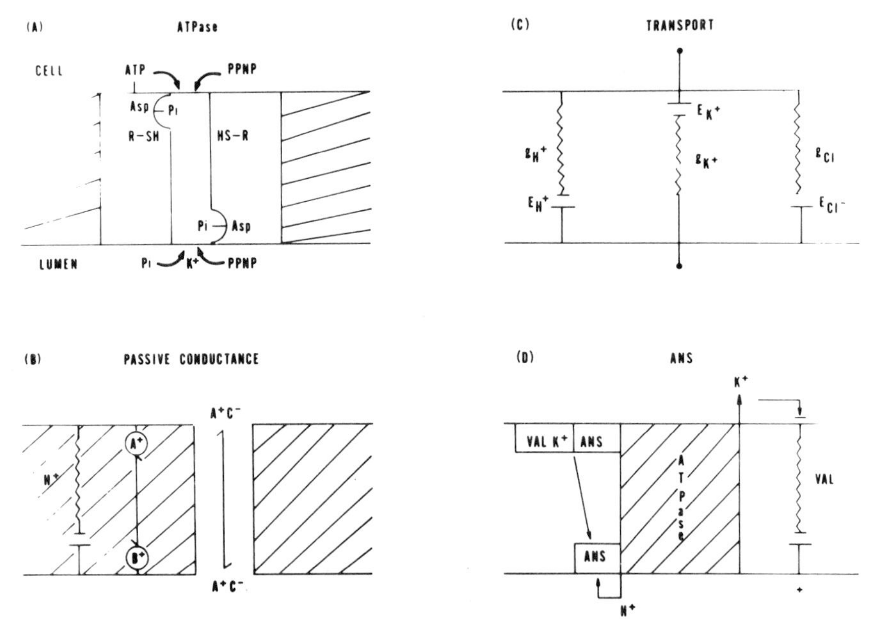

Figure 15. Models to illustrate the findings in gastric vesicles: *A*, a dimer of the ATPase symmetric with respect to p_{NPP} hydrolysis but asymmetric with respect to ATP binding forming β-aspartyl phosphate intermediate; *B*, the passive conductance pathways of the vesicle membrane comprising a free diffusion path, a cation exchange pump with cation selectivity as for pump, and an H^+ conductance; *C*, an equivalent circuit illustrating the presence of an active H^+ and K^+ component, and a postulated Cl^- conductance to account for the effects of anion substitution; *D*, a model to illustrate the ANS fluorescence enhancement found with the addition of ATP to gastric vesicles only in the presence of valinomycin and in the absence of a large H^+ conductance.

mobile and can react with *p*-NPP in either conformation, but with ATP only in one. Moreover, *p*-NPPase activity shows evidence of allosteric regulation with a Hill coefficient of 1.8 with respect to Mg *p*-NPP. This also suggests that a dimer of this enzyme may be involved.

Hence, evidence is accruing for a model consisting of a phosphorylating transport system with the site involved in phosphorylation (presumably a β-aspartyl carboxyl group) being mobile across the membrane and two such sites being involved in each pump unit (Figure 15).

Addition of valinomycin and ANS to these vesicles followed by ATP results in a quantum yield increase in ANS fluorescence dependent on the generation of both an H^+ and K^+ gradient. The ANS site responsible for these changes is located on the membrane on the opposite side from the ATP binding site, i.e., the ANS fluorescence changes only in the inside out vesicles (Figure 15). This ANS change shows either a localized change of PD due to the ATP-generated K^+ gradient or a gradient-dependent conformational change in the ATPase (184).

This ANS effect is similar to that found with Ca^{2+}-ATPase (185) and mitochondria (186).

A K^+ inward gradient results in phosphorylation of the protein from inorganic [^{32}P]phosphate. This is a vectorial reaction showing coupling of the cation gradient to partial reversal of the ATPase, again as found for Ca^{2+}- and ($Na^+ + K^+$)-ATPase (187, 188). Hence, there is a strong similarity in this vesicular preparation to vesicles containing the other cation transport ATPases, in contrast to the H^+ transport systems found, for example, in mitochondria.

In comparing the transport reaction with the properties of the K^+-ATPase, the K_A for K^+ for the ATPase is 7×10^{-3} M under the conditions described above, i.e., pH 6.1 and room temperature. In contrast, the K_A with a zero cation gradient for K^+ in terms of H^+ uptake is 3.2×10^{-2} M. This may suggest some difference between the "transport" site and the "hydrolytic" site.

This discussion emphasizes the detail in which transport can be defined in a vesicular system free of contaminating side reactions, as has already been established for sarcoplasmic reticulum vesicles (189). However, it is still a far cry from studying a system in a test tube at pH 6.1 to justifying its properties as directly relevant to H^+ secretion by the intact mucosa. Just as there are some similarities, such as a K^+ requirement or inhibition by SCN^-, so are there many differences. Most important is the pH optimum of transport, pH 6.1 (for H^+, not K^+). Other features, such as the lack of absolute K^+ specificity and the insensitivity of intact mucosa to ionophores, make one hesitate to accept this model as more than an assessment of the transport properties of the K^+-ATPase of gastric mucosa. At the least, however, this will be an informative addition to our growing arsenal of transport ATPases capable of being analyzed in considerable detail.

SUMMARY

This review has attempted to cover some of the findings that have been made in the mechanism of gastric secretion in recent years. It is hard to offer any firm conclusions, whether at the level of stimulus, metabolism, or the terminal process of secretion. However, some generalizations may be possible. At least amphibian gastric secretion is stimulated by cAMP as a second messenger, with histamine presumably acting as the primary messenger. The resultant metabolic change is due largely to a direct stimulation of catabolism, which in dog appears to be the metabolism of hexose, through the glycolytic process, the hexose monophosphate shunt, and the Krebs' cycle with cytoplasmic reduction and mitochondrial oxidation of pyridine nucleotides. No evidence could be obtained for changes in high energy phosphate or for lipolysis. One would expect gastric mucosal membranes during secretion to contain an anion-restricted electrogenic H^+ pump, but they in fact contain an ATPase stimulated by monovalent cations and are insensitive to ouabain. In addition, hog or dog gastric membranes have the vectorial properties of H^+ absorption, Rb^+ extrusion, and ANS fluorescence enhancement with the addition of ATP, as well as protein phosphorylation by ^{32}P dependent on a K^+ gradient.

REFERENCES

1. Forte, J. G. (1971). Hydrochloric acid secretion by gastric mucosa. *In* E. E. Bittar (ed.) Membranes and Ion Transport, Ch. 6, pp. 111–165. Wiley, New York.
2. Rehm, W. S. (1972). Proton Transport. *In* D. M. Greenberg (ed.), Metabolic Pathways, 3rd Ed. Ch. 6, pp. 187–241. Academic Press, New York.
3. Hersey, S. J. (1974). Interactions between oxidative metabolism and acid secretion in gastric mucosa. Biochim. Biophys. Acta 244:157.
4. Durbin, R. P. (1973). Secretory events in gastric mucosa. Current Top. Membr. Transp. 4:305.
5. Forte, J. G. (1973) Pharmacology of isolated amphibian gastric mucosa. Int. Rev. Exp. Pharmacol. Therap. 1:195.
6. Walsh, J. H., Isenberg, J. I., Ausfield, J., and Maxwell, D. J. Clin. Invest. In press.
7. Kasbekar, D. K., Ridley, H. A., and Forte, J. G. (1969). Pentagastrin and acetylcholine relation to histamine in H^+ secretion by gastric mucosa. Am. J. Physiol. 216:961.
8. Bovet, D. (1950). Introduction to antihistamine agents and antergan derivatives. Ann. N.Y. Acad. Sci. 50:1089.
9. Code, C. F. (1951). The inhibition of gastric secretion: a review. Pharmacol. Rev. 3:59.
10. Kier, L. B. (1968). Molecular orbital calculations of the preferred conformations of histamine and a theory on its dual activity. J. Med. Chem. 11:441.
11. Ganellin, C. R. (1973) Conformation of histamine derivatives. 3. A relationship between conformation and pharmacological activity. J. Med. Chem. 16:620.
12. Black, J. W., Duncan, W. A. M., Durant, C. J., Ganellin, C. R. and Parsons, E. M. (1972). Definition and antagonism of histamine H_2-receptors. Nature 236:385.
13. Pounder, R. E. Williams, J. E., Misiewicz, J. J., and Milton-Thompson, G. J. (1975). The 24-hour control of intragastric pH by cimetidine, a new H_2-receptor antagonist, in normal subjects and in patients with duodenal ulcer. Gut 16:831 (Abstr.).
14. Butcher, R. W., Baird, C. E., and Sutherland, E. W. (1968). Effects of lipolytic and antilioplytic substances on adenosine 3′5′-monophosphate levels in isolated fat cells. J. Biol. Chem. 243:1705.
15. Haverback, R. J., Stubrin, M. I., and Dyce, B. J. (1965). Relationship of histamine to gastric and other secretagogues. Fed. Proc. 24: 1326.
16. Jacobson, E. D. (1968). Clearances of the gastric mucosa. Gastroenterology 54:434.
17. Blum, A. L. (1971). Kinetics of inhibition of canine gastric secretion by vasopressin. Gastroenterology 61:461.
18. Sachs, G., Rabon, E., and Sarau, H. M. Ann. N.Y. Acad. Sci. In press.
19. Shoemaker, R. L., Buckner, E., Spenney, J. G., and Sachs, G. (1974). Action of burimamide, a histamine antagonist, on acid secretion in vitro. Am. J. Physiol. 226:898.
20. Kasbekar, D. K. (1972). Secretagogue-induced tachyphylaxis of gastric H^+ secretion and its reversal. Am. J. Physiol. 223:294.
21. Harris, J. B., and Alonso, D. (1965). Stimulation of the gastric mucosa by adenosine-3′5′-monophosphate. Fed. Proc. 24:1368.
22. Alonso, D., and Harris, J. B. (1965). Effect of xanthines and histamine on ion transport and respiration by frog gastric mucosa. Am. J. Physiol. 208:18.
23. Harris, J. B., Nigon, K., and Alonso, D. (1969). Adenosine-3′5′-monophosphate: intracellular mediator for methyl xanthine stimulation of gastric secretion. Gastroenterology 57:377.
24. Nakajima, S., Shoemaker, R. L., Hirschowitz, B. I., and Sachs, G. (1970). Comparison of actions of aminophylline and pentagastrin on *Necturus* gastric mucosa. Am. J. Physiol. 219:1259.

25. Kasbekar, D. K. (1967). Studies of resting isolated frog gastric mucosa. Proc. Soc. Exp. Biol. Med. 125:267.
26. Nakajima, S., Hirschowitz, B. I., and Sachs, G. (1971). Studies on adenyl cyclase in *Necturus* gastric mucosa. Arch. Biochem. Biophys. 143:123.
27. Ray, T. K., and Forte, J. G. (1971). Studies on oxyntic cell membranes of gastric mucosa. Fed. Proc. 30:477. (Abstract).
28. Amer, M. S. (1974). Cyclic GMP and gastric acid secretion. Am. J. Dig. Dis. 19:71.
29. Roth, J. A., and Ivy, A. C. (1944). The effect of caffeine upon gastric secretion in the dog, cat and man. Am. J. Physiol. 141:456.
30. Ridley, P. T. Unpublished observations.
31. Levine, R. A., and Wilson, D. E. (1971). The role of cyclic AMP in gastric secretion. Ann. N.Y. Acad. Sci. 185:363.
32. Bieck, P. R., Oates, J. A., Robison, G. A., and Adkins, R. B. (1973). Cyclic AMP in the regulation of gastric secretion in dogs and humans. Am. J. Physiol. 224:158.
33. Sung, C. P., Jenkins, B. C., Burns, L. R., Hackney, V., Spenney, J. G., Sachs, G., and Wiebelhaus, V. D. (1973). Adenyl and guanyl cyclase in rabbit gastric mucosa. Am. J. Physiol. 225:1359.
34. Perrier, C. V., and Laster, L. (1970). Adenyl cyclase activity of guinea pig gastric mucosa: stimulation by histamine and prostaglandins. J. Clin. Invest. 49:73a (Abstr.).
35. Shimizu, H., Creveling, C. R., and Daly, J. W. (1970). The effect of histamines and other compounds on the formation of adenosine 3′5′-monophosphate in slices from cerebral cortex. J. Neurochem. 17:441.
36. Mao, C. C., Shanbour, L. L., Hodgins, D. S., and Jacobson, E. D. (1972). Adenosine 3′5′-monophosphate (cyclic AMP) and secretion in the canine stomach. Gastroenterology 63:427.
37. Dozois, R. R., Wollin, A. and Rettmann, R. D. (1975). Stimulation of adenylate cyclase from dog gastric mucosa by histamine and its blockade by metiamide. Physiologist 18:196 (Abstr.).
38. Shoemaker, R. L., Sachs, G., and Hirschowitz, B. I. (1966). Secretion by guinea pig gastric mucosa *in vitro*. Proc. Soc. Exp. Biol. Med. 123:824.
39. Forte, J. G., Forte, T. M. and Machen, T. E. (1975). Histamine stimulated hydrogen ion secretion by *in vitro* piglet gastric mucosa. J. Physiol. (Lond.) 244:15.
40. Sung, C. P., Wiebelhaus, V. D., Jenkins, B. C., Adlercreutz, P., Hirschowitz, B. I., and Sachs, G. (1972). Heterogeneity of 3′5′-phosphodiesterase of gastric mucosa. Am. J. Physiol. 223:648.
41. Ray, T. K. and Forte, J. G. (1973). Studies on phosphodiesterase from oxyntic cells of bullfrog gastric mucosa. Arch. Biochem. Biophys. 155:24.
42. Miyamoto, E., Kuo, J. F., and Greengard, P. (1969). Adenosine 3′5′-monophosphate-dependent protein kinase from brain. Science 165:63.
43. Schwartz, I. L., Schlatz, L. J., Kinne-Saffran, E., and Kinne, R. (1974). Target cell polarity and membrane phosphorylation in relation to the mechanism of action of antidiuretic hormone. Proc. Nat. Acad. Sci. USA 71:2595.
44. Reimann, E. M., and Rapino, N. G. (1974). Partial purification and characterisation of an adenosine 3′5′-monophosphate-dependent protein kinase from rabbit gastric mucosa. Biochim. Biophys. Acta 350:201.
45. Ray, T. K., and Forte, J. G. (1974). Soluble and bound protein kinase of rabbit gastric secretory cells. Biochem. Biophys. Res. Commun. 61:1199.
46. Goldberg, N. D., O'Dea, R. F., and Haddox, M. K. (1973). Cyclic GMP. Adv. Cyclic Nucleotide Res. 3:155.
47. Forte, J. G., and Nauss, A. H. (1963). Effects of calcium removal on bullfrog gastric mucosa. Am. J. Physiol. 205:631.
48. Jacobson, A., Schwartz, M., and Rehm, W. S. (1965). Effects of removal of calcium from bathing media on frog stomach. Am. J. Physiol. 209:134.
49. Kasbekar, D. K. Am. J. Physiol. In press.
50. Sedar, A. W. (1965). Fine structures of the stimulated oxyntic cell. Fed. Proc. 24:1360.
51. Helander, H. F. (1962). Ultrastructure of fundus glands of the mouse gastric mucosa.

An electron microscopical study in fasted and refed animals with observations on ultrastructure after different fixation and embedding. J. Ultrastruct. Res. Suppl. 4:1.
52. Helander, H. G., and Hirschowitz, B. I. (1972). Quantitive ultrastructural studies on gastric parietal cells. Gastroenterology 63:951.
53. Forte, T. M., Machen. T. E., and Forte, J. G. (1975). Ultrastructural and physiological changes in piglet oxyntic cells during histamine stimulation and metabolic inhibition. Gastroenterology 69:1208.
54. Berridge, M. J., and Oschman, J. L. (1969). Structural basis for fluid secretion by Malpighian tubules. Tissue Cell 1:247.
55. Anderson, E., and Harvey, W. R. (1966). Active transport by the cecropia midgut. II. Fine structure of the midgut epithelium. J. Cell Biol. 31:107.
56. Ito, S., and Schofield, G. C. (1974). Studies on the depletion and accumulation of microvilli and changes in the tubulovesicular compartment of mouse parietal cells in relation to gastric acid secretion. J. Cell Biol. 63:364.
57. Kasbekar, D. K., Forte, G. M., and Forte, J. G. (1968). Phospholipid turnover and ultrastructural changes in resting and secreting bullfrog gastric mucosa. Biochim. Biophys. Acta 163:1.
58. Hokin, L. E., and Hokin, M. R. (1963). Biological transport. Annu. Rev. Biochem. 32:553.
59. Mitchell, P. (1966). Chemiosmotic coupling in oxidative and photosynthetic phosphorylation. Biol. Rev. 41:445.
60. Oesterhelt, D., and Stoeckenius, W. (1973). Functions of a new photoreceptor membrane. Proc. Natl. Acad. Sci. USA 70:2853.
61. Oesterhelt, D. (1974). Bacteriorhodopsin as a light driven proton pump. *In* G. F. Azzone et al. (eds.), Membrane Proteins in Transport and Phosphorylation, pp. 79–84. Elsevier, New York.
62. Racker, E., and Stoeckentius, W. (1974). Reconstitution of purple membrane vesicles catalyzing light-driven proton uptake and adenosine triphosphate formation. J. Biol. Chem. 249:662.
63. Rehm, W. S. (1975). Ion transport and short circuit technique. Current Top. Membr. Transp. 7:217.
64. Blum, A. L., Shah, G., Wiebelhaus, V. D., Brennan, F. T., Helander, H. F., Ceballos, R., and Sachs, G. (1971). Pronase method for isolation of viable cells from *Necturus* gastric mucosa. Gastroenterology 61:189.
65. Croft, D. N., and Ingelfinger, F. J. (1969). Isolated gastric parietal cells: oxygen consumption, electrolyte content and intracellular pH. Clin. Sci. Mol. Med. 37:491.
66. Blum, A. L., Hirschowitz, B. I., Helander, H. G., and Sachs, G. (1971). Electrical properties of isolated cells from *Necturus* gastric mucosa. Biochim. Biophys. Acta 241:261.
67. Forte, J. G., Ray, T. K., and Poulter, J. L. (1972). A method for preparing oxyntic cells from frog gastric mucosa. J. Appl. Physiol. 32:714.
68. Romrell, L. J., Coppe, M. R., Munro, D. R., and Ito, S. (1975). Isolation and separation of highly enriched fractions of viable mouse gastric parietal cells by velocity sedimentation. J. Cell Biol. 65:428.
69. Lewin, M., Cheret, A. M., Soumarmon, A., and Girodet, J. (1974). Methode pour l'isolement et le tri des cellules de la muqueuse fundique de rat. Biol. Gastroentenol. 7:139.
70. Michelangeli, F. *In* D. K. Kasbekar, G. Sachs, and W. S. Rehm, (eds.) Advances in Gastric H^+ Secretion. In press.
71. Kasbekar, D. K. In press.
72. Forte, J. G., Forte, T. M., and Ray, T. K. (1972). Membranes of the oxyntic cell: their structure, composition and genesis. *In* G. Sachs, E. Heinz, and K. J. Ullrich (eds.), Gastric Secretion, pp. 37–67. Academic Press, New York.
73. Wiebelhaus, V. D., Sung, C. P., Helander, H. G., Shah, G., Blum, A. L., and Sachs, G. (1971). Solubilization of anion ATPase from *Necturus* oxyntic cells. Biochim. Biophys. Acta 241:49.
74. Sachs, G., Shah, G., Strych, A., Cline, G., and Hirschowitz, B. I. (1972). Properties of ATPase of gastric mucosa. III. Distribution for HCO_3-stimulated ATPase in gastric mucosa. Biochim. Biophys. Acta 266:625.

75. Ray, T. K., and Forte, J. G. (1974). Adenyl cyclase of oxyntic cells: its association with different cellular membranes. Biochim. Biophys. Acta 363:320.
76. Forte, J. G., Ganser, A., Beesly, R., and Forte, T. M. (1975). Unique enzymes of purified microsomes from pig fundic mucosa; K^+-stimulated adenosine triphosphatase and K^+-stimulaged *p*NPPase. Gastroenterology 69:175.
77. Saccomani, G., Lewin, M., Stewart, B., Shaw, D., Dailey, D., and Sachs, G. Biochim. Biophys. Acta (Submitted for publication.)
78. Ganser, A. L., and Forte, J. G. (1973). K^+-stimulated ATPase in purified microsomes of bullfrog oxntic cells. Biochim. Biophys. Acta 307:169.
79. Koenig, C., and Vial. J. D. (1970). A histochemical study of adenosine triphosphatase in the toad (Bufo spinulosus) gastric mucosa. J. Histochem. Cytochem. 18:340.
80. Cross. S. A. M. (1970) Ultrastructural localisation of carbonic anhydrase in rat stomach parietal cells. Histochemie 22:219.
81. Hersey, S. J. Personal communication.
82. Saccomani, G., Shah, G., Spenney, J. G., and Sachs, G. (1975). Characterisation of gastric mucosal membranes. VIII. The localisation of peptides by iodination and phosphorylation. J. Biol. Chem. 250:4802.
83. Durbin, R. P., and Kircher, A. B. (1973). A surface *p*-nitrophenyl phosphate of frog gastric mucosa. Biochim. Biophys. Acta 321:553.
84. Fleit, H., Conklyn, M., Stebbins, R. D., and Silber, R. (1975) Function of 5'-nucleotidase in the uptake of adenosine from AMP by human lymphocytes. J. Biol. Chem. 250:8889.
85. Racker, E. (1974). Mechanism of ATP formation in mitochondria. *In* L. Ernster, R. W. Estabrook and E. C. Slater (eds.), BBA Library Cynamics of Energy Transducing Membranes, pp. 269–282. Elsevier, New York.
86. Scholes, P., Mitchell, P., and Moyle, J. (1969). The polarity of proton translocation in some photosynthetic micro-organisms. Eur. J. Biochem. 8:450.
87. Mitchell, P., and Moyle, J. J. (1971). Activation and inhibition of mitochondrial adenosine triphosphatase by various anions and other agents. J. Bioenerg. 2:1.
88. Blum, A. L., Shah, G., St. Pierre, T., Sung, C. P., Wiebelhaus, V. D. and Sachs, G. (1971). Properties of solble ATPase of gastric mucosa. II. Feffect of HCO_3^-. Biochim. Biophys. Acta 249:101.
89. Makinose, M. (1969). The phosphorylation of the membranal protein of the sarcoplasmic vesicles during active calcium transport. Eur. J. Biochem. 10:74.
90. Skou, J. C. (1965). Enzymatic basis for active transport of Na^+ and K^+ across cell membrane. Physiol. Rev. 45:596.
91. Uesugi, S., Dulak, N. C., Dixon, J. F., Hexum, T. D., Dahl. J. L., Perdue, J. F., and Hokin, L. E. (1971). Studies on the characterisation of the sodium-potassium transport adenosine triphosphatase. VI. Large scale partial purification and properties of a lubrolsolubilized bovine brain enzyme. J. Biol. Chem. 246:531.
92. Racker, E., and Eytan, E. (1973). Reconstitution of an efficient calcium pump without detergents. Biochem. Biophys. Res. Commun. 55:174.
93. Blostein, R. (1975). Na^+-ATPase of the mammalian erythrocyte membrane: reversibility of phosphorylation at 0°. J. Biol. Chem. 250:6118.
94. Taniguchi, K., and Post, R. L. (1975). Synthesis of adenosine triphosphate and exchange between inorganic phosphate and adenosine triphosphate in sodium and potassium ion transport adenosine triphosphates. J. Biol. Chem. 250:3010.
95. Mitchell, P. (1961). Conductance of protons through the membranes of mitochondria and bacteria by uncouplers of oxidative phosphorlation. Biochem. J. 81:24 (Abstr.).
96. Hokin, L. E., and Dahl, J. L. (1972). The sodium potassium adenosine triphosphatase. *In* D. M. Greenberg (ed.), Metabolic Pathways, 3rd Ed., Academic Press, New York. Ch. 6, pp. 269–315.
97. Martonosi, A. (1969). Sarcoplasmic reticulum. VII. Properties of a phosphoprotein intermediate implicated in calcium transport. J. Biol. Chem. 244:613.
98. Askari, A., and Rao, S. N. (1971). Studies on the partial reactions catalyzed by the (Na^++K^+) activated ATPase. III. Relation of K^+-dependent *p*-nitrophenyl-phosphatase to Na^+ transport in red cell ghosts. Biochim. Biophys. Acta 241:75.
99. Albers, R. W., and Koval, G. J. (1966). Sodium potassium activated adenosine

triphosphatase of Electrophorus electric organ. III. An associated potassium activated neutral phosphatase. J. Biol. Chem. 241:1896.
100. De Meis, L. (1969). Activation of Ca^{2+} uptake by acetyl phosphate in muscle microsomes. Biochim. Blophys. Acta 172:343.
101. Kasbekar, D. K., and Durbin, R. P. (1965). An adenosine triphosphatase from frog gastric mucosa. Biochim. Biophys. Acta 105:472.
102. Davenport, H. W. (1940). The inhibition of carbonic anhyrase and of gastric acid secretion by thiocyanate. Am. J. Physiol. 129:505.
103. Sachs. G., Mitch, W. E., and Hirschowitz, B. I. (1965). Frog gastric mucosal ATPase. Proc. Soc. Exp. Biol. Med. 119:1023.
104. Sachs, G., Collier, R. H., Pacifico, A., Shoemaker, R. L., Zweig, R. A., and Hirschowitz, B. I. (1969). Action of thiocyanate on gastric mucosa in vitro. Biochim. Biophys. Acta 173:509.
105. Sachs, G., Collier, R. H., and Hirschowitz, B. I. (1970). Action of SCN^- on rat liver mitochondria. Proc. Soc. Exp. Biol. Med. 133:456.
106. Sachs, G., Wiebelhaus, V. D., Blum, A. L., and Hirschowitz, B. I. (1972). Role of ATP and ATPase in gastric acid secretion. *In* G. Sachs, E. Hein, and K. J. Ullrich (eds.), Gastric Secretion, pp. 321–343. Academic Press, New York.
107. Soumarmon, A., Lewin, M., Cheret, A. M., and Bonfils, S. (1974). Gastric HCO_3^- stimulated ATPase: evidence against its microsomal localisation in rat fundus mucosa. Biochim. Biophys. Acta 339:403.
108. Durbin, R. P., and Kasbekar, D. K. (1965). Adenosine triphosphate and active transport by the stomach. Fed. Proc. 24:1377.
109. Du Pont, J. J., Hansen, T., and Bonting, S. L. (1972). An anion sensitive ATPase in lizard gastric mucosa. Biochim. Biophys. Acta 274:189.
110. Hill, B. S., and Hill, A. E. (1973). Cl transport in Limonia. *In* W. P. Anderson (ed.), Ion Transport in Plants, pp. 379–384. Academic Press, New York.
111. Harris, J. B., Frank, H., and Edelman, I. S. (1958). Effect of potassium on ion transport and bioelectric potentials of frog gastric mucosa. Am. J. Physiol. 195:499.
112. Davis, T. L., Rutledge, J. R., Keesee, D. C., Bajandas, F. J., and Rehm, W. S. (1965). Acid secretion potential and resistance of frog stomach in K^+-free solutions. Am. J. Physiol. 209:146.
113. Sedar, A. W., and Wiebelhaus, V. D. (1972). K^+ effect on acid secretion and ultrastructure of the amphibian oxyntic cell. Am. J. Physiol. 223:1088.
114. Conway, E. J. (1953). The Biochemistry of Gastric Acid Secretion. Charles C Thomas, Springfield, Ill.
115. Conway, E. J. (1954). Some aspects of ion transport through membranes. Symp. Soc. Exp. Biol. 8:297.
116. Seaston, A., Inkson, C., and Eddy, A. A. (1973). The absorption of protons with specific amino acids and carbohydrates by yeast. Biochem. J. 134:1031.
117. Azzone, G. F., and Massari, S. (1973). Active transport and binding in mitochondira. Biochim. Biophys. Acta 301:195.
118. Forte, J. G., Forte, G. M., and Saltman, P. (1967). K^+-stimulated phosphatase of microsomes from gastric mucosa. J. Cell. Physiol. 69:293.
119. Forte, J. G., and Forte, G. M. (1966). K^+-stimulated phosphatase of gastric mucosal microsomes. Fed. Proc. 25:514 (Abstr.).
120. Sachs, G., Rose, J. D., Shoemaker, R. L., and Hirschowitz, B. I. (1966). Phosphatase reactions of transport ATPases. Physiologist 9:281 (Abstr.).
121. Spenney, J. G., Shah, G., and Sachs, G. Unpublished observations.
122. Forte, J. G., Ganser, A. L., and Tanisawa, A. S. (1974). The K^+-stimulated ATPase system of microsomal from gastric oxyntic cells. Ann. N.Y. Acad. Sci. 242:255.
123. Pickett, W. J. III (1974). Acid secretion in saccharomyces cerevisiae. Master's thesis, University of Alabama.
124. Britten, J. S., and Blank, M. (1968). Thallium activation of the (Na^+ + K^+) activated ATPase of rabbit kidney. Biochim. Biophys. Acta 159:160.
125. Saccomani, G., Rabon, E., and Sachs, G. In preparation.
126. Sachs, G., Saccomani, G., Lewin, M., and Goodall, M. C. (1975). Transport properties of a membrane fraction of gastric mucosa. Vth Int. Biophys. Congress.

127. Lee, J., Simpson, E., and Scholes, P. (1974). An ATPase from dog gastric mucosa: changes of outer pH in suspensions of membrane vesicles accompanying ATP hydrolysis. Biochem. Biophys. Res. Commun. 60:825.
128. Ganser, A. L., and Forte, J. G. (1973). Ionophoretic stimulation of K^+-ATPase of oxyntic cell microsomes. Biochem. Biophys. Res. Commun. 54:690.
129. Rehm, W. S., Sanders, S. S., Rutledge, J. R., David, T. L., Kurfees, J. F., Keesee, D. C., and Bajandas, F. J. (1966). Effect of removal of external K^+ on frog's stomach in Cl^- free solutions. Am. J. Physiol. 210:689.
130. Moody, F. G. (1968). Oxygen consumption during thiocyanate inhibition of gastric acid secretion in dogs. Am. J. Physiol. 215:127.
131. Alonso, D., Nigon, K., Dorr, I., and Harris, J. B. (1967). Energy sources for gastric secretion: substrates. Am. J. Physiol. 212:992.
132. Harris, J. B., Alonso, D., Park, O. H., Cornfield, D., and Chacin, J. (1975). Lipoate effect on carbohydrate and lipid metabolism and gastric H^+ secretion. Am. J. Physiol. 228:964.
133. Sernka, T. J., and Harris, J. B. (1972). Pentose phosphate shunt and gastric acid secretion in the rat. Am. J. Physiol. 222:25.
134. Sachs, G., Shoemaker, R. L., and Hirschowitz, B. I. (1965). Action of 2-deoxy-D-glucose on frog gastric mucosa. Am. J. Physiol. 209:461.
135. Hersey, S. J. Discussed in Ref. 3.
136. Kasbekar, D. K. Am. J. Physiol. In press.
137. Sachs, G., Shoemaker, R. L., and Hirschowitz, B. I. (1967). The action of amytal on frog gastric mucosa. Biochim. Biophys. Acta 143:522.
138. Sachs, G., Collier, R. H., Shoemaker, R. L., and Hirschowitz, B. I. (1968). The energy source for gastric H^+ secretion. Biochim. Biophys. Acta 162:210.
139. Bannister, W. H. (1966). The effect of oligomycin and some nitrophenols on acid secretion and oxygen uptake by gastric mucosa of the frog. J. Physiol. (Lond.) 186:89.
140. Hersey, S. J. Ref. 3 and unpublished observations.
141. Hersey, S. J. (1971). The energetic coupling of acid secretion in gastric mucosa. Philos. Trans. R. Soc. Lond. (Biol.) 262:261.
142. Kidder, G. W., Curran, P. F., and Rehm, W. S. (1966). Interactions between cytochrome system and H ion secretion in bullfrog gastric mucosa. Am. J. Physiol. 211:153.
143. Hersey, S. J., and High, W. L. (1971). On the mechanism of acid secretory inhibition by acetazolamide. Biochim. Biophys. Acta 233:604.
144. High, W. L., and Hersey, S. J. (1974). Mechanism of theophylline stimulation of acid secretion by frog gastric mucosa. Am. J. Physiol. 226:1408.
145 Sarau, H. M., Foley, J., Moonsamy, G., Wiebelhaus, V. D., and Sachs, G. (1975). Metabolism of dog gastric mucosa. I. Nucleotide levels in parietal cells. J. Biol. Chem. 250:8321.
146. Durbin, R. P., Michelangli, F., and Michel, A. (1974). Active transport and ATP in frog gastric mucosa. Biochim. Biophys. Acta 367:177.
147. Sarau, H. M., Foley, J., Moonsamy, G., and Sachs, G. J. Biol. Chem. In press.
148. Opie, L. H., Mansofrd, K. R. L., and Owen, P. (1971). Effects of increased heart work on glycolysis and adenine nucleotides in the perfused heart of normal and diabetic rats. Biochem. J. 124:475.
149. Greenbaum, A. L., Gumao, K. A., and McLean, P. (1971). The distribution of hepatic metabolites and the control of the pathways of carbohydrate metabolism in animals of different dietary and hormonal status. Arch. Biochem. Biophys. 143:617.
150. O'Callaghan, J., Sanders, S. S., Shoemaker, R. L., and Rehm, W. S. (1974). Barium and K^+ on surface and tubular cell resistances of frog stomach with microelectrodes. Am. J. Physiol. 227:273.
151. Spenney, J. G., Shoemaker, R. L., and Sachs, G. (1974). Microelectrode studies of fundic gastric mucosa: cellular coupling and shunt conductance. J. Membr. Biol. 19:105.
152. Flemstrom G., and Sachs, G. (1975). Ion transport by amphibian antrum in vitro. I. General characteristics. Am. J. Physiol. 228:1188.

153. Spenney, J. G., Flemstrom, G., Shoemaker, R. L., and Sachs, G. (1975). Quantitation of conductance pathways in antral gastric mucosa. J. Gen. Physiol. 65:645.
154. Frizzell, R. A., and Schultz, S. G. (1972). Ionic conductances of extracellular shunt pathway in rabbit ileum; influence of shunt on transmural sodium transport and electrical potential differences. J. Gen. Physiol. 59:318.
155. Sachs, G., Shoemaker, R. L., Blum, A. L., Helander, H. F., Makhlouf, G. M., and Hirschowitz, B. I. (1971). Microelectrode studies of gastric mucosa and isolated gastric cells. *In* G. Giebisch (ed.), Symp. Medica Hoechst, pp. 257–279. Schattauer.
156. Spangler, J. G., and Rehm, W. S. (1968). Potential responses of nutrient membrane of frog's stomach to step changes in external K^+ and Cl^- concentrations. Biophys. J. 8:1211.
157. Bajaj, S., Spenney, J. G., and Sachs, G. (1975). Modification of permeability properties of gastric antrum. Gastroenterology 68:978 (Abstr.).
158. Machen, T., and Forte, J. G. (1975). Active and passive Na transport by isolated mammalian stomach. Vth Int. Cong. Biophys. 1:15 (Abstr.)
159. Sanders, S. S., O'Callaghan, J., Butler, C. F., and Rehm, W. S. (1972). Conductance of submucosal-facing membrane of frog gastric mucosa. Am. J. Physiol. 222:1348.
160. Loewenstein, W. R., Socolar, S. J., Higashino, S., Kanno, Y., and Davidson, N. (1965). Intercellular communication: renal, urinary bladder, sensory, and salivary gland cells. Science 149:295.
161. Hogben, C. A. M. (1968). Observations of ionic movement through the gastric mucosa. J. Gen. Physiol. 51:240.
162. Rehm. W. S., and Dennis, W. H. (1957). A discussion of theories of hydrochloric acid formation in the light of electrophysiological findings. *In* Q. R. Murphy (ed.), Metabolic Aspects of Transport Across Cell Membranes, pp. 303–330. University of Wisconsin Press, Madison.
163. Rehm, W. S. (1950). A theory of the formation of HCl by the stomach. Gastroenterology 14:410.
164. Rangachari, P. K. (1975). Ba^{2+} on the resting frog stomach: effects on electrical and secretory parameters Am. J. Physiol. 228:511.
165. Harris, J. B., and Edelman, I. S. (1964). Chemical concentration gradients and electrical properties of gastric mucosa. Am. J. Physiol. 206:769.
166. Sachs, G. Unpublished observations.
167. Villegas, L. (1962). Cellular location of the electrical potential difference in frog mocosa. Biochim. Biophys. Acta 64:359.
168. Hogben, C. A. M. (1972). The reverse Rehm effect. *In* G. Sachs, E. Heinz, and K. J. Ullrich (eds.), Gastric Secretion, pp. 111–130. Academic Press, New York.
169. Rehm, W. S. (1965). Electrophysiology of the gastric mucosa in chloride free solutions. Fed. Proc. 24:1387.
170. Sachs, G. Unpublished observations.
171. Nicholls, D. G. (1974). The influence of respiration and ATP hydrolysis on the proton electrochemical gradient across the inner membrane of rat liver mitochondria as determined by ion distribution. Eur. J. Biochem. 50:305.
172. Witt, H. (1974). Primary acts of energy conservation in he functional membrane of photosynthesis. Ann. N.Y. Acad. Sci. 227:203.
173. Kaback, H. R. (1974). Transport studies in bacterial membrane vesicles. Science 186:882.
174. Hasselbach, W. (1964). Relaxing factor and the relaxation of muscle. Prog. Biophys. Mol. Biol. 14:167.
175. Sweadner, K. J., and Goldin, S. M. (1975). Reconstitution of active ion transport by the sodium and potassium ion stimulated adenosine triphosphatase from canine brain. J. Biol. Chem. 250:4022.
176. Racker, E., and Kandrach, A. (1971). Reconstitution of the third site of oxidative phosphorylation. J. Biol. Chem. 246:7069.
177. Bretscher, M. S. (1971). Major human erythryocyte glycoprotein spans the cell membrane. Nature (New Biol.) 231:229.
178. Eytan, G. D., Carroll, R. C., Schatz, G., and Racker, E. (1975). Arrangement of the subunits in solubilized and membrane bound cytochrome c oxidase from bovine heart. J. Biol. Chem. 250:8598.

179. Racker, E., and Eytan, E. (1975). A coupling factor from sarcoplasmic reticulum required for the translocation of Ca^{2+} ions in a reconstituted Ca^{2+} ATPase pump. J. Biol. Chem. 250:7533.
180. Shamoo, A. E., and MacLennan, D. H. (1974). A Ca^{2+} dependent and selective ionophore as part of the Ca^{2+} + Mg^{+}-dependent adenosine triphosphatase of sarcoplasmic reticulum. Proc. Natl. Acad. Sci. USA 71:3522.
181. Skulachev, V. P. (1975). Energy coupling in biological membranes. *In* E. Racker, (ed.), MTP International Review of Science, Vol. 3, pp. 31–74. Butterworth, London.
182. Sachs, G., Lewin, M., Rabon, E., Hung, P., Schackmann, R., and Saccomani, G. J. Biol. Chem. In press.
183. Pullman, M. E., and Monroy, G. C. (1963). A naturally occurring inhibitor of mitochondrial adenosine triphosphatase. J. Biol. Chem. 238:3762.
184. Lewin, M., and Sachs, G. (1975). ANS as a probe of ATP dependent H^{+}/K^{+} transport of gastric membrane vesicles. Physiologist 18:290a (Abstr.).
185. Vanderkooi, J. M., and Martonosi, A. (1971). Sarcoplasmic reticulum. XII. The interaction of 8-anilino-l-naphthalene sulfonate with skeletal muscle microsomes. Arch. Biochem. Biophys. 144:87.
186. Azzi, A., Gherardini, P., and Santato, M. (1971). Fluorochrome interaction with the mitochondrial membrane: the effect of energy conservation. J. Biol. Chem. 246:2035.
187. Garrahan, P. J., and Glynn, I. M. (1966). Driving the sodium pump backwards to form adenosine triphosphate. Nature 211:1414.
188. Barlogie, B., Hasselbach, W., and Makinose, M. (1971). Activation of calcium efflux by ADP and inorganic phosphate. FEBS Lett. 12:267.
189. Knowles, A. F., and Racker, E. (1975). Properties of a reconstituted calcium pump. J. Biol. Chem. 250:3528.
190. Reuss, L., and Finn, A. (1974). Passive electrical properties of toad urinary bladder epithelium. J. Gen. Phys. 64:1.
191. Frömter, E. (1972). The route of passive ion movement through the epithelium of Necturus gallbladder. J. Membr. Biol. 8:259.

International Review of Physiology
Gastrointestinal Physiology II, Volume 12
Edited by Robert K. Crane

6 The Exocrine Pancreas

H. SARLES

Institut National de la Sante et de la Recherche Mèdicale
Unite of Recherches de Pathologie Digestive
Marseille, France

ELECTROLYTE SECRETION 174
Water 174
Bicarbonate 175
Na^+ 175
Cl^- 176
Site of Water and Electrolyte Transport 176
Relation between Water, Electrolyte, and Protein Secretion 177
Secretion of Calcium and Other Divalent Cations 177

BIOSYNTHESIS AND INTRACELLULAR TRANSPORT OF EXPORTABLE PROTEINS 178
Synthesis of Secretory Proteins 179
Transport 179

STIMULUS-SECRETION COUPLING 181
Water and Electrolyte Secretion–Duct Cell 182
Protein Secretion–Acinar Cell 182
cAMP versus cGMP 182
Membrane Depolarization–Na^+-K^+ 183
Role of Ca^{2+} in Protein Secretion 184
Role of Other Cell Components 186

NERVOUS REGULATION 187
Cholinergic Nerves 187
Adrenergic Nerves 188

PROSTAGLANDINS 190

HORMONAL REGULATION 190
Secretin 190
CCK 191

Gastrin 192
Other Gastrointestinal Hormones 192
Endocrine-Exocrine Interaction in the Pancreas 192
Other Hormonal Effects 193

REGULATION BY DIGESTIVE SECRETION AND FOOD 195
Bile 195
Pancreatic Juice 195
Other Effects 196
Adaptation to Diet 197
Trophic Effects of Hormones 198

DEVELOPMENT OF PANCREATIC FUNCTION 198

PHARMACOLOGICAL EFFECTS 198

VESSELS AND DUCTS 199

Reviewed are papers published from 1972 to August 1975 with some necessary references from previous years. Work prior to 1972 has been reviewed by Preshaw (1).

ELECTROLYTE SECRETION

Electrolyte secretion by the pancreas requires energy (2, 3). Omission of glucose from the perfusate of an isolated pancreas decreases flow nearly to zero. Glucose may be replaced by lactate (2) but not by α-ketoglutarate or by fumarate (3).

Water

Diamond (4, 5) has pointed out that fluid-transporting epithelial structures are characterized by narrow blind-ending channels. In his standing osmotic gradient model, solutes secreted into these channels diffuse toward the open end, and water enters across the channel walls to reduce osmolality. A mathematical analysis for channels with an internal radius less than 1 μm and length less than 200 μm shows that this water-solute coupling could easily provide flows of the required velocity. In the pancreas, such a system could reside either in the duct system or in the potential intercellular spaces between secretory cells. Ullrich (6) has reported the opening of spaces between pancreatic cells during fluid transport and that the morphological substrate for osmotic equilibration is most prominent along intercellular infoldings or canaliculi. If this is correct, little or no osmotic gradient would be predicted within the duct system proper.

Bicarbonate

Case et al. (7) have shown that pancreatic juice bicarbonate is derived mainly from plasma bicarbonate and have assumed that the primary event is the splitting of water into H^+ and OH^-. H^+ then moves into the plasma where it reacts with HCO_3^- to yield CO_2. CO_2 then moves into the cell to combine with OH^- under the influence of carbonic anhydrase forming HCO_3^-, which is actively transported into the duct lumen.

Different mechanisms have been proposed by Schulz et al. (3, 8–10) and by Swanson and Solomon (11), who have used micropuncture techniques to study electrolyte secretion by the in vitro rabbit pancreas. Most active HCO_3 transport is found in the smallest ducts near the acini (diameter 25–60 μm) coincident with the highest activity of the HCO_3-activated ATPase described by Simon et al. (12).

Omission of bicarbonate from the bathing fluid abolishes water and electrolyte secretion, but other weak acids can replace bicarbonate and partially restore pancreatic secretion (11, 13, 14). The similar characteristics exhibited by HCO_3 and acetate (11) or sulfamerazine (8) suggest that they are secreted by a common mechanism involving active H^+ transport. The secretion rate is controlled by the HCO_3^- concentration and pH. HCO_3 concentration in the primary secretion is controlled by the environmental HCO_3 concentration and to a lesser extent by pCO_2 (9, 11). These results suggest that a proton pump is located not just at the luminal side (9) or at the serosal (14) side but at both sides (11). Active fluxes of H^+ are coupled with influx of Na^+ (7, 11, 14). The serosal Na-H pump could be linked to metabolism through the Na-K pump. The mucosal Na-H pump requires direct metabolic coupling (11). The ability of a weak acid to restore flow in the HCO_3-free perfused preparation would depend on its ability to reach the site of OH^- formation, and there is evidence that nonionic diffusion of the lipid-soluble weak acid is critical in this process. It is assumed by Schulz et al. (9) that HCO_3^- as well as CO_2 can penetrate the interstitial cell side and buffer the hydrogen ions which are actively transported into the cell, forming carbonic acid which dissociates into CO_2 and water. CO_2 diffuses into the lumen and forms the luminal HCO_3^- with OH^- ion coming from the dissociated water. Carbonic anhydrase is assumed to participate.

Na^+

The presence of a (Na^+-K^+)-ATPase in pancreas homogenates, reported by Ridderstap and Bonting (15, 16) has been interpreted as proof that a Na^+ pump actively extrudes Na^+ from the cell and that it is a rate-limiting step in exocrine pancreatic secretion. However, in the small pancreatic ducts which are the predominant site of electrolyte secretion there is almost no (Na^+-K^+)-ATPase activity (17). Nevertheless, ouabain, frusimide, and ethacrynic acid inhibit pancreatic electrolyte secretion (2, 18, 19). The effects of nigericin, valinomycin, and amiloride on pancreatic secretion correspond to the effects of these substances not only on HCO_3-activated, but also on (Na^+-K^+)-ATPase (3). This may indicate that the (Na^+-K^+)-ATPase has an auxiliary function in secretion. However, Case and Scratcherd (2) suggest that the pancreas may transport Na^+ actively. During partial isosomotic replacement of perfusate sodium with sucrose, large concentration gradients of sodium can be established between secretion and perfusate. Lithium, which has a low affinity for the sodium pump, is transported at a low rate which is partially blocked by ouabain. The effect of K^+ omission and replacement by Rb^+ and Cs^+ also supports the concept of a sodium pump. Active transport of sodium into the lumen could be localized at both the luminal cell side (9) and at the interstitial side (14) and be coupled to H^+ transport (11).

There are two plateaux in $^{22}Na^+$ uptake by the rabbit pancreas in vitro. The first might be due to retardation by the cell, the second one to Na^+ equilibration within the ducts. In this model Na^+ transport time through pancreas is 3.5 min because of cell transit (20).

In man, furosemide increases water and electrolyte secretion, suggesting that a part of Na^+ is normally reabsorbed in the ducts (21). However, this effect is opposite to that found in vitro (2).

Cl^-

Some experiments favor the existence of a Cl^--rich primary secretion. In man and dog, Ribet et al. (22, 23) have shown that increases in the injected quantity of a cholecystokinin-containing preparation of secretin (Boots secretin) above those giving a maximal response for bicarbonate output increases flow rate and decreases bicarbonate concentration. This effect, which may be pharmacological, can only be explained by the secretion of a Cl^--rich juice (24). However, this effect may explain the decline in HCO_3^- concentration found in humans at high rates of secretion (25) and the observation in rats that cholecystokinin (CCK) is able to induce the secretion of a large volume of juice poor in bicarbonate and rich in Cl^-. In the rat, this phenomenon is physiological and not pharmacological, as it is reproduced by feeding (26, 27). The admixture theory is not to be rejected.

Site of Water and Electrolyte Transport

That water and bicarbonate secretion originate from the ducts has been definitively shown by Fölsch et al. (28). A copper-deficient diet supplemented with penicillamine destroys the acinar cells leaving the duct cells intact. In this model, secretin but neither CCK nor gastrin induces an almost normal hydroelectrolytic secretion, though the exact localization of secretion and reabsorption in the ducts is still unclear. Mangos et al. (29) have used micropuncture of interlobular ducts and the split oil drop technique in the rat and found that all water is secreted in the acinus and neither secreted nor reabsorbed in the ducts. They also found that rat pancreatic secretion is rich in Cl^- and poor in HCO_3^-, in contrast to what has been found in rabbits, dogs, cats, and man. Nevertheless, Cl^- and HCO_3^- are more concentrated in the acinus lumen than in plasma. In the small ducts, and only in case of secretion, there is an exchange of Cl^- for bicarbonate. But the idea that secretin in the rat, contrary to observations in other species, induces a Cl^--rich, bicarbonate-poor secretion has been repeatedly disproved (see below). According to Swanson and Solomon (30), primary secretion in the rabbit is produced in intralobular ducts during spontaneous secretion and in small extralobular ducts in response to secretin. A flow-dependent Cl-HCO_3 exchange takes place in the main collecting duct. However, the results of Reber et al. (31, 32), using microcannulation and micropuncture techniques in cats at different levels of the extralobular duct system, suggest that fluid at the intercalated duct level has an anion composition similar to the extracellular fluid and that subsequent modifications occur in the duct. In this case, extralobular ducts should have a major role in pancreatic exocrine secretion by secreting a bicarbonate-rich fluid and being the place of HCO_3^-/Cl^- exchange at the duct epithelium. As assumed by Schulz et al. (9, 17), it is much more probable that the whole duct system participates in HCO_3^- secretion with the highest transport rates located in the ducts near the acini. Greenwell (35) has shown that duct cells, regardless of their position in the duct system, function electrophysiologically in identical fashion. When flow rate is low, a one-to-one exchange of intraductal HCO_3^- for interstitial Cl^- takes place in the ducts. That this exchange is possible even in the main duct of the gland has been proved (33, 34) by perfusing a simulated pancreatic juice through the main duct.

Relation between Water, Electrolyte, and Protein Secretion

It is generally believed that water and bicarbonate secretion is necessary to flush released enzymes through the ducts. Greenwell and Scratcherd (18) find that the output of amylase in response to short duration stimulation occurs in the cat in a constant volume of approximately 0.5 ml, regardless of background electrolyte secretion. This is explained by a rapid secretion of enzymes into the duct system from the acini with subsequent wash-out by secretin-stimulated fluid secretion. Enzymes are packaged in zymogen granules and must be solubilized in the acinus and duct lumen by the electrolyte secretion, whatever its origin. Therefore, a minimum equilibrium between the two pancreatic secretions is necessary. The displacement of this equilibrium probably leads to pathological changes; in chronic alcoholic dogs where the release of gastrin (36) and CCK (37) is increased and the release of secretin (38) diminished, pancreatic juice protein precipitates in the ducts (39). The same phenomenon is observed during the course of increased protein secretion provoked in the dog by intravenous Ca^{2+} infusion (40). These two secretory abnormalities lead to the formation of pancreatic stones and chronic pancreatitis.

Secretion of Calcium and Other Divalent Cations

Goebell et al. (41) and Argent et al. (42) have shown that secretion of calcium in the pancreatic juice is a complex phenomenon which involves at least two distinct mechanisms. After secretin, the Ca^{2+} concentration is low (0.4 to 0.6

mEq/liter) and calcium secretion is independent of protein secretion. The results of Ceccarelli et al. (43) suggest that this calcium fraction may originate from the extracellular fluid, possibly by simple diffusion. On the contrary, Ca^{2+} secretion after CCK or octapeptide-CCK (OP-CCK) or caerulein injection increases in parallel with protein secretion (41, 42, 44) and is independent of blood Ca^{2+} level (45, 46). Ca^{2+} is first stored within zymogen granules of the acinar cell and then released by exocytosis; ^{45}Ca appears faster than L-[^{3}H]leucine labeled proteins in the zymogen granules isolated from pancreatic homogenates when the two tracers are injected concomittantly. This suggests that the zymogen granule-associated Ca^{2+} joins the exportable proteins sometime after their synthesis, possibly in the Golgi complex or in condensing vacuoles (43). The threshold dose of OP-CCK is higher for stimulating amylase secretion than for stimulating Ca^{2+} (and Zn^{2+}) secretion (48). The ^{45}Ca:protein ratios are constantly higher in caerulein-stimulated juice than in isolated zymogen granule preparations (43). Scratcherd has shown (48) that nicotinic acid superimposed on secretin infusion elicits a pancreatic secretion which is very poor in protein but rich in Ca^{2+} and Mg^{2+}, making it likely that a third source contributes to the Ca^{2+} of pancreatic juice (43). The data of Gullo et al. (49) collected in normal and diseased humans can only be explained by the assumption of a third source of Ca^{2+}. Zn^{2+} secretion, like Ca^{2+} secretion, is increased by vagal excitation, pentagastrin, and CCK. Mg^{2+} secretion is increased by pentagastrin and vagal stimulation but not by CCK (50).

BIOSYNTHESIS AND INTRACELLULAR TRANSPORT OF EXPORTABLE PROTEINS

Pancreatic protein secretion is dependent upon ATP and upon glucose oxidation (51), though Danielson and Sehlin (52) suggest that amino acid is preferred to D-glucose during stimulation by CCK. Pilocarpine stimulation modifies the intracellular level of amino acids, suggesting that neural stimulation is coupled with changes in oxidation-reduction potential and oxidative phosphorylation. The same phenomenon is observed with CCK (53).

Palade and his group have shown the different stages of the secretory process in the pancreas of guinea pig (54–56). The same process occurs in the frog *Rana esculenta* (57).

During intracellular transport, the enzymes and proenzymes are generally considered to be completely separated from the cytosol in a membrane-bound system. This, together with the presence of secretory anti-enzymes, is believed to protect the cell against autodigestion. However, it has been recently shown that proenzymes have inherent proteolytic activities even before being transformed into "active" enzymes. This ability for autoactivation is increased by increasing their concentration (58). Synthesis, intracellular transport, and exocytosis can be dissociated from one another. Biosynthesis is regulated in the long term by the composition of the diet (adaptation, see below) but has been considered by many authors to be independent of secretion and secretory stimuli such as acetylcholine and CCK (56, 59–64). Intracellular transport of secretory proteins from rough endoplasmic reticulum (RER) to zymogen granules (ZG) can be dissociated from protein synthesis and is considered to be unaffected by secretory stimuli (55, 56, 65). According to the Palade group, the different secretory proteins follow a parallel course from the RER to the acinus lumen and the ducts.

Synthesis of Secretory Proteins

During the course of secretion stimulated by food, CCK, or cholinergic drugs, it has been shown, contrary to previous work, that there is a link between secretion and biosynthesis. Stimulation is indeed followed by a short period of decreased biosynthesis lasting 15–60 min. After this time, biosynthesis increases above basal levels (59, 66–71). Infusion of caerulein for 3 days in rats increases incorporation of [^{3}H]leucine; the effects start at 12 hr, become maximal at 24 hr, and decrease to normal values at 48 to 72 hr (72). When secretion is inhibited by atropine (73) or by eliminating Ca^{2+} from the perfusate (59), the immediate decrease in protein synthesis following stimulation is still observed. Webster et al. (66) have shown that the relative importance of post-stimulus decrease or increase of synthesis is determined by the intensity of the stimulus; 1–2 mg kg^{-1} bethanechol increase biosynthesis, 4–12 mg kg^{-1} diminish it. They assume, like Mongeau et al. (69), that this could be ascribed to a regulated distribution of the energy supply available within the cell, protein synthesis having a priority second to secretion.

What explains the transmission of the stimulus from the cell membrane to the RER ribosomes? Cell penetration by the first messenger acetylcholine or CCK seems improbable. According to Reggio and Cailla (71), transcription is not responsible; when RNA synthesis is prevented by actinomycin D, CCK is still able to increase biosynthesis. Nevertheless, CCK, bethanechol, or a meal increase the synthesis of RNA at a moment when protein synthesis is decreased (74) and increase RNA polymerase 1 and 2 (75). After stimulation, two peaks of cAMP are observed, the second of which could be related to biosynthesis (76). Increased biosynthesis in response to bethanechol seems to be mediated by changes occurring mainly in the cytoplasm. On the contrary, long term responses observed when fasting rats are refed are mediated by both soluble and particulate components (77–79).

Venroij et al. (80) have shown that the amino acids which are incorporated into proteins come from the extracellular milieu and not from the intracellular pool. Transport of the amino acids into the cell is sensitive to extracellular Na^+ (81) but not to CCK (51, 71). The only proteins synthesized at appreciable rates are the secretory proteins (82).

Transport

In the unstimulated rat pancreas, Kramer and Poort (83) have shown that 80% of the protein synthesized at a given moment is secreted into the acinus lumen within 12 hr. This unstimulated basal secretion involves two mechanisms. The first starts 20 min after pulse labeling and is considered to be the result of fortuitous contacts between young secretory granules and the apical cell membrane. The second consists of an orderly movement of the mass of secretory granules towards the apical cell membrane caused by the continuous formation of new granules. This takes 7–12 hr. It is blocked by the protein synthesis inhibitor cycloheximide but not by atropine. Intracellular destruction is unimportant (82, 83). The time necessary for newly synthesized protein to be transported from RER to ZG seemed not to be modified by acute or chronic stimulation of secretion (84–87). Nevertheless, according to Bieger et al. (88) at the 24th hr of a prolonged caerulein infusion in the rat, transport is considerably accelerated. Glycosylation (89, 90) and sulfation (91) take place during transport of enzymes, mainly in the smooth vesicles of the Golgi region. The concentration of proteins in condensing vacuoles is not energy dependent (93), suggesting that water leaves following a decrease in osmotic activity in the vacuoles, possibly as a result of the formation of macromolecular aggregates in which a sulfated polyanionic component, peptidoglycan, may play a role (89, 90). This could explain the remarkable stability of ZG in sucrose or urea solution (94). However, when the pH is increased in vivo, the different enzymes are released from ZG, and their release is not identical. Granule contents reaggregate when the suspension is again made acidic in the presence of the insoluble residue (95). According to Tartakoff et al. (82, 92) all of the major proteins originate in the RER and have a parallel course through the Golgi complex, its associated condensing vacuoles, and ZG prior to discharge. A quantitatively identical protein mixture is secreted when lobules are stimulated by different stimuli, e.g., carbamylcholine, caerulein, or KCl. The kinetics of intracellular transport of labeled proteins are similar for the different enzymes with the exception of amylase, which doubles during time. This deviation could be explained by the fact that amylase, being a glycoprotein, requires an additional step, viz., glycosylation, which may increase intracellular transport time. When carbamylcholine is used as a stimulant, "the kinetics of discharge (of the different enzymes) are similar" (82, 92). Nevertheless, discharge of amylase seems to be approximately twice as great at the end of the experiment than at the beginning. Although this work confirms a rough parallelism between enzyme transport and discharge in both the nonstimulated and stimulated states, minor deviations especially in the case of amylase are not ruled out. In the rat, by increasing the dose of injected caerulein, the maximal concentration of total protein is reached with lower doses than the maximal concentration of amylase (96). In the pig, excitation of the vagus nerve increases amylase activity more than total proteins (50). In man, after injection of secretin + CCK or caerulein, there is good correlation between trypsinogen (Tg) and chymotrypsinogen (ChTg), but the secretions of amylase and lipase are delayed (97). In the rat, the duration of amylase transport from RER to ZG is probably longer than that of lipase, but the transit time from ZG to lumen is the same for the different enzymes (98). This difference, at least, could be explained by the fact that amylase is a glycoprotein. However, when one increases pancreatic secretion in man by increasing the quantity of infused CCK or secretin, the ratio ChTg:Tg diminishes (99). In the rat, the apparent half-life of lipase is longer than the half-life of the other hydrolases, and an appreciable fraction of newly formed lipase could be reincorporated into the cell (100) following sequestration of ZG membranes into the cell. In the rat during prolonged perfusion with CCK the plateau of amylase secretion is stable. However, Tg and ChTg tend to diminish with time (101), possibly due to a dissociation in biosynthesis rather than transport. Kraehenbuhl and Jamieson (102) have shown that Zg are not specialized in the transport of enzymes; each ZG contains trypsinogen, chymotrypsinogen, procarboxypeptidases A, ribonuclease, and deoxyribonuclease. Therefore, a nonparallel course of secretion for the different enzymes, if it exists, is still difficult to explain. These deviations from the model proposed by Palade et al. are minor. However, the ideas of Rothman are completely opposed. They are as follows: (*a*) Intracellular transport of the different enzymes is dissociated. The secretion of trypsinogen, for instance is selectively induced by intraduodenal perfusion of 10^{-3} M lysine (103) as part of a "bond-specific, short term regulation of digestive enzyme secretion" suggested much earlier by Walther (104). (*b*) Secretory proteins are derived from not one but two intracellular pools: the ZG (a slowly equilibrating pool), the secretion of which is preferentially stimulated by CCK and metacholine, and the cytoplasm (a rapidly equilibrating pool), which contains considerable amounts of enzymes (105). The observations on which this is based could be a mere artifact (82, 92) but, if not, they could explain the persistence of secretion after repeated injections of metacholine have produced an almost complete disappearance of ZG (106). Arguments in favor of the transport of amylase via the cytosol and its excretion across the intact apical membrane in pigeon pancreas have recently been given by Dandrifosse and Simar: ZG are not fragile, at least within certain limits, thus their destruction during preparation does not seem to explain the high level of amylase in the cytosol (107). (*c*) The passage of enzymes from one of the two assumed intracellular pools to the other is possible as enzymes are able to permeate cellular membranes (108). At least it has been reported that some digestive enzymes undergo an enteropancreatic circulation; that is, intact digestive enzymes are absorbed by the intestine and subsequently secreted by the pancreas (109) (see also chymodenin, below).

Rothman's data might possibly be rationalized with Palade's concept by assuming the presence of two different types of acinus cell.

STIMULUS-SECRETION COUPLING

Cyclic adenosine 3′:5′-monophosphate (cAMP), cyclic guanosine 3′:5′-monophosphate (cGMP), and Ca^{2+} have been proposed as second messengers for the pancreatic stimulus-secretion coupling of water and electrolyte or protein secretion; pancreatic tissue contains the necessary enzymatic equipment. Adenylate cyclase has been isolated in the particulate fraction of pancreas homogenates (110–112). Two cAMP phosphodiesterase activites and two cGMP phosphodiesterase activities (111, 113, 114) with different K_m have been found mainly in the cytosol. Three protein kinase activities have been found in pancreas; one or two are activated by cAMP and one by cGMP (115, 116). However, the pancreas is a complex organ containing different types of endocrine cells and at least two types of exocrine cells. Thus far, none of these enzyme activites have been localized to any type of secretory cells. This explains the confusion of our knowledge on stimulus-secretion coupling in exocrine pancreatic secretion.

Water and Electrolyte Secretion–Duct Cell

There are three lines of evidence that cAMP is the second messenger. First, Greenwell (35) has shown that secretin, but not CCK, hyperpolarizes the ductal cell. The onset of the secretory response is faster than the onset of the membrane potential response, suggesting that the potential changes may be secondary to secretion. The electrical response to secretion is much longer than that to CCK or acetylcholine (35), in keeping with the longer secretory response to secretin. Secretin increases cAMP concentration in pancreas tissue (117, 118), and a low concentration (10^{-8} M) gives half-maximal activity of adenylate cyclase (110, 119, 120). Tissue concentrations of cAMP and pancreatic juice levels of cAMP and bicarbonate correlate with the dose of exogenous secretin either injected or released in conscious dogs (121). Second, dibutyryl cAMP increases water and electrolyte output (122). Third, cholera toxin, which activates adenylate cyclase, induces the secretion of water and bicarbonate (123, 124). Methylxanthines (aminophylline, caffeine) and papaverine, which inhibit the destruction of cAMP by phosphodiesterase, potentiate the action of secretin. On the contrary, alloxan, which inhibits membrane-bound adenylate cyclase, inhibits pancreatic secretion (47, 118, 121, 125). It is therefore very probable that cAMP is the second messenger in water and electrolyte secretion elicited by secretin. On the other hand, Ca^{2+} is probably not directly involved (126). It has been suggested that superficially located –SH groups are involved in the secretin cell receptor (10).

Protein Secretion–Acinar Cell

cAMP versus cGMP Numerous arguments against the role of cAMP have been published in recent years. First, the concentration of CCK, which is necessary for stimulating adenylate cyclase activity, has been reported to be far higher than physiological doses (110) and the concentrations necessary for secretin or VIP to stimulate the same enzymatic activity (110, 112, 119, 120). Acetylcholine and isoproterenol, which both stimulate enzyme secretion in the rat, do not stimulate adenylate cyclase in broken cell membranes of the pancreas (9). Phosphodiesterase activities are not modified by pilocarpine (114). It has been reported that acetylcholine and carbachol have a biphasic effect on cAMP levels, i.e., stimulation followed by inhibition (127), or no effect, CCK being also inefficient (9), or even that they reduce cAMP levels (127). Second, dibutyryl cAMP has no action on the cell membrane potential or resistance (128), and it has been possible with pure CCK to increase tissue level of cAMP without increasing enzyme release (117). Dibutyryl cAMP does not increase pancreatic protein secretion (129). Also, cAMP added to isolated pig pancreatic zymogen granules leads to an increased lysis of the granules, but the slowness of this effect makes its physiological significance dubious (130). Last, theophylline (129), alloxan, nicotinic acid (47), and cholera toxin (123, 124) have no action on protein secretion.

On the other hand, there is some evidence that cGMP plays an important role in stimulus-secretion coupling in the acinar cells. First, CCK and OP-CCK produce a rapid and marked rise in cGMP in isolated guinea pig pancreatic slices (131, 132) and in the rat pancreas in vivo (133), whereas secretin fails to increase cGMP (131, 132). However, Petersen and Ueda observed that cGMP had no effect on cell membrane potential resistance (128), and Heisler and Grondin (134) found that dibutyryl cGMP did not affect basal or carbachol-stimulated secretion of α-amylase from in vitro rat pancreas. Second, the pancreas contains a cGMP-dependent protein kinase (133). cGMP is able to induce the phosphorylation of a membrane peptide (135), and, indeed, pancreatic stimulation is associated with ZG phosphorylation (136, 137).

A role for cAMP, however, cannot be completely excluded. Bonting and de Pont (138) have shown that the CCK receptor site is sensitive to lipolytic (and possibly proteolytic) attack. Addition of phospholipids, especially phosphatidylcholine and phosphatidylserine, and also bovine albumin during the adenylate cyclase assay decreases the molar half-maximally stimulating CCK concentration to that of secretin (1.5×10^{-7} M). Therefore, the presumed different sensitivity of adenylate cyclase to secretin and CCK could be an artifact. The fact that secretin and VIP increase cAMP production in "a pure preparation of acinar cells" (139) does not help with understanding the problem. Theophyline, 10 mM, and papaverine stimulate enzyme secretion and simultaneous addition of CCK and theophylline gives a larger stimulation than the sum of the individual effects (138). Tetracaine inhibits the secretory effect of CCK and urecholine but increases the secretion of protein provoked by dibutyryl cAMP possibly because it increases cell membrane permeability to cAMP (141). CCK, 1.10^{-8} M, carbamylcholine, and dibutyryl cAMP, but not secretin, provoke a 30–40% increase in protein phosphorylation in the rat pancreas. The phosphorylation is greater in the ZG fraction (115).

Membrane Depolarization–Na^+ K^+ Although Kanno (141, 142) has found opposite results, it has been shown that acetylcholine, CCK, gastrin, and stimulation of perivascular pancreatic nerves depolarize the acinar cell (128, 143, 144), increase Ca^{2+} efflux, and raise amylase output with similar dose dependence (144). Ca^{2+} efflux precedes or coincides with the increase in amylase output (50, 145). Nevertheless, depolarization is not an adequate stimulus for amylase release. When K^+ is increased 10-fold in the presence of atropine, depolarization is observed, although there is no amylase secretion (146–148) and the acinar cell membrane potential returns to basal level during maintained stimulation with CCK (35). This could nevertheless result from the fact that the response to CCK is an extremely transient phenomenon (18, 35). Depolarization is accompanied by a marked decrease in membrane resistance to Na^+ and K^+ (145). The secretory effect of CCK and acetylcholine is suppressed during Na^+ substitution by Li^+ (50), varies with the concentration of Na^+ in the perfusate (149), and depolarization of the acinar cell is provoked by increasing K^+ concentration in the perfusate and reduced by the local anesthetic agent tetracaine. This suggests that the resting potential of pancreatic acinar cell is dependent on the concentration gradient across the cell membrane and that the primary action of CCK or acetylcholine may be to increase the influx of Na^+ by changing cell permeability (50, 144, 148). According to Petersen (150), an (Na^+-K^+)-requiring ATPase activated by Na^+ at the inside and K^+ at the outside of cell membrane could play a role in the movement of K^+ and Na^+ and in the modifications of the cell membrane potential (150). According to Kanno (141), acetylcholine increases Na influx, and the subsequent increase in the intracellular Na^+ activates the Na^+ pump. The action of acetylcholine is abolished by atropine but the action of CCK is not, suggesting that the receptors for these two agents are different (143).

Role of Ca^{2+} in Protein Secretion Reviews on this subject have been published by Case (126, 151). Ca^{2+} plays an important role in enzyme secretion, but this action is very complex: there is an indirect action through nervous and hormonal mechanisms and a direct action on the acinar cell.

Indirect Actions In man, hypocalcemia induced by EDTA reduces volume and enzyme secretion (45, 152). In man (45, 152–156) and in dog (40), an intravenous infusion of Ca^{2+} increases volume and enzyme output in both basal and secretin stimulated secretion. Secretion stimulated by submaximal doses of CCK can be increased (155) or not modified (45). Atropine reduces the effect of Ca^{2+} infusion (40, 42). If smaller Ca^{2+} doses (1.8 mg kg^{-1} hr^{-1}) are infused intraduodenally in the conscious dog (157), volume, bicarbonate output, and protein concentration and output rise during the first 20 min of calcium infusion, but, during the next 20 min, protein returns to the plateau levels found before Ca^{2+} infusion and volume and bicarbonate are inhibited. After stopping the Ca^{2+} infusion, protein secretion again increases (157). These complex results illustrate that in addition to a possible direct action on the acinar cell, Ca^{2+} probably acts through nervous (cholinergic) and hormonal mechanisms (126). In the conscious rat, neither hypercalcemia nor hypocalcemia is able to modify pancreatic secretion (45), although, in vitro, Ca^{2+} does so.

Direct Action on Acinar Cell In the isolated, perfused cat pancreas a Ca^{2+}-rich perfusate has no effect on water and electrolyte secretion, but potentiates submaximally stimulated amylase secretion. When the perfusate is deprived of Ca^{2+} and contains EGTA, there is a progressive inhibition of pancreatic electrolyte secretion which becomes apparent after 50–70 min (42). An injection of $CaCl_2$ into the arteries of the perfused cat pancreas elicits enzyme secretion which is directly correlated to the effluent Ca^{2+} concentration (158), but atropine was not present to block acetylcholine release. This effect is potentiated by CCK, acetylcholine, and cAMP or cGMP. Pancreatic protein secretion is inhibited by high concentrations of Mg^{2+} (147, 159, 160). In the presence of ionophore A23187 which increases Ca^{2+} influx into cells, Ca^{2+} added to the incubation milieu of pancreas slices is able to stimulate the release of amylase. Dinitrophenol inhibits Ca^{2+}-stimulated release, indicating a requirement for metabolic energy. Tetracaine at concentrations which block the enzyme secretion provoked by cholinergic stimulus also blocks the secretory effect of Ca^{2+} in the presence of A23187 but atropine does not (161, 162). EGTA inhibits the secretion in response to carbachol or dibutyryl cAMP. La^{3+}, which blocks the carbachol-stimulated uptake of Ca^{2+} by the pancreas, also inhibits the secretion in response to the same agent (163). Replacement of Na^+ by Li^+ in the incubation medium of pieces of rat pancreas increases the release of $^{45}Ca^{2+}$ and amylase, but only in the presence of extracellular Ca^{2+}, and increases $^{45}C^{2+}$ uptake (50). According to Schulz et al. (158), carbamylcholine and CCK increase Ca^{2+} uptake significantly, thus extracellular Ca^{2+} could be the source of that fraction of Ca^{2+} responsible for stimulating secretion as in pancreatic islets. However, in isolated pancreatic acinar cells, neither CCK nor carbamylcholine modify Ca^{2+} influx (164), and according to Petersen, there is little sign of an acetylcholine-evoked Ca^{2+} influx (145, 165). Case and Clausen (51) suggest that the irreversible inhibition of secretion in Ca^{2+}-free media containing EGTA, very probably results from lesions of the cell membrane. Natural stimuli of pancreatic enzyme secretion modify $^{45}Ca^{2+}$ distribution in the cell by a process which is independent of extracellular Ca^{2+} and which is associated with amylase release (51, 163). The assumed increase in Ca^{2+} concentration at those sites where it is necessary for stimulating secretion, with the exception of artificial conditions (51, 161, 162), probably does not originate from extracellular sources but from intracellular ones. The Ca^{2+} efflux associated with stimulation (51, 144, 163) probably results from this movement. Extracellular Ca^{2+} may only regulate the release process which could explain the decreased secretion of rat pancreatic slices in a Ca^{2+}-free medium.

Nevertheless, $^{45}Ca^{2+}$ efflux from the pancreas is not precisely coupled to the amylase secretion; a plateau in the rate of loss of ^{45}Ca from the tissue is observed after 10–20 min of incubation, the release of amylase being linear up to 60 min (163). ^{45}Ca efflux from the cell is not derived solely from a zymogen pool; carbachol is able to stimulate ^{45}Ca efflux in the presence of 1 mM EGTA, a concentration which inhibits the protein secretory response (163). Therefore, other subcellular fractions must play a role. Ca^{2+} is found in the particulate subcellular fractions of acinar cells, namely mitochondria, ZG, plasmalemma, and smooth but not rough microsomes (166). In general, Ca^{2+} is found in membranes where it is bound to proteins of molecular weight of about 12,000 and not to phospholipids (167). Three distinct Ca-ATPases have been detected, one located in mitochondria, another in light microsomes (168) and a third in ZG membranes (169). Isolated mitochondria but not light microsomes demonstrate ATP-stimulated accumulation of Ca^{2+}, small amounts of which can be released by cAMP, cGMP, and sodium ions (168). Ca^{2+} might be the second messenger of stimulus-secretion coupling in the acinar cell (162). Ca^{2+} binds superficially to ZG membranes and could have a direct role in the fusion of the ZG and plasma membrane preceding exocytosis. According to Dean (170), Ca^{2+} has a physiological function of surface charge neutralization to diminish the electrostatic repulsion energy barrier between ZG and plasmalemma.

Role of Other Cell Components The presence of exogenous lysolecithin in the incubation medium of in vitro rat pancreas elicits a massive release of amylase (650% of basal secretion) (171). Nevertheless, the role of a phospholipase which would result in the formation of lysophospholipids has been rejected (130). The possible role of myoinositol, cyclic1,2-phosphate as a second messenger (172) has also been rejected (173).

Stimulation of enzyme secretion is associated with an increased synthesis of phosphatidylinositol, viz, the phosphatidylinositol effect described by Hokin (159). This synthesis is preceded by a breakdown of phosphatidylinositol in response to acetylcholine and to high concentrations of CCK (173). During carbachol stimulation, incorporation of [^{14}C]acetate into total lipids is diminished but increased in phosphatidylinositol (175). According to Gerber et al. (174), the major synthesis is confined to microsomes, which would suggest that there is a flow of microsomal membrane to the Golgi and ZG membranes. However, according to de Camilli and Meldolesi (176), incorporation of [2-^{3}H]myoinositol in membranes following acetylcholine + eserine is of the same order (4 × values before stimulation) in other membrane fractions. It is generally agreed (174, 175, 177) that there is a dissociation between the phosphatidylinositol effect and protein secretion. The physiological significance of this effect is therefore not clear. However, the phosphotidylinositol effect always appears when the rate of protein secretion is at least 2.5 times greater than the prestimulated values (177).

It has been proposed by Morre et al. (178) that the different cytoplasmic membranes are interconnected biogenetically, the membrane of the RER being the origin of the others. If this concept is correct, acinar cells must have a means by which they can remove the enormous quantity of extra membrane contributed to the cell membrane by the fusion of the ZG (179, 180). Kramer and Geuze suggest that redundant membranes are catabolized by lysosomes and reutilized (180). Meldolesi has criticized this point of view (181–183) since the turnover of different cell membranes is asynchronous and slower than the turnover of secretory proteins. Also, the composition of the different cytoplasmic membranes is different and there is no room for a sufficiently large pool of membranes in the cell. Applied to the pancreatic acinar cell, the freeze fracture technique shows that the luminal portion of the plasma membrane is similar to the ZG membranes. Finally, Meldolesi proposes that "intracellular transport occurs through the non-random fusion, fission and recycling of the various membranes involved. At any boundary between cell compartments, vesicles would bud off the membrane of the second compartment, discharge their content and then shuttle back for further loading."

A role for microfilaments and microtubules in the secretion of pancreatic proteins has been recently assumed (115, 184–187).

Some models have been proposed to explain the entire process of stimulus-secretion coupling, as, for example, by Matthews et al. (144), Beaudoin et al. (140), and Christophe et al. (115). The recent review by Case (76) should also be consulted.

NERVOUS REGULATION

Cholinergic Nerves

Acetylcholine increases enzyme secretion in all species but induces different actions on hydroelectrolytic secretion in different species: in pig, rat, and rabbit, contrary to cat, dog, and man, pancreatic secretion of water and bicarbonate is increased by acetylcholine to a level two to four times basal secretion, or about 30% of the secretion with maximal doses of secretin in the rat (192, 193) and 20% in isolated rabbit pancreas (195). In the cat, a rapid intravenous injection of acetylcholine is responsible for a secretion of pancreatic enzymes associated with a first peak of increased electrical conductance across the gland. A second peak of conductance probably corresponds to vasodilatation (18). In the dog (195), sub-threshold doses of urecholine increase pancreatic response to CCK at all dose levels.

Vagal stimulation of the cat pancreas provokes a secretion of fluid and electrolyte up to 20% of the maximum response observed with secretin (196). In this species, a short excitation of the vagus nerves has the same stimulatory effect as a rapid intravenous injection of acetylcholine on enzyme secretion and conductance (18). In the dog, electrical stimulation of the distal end of the right vagus nerve (but not of the left one) (197) increases volume and protein output even after stomach denervation. Stimulation of the intact vagus nerve is not effective on the pancreas but provokes a gastric secretion of H^+. This is interpreted as suggesting the possibility of vagal inhibitory fibers. However, Bourde et al. (198) with a very sophisticated technique found that stimulation of the intact vagus caused pancreatic secretion in four out of six dogs.

In dog, the stimulatory effect secondary to insulin or 2-deoxyglucose injection (199, 200) is mainly exerted through the release of gastrin, secretin, and CCK. Eighty percent of this effect is suppressed by antrectomy. The response to 2-deoxyglucose is probably not completely vagally mediated (201). In man, enzyme secretion is increased by hypoglycemia (202). This effect is suppressed by truncal vagotomy but not by highly selective gastric vagotomy (203). Atropine does not prevent the stimulatory effect of hypoglycemia either in the pig (204) or in man (206).

In the dog, extragastric or truncal vagotomy has little or no effect on the secretory response to secretin, to CCK, or to duodenally perfused HCl (201, 205). The response to intraduodenal amino acids and oleate perfusion is markedly depressed as is basal secretion (201). Tonic vagal activity contributes to basal pancreatic secretion but has little effect on the sensitivity of dog pancreas to secretin and CCK. Nevertheless the potentiating effect of exogenous CCK on secretin-induced bicarbonate secretion is reduced (207). Vagotomy decreases pancreatic response to a meal by decreasing the release of gastrin and CCK and, to a much less extent, of secretin (201, 208). Protein output falls by 77% following selective gastric vagotomy and by 15% following selective extragastric vagotomy in the dog. This demonstrates that vagal release of gastrin is the major element in vagally stimulated enzyme production (208a). The claim that in man vagotomy renders the pancreas hypersensitive to submaximal CCK doses (209) has been criticized (210). Thambugala and Baron (211) explain the earlier discrepancies on the results of vagotomy (191) by the fact that immediately after operation pancreatic secretion is reduced, but later the response to secretin and CCK increases above pre-vagotomy values.

Confirming the ideas of Thomas (212), the importance of duodenopancreatic reflexes has been underlined. In conscious dogs, xylocaine spray on the duodenal papilla decreases bicarbonate and protein secretion during the course of a prolonged perfusion of secretin (211) but increases protein secretion induced by CCK (213). This last action is more marked in chronic alcoholic dogs (213), which are known to have an increased pancreatic cholinergic tone (39).

Atropine suppresses the protein secretion due to intravenous injection of acetylcholine (18, 193) and to vagal stimulation (18, 196). In the dog, atropine infusion markedly reduces the secretion of proteins but not of water and bicarbonate when the animal is submitted to an infusion of submaximal doses of secretin + CCK. The ganglion blocker, pentolinium, has the same effect (205). These results are in agreement with most of the previous experiments (1). Neither pancreatic vasodilation secondary to excitation of the vagus nerves (18) nor the secretory response to hypoglycemia (202) nor 2-deoxyglucose (201) are suppressed by atropine confirming the existence of atropine resistant fibers in the vagus nerves.

The fact that atropine and ganglion blockers, but not vagotomy, decrease CCK-induced secretion of proteins in intact animals is probably due to the fact that after vagotomy the pancreatic ganglion cells are able to function independently and, after a transitory inhibition, to synthesize acetylcholine (214). In the rat, the cephalic and gastric phases of pancreatic secretion are probably of minor importance (215).

Adrenergic Nerves

In the rat, in contrast to other animals, isoproterenol has a potent stimulatory effect on pancreatic water and bicarbonate secretion (216), with a linear dose-response relationship between 0.1 and 10 μg kg^{-1} min^{-1}. The secretion of enzymes is not stimulated. Adrenaline and noradrenaline have the same effect with doses 20–40 times greater. Propranolol slightly diminishes basal secretion and irregularly diminishes the stimulatory effect of isoproterenol. Phenoxybenzamine increases basal secretion and the stimulatory effect of isoproterenol (216). In the perfused cat pancreas isoproterenol stimulates enzyme secretion probably by releasing acetylcholine from nerve endings (9). In the isolated rabbit pancreas, norepinephrine (noradrenaline) and epinephrine (adrenaline) inhibit the pancreatic secretion of water and bicarbonate, whereas phenoxybenzamine stimulates hydroelectrolyte secretion in vitro and blocks the inhibitory effect of nicotine and catecholamine on stimulated water and bicarbonate secretion (217). In the dog, isoproterenol inhibits the pancreatic secretion of protein. Stimulation of water and electrolyte secretion is either weak (218) or absent (219). This effect is blocked by propranolol (218). Phenoxybenzamine and, to a lesser degree, propranolol, increase volume and protein output (220). In man propranolol does not modify pancreatic secretion (221).

In the cat, the stimulation of splanchnic nerves produces, first, a phase of vasoconstriction abolished by α-blocking agents and, second, a phase of vasodilatation, abolished by β-blocking agents, which develops when stimulation is stopped (222). The rate of secretion is reduced partly by vasoconstriction and partly by a direct inhibition of the secretory cells which is not abolished by phenoxybenzamine. Noradrenaline has the same effect. Amylase concentration and output are augmented when bretylium tosylate, which prevents the release of catecholamine from nerves endings, is injected simultaneously. This effect is suppressed by atropine and is not reproduced by noradrenaline. It could be explained if one accepts the old assumption (195) that parasympathetic efferent fibers in dorsal roots reach the pancreas by way of the splanchnic nerves (223).

The most important action of the adrenergic nerves on the pancrease is probably through the release of dopamine, which has a stimulatory effect on volume and bicarbonate secretion in perfused dog pancreas (224–226). This effect is abolished by prostaglandin $F_{2\alpha}$ (227). The increased secretion induced by infusion of L-dopa is completely antagonized by RO 4.4602, a dopa decarboxylase inhibitor. The secretagogue effect of dopamine is enhanced by fusaridic acid, a dopamine-β-hydroxylase inhibitor, and by nialamide, a monoamine oxidase inhibitor. It is not affected by β- or α-adrenergic blocking agents or by atropine; hence, the dog pancreas may have a specific receptor for dopamine (225).

The role of adrenergic nerves in modulating the pancreatic secretory response seems therefore to be more important than has been thought, even though the effect of sympathomimetic agents is not always the same in different species. α-Adrenergic receptor stimulation probably decreases pancreatic secretion of water and bicarbonate. β-adrenergic receptor stimulation decreases pancreatic secretion in dogs (and rabbits) but is ineffective in man and induces an abundant secretion of water (and protein) in rats. Dopamine-receptor stimulation induces electrolyte and water secretion in the dog. Other reviews which may be consulted are references 188, 189, 190, and 191.

PROSTAGLANDINS

In the rat, prostaglandin E_1 (1 mg/liter) provokes a sustained hypersection of amylase from pancreas fragments and does not affect Ca efflux (115). In the anesthetized cat, intra-arterial injections of PGE_1 or E_2 and to lesser degree of $PGF_{1\alpha}$ or $F_{2\alpha}$, decrease arterial blood pressure and pancreatic blood flow. This could be responsible for the inhibition of secretin-stimulated electrolyte secretion, which is also observed. In effect, in the isolated perfused cat pancreas, prostaglandins stimulate water and electrolyte secretion. Nevertheless, the decrease of blood pressure and flow after prostaglandin are much less marked if phenoxybenzamine is injected but the secretory inhibition is not reduced (228). In the dog, 16-dimethyl-PGE_2, which is active by the oral route, inhibits hydroelectrolyte secretion and increases enzyme secretion (229). $PGF_{2\alpha}$ abolishes the secretory effect of dopamine (227). When blood pressure is kept constant, PGF_1 inhibits water and electrolyte secretion but does not modify dopamine-induced secretion (227). However, according to Heisler (230) and Bauduin (177), PGE_1, E_2, and $F_{2\alpha}$ do not modify basal or stimulated protein secretion of in vitro rat pancreas.

HORMONAL REGULATION

Secretin

The half-life of the response to secretin in the dog is 3.2 min according to Lehnert et al. (232). In the rat, subcutaneous or intraperitoneal injection of secretin and CCK gives results identical to intravenous injection (193). In man, administration via the nasal mucosa (snuff) is effective (233, 234). In man, the response to an intravenous infusion of 75 CU in 1 hr is higher and more sustained than the response to a single injection of the same dose (235). The dose of porcine secretin required for half-maximal flow rate in the bird (5.9 $\mu g\ kg^{-1}$) is 180 times higher than in mammals (0.033 $\mu g\ kg^{-1}$) (236). In dog (191), cat, pig, and perhaps in man, there is little or no spontaneous flow of juice, but following adequate secretin stimulation, HCO_3^- concentration approaches an asymptote well in excess of 100 mM[2]. In rabbit, calf, horse, and particularly rat (26, 27, 237, 238), there is a considerable resting flow and although secretin augments it, HCO_3^- concentration is less than 100 mM. However, according to Mangos et al. (29) secretin in the rat increases, at the same time, flow rate and Cl^- secretion and decreases bicarbonate. These results raise the point of the technical difficulties of rat physiology. According to de Waele et al. (239), experiments must be realized 48 hr or more after operation, since surgical trauma decreases pancreatic secretion. On the other hand, anesthesia by itself has only a minor effect if the pylorus is ligated because anesthesia substantially decreases gastric secretion. Body temperature must be stabilized at 38°C because hypothermia, which is usual when rats are immobilized, reduces pancreatic secretion. According to Sewell and Young (27) the results of Mangos et al. (29) are explained by the fact that rats were not fasted, the pylorus was not ligated, and an impure preparation of secretin rich in CCK has been used (see CCK), therefore the rats were submitted to the effects of both secretin and CCK; in the rat, CCK is responsible for an abundant secretion of juice with a plasma-like HCO_3^- concentration. Following an injection of both CCK and secretin, flow rate increases far above maximal secretion with secretin and HCO_3^- concentration is low. Since the half-life of secretin is longer than that of CCK, the CCK-evoked pancreatic secretion fades out sooner than the HCO_3^--rich secretin-evoked juice (27). This does not however explain the fact that the peak bicarbonate response frequently occurs after stopping an intravenous infusion of secretin (235). Secretin does not modify the secretion of K^+ in the dog or rabbit. It increases it slightly in the cat, but in the rat there is an increased K^+ concentration up to 4 mM as long as the secretin- (but not CCK) stimulated flow rate persists (27). In the isolated perfused dog pancreas, Stock et al. (240) have shown that pure synthetic secretin is able to stimulate protein secretion. In cannulated dogs, secretin increases protein secretion by 40–60% (191). In the rat, pure secretin has been reported both to have no effect on protein secretion (101, 193) and to stimulate protein secretion (123). In man, protein secretion is not modified (241).

CCK

According to Lehnert, the half-life of the response to CCK in the dog is 1.8 min (232). The COOH-terminal octapeptide of CCK (OP-CCK) is 13–33 times (by weight) more active than the tetrapeptide (46). The relative potency for Ca^{2+} release from isolated pancreatic acinar cells is octapeptide > heptapeptide > CCK > hexapeptide = gastrin (164). Those peptides in which the tyrosine residue is sulfated and is at position 7 from the COOH-terminal are considerably more potent (164, 242). Caerulein is 5 times more active than CCK and 5,000 times more active than carbamylcholine in equimolar concentration (243). In man, the maximum secretion of protein is reached with an infusion of 2 Ivy Dog Units kg^{-1} hr^{-1}. The secretory plateau is higher than the plateau which is obtained with a maximal dose of caerulein (244). In the rat, the maximum concentration of protein is obtained with 330 ng kg^{-1} and the maximum concentration of amylase with 110 ng kg^{-1} caerulein (96). In the same animal, the maximum secretion of amylase is obtained with 60 Ivy Dog Units kg^{-1} hr^{-1} CCK and the maximum secretion of trypsin with 120 U (245). Submaximal doses are potentiated by secretin (245). Supramaximal doses of CCK, caerulein and OP-CCK cause an irreversible reduction in flow rate and protein output (27, 96, 191, 245), which may be due to excessive viscosity of the juice secondary to high protein content (96) and complicates the choice of maximal doses. Pancreatic secretion is poorly sustained during constant stimulation at all doses of hormone (245, 246). Whereas the cat pancreas does not secrete water in response to CCK, the dog pancreas responds vigorously to CCK, caerulein, and OP-CCK (247). The same is observed with the rat, which responds even more vigorously to CCK than to secretin. In the juice secreted in response to CCK, the bicarbonate concentration is similar to the plasma (27), and therefore Cl^- concentration should be high. In man, the smallest quantity of CCK able to contract the gall bladder is higher (0.025 Crick Harper Raper Unit) than that able to initiate pancreatic secretion (0.0078 U) (248).

Gastrin

In man, pentagastrin has no effect, or an insignificant effect, on the resting or secretin-stimulated secretion except at doses higher than those producing maximal gastric acid response (249–252). According to Duffaut et al. (253), 1 $\mu g\ kg^{-1}\ hr^{-1}$ pentagastrin does not modify the electrolyte secretion evoked by 2 U $kg^{-1}\ hr^{-1}$ secretin and 0.5–2.4 U $kg^{-1}\ hr^{-1}$ CCK, but inhibits enzyme secretion noncompetitively (inhibition being greater with higher doses of CCK). The trophic effect of gastrin will be discussed below (see section "Trophic Effects of Hormones," below).

Other Gastrointestinal Hormones

Gastric inhibition peptide (GIP) has been shown to have no effect on pancreatic secretion (254). Vasoactive intestinal peptide (VIP) stimulates electrolyte and water secretion by the pancreas. It also augments the response to secretin and to OP-CCK (255). In conscious dogs (256), maximal bicarbonate response to VIP is about 17% of that of secretin. Caerulein and endogenous CCK released by a peptone meal potentiate pancreatic bicarbonate response to VIP. VIP is a typical competitive inhibitor of secretin-induced pancreatic secretion. VIP is a close hormolog of chicken secretin and may have pronounced secretin like effects in nonmammalian species (257).

Endocrine-Exocrine Interaction in the Pancreas

In humans, glucagon inhibits CCK-stimulated enzyme secretion and reduces volume, whereas bicarbonate secretion is not influenced (259). In the dog, glucagon (6–60 $\mu g\ kg^{-1}\ hr^{-1}$) competitively inhibits the pancreatic secretion evoked by CCK infusion or intraduodenal infusion of amino acids (260, 261). In isolated dog pancreas, 1 mg of glucagon inhibits the pancreatic secretion evoked by an infusion of either secretin or CCK. However, when glucagon is superimposed on secretin + CCK given together, flow rate and bicarbonate and enzyme secretion are increased. This is not reproduced by a second injection (262).

In the cat, glucagon inhibits bicarbonate and protein secretion due to endogenous and exogenous secretin and CCK (263). In the perfused cat pancreas, one observes an identical inhibition of CCK-induced secretion (264). In the fasting rat, glucagon diminishes flow rate and bicarbonate and protein secretion. On the contrary, during a meal, glucagon increases flow rate and bicarbonate secretion, though modifications of blood glucose concentration are not responsible for this effect (265). However, according to Adler and Kern (266), stimulated pancreatic secretion is not influenced. Glucagon has no effect on amylase synthesis or release in vitro (267). In summary, glucagon as a rule inhibits pancreatic secretion, but the interplay between glucagon and gastrointestinal hormones is not well understood.

Insulin has no effect on the secretion of the perfused cat pancreas (264). In the isolated mouse pancreas, insulin (100 μg/ml) depresses amylase secretion and stimulates glucose oxidation but has no effect on amino acid uptake or oxidation (51). Most of the in vivo effects of insulin on the exocrine pancreas are exerted through the vagus nerves (268) or gastrin secretion (269). However, insulin seems to play an important role in adaptation to food (see below).

Other Hormonal Effects

Hypophysectomy results in a characteristic atrophy of the pancreas which is reversed by chronic administration of pentagastrin or a mixture of growth hormone, thyroxin, and corticosterone (270). In the rat, thyroidectomy decreases enzyme output (271, 272). Oral administration of thyroxine in dogs increases pancreatic secretion (273), and thyroidectomy reduces hormonally stimulated enzyme secretion (272). Vasopressin decreases flow rate and bicarbonate secretion in the rabbit. It abolishes or diminishes the secretory response to secretin in the cat and dog and decreases the output of pancreatic fistulae in man (274). Antidiuretic hormone also decreases flow rate and bicarbonate output and concentration when secretin is infused (275, 276). Somatostatin mainly decreases protein secretion stimulated by CCK in man (277), but also secretin-induced bicarbonate concentration though not the volume (278). In the anesthetized cat, somatostatin causes only a slight inhibition of amylase secretion but potentiates electrolyte secretion. In the perfused cat pancreas, however, amylase secretion is inhibited only at the highest doses and electrolyte secretion is unaffected (279). In the pig, somatostatin has no effect on endogenous or exogenous secretin action (280). In conscious dogs, it decreases pancreatic secretion in response to intraduodenal HCl, mainly by suppressing secretin release (281).

The binding of estriol, estradiol, and diethylstilbestrol is more important in the pancreas of the dog than in any other organ of this animal. One protein (3.7–3.9 S) of the cytosol seems to play a specific role in this binding (282).

In humans (45, 152, 156, 283, 284), calcitonin decreases only protein secretion induced by CCK; secretion induced by hypoglycemia or carbamylcholine is not influenced (45). Bicarbonate secretion is not influenced during the course of secretin infusion but is reduced during the course of CCK infusion (285). The maximum inhibitory effects are obtained with doses 30% of those necessary for the stomach (45). Contrary to an acute increase of blood calcitonin, a chronic increase (45), as, for example, from medullary tumors of the thyroid gland or repeated injections over 14 days, does not modify pancreatic secretion. The effect of calcitonin in man is probably more pharmacological than physiological. In the rat, pancreatic secretion is not influenced by calcitonin, but calcitonin decreases both components of pancreatic secretion in cats (286).

Among the candidate hormones of the gastrointestinal tract, bovine pancreatic peptide (BPP), a 36-amino-acid peptide (287, 288), increases water and electrolyte secretion evoked by secretin and inhibits enzyme secretion and the effect of CCK on water and electrolyte secretion.

According to Dick and Felber (289, 290), an extract of duodenal mucosa of rat made 30 min after glucose ingestion contains a factor which, given intravenously to the same animal, increases the pancreatic secretion of amylase but not of trypsinogen and chymotrypsinogen. However, 5 and 45 min after ingestion of lactalbumin, the duodenal mucosa contains a factor which increases the secretion of trypsinogen. If these reports are verified, they would be a strong argument against the parallel secretion of enzymes.

Chymodenin (291, 292) enhances rat pancreatic fluid flow and the release of digestive enzymes from isolated zymogen granules. In the anesthetized rabbit it increases ChTg secretion 3-fold but protein output only moderately (+ 40%) and lipase secretion not at all.

In the dog, during gravidity and puerperium, basal electrolyte and CCK-induced protein secretion are increased, but the response to secretin is not modified (293). During lactation in the rat, the pancreas weight is maximal 2 weeks post partum and regresses after weaning. The DNA content of the pancreas tissue, flow rate, and protein secretion are increased, as well as the sensitivity of the pancreas to CCK (294).

An interaction occurs between the gastrointestinal hormones and the nerves controlling alimentary gland secretion (295–297), but the mechanisms of these interactions are still unclear. The two-receptor theory of Grossman is still insufficiently tested to be accepted or rejected (298). The transposition of the Michaelis-Menten laws for enzyme-substrate complex formation to the relation between injected hormone (299) and magnitude of secretion certainly furnishes useful curves, but the data on dose 50 and maximum secretion drawn from these curves have never been proved to correspond to the real kinetics of secretion. There are many more intermediary steps between the intravenous injection of a hormone and the secretion which follows during a certain lapse of time than between substrate concentration and enzymic action. This explains why the conclusions on competition, facilitation, effication, and potentiation drawn from experiments in intact animals are often contradictory (298). Convincing evidence for Grossman's hypothesis will require isolation of pure pancreatic or ductal cells and kinetic studies on the specific hormonal receptors using isolated cells. Christophe et al. (300) have shown that VIP binds to two distinct classes of membrane receptors, one with high affinity and the other with low affinity. Binding of ^{125}I-VIP is inhibited competitively by high concentrations of secretin or secretin fragments 5–27 and 14–27 but not by secretin fragment 1–14 nor glucagon. In isolated pancreatic cells, secretin, glucagon, and VIP fail to alter Ca^{2+} outflux and do not affect stimulation by OP-CCK or carbamylcholine (164). It is generally agreed that CCK or its analogs (caerulein, OP-CCK, or gastrin) "potentiate" the action of secretin on water and electrolyte secretion. This potentiation has been found in dog (301), man (302), cat (303), and rat (304). However, in rats, the maximum rates of secretion of HCO_3^- in response to secretin and caerulein are additive when these hormones are given together. This suggests that the two hormones have independent actions on independent receptors (27).

Other reviews which may be consulted are references 189, 191, 231, 258.

REGULATION BY DIGESTIVE SECRETION AND FOOD

Bile

Confirming previous observations in cat by Mellanby (305), Forell, and co-workers (306–308) found that in man introduction of ox bile into the duodenum or emptying of the gall bladder by means of hypophysin stimulated pancreatic secretion (mainly enzymes). This effect is significantly reduced by atropine, suggesting that bile acts through hormone release, possibly by the way of a local nervous reflex. The sodium salt of 3α, 12α-dihydroxy-5β-cholanic acid is the most active stimulant of pancreatic secretion in man, but pure glycocholate evokes no secretion. When the dose of bile is sufficiently large, the increase in pancreatic secretion is as large as after intravenous injection of 1 U/kg secretion + U/kg of CCK. This probably explains the slight modification of pancreatic secretion observed in man after cholecystectomy in the third 20-min period following secretin + CCK injection (309). Malagelada et al. (310) have shown that intraduodenal, but not intrajejunal, infusion of taurocholate, taurochenodeoxycholate, or taurodeoxycholate (5–10 mM), in man inhibits the pancreatic secretion evoked by intraduodenal infusion of mono-olein or essential amino acids but not secretion evoked by intravenous injection of CCK. The effect of bile salts is attributed to an inhibition of CCK release.

In dogs, reintroduction of bile into the duodenum and jejunm, but not the ileum, increases basal secretion of protein and, to a lesser extent, of bicarbonate; the response to secretin, caerulein, or a meal is not modified (311). Intraperitoneal administration of bile salts and biliary obstruction increases volume and bicarbonate (312). In the rat, ligation of the main bile duct diminishes the stimulated and unstimulated pancreatic secretion of trypsin, but this inhibition is not overcome by reintroduction of the lost bile into the duodenum or by an intraduodenal infusion of taurocholate (313).

Pancreatic Juice

Since the work of Annis and Hallenbeck (314), it has been shown that in dogs water and bicarbonate secretion, but not protein secretion, were controlled by the presence of bicarbonate in the duodenum. In contradiction to Shaw and

Heath's results (237), it has been repeatedly found that in the rat and in the pig, when the pancreatic duct is cannulated and pancreatic juice is prevented from entering the duodenum, pancreatic secretion of protein is increased (313, 315, 316). The specific inhibition of intraluminal trypsin by oral administration of soybean trypsin inhibitor (SBTO) is also able to increase pancreatic secretion. This is consistent with the fact that, in the rat, the presence of trypsin or pancreatic juice in the duodenum inhibits pancreatic secretion by a feedback mechanism (315–317). Laporte and Tremoliere (316) have proposed that trypsin passes through the duodenal wall and binds CCK in the blood. But in recent experiments, Schneemann and Lyman (317) have found that depression of enzyme secretion by trypsin, as well as its stimulation by SBTI, occurs only in the upper part of the small intestine. A bicarbonate solution or a solution of ChTg injected into the duodenum are ineffective, as is an intravenous infusion of trypsin, ChTg, or SBTI. The effect of trypsin is more likely to be due to a direct effect of the active enzyme molecule. According to these authors, trypsin inhibits CCK release. Intraduodenal infusion of protein is followed by secretion of pancreatic juice but infusion of protein hydrolysate is not (318), which may be explained by the fact that protein occupies the active center of trypsin. Coring (319) has confirmed that pancreatic juice or trypsin, even if injected at acid pH, but not bicarbonate buffer, depresses pancreatic secretion in the pig when pancreatic juice has been diverted from the duodenum. In the rat, when the pylorus is not ligated and both bile and pancreatic juice are diverted, the increase of protein secretion is 600%, according to Green and Lyman (315). When the pylorus is ligated, the reintroduction of pancreatic juice alone decreases protein output by only 18% (239). Both bile (see above) and the stomach could therefore explain the intensity of the inhibition observed by Green and Lyman. Indeed, pancreatic secretion of trypsinogen is also probably able to modify pancreatic secretion indirectly by way of an inhibition of the effects of gastrin. Abita et al. (320) and Palasciano et al. (321) have shown that the injection into the duodenum of the trypsin inhibitory hexapeptide, which is liberated from trypsinogen by enterokinase and trypsin, is able to inhibit the action of gastrin on oxyntic cells.

Other Effects

Hage et al. (322) have shown that, in conscious dogs, intracolonic but not intraileal injection of oleic acid inhibits pancreatic secretion, and Harper et al. (323) have prepared an extract of cat ileal and colonic mucosa that inhibit pancreatic volume and enzyme output. Another possible inhibitory effect has been described by Rothman (103), namely, perfusion of the rabbit duodenum with 1×10^{-3} M lysine depresses protein secretion. Portal vein blood from these animals, also depresses protein secretion relative to controls when injected into recipient animals.

When bovine trypsin lung inhibitor or SBTI is given orally to rats for 20 days, increases in pancreas weight, protein percentage content of the pancreatic tissue, and the ratio trypsinogen to amylase and trypsinogen to lipase are observed (324, 325). However, after 3 weeks, the only modification is an increased specific activity of trypsinogen (325). Fölsh et al. (326) observed similar results after 3 weeks, with the exception that the specific activities of the different enzymes were not modified. The response to submaximal and supramaximal doses of CCK was increased, but the response to a maximal dose was not modified.

Adaptation to Diet

Palla, Ben Abdeljlill, and Desnuelle (327) were the first to show that the increased amylase biosynthesis of the acinar cell in response to a prolonged carbohydrate diet was probably mediated by an increased release of insulin induced by glucose. This has been confirmed. Intravenous infusion of glucose, as well as oral administration of starch (328), fructose, or sucrose, but not galactose (329), have been shown to induce amylase synthesis and the role of insulin has been equally confirmed (268, 329). Increased levels of circulating insulin, associated either with hyperglycemia (glucose infusion) or hypoglycemia (glibenclamide), increase the amylase content of the pancreas, total protein synthesis, and the baseline discharge of amylase; pancreatic levels and basal discharge of lipase and ChTg are decreased. Basal or stimulated discharge of newly synthesized proteins are not modified. On the other hand, diabetes with low levels of blood insulin leads to a disappearance of secreted amylase and a decrease of biosynthesis of protein (330). However, the presence of glucose or insulin during in vitro incubation has no effect on the synthesis of amylase and total proteins (368). Adaptation of amylase and lipase to diet is unaffected by 7 days of glucagon administration. Adrenalectomy tends to decrease and hydrocortisone tends to increase the level of amylase in response to a starch diet (331). Lavau et al. (328) have shown that just as intravenous glucose induces amylase synthesis, intravenous oleic acid infusion induces lipase synthesis, with the regulation of these two enzymes being controlled by the final products of the hydrolysis of their specific substrates. Whereas oral casein increases the intrapancreatic level of lipase and proteases, oral or intravenous amino acids are ineffective.

Twenty-four hours after swallowing a prey, the specific activity of ChTg, carboxypeptidase, and elastase in pancreas homogenates of different snakes are three times those of snakes fasted for 1 month. Two weeks after feeding, enzyme activites are returned to basal levels. Modifications of biosynthesis explain modifications of enzyme levels (332).

Twenty-four hours fasting diminishes amylase and protein biosynthesis (73, 267) and induces a reduction in polysome content and the capacity of the polysomes to synthesize proteins (73, 78, 267, 333). RNAse activity is increased along with an increased degradation of RNA (77). The number of ZG markedly decreases after 1 day of fasting, and numerous small secretory granules approximately 0.2–0.3 μm in diameter appear in the apical cytoplasm (334). The effect of fasting is suppressed by the intravenous injection of glucose, an action which is in turn abolished by the concommitant injection of diasoxide, a potent inhibitor of glucose-induced insulin release (78). The decrease in intrapancreatic amylase is parallel to the blood level of insulin (333). Diabetes decreases the amylase level (see above) but increases the lipase level (327, 329).

Trophic Effects of Hormones

It has been repeatedly confirmed that gastrin has a trophic effect on the exocrine pancreas. Pancreatic atrophy secondary to parenteral nutrition (335) or hypophysectomy (270) is prevented by continuous intravenous infusion of pentagastrin. A 20 U/kg^{-1} injection of CCK twice a day for 5 days increases the size of the cells and the number of cells (hyperplasia), while 6 mg kg^{-1} day^{-1} bethanechol only increases the cell mass (hypertrophy) (336).

DEVELOPMENT OF PANCREATIC FUNCTION

As previously shown by Yalovski et al. (337), the second phase of development of the embryo pancreas is controlled by corticosteroids; production of amylase (337) and chymotrypsinogen (338) but not of nonspecific esterase (339) is increased by cortisone in rat or chicken embryos. Actinomycin prevents this effect when it is given at the same time as hydrocortisone but not if given 1 hr later (338). In the rat, intrapancreatic levels and secretion of enzymes decrease for some days after birth, then increase until adulthood. The decrease is prevented by hydrocortisone (340–343). This development is partly controlled by diet. An increase of trypsinogen is induced by casein, of lipase by lard, and of nonspecific esterase by oil. Amylase increases only after weaning under the influence of dietary starch (339, 344). In the pig, also, the amylase increase is late (345). In human pancreatic juice collected from the age of 2 months, lipase, amylase, trypsin, chymotrypsin, and also bicarbonate concentration and output increase significantly from birth to the age of 2–3 years. The increase in amylase is most obvious inasmuch as this enzyme is frequently lacking in the newborn (346). Modifications in the secretion of this enzyme are at least partly a result of adaptation to starch ingestion. Indeed, Zoppi et al. (347) have shown that in newborn infants administration of a small quantity of starch stimulates amylase secretion. A diet rich in protein stimulates trypsin and lipase but a fat-rich diet has no specific effect. However, the increase of the other enzymes and bicarbonate shows that the human pancreas is not mature until about 2 years of age (346). The decrease in pancreatic function in old age is not sufficient to influence digestion (348). In the dog, exercise decreases water and bicarbonate secretion (349).

PHARMACOLOGICAL EFFECTS

Nicotine decreases flow rate, bicarbonate secretion, and the response to secretin in rabbit (217, 350) and dog (351). Phenoxybenzamine blocks this effect (217, 350). Protein secretion is not modified (351). However, a dose of 1.6 mg of nicotine increases by 4-fold the protein output of an isolated rabbit pancreas (350). Cocaine and desmethylimipramine enhance L-dopa or dopamine-induced secretion but not CCK- or secretin-induced secretion (226). Cigarette smoking has been reported both to decrease flow rate and bicarbonate secretion in man (352) and to decrease basal secretion but to increase the secretory response to secretin (353). In heavy smokers, pancreatic secretion is diminished even when tobacco is stopped (353). Phenobarbital given for 3 days to rats increases pancreatic weight, microsomal proteins, in vivo and in vitro incorporation of L-[^{14}C]-leucine into proteins, and enzyme secretion. This effect is not abolished by atropine (354). Anesthesia with pentobarbital and urethane in rats depresses flow rate and enzyme output by 50% (193), but, when the pylorus is ligated, the effect is insignificant (239). In the dog, fluothane anesthesia depresses flow rate and bicarbonate output (205). The action of ethanol on the dog pancreas differs according to whether the animal is or is not accustomed to the regular consumption of ethanol. In the "non-alcoholic" dog an intravenous injection of ethanol inhibits pancreatic secretion mainly of protein but also of water and bicarbonate. This action is suppressed by atropine (355), pentolinium, a ganglion blocker and decreased by vagotomy (356), but it is not modified by reserpine (357). When the animal is accustomed to the daily consumption of 2 g kg^{-1} ethanol ("alcoholic dogs") the effect of intravenous infusion of ethanol is reversed to one of stimulation of pancreatic secretion, again mainly of protein (358). The effect is suppressed by atropine but not by pentolinium or vagotomy (359). At the time of alcohol consumption, the sensitivity to CCK of the acinus cell is not modified (360). The mechanism of this action may be due to an ethanol-induced cholinergic impulse originating in the muscarinic receptor or intrapancreatic ganglion cells of the vagus nerves.

VESSELS AND DUCTS

If injected into the superior pancreatico-duodenal arteries, secretin, CCK, and histamine increase blood flow through the arteries and the pancreatic tissue. The relationship between the logarithm of the drug dose and its effect is linear (361). This action has been verified with CCK on the isolated perfused dog pancreas. It potentiates the effect of gastrointestinal hormones on the pancreas (362, 363). Similarly to an intravenous injection of CCK, gastrin, and acetylcholine, a rapid excitation of the vagus nerves of the cat is followed by two peaks of conductance. The first peak corresponds to secretion and the second to vasodilatation. The second peak is suppressed by atropine in the case of CCK, gastrin, or acetylcholine injection but not of vagus nerve excitation. Excitation of the left celiac ganglion in the dog diminishes the quantity of blood perfusing the pancreas in proportion to the intensity of electrical excitation (364). PGE_1 and E_2, and to a lesser extent PGF_{α} and $PGF_{2\alpha}$ diminish arterial blood pressure and pancreatic secretion. This effect is reduced by phenoxybenzamine. However, a direct constrictive action of prostaglandins on the vasculature of the pancreas is unlikely as these products have no effect on the perfusion of the isolated pancreas, suggesting that vasomotor effects represent part of the reflex homeostatic response to the fall in systemic blood pressure (298). PGB_1 and B_2 increase vascular resistance in the anesthetized rat pancreas (365). The protein secretory response of in vitro rabbit pancreas is more regular when CCK or acetylcholine are given arterially than when placed in the bathing media (11).

According to Lenninger (366) methacholine may increase or decrease the caliber of the cat pancreatic duct, and both the increase and decrease can be explained by smooth muscle contraction of duct walls. Although these variations of caliber are small, they could, according to the law of Poiseuille, markedly increase resistance to flow. Histochemical results show that the pancreas duct of the rat is cholinergically innervated (367).

Amylase output into the portal vein is observed with high doses of CCK which produce a dissociation of the intercellular connections of acinar cells. This effect is probably more pharmacological than physiological (368).

ADDENDUM ADDED IN PROOF

The following important observations have been made in the interval since the manuscript was completed:

Electrolyte Secretion

In agreement with the results of Case et al. in cats (7), it has been found in the same animal (369) that only 6% of the secreted HCO_3 could originate from cell metabolism (369). However, Pascal et al. (370) have found this percentage to be 25% in the isolated perfused dog pancreas. According to Simon and Knauf (371), HCO_3-sensitive ATPase is inhibited by phospholipase and this inhibition is removed by the addition of phospholipids. Being activated not only by bicarbonate but by other oxybases, this enzyme acts as a proton carrier. It is probably located, as in the stomach, at the luminal cell side, which on the contrary to the serosal cell side, is not permeable to HCO_3. In the model proposed by these authors, CO_3 functions as the proton acceptor. It is assumed to be bound to a carrier by phosphatidylethanolamine. It accepts a proton from dissociated water, which does not require energy. Coupling of the carrier with carbonic buffer or phosphatidylethanolamine changes its configuration and allows influx of H^+. The carrier is regenerated at the intracellular side of the plasma membrane by reaction with ATP. Carbonic anhydrase which is bound to the membrane at the same side as the HCO_3-sensitive ATPase could play a role in this mechanism.

Biosynthesis and Transport of Proteins

Sequence determination up to 24 NH_2-terminal residues of several putative percursors of dog pancreas secretory proteins synthesized in vivo have revealed sequence homology in the 16 NH_2-terminal residues: this sequence could constitute a metabolically short-lived peptide which precedes the NH_2-terminal sequences of all pancreatic secretory proteins and might function in the transfer of protein across the microsomal membrane (372).

Dagorn et al. (373, 374) compared in rats the amylase and chymotrypsinogen response to secretory stimulus. The comparison concerned the rates of biosynthesis of these two enzymes, their intrapancreatic levels of storage, and their rates of excretion in the juice. In each of these three steps, stimulation induced a nonparallel response of amylase versus trypsinogen. The nonparallelism in the rates of biosynthesis is unable to explain entirely the nonparallelism in the rates of excretion. These results suggest that the mechanism which controls the proportion of the different enzymes in the juice is different from that which monitors the rate of individual enyme biosynthesis.

On the contrary when isolated guinea pig lobules are used, carbachol- or caerulein-induced secretion of several different enzymes is parallel (375). Discrepancies between isolated lobules studied as an in vivo preparation of perfused pancreas or as in vitro perfused pancreas fragments could be explained by the finding by Malaisse-Lagae et al. (376) that peri-insular acini have different enzyme proportions and respond later to stimulation that tele-insular acini.

Although neither dibutyryl-cAMP nor dibutyryl-cGMP seems to modify Ca uptake by isolated rat cells (377), Haymovits and Scheele confirmed the role of cGMP (and possibly cAMP) as possible messenger(s) of protein secretion (378). In isolated guinea pig pancreatic lobules, optimal doses of carbachol, CCK, or caerulein increase by 20-fold cGMP levels in the 2 min following stimulation. Atropine blocks the effect of carbachol but not of caerulein. The effect on cGMP levels are dose related to protein secretion and doses of carbachol. High concentrations of dibutyryl-cAMP both increase cGMP and stimulate protein secretion. Contrary to previous works (see above) high concentrations of dibutyryl-cGMP increase protein secretion but not cAMP levels. For Bauduin (379) cAMP plays a role in the last stages of exocytosis: high concentrations of it increase protein secretion and induce the formation of emiocytosis figures. Inhibition of oxydative phosphorylation by antimycin does not decrease this action of cAMP but inhibits carbachol- and caerulein-induced secretion. For Gardner (380) CCK-P increases cAMP levels only in plasma membranes.

Petersen and Ueda (381) showed that when two microelectrodes impaling acinar cells are at a distance of less than 50 μm, these two cells are electrically coupled and Ach induces the same depolariation in both cells. This shows that the acinus is a functional unit.

When in incubation media of pancreas fragments Na is replaced by Tris (382, 381) or Cl by sulfate, Ach-, bethanecol-, or caerulein-induced secretion is abolished. By reintroducing Na or Cl, the secretion normally induced by Ach is obtained (381). For these authors, Na influx would induce the secretion of fluids necessary for driving out secretory protein. However, protein secretion itself should therefore be independant from Na movements (nevertheless Na influx into the cell induces Ca release from mitochondria (383)).

Input of Na would be caused by a decreased specific resistance of the membrane to Na and this in turn would result from the release of Ca from the same membrane (381).

The role of an increased secretion of Ca in the cytosol during stimulation is fairly well established and it is generally admitted that this Ca originates from intracellular stores (see above, 384). According to Clemente and Meldolesi (383) after caerulein stimulation of isolated lobules the influx of Ca into cytosol originates from mitochondria, other organelles having no important part in this phenomenon. The role of extracellular Ca is nevertheless not excluded. The Ca concentration in the incubation media of pancreas fragments has an influence on acinus cell sensitivity to Ach which respectively increases or decreases with high or low calcium concentrations (381). The effect of Ca an amylase sensitivity is independant from Na. According to Petersen and Ueda, uptake of extracellular calcium probably happens during prolonged stimulation and is necessary for a sustained secretion, but cannot explain the response to short stimulation (381).

Nervous Regulation

It has been confirmed by Vaysse et al. (385) using the isolated perfused dog pancreas that carbamylcholine increases the secretory response to small, submaximal doses of secretin, responses to maximal doses being not modified or slightly increased. This effect is not (or not only) explained by vasodilatation. As the slope of the dose action response curve to secretin is modified by carbomyl choline, this drug should act on a different receptor than secretin.

Contrary to the pancreas of the intact animal but similarly to isolated pancreas fragments (see above) in the isolated perfused dog pancreas, the secretory response to secretin or to secretin + CCK-PZ is not affected by atropine (385) which is fairly well explained by the fact that in these two last preparations Ach would no more be released by gangolion cells. The protein secretory response of all types of preparations to Ach (or derived products) is blocked by atropine (see above, 385) but not by nicotin (386) which proves that the receptors are muscarinic.

The action of different catecholamines has been studied by Bastie et al. (387) in the isolated perfused dog pancreas. Dopamine increases volume with a linear dose action relation. This is antagonized by Haloperidol (Ro 4.4602) but not by phenoxybenzamine and propranolol. The vascular action of dopamine is variable and is not responsible for secretion. Dopamine and secretin action are additive. Dopamine potentiates the protein secretory response to CCK, probably by washing out. Haloperidol has no effect on secretin-induced secretion. This secretory action of noradrenalin which is not blocked by phenoxybenzamin is probably exerted on a dopamine receptor. This action is weaker than the response to dopamine. The opposite action of the different catecholamines probably explains the fact that in the nonanesthetized dog, when all types of catecholamines have been depleted by repeated injection of reserpine (388), fluid and bicarbonate output is decreased while protein output is increased.

Hormonal Regulation

The isolated cat pancreas perfused with theophylline is a very sensitive preparation for secretin bioassay (389). A nice experimental model has been realized in the monkey by Gardiner (390). This model allows separate collection of bile and pancreatic juice and simultaneous reintroduction into the intestine of 95% of the secretion. The maximum secretion of fluid, bicarbonate and protein is obtained with 2.3 U kg^{-1}; with 5 U kg^{-1}, the protein secretory response is inhibited. Domschke et al. (391) and Cotton et al. (392) showed in men that when pancreatic juice is collected by endoscopic retrograde catheterization of the papilla, secretin doses necessary for obtaining a maximal rate of secretion are small: that is, 1 CU (392), 129 ng kg^{-1} hr^{-1}, which is lower than the secretin dose required when duodenal juice is collected by a tube (391). Secretin has no direct effect on protein secretion. cAMP is secreted in parallel with bicarbonate (391).

According to Ueda et al. (393) the maximum effect of CCK-PZ on fluid secretion in the isolated perfused rat pancreas as well as that of caerulein, gastrin, and Ach is greater than that of secretin. If bicarbonate is not present in the perfusate, the effect on flow rate of secretin but not of caerulein is abolished. If bicarbonate is replaced by sulfate and Na by Li, caerulein action is inhibited.

In the cat, dose response curve to VIP is the same as to secretin, the action of these two hormones when combined being additive. In the cat and in the dog, the dose response curve, that is fluid and bicarbonate in response to intraduodenal oleic acid infusion, is similar to the one in response to VIP in the same species (394).

To a lesser extent than calcitonin, parathormone inhibits Ach or CCK-PZ-induced protein secretion but not fluid and electrolyte secretion in the isolated perfused cat pancreas (395).

REFERENCES

1. Preshaw, R. M. (1974). Pancreatic exocrine secretion. Gastrointestinal Physiology, Series one, University Park Press, Baltimore.
2. Case, R. M., and Scratcherd, T. (1974). The secretion of alkali metal ions by the perfused cat pancreas is influenced by the composition and osmolality of the external environment and by inhibition of metabolism and Na^+ K^+-ATPase activity. J. Physiol. (Lond.) 242:415.
3. Wizemann, V., and Schulz, I. (1973). Influence of amphetamin, amiloride, ionophores

and 2-4-dinitrophenol on the secretion of the isolated cat's pancreas. Pflügers Arch. 339:317.

4. Diamond, J. M. (1971). Standing gradient model of fluid transport in epithelia. Fed. Proc. 30:6.
5. Diamond, J. M., and Brossert, W. H. (1967). Standing gradient osmotic flow. A mechanism for coupling of water and solute transport in epithelia. J. Gen. Physiol. 50:2061.
6. Ullrich, K. J. (1974). The acinar ion and water transport processes in exocrine glands. *In* N. A. Thorn and O. H. Petersen (eds.), Secretory Mechanism of Exocrine Glands, pp. 423–431. Munksgaard, Copenhagen.
7. Case, R. M., Scratcherd, T., and Wynne, R. D. (1970). The origin and secretion of pancreatic juice bicarbonate. J. Physiol. (Lond.) 210:1.
8. Schulz, I., Yamagata, A., and Weske, M. (1969). Micropuncture studies of the pancreas of the rabbit. Pflügers Arch. 308:277.
9. Schulz, I., Pederson, R., Wizemann, V., and Konmo, S. (1974). Stimulatory process of the exocrine pancreas and their inhibition by a non-penetrating SH reagent. *In* N. A. Thorn and O. M. Petersen (eds.), Secretory Mechanism of Exocrine Glands, pp. 88–90. Munksgaard, Copenhagen.
10. Wizemann, V., Schulz, I., and Simon, B. (1973). SH-groups on the surface of pancreas cells involved in secretion stimulation and glucose mediated secretion. Biochim. Biophys. Acta 307:366.
11. Swanson, C. H., and Solomon, A. K. (1975). Micropuncture analysis of the cellular mechanisms of electrolyte secretion by the *in vitro* rabbit pancreas. J. Gen. Physiol. 65:22.
12. Simon, B., Kinne, R., and Sachs, G. (1972). The presence of a bicarbonate stimulated ATPase in pancreatic tissue. Biochim. Biophys. Acta 282:293.
13. Noda, A., Toda, Y. T., Hakayama, T., and Nakajima, S. (1973). The excretion of 5,5-dimethyl-3,4-oxazolidinedione from the canine pancreas and bile. Am. J. Dig. Dis. 18:498.
14. Swanson, C. H., and Salomon, A. K. (1972). Evidence for Na–H exchange in the rabbit pancreas. Nature (New Biol.) 236:183.
15. Ridderstap, A. S., and Bonting, S. L. (1969). Na–K activated adenosine triphosphatase and pancreatic secretion in the dog. Am. J. Physiol. 216:547.
16. Ridderstap, A. S., and Bonting, S. L. (1969). Na^+–K^+ activated ATPase and exocrine pancreatic secretion *in vitro*. Am. J. Physiol. 217:1721.
17. Wizemann, V., Christian, A. L., Wiechmann, J., and Schulz, I. (1974). The distribution of membrane-bound enzymes in the acini and ducts of the cat pancreas. Pflügers Arch. 347:39.
18. Greenwell, J. R., and Scratcherd, T. (1974). The kinetics of pancreatic amylase secretion and its relationship to flow and electrical conductances in the anaesthetized cat. J. Physiol. (Lond.) 239:443.
19. Guelrud, M., Rudick, J., and Janowitz, A. D. (1972). Effects of some inhibitors of sodium transport (adenosinetriphosphatase inhibitors) on pancreatic secretion. Gastroenterology 62:540.
20. Rossier, M., and Rothman, S. S. (1975). Kinetics of Na^+ uptake and transcellular transit by the pancreas. Am. J. Physiol. 228:1199.
21. Thomas, F. B., Falko, J. M., Caldwell, H. H., and Hekadtian, H. S. (1975). Effect of furosemide on pancreatic exocrine function. Gastroenterology 68:A141.
22. Ribet, A., Pascal, J. P., and Sannou, N. (1967). Etude de la fonction exocrine du pancréas humain par les perfusions continues de secretine I. Effet des doses croissantes sur la sécrétion hydroelectrolytique. Arch. Fr. Mal. App. Dig. 56:677.
23. Ribet, A., Pascal, J. P., Vaysse, N., and Boucard, J. P. (1968). Relationship between bicarbonate, chloride and volume flow at high secretory rate in the pancreatic juice of the dog. Scand. J. Gastroenterol. 3:401.
24. Pascal, J. P., Vaysse, N., Louis, A., Augier, D., and Ribet, A. (1969). Etude des variations du contenu electrolytique du suc pancreatique (Interprétation en fonction de la théorie des deux composants). Biol. Gastroenterol. (Paris) 1:59.

25. Wormsley, K. G. (1968). Response to secretin in man. Gastroenterology 54:197.
26. Sewell, W. A., and Young, J. A. (1975). Secretion of eletrolyte by the pancreas of the anaesthetized rat. J. Physiol. (Lond.). In press.
27. Sewell, W. A., Coroneo, M., and Young, J. A. (1975). Bicarbonate secretion by the pancreas of the anaesthetized rat. Digestion 13:123.
28. Fölsch, U. R., and Creutzfeldt, W. (1976). Electrolyte secretion of a pancreatic duct model in the rat in vivo and accumulation of cyclic adenosine 3′5′-monophosphate in vitro in response to gastrointestinal hormones. *In* R. M. Case and H. Goebell (eds.), Stimulus Secretion Coupling in the Gastrointestinal Tract, pp. 381–386. MTP Press, Lancaster, England.
29. Mangos, J. A., McSherry, N. R., Nousia-Arvanitakis, S., and Scilling, R. F. (1974). Transductal flux of anions in the rat pancreas. Proc. Soc. Exp. Biol. Med. 146:325.
30. Swanson, C. H., and Solomon, A. K. (1973). A micropuncture investigation of the whole tissue mechanism of electrolyte secretion by the *in vitro* rabbit pancreas. J. Gen. Physiol. 62:407.
31. Reber, H. A., Lightwood, R., and Zakula, S. (1975). Micropuncture study of electrolyte secretion in the cat pancreas. Gastroenterology 68:A115/972.
32. Reber, H. A., Wolf, C. J., and Lee, S. P. (1969). Role of the main duct in pancreatic electrolyte secretion. Surg. Forum 20:382.
33. Case, R. M., Harper, A. A., and Scratcherd, T. (1969). The secretion of electrolytes and enzymes by the pancreas of the anaesthetized cat. J. Physiol. (Lond.) 201:177.
34. Moqtaderif Himal, H. S., Rudick, J., and Dreiling, D. A. (1972). Pancreatic transductal electrolyte flux. Am. J. Gastroenterol. 58:177.
35. Greenwell, J. R. (1975). The effects of cholecystokinin pancreozymin, acetylcholine and secretin on the membrane potential of mouse pancreatic cell *in vitro*. Pflügers Arch. 353:159.
36. Treffot, M. J., Tiscornia, O., Palasciano, G., Hage, G., and Sarles, H. (1975). Chronic alcoholism and endogenous gastrin. Am. J. Gastroenterol. 63:29.
37. Palasciano, G., Tiscornia, O., Hage, G., Sarles, H., Devaux, M. A., Michel, G., and Grimaud, R. (1974). Chronic alcoholism and endogenous CCK PZ. Biomedicine 21:94.
38. Bretholz, A., Levesque, D., Voirol, M., Laugier, R., Tiscornia, O., Sarles, H., and Blooms, S. Secretin release in chronic alcoholic dogs. To be published.
39. Sarles, H., and Tiscornia, O. (1974). Ethanol and chronic calcifying pancreatitis. Med. Clin. North Amer. 58:1333.
40. Tiscornia, O., Palasiano, G., Dzieniszewski, J., Verine, H., and Sarles, H. (1975). Analysis of the mechanism of action of calcium-induced exocrine pancreatic secretory changes in the dog. Am. J. Gastroenterol. 63:293.
41. Goebell, H., Steffen, C., and Bode, C. (1972). Stimulatory effect of pancreozymin–cholecystokinin on calcium secretion in pancreatic juice of dogs. Gut 13:477.
42. Argent, B. E., Case, R. M., and Scratcherd, T. (1973). Amylase secretion by the perfused cat pancreas in relation to the secretion of calcium and other electrolytes and as influenced by the external ionic environment. J. Physiol. (Lond.) 230:575.
43. Ceccarelli, R., Clemente, F., and Meldolesi, J. (1975). Secretion of calcium in pancreatic juice. J. Physiol. (Lond.) 245:617.
44. Schreurs, V. V. A. M., Swarts, H. G. P., De Pont, J. J. H. M., and Bonting, S. L. (1975). Role of calcium in exocrine pancreatic secretion. I. Calcium movements in the rabbit pancreas. Biochim. Biophys. Acta 404:257.
45. Goebell, H., Baltzer, G., Shlott, K. A., and Bode, C. H. (1973). Parallel secretion of calcium and enzymes by the human pancreas. Digestion 8:336.
46. Hotz, J. (1975). Beziehungen zwischen calcium homöostase U. Sekretion von Magen U. Bauchspeichel drüse, Habilitations Shift zur Erlangung der venia legendi für das Fach Innere Medizin der Hohen Klinisch-Medizinischen Fakultät. Germany, Universität Ulm.
47. Nakajimu, S. (1973). The action of the C-terminal octapeptide of cholecystokinin and related peptides on pancreatic exocrine secretion. Gut 14:607.
48. Scratcherd, T. (1974). The action of some inhibitors and stimulators of phosphodi-

esterase and adenylate cyclase on electrolyte and enzyme secretion by the pancreas. *In* N. A. Thorn and O. H. Petersen (eds.), Secretory Mechanisms of Exocrine Glands, pp. 379–388. Munksgaard, Copenhagen.
49. Gullo, L., Sarles, H., Mott, C. De Barros, Tiscornia, O., Pauli, A. M., and Pastor, J. (1974). Pancreatic secretion of calcium in healthy subjects and various diseases of the pancreas. Rendic. Gastroenterol. 6:35.
50. Sullivan, J. F., Burch, R. E., and Magee, D. F. (1974). Enzymatic activity and divalent cation content of pancreatic juice. Am. J. Physiol. 226:1420.
51. Case, R. M., and Clausen, T. (1973). The relationship between calcium exchange and enzyme secretion in the isolated rat pancreas. J. Physiol. (Lond.) 235:75.
52. Danielson, A., and Sehlin, J. (1974). Transport and oxidation of amino acids and glucose in the isolated exocrine mouse pancreas: effects of insuline and pancreozymin. Acta Physiol. Scand. 91:557.
53. Robberecht, P., Deschodt-Lanckman, M., Camus, J., Kutzner, R., and Christophe, J. (1974). Amino acid levels in rat pancreas after pilocarpine and pancreozymin. Am. J. Physiol. 224:1309.
54. Palade, G. E., Siekevitz, P., and Caro, L. G. (1962). Structure, chemistry and function of the pancreatic exocrine cell. *In* C. A. V. S. de Reuck and M. P. Cameron (eds.), The Exocrine Pancreas, CIBA Foundation Symposium, p. 23. J. and A. Churchill, London.
55. Caro, L. G., and Palade, G. E. (1964). Protein syntheses, storage, and discharge in the exocrine cell: an autoradiographic study. J. Cell Biol. 20:473.
56. Jamieson, J. D., and Palade, G. E. (1971). Synthesis, intracellular transport, and discharge of secretory proteins in stimulated pancreatic exocrine cells. J. Cell Biol. 50:135.
57. Slot, J. N., Geuze, J. J., and Poort, C. (1974). Synthesis and intracellular transport of protein in the exocrine pancreas of the frog (*Rana esculenta*) I. An ultrastructural and autoradiographic study. Cell Tissue Res. 155:135.
58. Kassel, B., and Kay, J. (1973). Zymogens of proteolytic enzymes. These enzyme precursors, formerly thought to be inert substances, have inherent proteolytic activity. Science 180:1022.
59. Bauduin, H., Tondeur, T., Vansande, J., and Vincent, D. (1973). Secretion and protein metabolism in the rat pancreas *in vitro*. Biochim. Biophys. Acta 304:88.
60. Danielson, A. (1976). Effects of glucose, secretin and glucagon on amylase secretion from incubated mouse pancreas. Plügers Arch. In press.
61. Irwin, C., and Tenenhouse, A. (1974). Effect of carbachol on amino acid incorporation into protein of rat pancreas *in vitro*. Can. J. Physiol. Pharmacol. 52:632.
62. Hokin, L. E. E., and Hokin, M. R. (1962). The synthesis and secretion of digestive enzymes by pancreas tissue *in vitro*. *In* C. A. V. S. De Reuck and M. P. Cameron (eds.), The Exocrine Pancreas, CIBA Foundation Symposium, p. 186. J. and A. Churchill, London.
63. Dickman, S. R., Holtzer, R. L. and Gazzinelli, G. (1962). Protein systhesis by beef pancreas slices. Biochemistry 1:574.
64. Amsterdam, A., and Jamieson, J. D. (1972). Structural and functional characterization of isolated pancreatic exocrine cells. Proc. Natl. Acad. Sci. USA 69:3028.
65. Jamieson, J. D., and Palade, G. E. (1968). Intracellular transport of secretory proteins in the pancreatic exocrine cell III. Dissociation of intracellular transport from protein synthesis. J. Cell Biol. 39:580.
66. Webster, P. D., Black, O., and Morisset, J. (1974). Effect of bethanechol chloride on protein synthesis in rat pancreas. Am. J. Dig. Dis. 19:167.
67. Webster, P. D., and Tyor, M. P. (1966). Effect of intravenous pancreozymin on amino acid incorporation *in vitro* by pancreatic tissue. Am. J. Physiol. 211:157.
68. Dragon, N., Beaudoin, A. R., and Morisset, J. (1975). Effect of tetracaine on pancreatic protein synthesis. Proc. Soc. Exp. Biol. Med. 149:278.
69. Mongeau, R., Couture, Y., Dunnigan, J., and Morisset, J. (1974). Early dissociation of protein synthesis and amylase secretion following hormonal stimulation of the pancreas. Can. J. Physiol. Pharmacol. 52:198.

70. Mongeau, R., Dagorn, J. C., and Morisset, J. (1975). Baisse précoce de la synthèse protéique après stimulation hormonale *in vivo*. Biol. Gastroenterol. (Paris) 8:149.
71. Reggio, H., and Cailla, H. L. (1974). Effect of actinomycine D, pancreozymin and secretin on RNA synthesis and protein synthesis measured *in vivo* on rat pancreas. Biophys. Biochim. Acta 378:37.
72. Kern, H. F., and Bieger, W. (1975). Functional adaptation of rat exocrine pancreas to prolonged stimulation by caerulein *in vivo*. Digestion 13:113.
73. Morisset, J. B., and Webster, P. D. (1972). Effects of fasting and feeding on protein synthesis by the rat pancreas. J. Clin. Invest. 51:1.
74. Black, O., and Webster, P. D. (1974). Nutritional and hormonal effect on RNA polymerase enzyme activities in pancreas. Am. J. Physiol. 227:1276.
75. Rajotte, D., Dagorn, J. C., and Morisset, J. (1975). Augmentation de la synthèse des ARNs pancréatiques chez le rat après stimulation. Biol. Gastroenterol. (Paris) 8:149.
76. Case, R. M. (1973). Cellular mechanisms controlling pancreatic exocrine secretion. Acta Hepatogastroenterol. 20:435.
77. Morisset, J. A., Black, O., and Webster, P. D. (1972). Effects of fasting, feeding and bethanechol chloride on pancreatic microsomal protein synthesis *in vivo*. Proc. Soc. Exp. Biol. Med. 140:1308.
78. Black, J. O., and Webster, P. D. (1973). Protein synthesis in pancreas of fasted pigeons. J. Cell Biol. 57:1.
79. Morisset, J. A., Black, O., and Webster, P. D. (1972). Changes with fasting in pigeon pancreas alkaline and acid ribonuclease. Proc. Soc. Exp. Biol. Med. 139:562.
80. Venroij, W. J., Poort, C., Kramer, M. F., and Jansen, M. I. (1972). Relationship between extracellular amino acids and protein synthesis *in vitro* in the rat pancreas. Eur. J. Biochem. 30:427.
81. Cheneval, J. P., and Johnstone, R. M. (1974). Transport of amino acid in rat pancreas during development. Biochim. Biophys. Acta 345:17.
82. Tartakoff, A. M., Greene, L. J., Jamieson, J. D., and Palade, G. E. (1974). Parallelism in the processing of pancreatic protein. *In* B. Ceccarelli, F. Clemente, and J. Meldolesi (eds.), Advance in Cytopharmacology, Vol. 2, p. 177. Raven Press, New York.
83. Kramer, M. F., and Poort, C. (1972). Unstimulated secretion of protein from rat exocrine pancreas cells. J. Cell Biol. 52:147.
84. Singh, M. (1974). Effect of chronic administration of cholecystokinin-pancreozymin (CCK-PZ) on pancreatic macromolecular transport. Gastroenterology 66:A 223/877.
85. Singh, M., Black, O., and Webster, P. D. (1973). Effects of selected drugs on pancreatic macromolecular transport. Gastroenterology 64:983.
86. Singh, M., and Webster, P. D. (1975). Effect of hormones on pancreatic macromolecular transport. Gastroenterology 68:1536.
87. Lin, K. T. (1973). Ph.D. thesis, Department of Biochemistry. McGill University, Montreal, Quebec.
88. Bieger, W., Bassler, M., Martin-Acharo, A., and Kern, H. F. (1975). Functional adaptation of rat exocrine pancreas to prolonged stimulation by caerulein *in vivo*. European Pancreatic Club, VIIth Symposium, Toulouse, October 23–25, Abstract p. 67.
89. Ronzio, R. A. (1973). Glycoprotein synthesis in the adult rat pancreas. II. Characterization of Golgi rich fractions. Arch. Biochem. Biophys. 159:777.
90. Morre, D. J., Mollenhaver, H. A., and Bracker, C. E. (1971). *In* J. Reinert and H. Unsprung (eds.), Results and Problems in Cell Differentiation. Vol. 2, pp. 82–126. Springer-Verlag, New York.
91. Berg, N. B., and Young, R. W. (1971). Sulfate metabolism in pancreatic acinar cells. J. Cell Biol. 50:464.
92. Tartakoff, A., Greene, L. J., and Palade, E. (1974). Studies on the guinea pig pancreas. Fractionation and partial characterization of exocrine proteins. J. Biol. Chem. 249:7420.
93. Jamieson, T. D., and Palade, G. C. (1971). Condensing vacuole conversion and zymogen granule discharge in pancreatic exocrine cells: metabolic studies. J. Cell Biol. 48:503.

94. Hokin, L. E. (1955). Isolation of the zymogen granules of the dog pancreas and a study of their properties. Biochim. Biophys. Acta 18:379.
95. Rothman, S. S. (1971). The behaviour of isolated zymogen granules: pH-dependent release and reassociation of protein. Biochim. Gastroenterol. (Paris) 6:97.
96. Debray, C., Vaille, C., Latour, J., de Roze, C., Charlot, J., and Fox, A. (1973). Action de la caeruleine sur la secrétion pancréatique externe du rat, cinétique et relation dose–effet. Biol. Gastroenterol. (Paris) 6:97.
97. Vandermeers Piret, M. L., Vandermeers, A., Rathe, J., Christophe, J., Vanderhoeden, R., Wettendorf, P., and Delcourt, A. (1974). A comparison of enzyme activities in duodenal aspirates following injection of secretin, caerulein and cholecystokinin. Digestion 10:191.
98. Christophe, J., Camus, J., Deshodt Lankman, M., Rathe, J., Robberecht, P., Vandermeers Piret, M. L., and Vandermeers, A. (1971). Factors regulating biosynthesis, intracellular transport and secretion of amylase and lipase in the rat exocrine pancreas. Horm. Metab. Res. 3:393.
99. Goldberg, D. M., Sale, J. K., and Wormsley, K. G. (1973). Ratio of chymotrypsin to trypsin in human duodenal aspirates. Digestion 8:101.
100. Christophe, J., Vandermeers, A., Vandermeers Piret, M. C., Rathe, J., and Camus, J. (1973). The relative turnover time *in vivo* of the intracellular transport of five hydrolases in the pancreas of the rat. Biochim. Biophys. Acta 308:285.
101. Schmidt, H. P., Goebell, H., and Johannson, F. (1972). Pancreatic and gastric secretion in rats studied by means of duodenal and gastric perfusion. Scand. J. Gastroenterol. 7:47.
102. Kraehenbuhl, J. P., and Jamieson, J. D. (1972). Solid-phase conjugation of ferritin to Fab-fragments of immunoglobulin G for use in antigen localization on thin section. Proc. Natl. Acad. Sci. (USA) 69:1771.
103. Rothman, S. S. (1974). Molecular regulation of digestion: short term and bond specific. Am. J. Physiol. 226:77.
104. Walther, A. A. (1910). (Quoted in J. P. Pavlov.) The Work of the Digestive Glands. Charles Griffin and Co., London.
105. Rothman, S. S., and Isenman, L. D. (1974). Secretion of digestive enzymes derived from two parallel intracellular pools. Am. J. Physiol. 226:1082.
106. Rothman, S. S. (1975). Enzyme secretion in the absence of zymogen granules. Am. J. Physiol. 228:1828.
107. Dandrifosse, G., and Sinar, L. (1975). Transport of amylase across the apical membrane of the pancreatic exocrine cells. Quantitative analysis of zymogen granules. Pflügers Arch. 357:361.
108. Liebow, C., and Rothman, S. S. (1974). Transport of bovine chymotrypsinogen into rabbit pancreatic cells. Am. J. Physiol. 226:1077.
109. Liebow, C., and Rothman, S. S. (1975). Entero-pancreatic circulation of digestive enzymes. Science 189:472.
110. Rutten, W. J., De Pont, J. J. H. H. N., and Bonting, S. L. (1972). Adenylate cyclase in the rat pancreas: properties and stimulation by hormones. Biochim. Biophys. Acta 274:201.
111. Lemon, M. J. C., and Bhoola, K. D. (1975). Excitation-secretion coupling in exocrine gland. Properties of cyclic AMP (phosphodiesterase and adenylate cyclase from the submaxillary gland and pancreas). Biochim. Biophys. Acta 385:101.
112. Marois, C., Morisset, J., and Dunnigan, J. (1972). Presence and stimulation of adenyl cyclase in pancreas homogenate. Rev. Can. Biol. 31:253.
113. Rutten, W. J., Schoot, B. M. C., Depont, J. J. H. H. M., and Bonting, S. L. (1973). Adenosine 3′-5′-monophosphate diesterase in rat pancreas. Biochim. Biophys. Acta 315:384.
114. Robberecht, P., Deschodt-Lanickman, M., Deneef, P., and Christophe, J. (1974). Hydrolysis of the cyclic 3′–5′-monophosphate of adenosine and guanosine by rat pancreas. Eur. J. Biochem. 41:585.
115. Christophe, J., Robberecht, P., Deschodt-Lanckman, M., Lambert, M., Vanleemput-Coutrez, M., and Camus, J. (1974). Molecular basis of enzyme secretion by the

exocrine pancreas. *In* B. Ceccarelli, F. Celement, and J. Meldolesi (eds.), Advances in Cytopharmacology, Vol. 2, pp. 47–61. Raven Press, New York.

116. Cenatiempo, Y., Mangeat, P., and Marchis-Mouren, G. (1975). Purification and properties of cyclic AMP dependent and independent protein kinases from rat pancreas. Biochimie 57:865.
117. Case, R. M., Johnson, M., Scratcherd, T., and Sherratt, H. S. A. (1972). Cyclic adenosine 3′-5′-monophosphate concentration in the pancreas following stimulation by secretin, cholecystokinin-pancreozymin and acetylcholine. J. Physiol. (Lond.) 233:669.
118. Bonting, S. L., Case, R. M., Kempen, H. J. M., Depont, J. J. H. H. N., and Scratcherd, T. (1974). Further evidence that secretin stimulates pancreatic secretion through adenosine 3′-5′-monophosphate (cyclic AMP). J. Physiol. (Lond.) 240:34.
119. Robberecht, P., Deschodt-Lanckman, M., Deneef, Ph., and Christophe, J. (1974). Interaction des hormones gastro-intestinales au niveau des cellules pancréatiques chez le rat. Arch. Int. Physiol. Biochim. 82:196.
120. Deschodt-Lanckman, M., Robberecht, P., Deneef, Ph., and Christophe, J. (1974). Teneur en AMP cyclique et GMP cyclique du pancreas de rat stimuli *in vivo* et *in vitro* par des hormones gastro-intestinales. Arch. Int. Physiol. Biochim. 82:180.
121. Domschke, S., Konturek, S. J., Domschke, W., Dembinski, A., Thor, P., Krol, R., and Demling, L. (1975). Bicarbonate and cyclic AMP–secretion in canine pancreas. Digestion 13:108.
122. Case, R. M., and Scratcherd, T. (1972). The actions of dibutyryl cyclic adenosine 3′-5′-monophosphate and methyl xanthines on pancreatic secretion. J. Physiol. (Lond.) 223:649.
123. Kempen, H. J. M., Depont, J. J. H. H. M., and Bonting, S. L. (1975). Rat pancreas adenylate cyclase. III. Its role in pancreatic secretion assessed by means of cholera toxine. Biochim. Biophys. Acta 392:276.
124. Smith, P. A., and Case, M. (1975). Effects of cholera toxin on cyclic adenosine 3′-5′-monophosphate concentration and secretory processes in the exocrine pancreas. Biochim. Biophys. Acta 399:277.
125. Porter, G. M. L., and Scratcherd, T. (1973). The effect of nicotinic acid and alloxan on the secretion of electrolyte and amylase by the isolated saline perfused pancreas of the cat. J. Physiol. (Lond.) 232:33.
126. Case, R. M. (1973). Calcium and gastrointestinal secretion. Digestion 8:269.
127. Heisler, S., Grondin, G., and Forget, G. (1974). The effect of various secretagogues on accumulation of cyclic AMP and secretion of α-amylase from rat exocrine pancreas. Life Sci. 14:631.
128. Petersen, O. H., and Ueda, N. (1975). Pancreatic acinar cells: effect of acetylcholine, pancreozymin, gastrin and secretin on membrane potential and resistance *in vivo* and *in vitro*. J. Physiol. (Lond.) 247:461.
129. Williams, J. A. (1974). Intracellular control mechanism resulating secretion by exocrine and endocrine glands. *In* N. A. Thorn and O. H. Petersen (eds.), Secretory Mechanism of Exocrine Glands, pp. 389–407. Munksgaard, Copenhagen.
130. Rutten, W. J., Depont, J. J. H. H. M., Bonting, S. L., and Daemen, F. J. M. (1975). Lyophospholipid in pig pancreatic zymogen granules in relation to exocytosis. Eur. J. Biochem. 54:259.
131. Albano, J., Harvey, R. F., and Bhoola, K. D. (1975). Studies on pancreozymin and acetylcholine mediated enzyme secretion and cyclic GMP in the pancreas. Digestion 13:104.
132. Harvey, R. F., Albano, J., Bhoola, K. D., and Read, A. E. (1975). Effects of cholecystokinin, secretin and acetylcholine on pancreatic enzyme secretion and tissue levels of cyclic GMP and cyclic AMP. British Society of Gastroenterology Handbook, p. 43.
133. Robberecht, P., Deschodt-Lanckman, M., Deneef, P., Borgeat, P., and Christophe, J. (1974). *In vivo* effect of pancreozymin, secretin, vasoactive intestinal polypeptide and pilocarpine on the levels of cyclic AMP and cyclic GMP in the rat pancreas. FEBS Lett. 43:135.

134. Heisler, S., and Grondin, G. (1975). Absence of effects of dibutyryl cyclic guanosine 3′5′-monophosphate on release of α-amylase ^{45}Ca efflux and protein synthesis in rat pancreas *in vitro*. Experientia 31:936.
135. MacDonald, R. J., and Ronzio, R. A. (1974). Phosphorylation of a zymogen granule membrane polypeptide from rat pancreas. FEBS Lett. 40:203.
136. Lambert, M., Camus, J., and Christophe, J. (1973). Phosphorylation des protéines dans le pancréas de rat: stimulation par la caeruléine. Arch. Int. Physiol. Biochim. 81:381.
137. Lambert, M., Camus, J., and Christophe, J. (1973). Pancreozymin and caerulein stimulate *in vitro* protein phosphorylation in the rat pancreas. Biochem. Biophys. Res. Commun. 52:935.
138. Bonting, S. L., and DePont, J. J. H. H. M. (1974). Adenylate cyclase and phophodiesterase in rat pancreas. *In* N. A. Thorn and O. H. Petersen (eds.), Secretory Mechanism of Exocrine Glands, pp. 363–378. Munksgaard, Copenhagen.
139. Robberecht, P. (1976). Role of cyclic nucleotides in pancreatic enzymes and electrolyte secretion. *In* R. M. Case and H. Goebell (eds.), Stimulus Secretion Coupling in the Gastrointestinal Tract, pp. 203–226. MTP Press Ltd., Lancaster, England.
140. Beaudoin, A. R., Marois, C., Dunnigan, J., and Morisset, J. (1974). Biochemical reaction involved in pancreatic enzyme secretion. I. Activation of the adenylate *cyclase* complex. Can. J. Physiol. Pharmacol. 52:174.
141. Kanno, J. (1974). Relation between amylase release and change in electrophysiological and morphological properties of cells in the exocrine pancreas. *In* N. A. Thorn and O. H. Petersen (eds.), Secretory Mechanism of Exocrine Glands, pp. 278–301. Munskgaard, Copenhagen.
142. Kanno, T. (1975). The electrogenic sodium pump in the hyperpolarizing and secretory effects of pancreozymin in the pancreatic acinar cells. J. Physiol. (Lond.) 245:599.
143. Dean, S. M., and Matthews, E. K. (1972). Pancreatic acinar cells: measurement of membrane potential and miniature depolarization potential. J. Physiol. (Lond.) 225:1.
144. Matthews, E. K., Petersen, O. H., and Williams, J. A. (1973). Pancreatic acinar cells: acetylcholine-induced membrane depolarization, calcium efflux and amylase release. J. Physiol. (Lond.) 234:689.
145. Nishiyama, A., and Petersen, O. H. (1974). Membrane potential and conductance measurements in mouse and rat pancreatic acinar cells. *In* N. A. Thorn and O. H. Petersen (eds.), Secretory Mechanism of Exocrine Gland, pp. 267–277. Munksgaard, Copenhagen.
146. Argent, B. E., Case, R. M., and Scratcherd, T. (1971). Stimulation of amylase secretion from the perfused cat pancreas by potassium and other alkaline metal ions. J. Physiol. (Lond.) 216:611.
147. Benz, L., Eckstein, B., Matthews, E. K., and Williams, J. A. (1972). Control of pancreatic amylase release *in vitro*: effects of ions, cyclic AMP and colchicine. Br. J. Pharmacol. 46:66.
148. Matthews, E. K., and Petersen, O. H. (1973). Pancreatic acinar cell: ionic dependence on the membrane potential and acetylcholine induced depolarization. J. Physiol. (Lond.) 234:283.
149. Petersen, O. H. (1971). Initiation of salt and water transport in mammalian salivary glands by acetylcholine. Philos. Trans. R. Soc. Lond. (Biol.) 262:307.
150. Petersen, O. H. (1974). Electrogenic sodium pump in exocrine gland cells. *In* N. A. Thorn and O. H. Petersen (eds.), Secretory Mechanism of Exocrine Glands, pp. 474–486. Munksgaard, Copenhagen.
151. Case, R. M. (1974). The role of calcium and cyclic AMP in pancreatic secretory process. *In* N. A. Thorn and O. H. Petersen (eds.), Secretory Mechanisms of Exocrine Glands, pp. 344–362. Munksgaard, Copenhagen.
152. Goebell, H., Hotz, J., and Ziegler, R. (1976). The secretion of the human exocrine pancreas as related to acute alterations in calcium homeostasis. *In* R. M. Case and H. Goebell (eds.), Stimulus Secretion Coupling in the Gastrointestinal Tract, pp. 255–268. MTP Press Ltd., Lancaster, England.
153. Goebell, H., Steffen, C. H., Baltzer, C., Schlott, K. A., and Bode, C. H. (1972).

Stimulierung der Enzymsekretion in pancreas durch acute Hypercalciämie. Dtsch. Med. Wochenschr. 97:300.
154. Goebell, H., Steffen, C. H., Baltzer, C., and Bode, C. H. (1973). Stimulation of pancreatic secretion of enzyme by acute hypercalcaemia in man. Eur. J. Clin. Invest. 3:98.
155. Malagelada, J. R., Holter-Muller, K. H., Sizemore, G. W., and Go, V. L. W. (1974). Potentiation of cholecystokinin–pancreozymin (CCK–PZ) action by hypercalcaemia in man. Gastroenterology 66:A84.
156. Hotz, J., Minne, H., and Ziegler, R. (1971). Der Einfluss der akuten Hyper- und Hypocalciämie auf die basale und stimulierte Pankreas sekretion des Menschen. 26 Tagg. Dtsch. Ges. Verdaungs und stoffwechselkrankh. Stuttgart, Abstrakt 41.
157. Tiscornia, O. M., Levesque, D., Voirol, M., Bretholz, A., and Sarles, H. (1975). The effects of intraduodenal calcium infusion on canine exocrine pancreatic secretion. Digestion 13:126.
158. Schulz, I., Kondo, S., Milutinovic, S., and Sachs, G. (1975). Ca^{++} coupling in the exocrine pancreas. Digestion 13:125.
159. Hokin, L. E. (1968). Effect of calcium omission on acetylcholine-stimulated amylase secretion and phospholipid synthesis in pigeons pancreas slices. Biochim. Biophys. Acta 115:219.
160. Robberecht, P., and Christophe, J. (1971). Secretion of hydrolase by perfused fragments of rat pancreas: effect of calcium. Am. J. Physiol. 220:911.
161. Williams, J. A., and Lee, M. (1974). Pancreatic acinar cells: use of a Ca^{++} ionophore to separate enzyme release from the earlier steps in stimulus–secretion coupling. Biochim. Biophys. Res. Commun. 60:542.
162. Selinger, Z., Eimerl, S., Savion, N., and Schramm, M. (1974). A Ca^{++} ionophore A 23187 stimulating hormone and neurotransmitter action in the rat parotid and pancreas glands. *In* N. A. Thorn and O. H. Petersen (eds.), Secretory Mechanisms of Exocrine Glands, pp. 68–87. Munskgaard, Copenhagen.
163. Heisler, S. (1974). Calcium efflux and secretion of α-amylase from rat pancreas. Br. J. Pharmacol. 52:387.
164. Gardner, J. A., Conlon, J. P., Klaeveman, H. L., Adams, T. D., and Ondetti, M. A. (1975). Action of cholecystokinin and cholinergic agents on calcium transport in isolated pancreatic acinar cells. J. Clin. Invest. 56:366.
165. Petersen, O. H. (1974). General discussion. *In* N. A. Thorn and O. H. Petersen (eds.), Secretory Mechanisms of Exocrine Glands, p. 401. Munskgaard, Copenhagen.
166. Clemente, F., and Meldolesi, J. (1975). Calcium and pancreatic secretion. 1. Subcellular distribution of calcium and magnesium in the exocrine pancreas of the guinea pig. J. Cell Biol. 65:88.
167. Meldolesi, J., and Ramellini, G. (1975). Calcium binding to cellular membranes of the exocrine pancreas of the guinea pig. Digestion 13:118.
168. Argent, B. E., Smith, R. K., and Case, R. M. (1975). The role of calcium in pancreatic enzyme and electrolyte secretion. Digestion 13:237.
169. Watson, E. L., Siegel, I. A., and Robinovitch, M. R. (1974). Ca^{++} ATPase activity in isolated secretory granule membranes. Experientia 30:876.
170. Dean, P. M. (1974). The electrokinetic properties of isolated secretory particles. *In* N. A. Thorn and O. H. petersen (eds.), Secretory Mechanisms of Exocrine Glands, pp. 152–167. Munksgaard, Copenhagen.
171. Marchand, C., Morisset, J., and Beaudoin, A. R. (1975). Les lipides et la sécrétion exocrine du pancréas. Biol. Gastroenterol. (Paris) 8:150.
172. Lapetina, E. G., and Michell, R. A. (1973). Phosphatidyl inositol metabolism in cells receiving extracellular stimulation. FEBS Lett. 31:1.
173. Hokin, M. R. (1974). Breakdown of phosphatidyl inositol in the pancreas in response to pancreozymin and acetylcholine. *In* N. A. Thorn and O. H. Petersen (eds.), Secretory Mechanisms of Exocrine Glands, pp. 101–115. Munksgaard, Copenhagen.
174. Gerber, D., Davies, M., and Hokin, L. (1973). The effect of secretagogues on the incorporation of [2-^{3}H]myoinositol into lipid in cytological fractions in the pancreas of the guinea pig *in vivo*. J. Cell Biol. 56:736.
175. Calderon, P., Fornelle, J., Winand, J., and Christophe, J. (1973). Marquage *in vitro*

des lipides du pancreas exocrine de rat par l'acétate radioactif. Arch. Int. Physiol. Biochim. 81:363.
176. De Camilli, P., and Meldolesi, J. (1975). Subcellular distribution of the PI effect in the pancreas of the guinea pig. Life Sci. 15:711.
177. Bauduin, H., and Cantraine, F. (1972). Phospholipid effect and secretion in the rat pancreas. Biochim. Biophys. Acta 270:249.
178. Morre, D. J., Keenan, J. W., and Huang, G. N. (1974). Membrane flow and differentiation: origin of Golgi apparatus membranes from endoplasmic reticulum. *In* B. Ceccarelli, F. Clementi, and J. Meldolesi (eds.), Advances in Cytopharmacology, Vol. 2, pp. 107–125. Raven Press, New York.
179. Geuze, J. J., and Poort, C. (1973). Cell membrane resorption in the rat exocrine pancreas cell after *in vivo* stimulation of the secretion, as studied by *in vitro* circulation with extracellular space markers. J. Cell Biol. 57:169.
180. Kramer, M. F., and Geuze, J. J. (1974). Redundant cell membrane regulation in the exocrine pancreas cells after pilocarpine stimulation of the secretion. *In* B. Ceccarelli, F. Clemente, and J. Meldolesi (eds.), Advances in Cytopharmacology, Vol. 2, pp. 87–97. Raven Press, New York.
181. Meldolesi, J., De Camilli, P., and Peluchetti, D. (1974). The membrane of secretory granules: structure, composition and turnover. *In* N. A. Thorn and O. H. Petersen (eds.), Secretory Mechanisms of Exocrine Glands, pp. 137–151. Munskgaard, Copenhagen.
182. Meldolesi, J. (1974). Secretory mechanisms in pancreatic acinar cells. Role of the cytoplasmic membrane. *In* B. Ceccarelli, F. Clemente, and J. Meldolesi (eds.), Advances in Cytopharmacology, Vol. 2, pp. 71–85. Raven Press, New York.
183. Meldolesi, J. (1974). Dynamics of cytoplasmic membranes in guinea pig pancreatic acinar cells. I. Synthesis and turnover of membrane proteins. J. Cell Biol. 61:1.
184. Stock, C., Vincent, D., and Bauduin, H. (1975). The involment of the microfilamentous system in the exocytosis of the pancreatic zymogen granules. Digestion 13:125.
185. Kern, H. F., Seybold, J., and Bieger, W. (1976). Inhibition of secretory processes in the rat exocrine pancreatic cells by microtubule inhibitors. *In* R. M. Case and H. Goebell (eds.), Stimulus Secretion Coupling in the Gastrointestinal Tract, pp. 79–84. MTP Press Ltd., Lancaster, England.
186. Beaudoin, A. R., Dunnigan, J., and Morisset, J. (1975). Les microfilaments, sont-ils impliqués dans la sécretion des enzymes du pancréas? Biol. Gastroenterol. (Paris) 8:156.
187. Nevalainen, T. J. (1975). Inhibition of pancreatic exocrine secretion by vinblastine. Res. Exp. Med. (Berl.) 165:163.
188. Konturek, S. J. (1974). Non-humoral control of pancreatic secretion. Acta Gastroenterol. Belg. 37:273.
189. Hubel, K. A. (1972). Secretin: a long progress note. Gastroenterology 62:318.
190. Harper, A. A. (1972). The control of pancreatic secretion. Gut 13:308.
191. Henriksen, F. W. (1974). Studies on the external pancreatic secretion. Scand. J. Gastroenterol. (Suppl.) 9:26.
192. Debray, C., Latour, J., Roze, C., and Souchard, M. (1972). Action de l'acetylcholine sur la sécretion pancreatique externe du rat: cinétique et relation dose–effet. Biol. Gastroenterol. (Paris) 5:61.
193. Hotz, J., Zwicker, M., Minne, H., and Ziegler, R. (1975). Pancreatic enzyme secretion in the conscious rat. Method and application. Pflügers Arch. 353:171.
194. Solomon, T. E., Solomon, N., Shanbour, L. L., and Jacobson, E. D. (1973). Direct effect of cholinergic stimulation on pancreatic secretion. Gastroenterology 64:A121.
195. Konturek, S. J., Tasler, J., and Obtulowicz, W. (1973). Effect of caerulein and endogenous cholecystokinin on urecholine-induced gastric and pancreatic secretion in dogs. Gastroenterology 65:235.
196. Lenninger, S., and Ohlin, P. (1971). The flow of juice from the pancreatic gland of the rat in response to vagal stimulation. J. Physiol. (Lond.) 216:303.
197. Kaminski, D. L., Ruwart, M. J., and Willman, V. L. (1975). The effect of electrical vagal stimulation on canine pancreatic exocrine function. Surgery 77:545.
198. Bourde, J., Lawrence, A., Robinson, B. A., Suda, Y., and White, T. T. (1970). Vagal

stimulation. II. Its effects on pancreatic secretion in conscious dogs. Ann. Surg. 171:357.

199. Rosemberg, I. R., Zambrano, V. J., Janowitz, H. D., and Rudick, J. (1973). Parasympathetic innervation and pancreatic secretion. The role of gastric antra. Gastroenterology 64:869.
200. Sugawara, I., Yanagisawa, J., and Eisenberg, M. M. (1973). Sustained vagal stimulation of the exocrine pancreas. Gastroenterology 64:807.
201. Debas, H. T., Konturek, S. J., and Grossman, M. I. (1972). Effect of extragastric and truncal vagotomy on pancreatic secretion in dog. Am. J. Physiol. 228:1172.
202. Wolfert, W., Hartmann, W., and Hotz, J. (1974). Exocrine pancreatic secretion and output of bile in humans under conditions of hypoglycemia and administration of atropine. Digestion 10:9.
203. Davis, M., Gupta, S., and Elder, J. B. (1973). The effect of vagotomy on the exocrine pancreatic secretory response to pentagastrin and to a 400 g meat meal in dogs. Br. J. Surg. 60:318.
204. Hickson, J. C. D. (1970). The secretion of pancreatic juice in response to stimulation of the vagus nerve in the pig. J. Physiol. (Lond.) 206:275.
205. Tiscornia, O. M., Brasca, A. F., Hage, G., Palasciano, G., Devaux, M. A., and Sarles, H. (1972). Les effets de l'atropine, du penthonium, de la vagotomie et du fluothane sur la secretion pancreatique du chien. Biol. Gastroenterol. (Paris) 5:249.
206. Dembinski, A., Konturek, S. J., and Thor, P. (1974). The role of vagal innervation in gastric and pancreatic responses to meals varying in pH. J. Physiol. (Lond.) 241:677.
207. Konturek, S., Radecki, T., Pawlik, W., Thor, P., and Biernat, J. (1974). The significance of vagus nerves in the regulation of pancreatic secretion. Acta Physiol. Pol. 25:423.
208. Konturek, S. J., Becker, H. D., and Thompson, J. C. (1974). Effect of vagotomy on hormones stimulating pancreatic secretion. Arch. Surg. 108:704.
208a. Beesley, W. H., Orbhood, R., Dutta, P., and Eisenberg, M. M. (1972). The role of vagal release of gastrin in pancreatic enzyme secretion. Br. J. Surg. 59:912.
209. Malagelada, J. R., Go, V. L. A., Magno, O. P., and Summerskill, W. J. H. (1973). Interactions between intraluminal bile acids and digestive products on pancreatic and gall bladder function. J. Clin. Invest. 52:2160.
210. Long, W. B. (1974). Effect of vagotomy on pancreatic secretion. Gastroenterology 67:768.
211. Thambugala, R. L., and Baron, J. H. (1971). Pancreatic secretion after selective and truncal vagotomy in the dog. Br. J. Surg. 58:839.
212. Thomas, J. E. (1967). Neural regulation of pancreatic secretion. *In* C. F. Code (ed.), Handbook of Physiology, Section 6, Vol. 11, Alimentary Canal, p. 955. Am. Physiol. Soc., Washington D.C.
213. Tiscornia, O. M., Palasciano, G., and Sarles, H. (1974). Pancreatic changes induced by chronic (two years) ethanol treatment in the dog. Gut 15:839.
214. Grossman, M. I., (1962). Nervous and hormonal regulation of pancreatic secretion. Ciba Foundation Symposium of the Exocrine Pancreas, pp. 208–219. J. and A. Churchill Ltd., London.
215. Shaw, H. M., and Heath, T. J. (1973). The phases of pancreatic secretion in rats. Q. J. Exp. Physiol. 58:229.
216. Roze, C., Debray, L., Latour, J., de Souchard, M., and Vaille, C. (1974). Action de quelques amines sympathomimétiques sur la secretion pancreatique externe chez le rat. J. Pharmacol. 5:155.
217. Solomon, T. E., Solomon, N., Shanbour, L. L., and Jacobson, E. D. (1974). Direct and indirect effect of nicotine on rabbit pancreatic secretion. Gastroenterology 67:276.
218. Rudick, J., Gonda, M., Rosenberg, R., Chapman, N. L., Dreiling, D. A., and Janowitz, H. B. (1973). Effects of a beta adrenergic receptor stimulant (isoproterenol) on pancreatic exocrine secretion. Surgery 74:338.
219. Kelly, G. A., Rose, R. C., and Nahrwold, D. L. (1974). Effect of isoproterenol on pancreatic responses to exogenous and endogenous cholecystokinin. Gastroenterology 66:A68/722.

220. Dzieniszewski, J., Tiscornia, O. M., Palasciano, G., and Sarles, H. (1974). Les effets du blocage alpha et beta adrénergique sur la sécretion pancreéatique exocrine du chien stimulée par la sécrétion et la cholecystokinine–pancréozymine (CCK–PZ). Biol. Gastroenterol. (Paris) 7:131.
221. Raptis, S., Dollinger, H., Chrissiku, M., Rothenbuchner, G., and Pfeiffer, E. F. (1973). The effects of the β-receptor blocker (propranolol) on endocrine and exocrine pancreatic function in man after the administration of intestinal hormone (secretin and cholecystokinin–pancreozymin). Eur. J. Clin. Invest. 3:163.
222. Barlow, T. E., Greenwell, J. R., Harper, H. A., and Scratcherd, T. (1974). The influence of the splanchnic nerves on the external secretion. blood flow and electrical conductance of the rat pancreas. J. Physiol. (Lond.) 236:421.
223. Kure, K., and Fujii, M. (1933). Spinal parasympathetic: influence of spinal parasympathetic on blood vessels and external secretion of pancreas. Q. J. Exp. Physiol. 22:323.
224. Hashimoto, K., Satoh, S., and Takeuchi, O. (1971). Effect of dopamine on pancreatic secretion in the dog. Br. J. Pharmacol. 43:739.
225. Furuta, Y., Hashimoto, K., Iwatsuki, K., and Takeuchi, O. (1973). Effects of enzyme inhibition of catecholamine metabolism and of haloperidol on the pancreatic secretion induced by L-dopa and by dopamine in dogs. Br. J. Pharmacol. 47:77.
226. Furuta, Y., Iwatsuki, K., and Hashimoto, K. (1974). Enhancement by cocaine of dopamine-induced pancreatic secretion. Jpn. J. Pharmacol. 24:S24.
227. Iwatsuki, K., Furuta, Y., and Hashimoto, K. (1973). Effects of prostaglandin $F_2\alpha$ on the secretion of pancreatic juice induced by secretin and by dopamine. Experientia 29:319.
228. Case, R. M., and Scratcherd, T. (1972). Prostaglandin action on pancreatic blood flow and on electrolyte and enzyme secretion by exocrine pancreas *in vivo* and *in vitro*. J. Physiol. (Lond.) 226:393.
229. Rosenberg, V., Robert, A., Gonda, M., Dreiling, D. A., and Rudick, J. (1974). Synthetic prostaglandin analogs and pancreatic secretion. Gastroenterology 66:A113/767.
230. Heisler, S. (1973). Effect of various prostaglandins and serotonin on protein secretion from rat exocrin pancreas. Experientia 29:1233.
231. Grossman, M. I., et al. (1974). Candidate hormones of the gut. Gastroenterology 67:730.
232. Lehnert, P., Strahleber, H., and Forell, M. M. (1972). Kinetics of exocrine pancreatic secretion. Digestion 6:9.
233. B'Hend, P., Hadorn, B., Haldeman, B., Kleb, M., and Lüthi, H. (1973). Stimulation of pancreatic secretion in man by secretin snuff. Lancet 1:509.
234. Tympner, F., Domschke, S., Domschke, W., Wünsch, E., Jaeger, E., and Demling, L. (1975). Pancreatic response to graded doses of synthetic secretin: comparison between intravenous and paranasal administration in man. Klin. Wochenschr. 53:713.
235. Petersen, H., and Berstad, A. (1972). Comparison of response to intravenous injection and infusion of secretin in man. Scand. J. Gastroenterol. 7:463.
236. Docrkay, G. J. (1975). Comparison of the action of porcine secretion and extracts of chicken duodenum on pancreatic exocrine secretion in the cat and turkey. J. Physiol. (Lond.) 244:625.
237. Shaw, H. M., and Health, J. J. (1973). Basal and post-prandial pancreatic secretion in rats. Q. J. Exp. Physiol. 58:335.
238. Desmul, A., De Waele, B., Wissocq, P., and Kiekens, R. (1974). Exogenous and endogenous secretin stimulation in the conscious rat. Digestion 11:39.
239. De Waele, B., Desmul, A., Wissocq, P., and Kiekens, R. (1974). La sécretion pancreatique chez le rat, influence de l'intervention chirurgicale, de la narcose, de l'hypothermie et de la dérivation du suc gastrique ou du suc pancréatique. Biol. Gastroenterol. (Paris) 7:253.
240. Stock, C., Stoebner, P., Kachelochofer, J., and Grenier, J. F. (1972). Perfusion de pancréas isolé chez le chien. Etudes biochimiques et ultra structurales de la secretion externe stimulée. Biol. Gastroenterol. (Paris) 5:163.
241. Domschke, S., Domschke, W., Rosch, W., Konturek, S. J., Wunsch, E., and Demling,

L. (1975). Bicarbonate and cyclic AMP secretion in pure pancreatic juice: dose–response curves to synthetic secretin in man. British Society of Gastroenterology Handbook, p. 42.
242. Dockray, G. J. (1973). The action of gastrin and cholecystokinin related peptides on pancreatic secretion in the rat. Q. J. Exp. Physiol. 58:163.
243. Fox, A., Van der Hoeden, R., and Delcourt, A. (1972). Influence de la caeruleine sur la fonction exocrine du pancreas humain. Biol. Gastroenterol. (Paris) 5:91.
244. Vaysse, N., Laval, J., Duffaut, M., and Ribet, A. (1974). Effect of secretin and graded doses of CCK–PZ on pancreatic secretion in man. Am. J. Dig. Dis. 19:887.
245. Folsch, V. R., and Wormsley, K. G. (1973). Pancreatic enzyme response to secretin and cholecystokinin–pancreozymin in the rat. J. Physiol. (Lond.) 234:79.
246. Robberecht, P., Cremer, M., Van der Meers, A., Van der Meers-Piret, M. C., Cotton, P., Deneef, P., and Christophe, J. (1975). Pancreatic secretion of total protein and three hydrolases collected in healthy subjects. Gastroenterology 69:374.
247. Debas, H. T., and Grossman, M. I. (1973). Pure cholecystokinin: pancreatic proteins and bicarbonate response. Digestion 9:469.
248. Malalelada, J. R., Go, V. L. W., and Summerskill, W. H. J. (1973). Different sensitivity of gallbladder and pancreas to cholecystokinin–pancreozymin (CCK–PZ) in man. Gastroenterology 64:950.
249. Vagne, M. (1970). Les hormones gastro-intestinales. I. La gastrine. Pathol. Biol. (Paris) 18:887.
250. Wormsley, K. G., Mahoney, M. P., and Ng, M. (1966). Effects of a gastrin-like pentapeptide (ICI 50, 123) on stomach and pancreas. Lancet I:993.
251. Petersen, H., Berstad, A., and Myren, J. (1971). Effects of pentagastrin on pancreatic secretion in man. *In* E. Hess-Thaysen (ed.), Gastrointestinal Hormones and other Subjects, pp. 72–73. Alfred Benzon Pub. I. Munksgaard, Copenhagen.
252. Petersen, H., and Berstad, A. (1973). The interaction between pentagastrin and cholecystokinin on pancreatic secretion in man. Scand. J. Gastroenterol. 8:257.
253. Duffaut, M., Vaysse, N., Laval, J., Louis, A., and Ribet, A. (1972). Action conjugée de la secretine cholecystokinine et pentagastrine sur le pancreas exocrine humain. Biol. Gastroenterol. (Paris) 5:71.
254. Brown, J. C., Dryburgh, J. R., and Pederson, R. A. (1974). Gastrin inhibitory polypeptide and motiline. *In* W. Y. Chey, F. P. Brooks, N. J. Thorofare, and C. B. Slack (eds.), Endocrinology of the Gut, pp. 76–82.
255. Schebalim, R., Said, S. I., and Makhlouf, G. M. (1974). Inhibition of gastric secretion by synthetic vasoactive intestinal peptide (VIP). Clin. Res. 22:23a.
256. Konturek, S. J., Thor, P., Dembinski, A., and Krol, R. (1975). Comparison of secretin and vasoactive intestinal peptide on pancreatic secretion in dogs. Gastroenterology 68:1527.
257. Dockray, G. J. (1972). Pancreatic secretion in the turkey. J. Physiol. (Lond.) 227:49.
258. Youngs, G. (1972). Hormonal control of pancreatic endocrine and exocrine secretion. Gut 13:154.
259. Dyck, W. P., Texter, E. C., Lasater, J. M., and Hightower, N. C. (1970). Influence of glucagon on pancreatic exocrine secretion in man. Gastroenterology 58:535.
260. Konturek, S. J., Tasler, T., and Obtulowicz, W. (1973). Characteristics of inhibition of pancreatic secretion by glucagon. Gastroenterology 64:755.
261. Konturek, S. J., Tasles, J., and Obtulowicz, W. (1974). Effect of glucagon on food-induced gastrointestinal secretion. Digestion 8:220.
262. Fitzgerald, O., McGeeney, K. F., and Murphy, J. J. (1974). An endocrine–exocrine interaction in the isolated canine pancreas. J. Physiol. (Lond.) 239:59.
263. Konturek, S. J., Dimitreschu, T., Radecki, T., Thorp, P., and Pucher, R. (1974). Effect of glucagon on gastric and pancreatic secretion and peptic ulcer formation in rats. Am. J. Dig. Dis. 19:557.
264. Wizeman, V., Weppler, P., and Mahrt, R. (1974). Effect of glucagon and insulin on the isolated exocrine pancreas. Digestion 11:432.
265. Werster, P. D., Singh, M., Tucker, P. C., and Black, O. (1972). Effect of fasting and feeding on the pancreas. Gastroenterology 62:600.
266. Adler, G., and Kern, H. F. (1974). Influence of islet hormones on the secretory

process of the exocrine pancreas in rats. VII Meeting Eur. Panc. Club., Dundee, Scotland, July, 1974, Abstract 18.
267. Danielson, Å, Marklond, S., and Stigbrand, T. (1974). Effects of starvation and islet hormones on on the synthesis of amylase in isolated exocrine pancreas of the mouse. Acta Hepatogastroenterol. 21:289.
268. Couture, Y., Dunnigan, J., and Morisset, J. (1972). Stimulation of pancreatic amylase secretion and protein synthesis by insulin. Scand. J. Gastroenterol. 7:257.
269. Gupta, S., Elder, J. B., and Kay, A. W. (1973). Exocrine secretory responses of the pancreas to insulin and to meat in dogs. Gut 14:54.
270. Mayston, P. D., and Borrowman, J. A. (1973). Influence of chronic administration of pentagastrin on the pancreas in hypophysectomized rats. Gastroenterology 64:391.
271. Belleville, J. (1973). Effets de la thyroidectomie sur les activités phospholipasique, lipasique, cholesterolesterasique et trypsique du suc pancreatique et du pancreas de rat. C.R. Acad. Sci. (Paris) 277:81.
272. Kayman, A., Ben-Ari, G. Y., and Dreiling, D. A. (1973). The effect of thyroidectomy on pancreatic exocrine secretion. Am. J. Gastroenterol. 59:336.
273. Tandon, G. S., Shukla, R. C., Schkla, S. N., and Singh, S. S. (1973). Effect of thyroid feeding on pancreatic secretion in dogs. Indian J. Physiol. Pharmacol. 17:257.
274. Schapiro, H., and Britt, L. G. (1972). The action of vasopressin on the gastrointestinal tract. A review of the literature. Am. J. Dig. Dis. 17:649.
275. Dyck, W. P. (1973). Influence of antidiuretic hormone on pancreatic exocrine secretion in man. Am. J. Dig. Dis. 18:33.
276. Schapiro, H. (1975). Inhibitory action of antidiuretic hormone on canine pancreatic exocrine flow. Am. J. Dig. Dis. 20:853.
277. Creutzfeldt, W., Lankisch, P. G., and Fölsch, U. R. (1975). Hemmung des Sekretin und Cholecystokinin–Pancreozymin–induzierten Saft und Enzymsekretion des Pankreas und der Gallenblasenkontraktion beim Menschen durch Somatostatin. Dtsch. Med. Wochenschr. 100:1135.
278. Dollinger, H. C., Raptis, S., Goebell, H., and Pfeiffer, E. F. (1975). Effect of somatostatin on gastrin, insulin, gastric and exocrine pancreatic secretion in man. Digestion 13:403.
279. Albinus, M., Case, R. M., Reed, J. D., Shaw, B., Smith, P. A., Gomez Pan, A., Hall, R., Besser, G. M., and Schally, A. V. (1975). Effects of growth hormone-release inhibiting hormone (GH-RIH or Somatostatin) on secretory processes in stomach and pancreas. Digestion 13:407.
280. Bloom, J. R., Joffe, J. N., and Dolack, J. M. (1975). Effect of somatostatin on pancreatic and biliary function. British Society of Gastroenterology Handbook, pp. 41–42.
281. Boden, G., Sivitz, M. C., Owen, O. E., Koumar, N. E., and Landor, J. S. (1975). Somatostatin suppresses secretin and pancreatic exocrine secretion. Science 190:163.
282. Kirdani, R. Y., Sandberg, A. A., and Murphy, G. P. (1973). Oestrogen binding to pancreas. Surgery 74:8490.
283. Goebell, H., Hotz, J., Steffen, C. H., and Ziegler, R. (1973). Pattern of glucagon and calcitonin induced inhibition of pancreatic secretion in the human. IV Meeting Eur. Panc. Club, Göteborg, Sweden, Abstract 45.
284. Hotz, J., Minne, H., and Ziegler, R. (1973). The influences of acute hyper- and hypocalcaemia and of calcitonin on exocrine pancreatic function in man. Res. Exp. Med. 160:152.
285. Schmidt, H., Hesch, R. D., Hufner, M., Paschen, K., and Creutzfeldt, N. (1971). Hemmung der exokrinen Pankreas Sekretion des Menschen durch Calcitonin. Dtsch. Med. Wochenschr, 96:1773.
286. Konturek, S. J., Radecki, T., Konturek, D., and Dimitrescu, T. (1974). Effect of calcitonin on gastric and pancreatic secretion and peptic ulcer formation in cats. Am. J. Dig. Dis. 19:235.
287. Lin, T. M., Chance, R., and Evans, D. (1973). Stimulatory and inhibitory action of a bovine pancreatic peptide on gastric and pancreatic secretion of dogs. Gastroenterology 64:865.
288. Lin, T. M., Evans, D. C., and Chance, R. E. (1974). Action of bovine pancreatic polypeptide 9BPP) on pancreatic secretion in dogs. Gastroenterology 66:A198/852.

289. Dick, J., and Felber, J. P. (1973). Study of the regulation of duodenal hormone regulating pancreatic amylase secretion. Acta Endocrinol. (Suppl.) 177:334.
290. Dick, J., and Felber, J. P. (1975). Specific hormonal regulation by food of the pancreas enzymatic (amylase and trypsin) secretion. Horm. Metab. Res. 7:161.
291. Adelson, J. W., and Ehrlich, A. (1972). The effect of porcine duodenal mucosa extract upon enzyme release from pancreatic zymogen granules *in vitro*. Endocrinology 90:60.
292. Adelson, J. W., and Rothman, P. M. (1974). II. Chymodenin. Gastroenterology 67:731.
293. Rosenberg, V., Rudick, J., and Dreiling, D. A. (1974). Pregancy and canine pancreatic exocrine secretion. Gastroenterology 66:A113.
294. Barrowman, J. A., and Mayston, P. D. (1973). Pancreatic secretion in lactating rat. J. Physiol. (Lond.) 229:41p.
295. Magee, D. F., Nakajima, S., and Odori, Y. (1968). On potentiation. Gastroenterology 55:648.
296. Grossman, M. I. (1969). Potentiation, a reply. Gastroenterology 56:815.
297. Way, L. W. (1969). Comment on potentiation. Gastroenterology 57:619.
298. Grossman, M. I. (1976). Interaction of gastrointestinal hormones on secretory processes. *In* R. M. Case and H. Goebell (eds.), Stimulus Secretion Coupling in the Gastrointestinal Tract, pp. 361–367. MTP Press Ltd., Lancaster, England.
299. Makhlouf, G. M. (1974). The neuroendocrine design of the gut. The play of chemicals in a chemical playground. Gastroenterology 67:159.
300. Christophe, J., Conlon, T. P., Robberecht, P., and Gardner, J. D. (1976). The specific binding and action of vasoactive intestinal peptide (VIP) on isolated pancreatic acinar cells. *In* R. M. Case and H. Goebell (eds.), Stimulus Secretion Coupling in the Gastrointestinal Tract, pp. 377–380. MTP Press Ltd., Lancaster England.
301. Henriksen, F. W., and Worning, H. (1967). The interaction of secretin and pancreozymin on the external pancreatic secretion in dogs. Acta Physiol. Scand. 70:241.
302. Wormsley, K. G. (1969). A comparison of the response to secretin, pancreozymin and a combination of these hormones in man. Scand. J. Gastrolenterol. 4:413.
303. Way, L. W., and Grossman, M. I. (1970). Pancreatic stimulation by duodenal acid and exogenous hormones in conscious cats. Am. J. Physiol. 219:449.
304. Shaw, H. M., Heath, T. J., and Stark, A. E. (1972). Response of the rat exocrine pancreas to secretin and cholecystokinin–pancreozymin. Can. J. Physiol. Pharmacol. 51:383.
305. Mellanby, J. M. (1926). The secretion of pancreatic juice. J. Physiol. (Lond.) 61:419.
306. Forell, M. M. (1972). Bile salts as stimulants of pancreatic secretion. *In* S. Anderson (ed.), Nobel Symposium XVI. Frontiers in Gastrointestinal Hormone Research, Stockholm, July 20–21, Almquist and Wiksell, Upsala.
307. Forell, M. M., Otte, M., Kohl, H. S., Lehnert, P., and Stahlheber, H. P. (1971). The influence of bile and pure bile salts on pancreatic secretion in man. Scand. J. Gastroenterol. 6:261.
308. Forell, M. M., Stahlheber, H., and Scholz, F. (1965). Galle als Reiz der Enzymsekretion des Pankereas. Dtsch. Med. Wochenschr. 90:1128.
309. Forell, M. M., Otte, M., Lechelt, B., Roder, O., Stahlheber, H., Thurmar, G. R., and Thurmayr, R. (1973). Einfluss der Cholezystektomie an die exkretorische Pankreas Funktion des Menschen. Dtsch. Med. Wochenschr. 98:930.
310. Malagelada, T. R., Dimagno, E. P., and Summerskill, W. H. J. (1973). Interaction between intraluminal bile acids and digestive products on pancreatic and gallbladder function. J. Clin. Invest. 52:2160.
311. Konturek, S. J., and Thor, P. (1973). Effect of diversion and replacement of bile on pancreatic secretion. Am. J. Dig. Dis. 18:971.
312. Tandon, G. S., Shukla, R. C., Shukla, S. N., and Shukla, N. (1972). Effect of bile salts and biliary obstruction on pancreatic secretion. Int. J. Physiol. Pharmacol. 16:167.
313. Geratz, J., and Lamb, J. C. (1974). Influence of bile duct ligation on exocrine pancreatic secretory activity in the rat. Am. J. Physiol. 227:119.
314. Annis, D., and Hallenbeck, A. H. (1951). Effect of excluding pancreatic juice from the duodenum on secretory response of the pancreas to a meal. Proc. Soc. Exp. Biol. Med. 77:383.

315. Green, G M., and Lyman, R. L. (1972). Feedback regulation of pancreatic enzyme secretion as a mechanism for trypsin induced inhibition of hypersecretion in rats. Proc. Soc. Exp. Biol. Med. 139:6.
316. Laporte, J. C., and Tremolieres, J. (1973). Action de la trypsine et des inhibiteurs trypsiques sur la secretion pancreatique. Nutr. Metab. 15:182.
317. Schneemann, B. O., and Lyman, R. L. (1974). Factors involved in the intestinal feedback regulation of pancreatic enzyme secretion in the rat. Proc. Soc. Exp. Biol. Med. 148:897.
318. Green, G. M., Olds, B. A., Matthews, G., and Lyman, R. L. (1973). Protein as a regulator of pancreatic enzyme secretion in the rat. Proc. Soc. Exp. Biol. Med 142:1162.
319. Corring, T. (1974). Régulation de la sécretion pancreatique par retro-action négative chez le porc. Ann. Biol. Anim. Biochim. Biophys. 14:487.
320. Abita, J., Moulin, A., Lazdunski, M., Hage, G., Palasciano, G., Brasca, A., and Tiscornia, O. (1973). A physiological inhibition of gastric secretion, the activation peptide of trypsinogen. FEBS Lett. 34:251.
321. Palasciano, G., Tiscornia, O., Dzienizewski, J., Sarles, H., Vincent, J. P., and Lazdunski, M. (1974). Influence of pancreatic secretion on gastrin secretion in dogs. Digestion 11:64.
322. Hage, G., Tiscornia, O., Palasciano, G., and Sarles, H. (1974). Inhibition of pancreatic exocrine secretion by intracolonic oleic acid infusion in the dog. Biomedicine 21:263.
323. Harper, A. A., Hood, A. J. C., and Mushens, J. Pancreotone: an inhibitor of pancreatic secretion in extract of ileal and colonic mucosa. Digestion. In press.
324. Arnesjö, B., Ihse, I., Lundquist, J., and Quist, I. (1973). Effect on exocrine and endocrine rat pancreatic function of bovine lung trypsine inhibitor administered perorally. Scand. J. Gastroenterol. 8:545.
325. Ihse, I., Arnesjö, B., and Lundquist, J. (1975). Studies on the reversibility of oral trypsin inhibition induced changes of rat pancreatic exocrine enzyme activity and insulin secretory capacity. Scand. J. Gastroenterol. 10:321.
326. Fölsch, U. R., Win Ler, K., and Wormsley, K. G. (1974). Effect of a soybean diet on enzyme content and ultrastructure of the rat pancreas. Digestion 11:161.
327. Palla, J. C., Ben Abdeljlil, A., and Desnuelle, P. (1968). Action de l'insuline sur la biosynthèse de l'amylase et de quelques autres enzymes du pancreas de rat. Biochim. Biophys. Acta 158:25.
328. Lavau, M., Bazin, R., and Herzog, J. (1974). Comparative effects of oral and parenteral feeding on pancreatic enzymes in the rat. J. Nutr. 104:1432.
329. Deschodt-Lanckman, M., Robberecht, P., Camus, J., and Christophe, J. (1971). Short-term adaptation of pancreatic hydrolases to nutritional and physiological stimulation in adult rats. Biochimie 53:789.
330. Adler, G., and Kern, H. F. (1975). Regulation of exocrine pancreatic secretion process by insulin *in vivo*. Horm. Metab. Res. 7:290.
331. Deschodt-Lankman, M., Robberecht, P., Camus, J., Baya, C., and Christophe, J. (1974). Hormonal and dietary adaptation of rat pancreatic hydrolases before and after weaning. Am. J. Physiol. 226:34.
332. Alcon, E., and Bdolah, A. (1975). Increase of proteolytic activity and synthetic capacity of the pancreas in snakes after feeding. Comp. Biochem. Physiol. (A) 50:627.
333. Danielsson, A. (1974). Effects of nutritional state and of administration of glucose, glibenclamide or diazoxide on the storage of amylase in mouse pancreas. Digestion 10:150.
334. Nevalainen, J. J., and Janigan, D. T. (1974). Degeneration of mouse pancreatic acinar cells during fasting. Wirchows, Arch. Abt. B. Zellpath. 15:105.
335. Johnson, L. R., Lichtenberger, L. M., Copeland, E. M., Dudrick, S. J., and Castro, G. A. (1971). Action of gastrin on gastrointestinal structure and function. Gastroenterology 68:1184.
336. Mainz, D. L., Black, O., and Webster, P. D. (1974). Hormonal control of pancreatic growth. J. Clin. Invest. 52:2300.

337. Yalovski, U., Heller, H., and Kulka, R. C. (1973). Accumulation of amylase by chick pancreas in organ culture. Exp. Cell Res. 80:322.
338. Cohen, A., and Kulka, R. G. (1974). Induction of chymotrypsinogen by hydrocortisone in embryonic chick pancreas *in vitro*. J. Biol. Chem. 249:122.
339. Machovich, R. (1974). Post-natal development of rat pancreatic esterase. Enzyme 17:265.
340. Deschodt-Lankmann, M., Robberecht, P., Camus, J., Baya, C., and Christophe, J. (1974). Hormonal and dietary adaptation of rat pancreatic hydrolases before and after weaning. Am. J. Physiol. 226:39.
341. Sesso, A., Abrahamsohn, P. A., and Tsanaclis, A. (1972). Acinar cell proliferation in the rat pancreas during early postnatal growth. Acta Physiol. Lat. Amer. 22:37.
342. Sesso, A., Carnero, J., Cruz, A. R., and Arruda Leite, J. A. (1973). Biochemical, cytochemical and electron microscopic observations on the enhancement of the pancreatic acinar cell secretory activity in the rat during early postnatal growth. Arch. Histol. Jpn. 35:343.
343. Larose, L., and Morisset, J. (1975). Les effets de l'âge sur la sensibilité des acini pancreatiques aux stimulations cholinergiques chez le rat. Biol. Gastroenterol. (Paris) 8:151.
344. Robberecht, P., Deschodt-Lankman, M., Camus, J., Bruylands, J., and Christophe, J. (1971). Rat pancreatic hydrolase from birth to weaning and dietary adaptation after weaning. Am. J. Physiol. 221:376.
345. Kitts, W. D., Bailey, C. B., and Wood, A. J. (1956). The development of the digestive enzyme system of the pig during its pre-weaning phase of growth. A pancreatic amylase and lipase. Can. J. Agr. Sci. 36:45.
346. Delachaume-Salem, E., and Sarles, H. (1970). Evolution en fonction de l'âge de la sécretion pancreatique humaine normale. Biol. Gastroenterol. (Paris) 2:135.
347. Zoppi, G., Andreotti, G., Pajno-Ferrara, F., Njai, D. M., and Gaburro, D. (1972). Exocrine pancreas function in premature and full-term neonates. Pediat. Res. 6:880.
348. Dietze, F., Ahlert, G., Kalbe, I., Reitzig, P., Schulz, H. J., and Wenlandt, H. (1972). Die exkretorische Pankreasfunktion im Alter. Dtsch. Ges. Wesen. 27:2361.
349. Konturek, S., Tasler, J., and Obtulowicz, W. (1973). Effect of exercise on gastrointestinal secretion. J. Appl. Physiol. 34:324.
350. Solomon, T. D., Solomon, N., Shambour, L. L., and Jacobson, E. D. (1973). Mechanism of inhibition of pancreatic secretion by nicotine. Gastroenterology 64:805.
351. Konturek, S. J., Dale, J., Jacobson, E. D., and Johnson, L. L. (1972). Mechanism of nicotine-induced inhibition of pancreatic secretion of bicarbonate in a dog. Gastroenterology 62:425.
352. Bynum, T. E. S., Solomon, T. E., Johnston, L. R., and Jacobson, E. D. (1972). Inhibition of pancreatic secretion in man by cigarette smoking. Gut 13:361.
353. Bochenek, W. J., and Koronczewski, R. (1973). Effect of cigarette smoking on bicarbonate and volume of duodenal contact. Am. J. Dig. Dis. 18:729.
354. Lavigne, J. G., and Marchand, P. C. (1972). Effect of phenobarbital pretreatment on rat pancreas. Am. J. Physiol. 222:360.
355. Tiscornia, O., Gullo, L., and Sarles, H. (1973). The inhibition of canine exocrine pancreatic secretion by intravenous ethanol. Digestion 9:231.
356. Tiscornia, O., Hage, G., Palasciano, G., Brasca, A. P., Devaux, M. A., and Sarles, H. (1973). The effects of pentolinium and vagotomy on the inhibition of canine exocrine pancreatic secretion by intravenous ethanol. Biomedicine 18:159.
357. Tiscornia, O., Palasciano, G., Dzieniszewski, J., and Sarles, H. (1975). Simultaneous changes in pancreatic and gastric secretion induced by acute intravenous ethanol infusion. Effects of atropine and reserpine. Am. J. Gastroenterol. 63:389.
358. Tiscornia, O., Palasciano, G., and Sarles, H. (1974). Effects of chronic ethanol administration on canine exocrine pancreatic secretion. Digestion 11:172.
359. Tiscornia, O., Palasciano, G., and Sarles, H. (1975). Atropine and exocrine pancreatic secretion in alcohol-fed dogs. Am. J. Gastroenterol. 63:33.
360. Tiscornia, O., Sarles, H., and Palasciano, G. (1976). Chronic alcholism and canine exocrine pancreas. A long follow-up study. To be published.

361. Papp, M., Varga, B., and Folly, G. (1973). Effect of secretin, pancreozymin, histamine and decholin ® on canine pancreatic blood flow. Pflügers Arch. 340:349.
362. Vaysse, N., Martinel, C., Lacroix, A., Pascal, J. P., and Ribet, A. (1973). Effet de la cholecystokinine–pancreozymine GIH sur la vasomotricité du pancreas isolé de chien. Relation entre l'effet vasomoteur et la résponse sécrétoire. Biol. Gastroenterol. (Paris) 6:33.
363. Pascal, J. P., and Vaysse, N. (1973). Rôle de la vasomotricité dans la régulation de la sécrétion pancréatique exocrine. Biol. Gastroenterol. (Paris) 6:33.
364. Varga, B., Folly, G., and Papp, M. (1974). L'effet de l'excitation électrique du ganglion coeliaque sur le débit sanguin du pancreas. Lyon Chir. 70:168.
365. Saunders, R. W., and Moser, C. A. (1972). Increased vascular resistance by prostaglandin B_1 and B_2 in the isolated rat pancreas. Nature (New Biol.) 237:285.
366. Lenninger, S. (1974). The autonomic innervation of the exocrine pancreas. Med. Clin. North Amer. 58:1311.
367. Eström, J., and Lenninger, S. (1973). Choline acetyltransferase and cholinesterase in the pancreatic duct of the cat. Acta Physiol. Scand. 87:78.
368. Saito, A., and Kanno, J. (1973). Concentration of pancreozymin as a determinant of the exocrine–endocrine partition of pancreatic enzyme. Jpn. J. Physiol. 23:477.
369. Hadi, N., Hotz, J., Scratcherd, T., and Wynne, R. D. (1976). The secretion of organic anions by the isolated perfused pancreas of the rat. J. Physiol. (Lond.) 259:56P.
370. Pascal, J. P., Roux, P., Vaysse, N., Lacroix, A., Martinel, C., and Ribet, A. (1976). Respiratory exchanges and acid-base balance during perfusion of ex-vivo isolated pancreas. Am. J. Dig. Dis. 21:381.
371. Simon, B., and Knauf, H. (1976). Die bedeutung der HCo^{-3}-ATPase in der H^+/HCo^-_3-sekretion. Klin. Wochenschr. 54:97.
372. Devillers-Thiery, A., Kindt, J., Scheele, L., and Blobel, G. (1975). Homology in amino terminal sequence of precursors to pancreatic secretory proteins. Proc. Natl. Acad. Sci. USA 72:5016.
373. Dagorn, J. C., Paradis, D., and Morisset, J. (1976). Non-parallel response of amylase and chymotrypsinogen biosynthesis following pancreatic stimulation: a possible explanation for observed non-parallelism in pancreatic secretion. A paraître dans Digestion.
374. Dagorn, J. C., and Michel, R. (1976). Nonparallel courses of intrapancreatic levels of exportable enzymes after a fatty meal. Proc. Soc. Exp. Biol. Med. 151:608.
375. Scheele, G. A., and Palade, G. E. (1975). Studies on the guinea pig pancreas. Parallel discharge of exocrine enzyme. J. Physiol. (Lond.), 250:2660.
376. Malaisse-Lagae, F., Ravazzola, M., Robberecht, P., Vandermeers, V., Malaisse, W. J., and Orci, L. (1975). Exocrine pancreas. Evidence for topography partition of secretory function. Science 190:795.
377. Kondo, S., and Schulz, I. (1976). Calcium ion uptake in isolated pancreas cells induced by secretagogues. Biochim. Biophys. Acta 419:76.
378. Haymovits, A., and Scheele, G. A. (1976). Cellular cyclic nucleotides and enzyme secretion in the pancreatic acinar cell. Proc. Natl. Acad. Sci. USA 73:156.
379. Bauduin, H., Stock, C., Vincent, D., and Potvliege, P. (1976). About the site of action of the dibutyryl derivative of 3′5′ cyclic adenosine monophosphate in the exocrine pancreas of the rat. Réunion du Club Européen du Pancréas (Toulouse 23–25 octobre 1975) in Biol. Gastroentérol. 9:80.
380. Gardner, J. D., Conlon, J. P., and Adams, T. D. (1976). Cyclic AMP in pancreatic acinar cells. Effects of gastrointestinal hormones. Gastroenterology 70:29.
381. Petersen, O. H., and Ueda, N. (1976). Pancreatic acinar cells: the role of calcium in stimulus-secretion coupling. J. Physiol. (Lond.) 254:583.
382. Williams, J. A. (1975). Na^+ dependence of in vitro pancreatic amylase release. Am. J. Physiol. 229:1023.
383. Clemente, F., and Meldolesi, J. (1975). Calcium and pancreatic secretion-dynamics of subcellular calcium pools in resting and stimulated acinar cells. Br. J. Pharmacol. 55:369.
384. Schreurs, V. V. A. M., Swarts, H. G. P., Depont, J. J. H. H. M., and Bonting, S. L. (1976). Role of calcium in exocrine pancreatic secretion. II. Comparison of the

effects of carbachol and the ionosphore A-23187 on enzyme secretion and calcium movements in rabbit pancreas. Biochim. Biophys. Acta 419:320.

385. Vaysse, N., Pascal, J. P., Roux, P., Martinel, C., Lacroix, A., and Ribet, A. (1975). Role of cholinergic mechanisms in the response to secretion of isolated canine pancreas. Gastroenterology 66:1269.

386. Williams, J. A. (1975). An in vitro evaluation of possible cholinergic and adrenergic reception affecting pancreatic amylase secretion. Proc. Soc. Exp. Biol. Med. 150:513.

387. Bastie, M., Vaysse, N., Pascal, J. P., and Ribet, A. (1976). Role of dopaminergic mechanism in the exocrine secretion of the pancreas. Réunion du Club Européen du Pancréas (Toulouse 23–25 octobre 1975) in Biol. Gastroentérol. 9:65.

388. Voirol, M., Tiscornia, O., Levesque, D., Dzieniszewski, J., Palasciano, G., Laugier, R., and Sarles, H. (1976). Evidence of a dissociation in pancreatic secretion when catecholamines storage is depleted. Réunion du Club Européen du Pancréas (Toulouse 23–25 octobre 1975) in Biol. Gastroentérol. 9:73.

389. Scratcherd, J., Case, R. M., and Smith, P. A. (1975). A sensitive method for the biological assay of secretin and substances with "secretin-like" activity in tissues and biological fluids. Scand. J. Gastroenterol. 10:821.

390. Gardiner, B. N., and Small, D. M. (1976). Simultaneous measurement of the pancreatic and biliary response to CCK and secretin: primate biliary physiology XIII. Gastroenterology 70:403.

391. Domschke, S., Domschke, W., Rosch, W., Konturek, S. J., Wunsch, E., and Demling, L. (1976). Bicarbonate and cyclic AMP content of pure human pancreatic juice in response to gradued doses of synthetic secretin. Gastroenterology 70:533.

392. Cotton, P. B., Heap, T. R., Reuben, A., Stern, R., Townson, J., Corns, C., Bloom, A., and Miller, A. (1976). Pure pancreatic juice response to lox dose secretin in man. Réunion du Club Européen du Pancréas (Toulouse 23–25 octobre 1975) in Biol. Gastroentérol. 9:61.

393. Ueda, N. (1976). Secretion of fluid and amylase by the perfused rat pancreas. Réunion du Club Européen du Pancréas (Toulouse 23–25 octobre 1975) in Biol. Gastroentérol. 9:70.

394. Konturek, S. J., Pucher, A., Radecki, T. (1976). Comparison of vasoactive intestinal peptide and secretin in stimulation of pancreatic secretion. J. Physiol. (Lond.) 255:497.

395. Heidbreder, E., Sieber, P., and Heidland, A. (1975). Exokrine pankreas-funktion und calciumhomöostase. Vergleichende tier experimentelle untersuchungen zur wirkung von parathormon, vitamin D3, 25-hydroxycholecalciferol, dihydrotachysterin und thyreocalcitonin. Res. Exp. Med. 166:147.

International Review of Physiology
Gastrointestinal Physiology II, Volume 12
Edited by Robert K. Crane

7
Biliary Secretion and Motility

A. GEROLAMI AND J.-C. SARLES

Institut National de la Sante et de la Recherche Mèdicale
Unite de Recherches de Pathologie Digestive
Marseille, France

BILE SECRETION: MORPHOLOGICAL BASIS AND GENERAL ASPECTS 224
Sites of Bile Formation 224

MECHANISMS INVOLVED IN BILE SECRETION 225
Water and Electrolytes 225
Bile Salt-dependent Flow 226
Bile Salt-independent Bile Flow 226
Interrelations between Bile Salt-dependent and Bile Salt-independent Flows 227
Bile Flow of Ductular Origin 227
Biliary Secretion of Organic Substances 228
Anion Secretion 228
Biliary Secretion of Bilirubin 228
Biliary Secretion of Bile Salts 230
Canalicular Transport 231
Biliary Lipid Secretion 231
Biliary Lecithin Secretion 232
Cholesterol Secretion 233
Control of Bile Saturation with Cholesterol 234
Quantitative Regulation of Bile Salt Secretion 234
Basic Mechanisms 234
Influences on Bile Secretion 235
Factors other than Bile Salts 237

BILIARY MOTILITY 237
Actions of Cholecystokinin 238
Action of Other Gastrointestinal Peptides 239

Editor's note: the section on Biliary Secretion was written by Professor Gerolami, the section on Biliary Motility by Dr. Sarles.

Action of Autonomic Nervous System 240
Other Factors 241

BILE SECRETION: MORPHOLOGICAL BASIS AND GENERAL ASPECTS

The stress in this review is different from that in other recent reviews (1, 2). In the liver, bile occupies a channel system formed by (*a*) the canalicular system and (*b*) the ductular system (3–5). The canalicular system is a network with a lumen of 0.6–1.2 μm diameter running between the membranes of two adjacent hepatocytes. This lumen is separated from the remaining intercellular space by a localized fusion of the adjacent membranes, the junctional complex or tight junction (6), which may prevent direct exchanges between canalicular bile and intercellular or perisinusoidal spaces. The canalicular membrane is part of the hepatocyte plasma membrane, but shows (4–7) numerous microvilli with a diameter of 0.13 μm and a length of 0.40 μm (8) which increase the surface area in contact with canalicular bile. Such structures in other tissues are generally associated with osmotic transport (9). Inside the cell, the canalicular membrane is surrounded by a pericanalicular ectoplasm where generally neither membranous structures nor secretion vacuoles can be seen (5). However, two intracellular structures are located near the canaliculus, the Golgi apparatus (10) and the lysosomes, both of which show early changes during cholestasis (11) although their role in bile secretion is not known. Recently, microfilaments have been observed in the pericanalicular ectoplasm. Perfusion with cytochalasin B, which abolishes the contractility of microfilaments, induces dilation of bile canaliculi. The reduction of bile flow correlates with the degree of canalicular change induced. Although other effects of cytocholasin B may contribute to cholestasis, these results suggest that microfilaments are necessary to maintain the morphology and functions of the canalicular system (12–15).

The canalicular system is connected to the ductular system, formed by channels bordered by true epithelial cells which converge to form the intrahepatic and ultimately the extrahepatic biliary tree. These morphological features suggest that the canalicular membrane plays a major role in bile formation and that secondary changes in bile composition may occur because of the action of the biliary duct epithelium.

Sites of Bile Formation

Substances such as bilirubin, BSP, Rose bengal, and bile salts may be excreted in bile in concentrations exceeding those in plasma. Since these substances often undergo metabolic transformation inside the hepatocyte and hepatocyte lesions impair their biliary secretion, they are most probably secreted through the canalicular membrane. Direct evidence for this has been obtained from studies using fluorescent dyes (16, 17). Evidence of canalicular or ductular origin of water and electrolytes is more difficult to obtain. Electrolyte and water secretion may be obtained from isolated bile ducts in vitro (18, 19), and it seems established (see below) in most species that secretin induces a ductular secretion of sodium bicarbonate and water. However, it seems probable that the major part of bile water and electrolytes is secreted into the canaliculi. This conclusion was drawn after the discovery that biliary clearance of nonmetabolized sugars, e.g., erythritol, mannitol, and inulin, allows an estimation of canalicular bile flow (20–24). First, erythritol and mannitol penetrated freely through the sinusoidal membrane of the hepatocytes (20, 25) and their biliary clearance increased as did bile flow during choleresis induced by bile salts (20, 22). These results suggest that the canalicular membrane, like the sinusoidal membrane, is permeable to these solutes, with possible slight species variations (21). However, direct secretion into canalicular spaces through the tight junctions has not been excluded and could possibly explain the fact that bile and plasma concentrations of inulin and sucrose equilibrate more rapidly than do liver and plasma concentrations (26). Second, erythritol and mannitol clearance are unchanged during secretin-induced choleresis (22), suggesting that epithelial cells are not permeable to such solutes except for individual exceptions (28). Erythritol clearance in isolated rat liver and in rats and rabbits is almost identical to bile flow, suggesting that bile flow in these species is entirely of canalicular origin (27, 29–32).

Recent experiments with perfused rat livers on the site of electrolyte secretion have measured the efflux into bile of ^{24}Na, K, ^{36}Cl, or ^{45}Ca accumulated in liver cells before a washout perfusion with a nonradioactive medium. The decrease of radioactivity in bile generally followed the same slope as intracellular decrease, indicating an intracellular origin. However, the initial slope, particularly for Na, decreased more rapidly in bile than in the cells, suggesting the existence either of a restricted pathway in cells, or the penetration of some Na from extracellular spaces through the tight junction (33, 34).

MECHANISMS INVOLVED IN BILE SECRETION

Water and Electrolytes

Biliary water secretion is probably secondary to the active transport of osmotically active substances: (*a*) bile secretion is almost independent of sinusoidal blood pressure (35) and can occur against a pressure exceeding sinusoidal pressure (36, 37). (*b*) Bile and plasma are nearly isotonic, and changes in plasma osmolarity are followed by similar changes in bile. Sodium concentration is, however, higher in bile than in plasma, presumably because bile salt micelles have a low osmotic activity in themselves and also reduce the osmotic activity of other electrolytes (38, 39). (*c*) Infusion of hypertonic solutions decreases bile flow (40), and (*d*) bile secretion is temperature- and oxygen-dependent (41, 42).

Bile Salt-dependent Flow Canalicular secretion of water is probably induced by osmotic activity of bile salts and sodium. Bile salt infusion generally increases bile flow in animals and man; as shown by Sperber (43), there is a linear relationship between bile flow and the rate of bile salt secretion (44). This choleretic effect is to be expected as secreted, ionized bile salts and accompanying sodium have an osmotic activity even if this activity is reduced by the association of bile salts and lecithin in mixed micelles. The osmotic activity probably varies as a function of the identity of the bile salts and their concentration as well as that of other bile constituents. However, some experimental results seem difficult to explain by the osmotic filtration hypothesis (1), though they could be explained by the variations in osmotic activity coefficients: (*a*) the volume of fluid secreted for a given amount of bile salts shows a wide species variations (44); for 1 μmol of bile salts, it is 8 μl in the dog (22), 15 μl in the rat (32), and 30 μl in the rabbit (31). (*b*) The slope of regression line between bile flow and bile salt secretion varies with the amount of bile salt infused; the slope decreases significantly when the infusion rate is increased (45), and (*c*) dehydrocholate, a bile salt which is in part metabolized by the hepatocytes (46, 47), increases bile flow more than taurocholate but not in every species (48). Indeed, as dehydrocholate does not form micelles, a choleretic effect greater than taurocholate should be expected.

Other organic anions have a choleretic action which is probably related to their osmotic activities (49). Some neutral substances may also increase bile flow by an osmotic effect (50).

Bile Salt-independent Bile Flow Biliary clearance of canalicular flow markers such as erythritol, mannitol, etc., is linearly related to bile salt output, but extrapolation to zero bile salt secretion shows that canalicular flow is not suppressed (22). In the same way, in perfused rat liver, bile secretion of canalicular origin continues at a near zero rate of bile salt secretion (51). This bile salt-independent flow exists in every species so far studied, including monkey (52) and man (53, 54). The importance of this fraction varies greatly; a 15- to 40-fold difference, for example, exists between dog and guinea pig (1).

Various studies suggest that this fraction depends on sodium transport: in isolated perfused rat liver, the presence of sodium in the perfusate is necessary to maintain bile secretion (55) and (Na^+-K^+)-ATPase, which is responsible for the active sodium pump in plasma membranes (56), is present in isolated canalicular membranes. Inhibitors of (Na^+-K^+)-ATPase, such as ouabain, ethacrynic acid, and amiloride, decrease canalicular bile salt independent flow in the rabbit (27).

In perfused rat liver, scillaren, another (Na^+-K^+)-ATPase inhibitor, also reduces bile flow. However, ethacrynic acid and ouabain have no inhibitory effect in the rat. On the contrary, they have a choleretic effect (57, 58) possibly related to an osmotic effect of the drugs, which are largely secreted in bile. The species-dependent effect of (Na^+-K^+)-ATPase inhibitors is consistent with studies showing that rat ATPase is less sensitive to ouabain than is bovine or dog (Na^+-K^+)-ATPase (59).

Participation of (Na^+-K^+)-ATPase was further assessed by experiments showing that bile secretion and canalicular (Na^+-K^+)-ATPase have the same temperature dependence, whereas the relation between Mg^+ATPase activity and temperature had a different form and was not related to bile flow variations (60). Similarly, when bile salt-independent canalicular flow decreased in thyroid deprived rats and increased during hypoerthyroidism, canalicular (Na^+-K^+)-ATPase (but not Mg^{2+}ATPase) followed exactly the same pattern (61). Also, some phthalein dyes lower bile flow and (Na^+-K^+)-ATPase activity, whereas phthalein dyes without action on ATPase do not influence bile flow (62, 63). 17-α-Ethinyl substituted steroids which inhibit (Na^+-K^+)-ATPase impair bile secretion. Other transport systems may be involved: (*a*) cAMP increases bile salt-independent water secretion in dog. Theophylline (64–66) and glucagon (67, 68), which increase the intracellular concentration of cAMP, have the same effect in dog or man, and (*b*) phenobarbital increases bile salt-independent flow of canalicular origin (32), though this effect is probably not related to the microsomal action of the drug, as other microsomal inducers did not increase bile flow (69). Phenobarbital seems to induce protein synthesis, with possible specific effects on low molecular weight proteins in the canalicular membranes (70). It does not seem to change canalicular (Na^+-K^+)-ATPase activity (63).

Interrelations between Bile Salt-dependent and Bile Salt-independent Flows Bile salts may influence bile salt-independent flow by two mechanisms: 1) they may have an action on sodium transport, and 2) bile salt micelles may reduce the osmotic activity of secreted Na and thus change the bile flow rate for the same electrolyte secretion. These two effects would clearly depend on the structure of the secreted bile salt as well as on its concentration and may be species dependent. Variations in these mechanisms may explain the various effects of bile salts on bile secretion: (*a*) taurodeoxycholate did not increase bile flow in rats (71) but increased it in dogs. Na secretion was increased by taurodeoxycholate, suggesting a reduction of osmotic activity in rats rather than inhibition of sodium transport, (*b*) taurochenodeoxycholate decreased bile salt-independent canalicular flow in perfused female rat liver (72); (*c*) monohydroxy bile salts, which are poorly water soluble, induced cholestasis by decreasing bile salt-independent flow at low doses possibly in relation to structural alterations of canalicular membranes (73–75). Sulfation of taurolithocholate increases its solubility, and sulfotaurolithocholate does not decrease bile flow (76, 77); (*d*) the relation between bile salt secretion and bile flow may not be linear at high bile salt secretion rates (45).

Bile Flow of Ductular Origin Isolated bile ducts (19) secrete an electrolyte solution isotonic with plasma, probably by active sodium transport. Bile duct secretion is enhanced by secretin. Secretin has a choleretic action in many species in vivo, including man (44, 78, 79), but it seems to be inactive in rabbit (78) and rat (1). Its effect in vivo is likely to be exerted on bile ducts, as the biliary washout volume is less during secretin choleresis than during bile acid choleresis, which suggests that secretin acts at a more distal site than bile salts

(80). The fluid secreted under secretin influence contains bicarbonate as the main cation (78) (see Chapter 6).

Gastrin, which increases bile flow in dogs, may also act on bile ducts (82). Insulin also increases bile flow possibly by vagal mechanisms and gastrin secretion inasmuch as the site if any of a direct action of insulin is unknown (83–85). Bile ducts have a reabsorptive capacity, and it is not known whether this action in vivo results in an increase or a decrease of bile volume. At low bile flow rates in the dog, a net absorption is suggested since, after erythritol perfusion, the bile concentration of erythritol may be higher than in plasma (22). Prostaglandins PGE_1 and PGE_2 have a choleretic activity in cats which is caused by a decrease in ductular reabsorption of bile (86), though net ductular secretion was occasionally observed. In cholecystectomized man, erythritol clearance studies suggest that part of bile water secretion is not of canalicular origin (54) and, therefore, that a basal ductular secretion exists.

Biliary Secretion of Organic Substances

Organic molecules normally found in bile are anions (bilirubin, bile salts) and lipids (cholesterol and phospholipids).

Anion Secretion For a given molecule, bile secretion involves at least three main steps: entry into hepatocytes, transfer to the canalicular membrane with possible biotransformations inside the hepatocyte, and biliary secretion. This general schema is well known for dyes (87); for example, infused fluorescein rapidly appears in sinusoidal blood. From 15 to 32 s later, fluorescence appears in the hepatocytes where its concentration increases in spite of rapid biliary secretion. The dye is finally concentrated in canaliculi. The precise pathways of transport are largely unknown, but they may be studied by Goresky's method which allows a calculation of maximal uptake rate, K_m, and the influences of other substances (89).

Compartmental analysis, with radioactive tracers assumed to equilibrate rapidly with nonradioactive material, has been used for dyes, bilirubin, and bile salts (89–92) and two- or three-compartment models generally fit well with the experimental results. This method has been proposed for studies of substrate competition, but the results obtained must be interpreted with caution. Analogies to Michaelis-Menten kinetics for rate-limiting steps are generally obtained, but these rate-limiting steps are not necessarily linked to the affinity of a molecule for a carrier (88).

Storage in the cell may be calculated by the multiple infusion method described by Wheeler (93). However, the theoretical basis of this method has been questioned (94). Direct measurement of BSP concentration in cell homogenates has been performed.

Canalicular transport seems to be the rate-limiting step for substances which are secreted in bile with a high bile to plasma concentration ratio.

Biliary Secretion of Bilirubin Bilirubin is formed by heme catabolism in the reticulo-endothelial system and by cytochrome metabolism in the hepatocytes.

Elimination of bilirubin therefore requires three steps: 1) hepatic uptake, 2) transport and conjugation inside the liver cells, and 3) canalicular secretion. These steps have been recently reviewed (2) and are briefly summarized.

Unconjugated plasma bilirubin is bound to albumin, and its uptake by liver seems to involve preferential binding to liver plasma membranes and/or cytoplasmic proteins (95). Two hepatic cytosol proteins, Y and Z, bind various organic anions and particularly dyes and bilirubin (96). Y, a dimeric protein with a molecular weight of 46,000, is identical to ligandin or GSH transferase B (97, 98). It represents 5% of cytosol proteins and is also present in the small intestine and kidney. It seems to be responsible for 80% of the binding capacity of cytosol (100). A role of cytoplasmic proteins in bilirubin uptake is suggested by many results (for a review, see Ref. 99): (*a*) dyes which compete with bilirubin for hepatic uptake in vivo competitively inhibit bilirubin binding to ligandin in vitro; (*b*) phenobarbital increases hepatic uptake of bilirubin and dyes and Y concentration in liver (100); (*c*) in the newborn, the liver concentration of Y and the hepatic uptake of organic anions increase simultaneously (101); and (*d*) the binding affinity of hepatic cytosol for bilirubin is higher than that of plasma (102). Hepatic cytosol proteins retain bilirubin in the presence of albumin (102), even though purified ligandin has lower affinity constants than albumin has for bilirubin binding (103). However, some other results have suggested a role of plasma liver membranes in bilirubin uptake: (*a*) Rifamycin impairs hepatic uptake of bilirubin and BSP but does not compete with BSP for binding to Y and Z proteins (104), and (*b*) during steady state bilirubin infusion in rats, plasma disappearance of a tracer dose of [^{3}H] bilirubin does not differ from that in saline-infused rat controls. These results seem consistent with a role of membrane carriers (105).

During cell transport, 98–99% of bilirubin is conjugated (106), mainly in the form of mono- and diglycuronides, although other conjugates have been observed during cholestasis (107). Some unconjugated hyperbilirubinemias, probably secondary to impairment of bilirubin glycuro-conjugation, are known (for a review see Ref. 2). Normally, enzyme activity is not a rate-limiting step for bilirubin excretion. Phenobarbital pretreatment enhances glycuronyltransferase activity but does not change the T_m of bilirubin (108). It is not known if locally formed bilirubin (109) and bilirubin of extrahepatic origin mix in a single homogeneous pool (109) (for a discussion see Ref. 2).

The kinetics of bilirubin and dye elimination suggest that biliary transport of conjugated bilirubin is the rate-limiting step (90, 92, 110) of canalicular secretion. The canalicular transport system can be saturated, and competition between bilirubin and BSP or cholescystographic agents occurs at this level. Furthermore, congenital transport defects exist in animals (Corriedale sheep) and man (Dubin, Johnson, and Rotor syndromes) where excretion of organic anions is impaired but bile salt secretion is normal (111). These studies suggest a carrier-mediated canalicular transport for bilirubin and other anions which is different from the bile salt carrier, although interactions between bile salt and bilirubin excretion are possible. For example, bile salts generally increase the T_m of bilirubin and BSP (see below).

Under physiological conditions, biliary secretion of bilirubin represents almost exclusively conjugated bilirubin, although significant unconjugated bilirubin secretion may occur during inhibition of glycuronyltransferase (112) or phototherapy in Gunn rats (113).

Biliary Secretion of Bile Salts Bile salt uptake has been studied by the Goresky multiple indicator dilution technique. In rat liver, uptake is sodium dependent (114, 116, 117) and follows Michaelis-Menten kinetics (115). Maximal uptake rates (V_{max}) were higher for cholate than for chemodeoxycholate (114, 118). Taurocholate and cholate showed the same V_{max} but the K_m was lower for the former. Similar results were obtained in dog liver; maximal rate and K_m were higher for taurocholate than for taurochenodeoxycholate (119, 120). These values of K_m and V_{max} are influenced by differences between bile salts in albumin binding (121) and the decrease, in intact liver, of the blood bile salt concentration during transfer from portal to centrolobular areas (119). However, taurocholate uptake by isolated rat liver cells yields roughly similar results. The plasma membrane has a very high binding capacity (3.8 nM/mg of cellular protein) for taurocholate. Uptake is temperature dependent and inhibited by ouabain (122) or the replacement of extracellular sodium by potassium. Taurocholate uptake is competitively inhibited by taurochenodeoxycholate (122). Altogether these results suggest that bile uptake is carrier mediated they do not prove that uptake occurs by means of a Na^+-dependent, gradient-coupled carrier (see Chapter 11).

The maximal uptake rate of bile salts greatly exceeds the excretory rate (T_m) and, during bile salt loading, the excretory T_m is the rate-limiting step for secretion. In physiological situations, however, the secretion rate does not approach the maximal capacity of hepatic transport systems. When physiological doses of various bile acids were infused in the bile fistula of the rat, taurocholate was most efficiently extracted, followed in decreasing order by glycocholate, cholate, deoxycholate, and chenodeoxycholate (123). Such differences could reflect either differences in affinity for albumin or in affinity for cellular uptake. However, since biliary secretion paralleled hepatic uptake, it is possible that hepatic uptake is a rate-limiting step in overall transport (123). Alternatively, the affinity of bile salts for hepatic uptake may parallel their affinity for cellular transport systems.

Intracellular and canalicular transport mechanisms are largely unknown. Some experimental results suggest that intracellular bile salts do not equilibrate in a single homogeneous pool (124). Canalicular transport can be saturated and is possibly carrier mediated. It is rate limiting. Transport mechanisms for bile salts are different from those involved in bilirubin secretion. Bile salts do not compete for secretion with dyes such as BSP or Rose bengal but, on the contrary, increase their maximal biliary excretion (125).

Conjugation is not essential for secretion of bile acids; taurine depletion in dogs results in secretion of unconjugated cholic acid (126). However, transfer to bile of unconjugated cholate in taurine-depleted dogs is significantly less than the transfer of cholate, as taurocholate, in non-taurine depleted dogs (127).

It is not known if sulfated bile salts are transported via the same pathways as are nonsulfated bile salts: the ester sulfate 3-α-hydroxy-5-cholinyltaurine infused in rats was shown to have a very low biliary T_m (0.28 μmol/min) and to have no effect on taurocholate secretion (76). However, maximal secretion rates were found to be similar for each natural bile salt and its corresponding sulfated derivative (128).

Interaction between bile salts and dyes during hepatic transport is shown by uptake experiments performed in situ. For example, the simultaneous injection of bilirubin and taurocholate did not alter their individual kinetic parameters for hepatic uptake (129), but cholic acid or dehydrocholic acid impaired hepatic uptake of BSP (130). Since inhibition of BSP uptake by taurocholate in isolated cells appears to be noncompetitive (122), the interaction is probably not at the membrane carrier level.

Canalicular Transport Many experimental studies showed that bile salt infusion increases T_m for BSP (125, 131–134) and bilirubin (135). This effect is not related to an increase of bile flow; other drugs, e.g., hydrocortisone (136) and theophylline (136, 137), which enhance bile flow, have no effect on BSP T_m. The effect is unexplained, although interaction at the level of the canalicular membrane has been postulated (125). Some results suggest that association with bile salt micelles increases the bilirubin T_m by sequestration. For example, during ultracentrifugation or Sephadex filtration of bile, bilirubin remains associated with a macromolecular complex formed chiefly by bile salts, lecithins, and cholesterol (138, 139). Also, in man, infusion of dehydrocholate, which does not form micelles, depresses the secretion of bilirubin (140). On the other hand, an increased T_m for dyes is not well explained by micellar sequestration; (*a*) high doses of dehydrocholate increase Rose bengal (110) as well as BSP, DBSP, and indocyanine green canalicular transport (141); (*b*) biliary transport of phenolphthalein-D-glucuronide was not increased during bile salt infusion, although its binding to micelles is similar to that of BSP and ICG (141); and (*c*) dehydrocholate increases the T_m of DBSP more than does taurocholate (131). Obviously, the exact experimental conditions are critical. During continuous dehydrocholate infusion, BSP T_m may be reduced rather than increased if BSP infusion starts after the achievement of a steady state of dehydrocholate secretion (130).

Biliary Lipid Secretion The two major lipids of bile are free cholesterol and phosphatidylcholine (142). Others are very minor; phosphatidylethanolamine constitutes only from 1–5% of bile phospholipids, sphingomyelin from traces to 2%, and lysolecithin from 1 to 3%. In vitro studies (143, 144) have shown that lecithin and cholesterol are solubilized in water as mixed micelles with bile salts.

Such micelles probably explain the apparent solubility of cholesterol in bile (145). Triangular coordinate diagrams can be used to calculate the limit of cholesterol solubility in bile salt-lecithin mixed micelles for concentrations resembling those of bile (144, 146) as a cholesterol saturation index (147). This method seems to predict the lithogenic potential of bile. First, cholesterol gallstones appear in people who have a bile supersaturated with cholesterol. Second, drugs which decrease cholesterol saturation promote gallstone dissolution (148, 149). It seems logical to suppose that such micellar associations not only solubilize bile lipids but also explain their secretion.

Biliary Lecithin Secretion The structure of bile lecithins differs from that of hepatic lecithins; 70% of bile lecithin is accounted for by 1-palmitoyl-2-linoleyl, 1 palmitoyl-2-oleoyl, and 1-stearyl-2-linoleyl phosphatidylcholines. Also, the specific radioactivity of linoleyl lecithins is higher in bile than in plasma or liver after administration of ^{32}P, [^{14}C] fatty acids, or [^{14}C] choline (150, 151) and is also higher in bile than in isolated canalicular membranes or microsomes (152). These results have suggested that biliary lecithins are synthesized in a specific compartment with a very rapid turnover (153). However, the various molecular species of linoleyl lecithins due to the presence of different fatty acids in the 1 position were not separated. More recent studies (154, 155), using mass spectrometry analysis of deuterium incorporation from [1,1-^{2}H] ethanol into the glycerol moieties of various molecular species of biliary and hepatic phosphatidylcholines, have shown that, in the rat, identical phosphatidylcholine molecular species had the same half-lives in bile as a liver. Thus, Curstedt and Sjovall concluded that the corresponding molecular species of biliary and hepatic lecithins are probably synthesized in the same pool (154). Differences between biliary and hepatic lecithins must then be explained by the mechanism of secretion.

It is well established that secretion of bile salts is necessary for lecithin secretion (156, 158). In rats (159), man (162), and dogs (28), the relationship is curvilinear; bile lecithin output reaches a plateau at high bile salt excretion. Also, bile salts increase lecithin synthesis (157). Most results favor a hypothesis that the action of bile salts on secretion is related to the formation of mixed bile salt-lecithin micelles; (*a*) dehydrocholate and similar salts decrease biliary lecithin secretion (159). Occasional increases in lecithin secretion observed in isolated perfused livers (156) or during chronic administration in mice (160) may be explained by dihydroxylated metabolites which are able to form mixed micelles. In the same way, derivatives which are secreted in bile and have physical properties similar to those of cholate increase the secretion of bile lecithins (161, 163); (*b*) in rat, mouse, and dog, the curvilinear relation between lecithin and bile salt secretion resembles a solubilization curve of micelles acting on a constant surface area of lecithin in analogy to the Langmuir isotherm (164); (*c*) in rat, extrapolation to low bile salt secretion shows that bile lecithin secretion is suppressed in accordance with the micelle hypothesis. In aqueous solutions, bile salt micelles appear when the bile salt concentration exceeds the critical micellar concentration (CMC) and the micelles remain in equilibrium with some bile salts in true, monomeric solution (165). The micelle hypothesis thus implies that secretion of bile salts below CMC can take place without accompanying lecithin secretion. Similarly, the choleretic, SC-2644, decreases bile lecithin and cholesterol secretion in the dog probably by lowering the bile salt concentration to or below CMC without influencing the bile salt secretion rate. Secretin which dilutes bile salts at a more distal site than the canaliculus does not have the same effect. Phenobarbital pretreated rats behave in the same way; bile flow increases due to a stimulation of bile salt-independent flow (32) by an action which is probably not related to the well known effect on microsomes but to a specific modification of canalicular membrane proteins (70). In phenobarbital pretreated rats, lecithin secretion was reduced nearly to zero for a bile salt secretion rate higher than in controls (166), as would be expected were the bile salt concentration reduced below CMC by the increased canalicular flow. Altogether these results would appear generally to explain lecithin secretion. Selective transport of lecithin, however, remains unexplained. Membrane phospholipids contain molecular species other than lecithins, and these species are able to combine in vitro with bile salts as mixed micelles (167). It is possible that there is an asymmetric distribution of phospholipids with preferential repartition in the external (canalicular) layer of the membrane (166), as has been shown for other membranes (168).

Cholesterol Secretion Bile cholesterol is nonesterified and seems to come from a pool different from the substrate pool used for bile acid synthesis, but there is probably not a preferential secretion of newly synthesized cholesterol (169, 170).

Biliary secretion of cholesterol seems to depend on bile salts in the same way as does lecithin secretion in rat (159) and dog (28). In man and other species, cholesterol secretion decreased at a lower rate than did lecithin secretion during biliary diversion (163). A linear relationship was observed between lecithin and cholesterol secretion (159). Extrapolation for zero lecithin secretion suggested the existence of cholesterol secretion "independent" of lecithin or even bile salt secretion (159). However, in the rat after phenobarbitone pretreatment (166) the secretion of biliary cholesterol decreased more sharply than did that of lecithin. The curvilinear relationship obtained indicated that cholesterol secretion was suppressed at zero lecithin secretion and suggested that cholesterol secretion, like lecithin secretion, requires the presence of bile salt micelles in the vicinity of the canalicular membrane. It seems possible, therefore, that bile salts act on a preformed complex of lecithin and cholesterol (144).

Biliary secretion of a preformed association of cholesterol and lecithins would explain how a bile supersaturated with cholesterol can be formed by the liver. In experimental or human cholelithiasis or during biliary drainage, triangular coordinate plots indicate a supersaturated bile which could be formed if the lecithin contained a large amount of cholesterol and if bile salts interact with lecithin independently of the amount of cholesterol.

It has been suggested that membrane proteins may be associated with lipids in bile (171). Consistent with this hypothesis are the following results: (*a*) extraction of bile with organic solvents yielded, in addition to lipids, a proteic fraction with a constant amino acid composition; (*b*) during ultracentrifugation, the lecithin-cholesterol complex of bile had a sedimentation coefficient higher than pure mixed micelles of similar composition (173); and (*c*) electron microscopy of bile showed structures similar to lipoproteins (172).

Control of Bile Saturation with Cholesterol

The close relationship between secretion of bile salts and of lecithin or cholesterol suggests that variations in bile salt secretion represent the main determining factor in bile cholesterol saturation.

Quantitative Regulation of Bile Salt Secretion In the absence of biliary obstruction or hepatic disease, bile salt secretion rate does not approach the T_m; it is limited by the enterohepatic circulation and bile salt synthesis. Compartmental models for man have been recently developed (174). The bile salt pool varies from 3 to 5 g (176, 178), and similar values are found in primates (177). In man, the pool consists of approximately identical levels of cholate, deoxycholate, and chenodeoxycholate (178) which circulate from bile to intestine to liver about 6–10 times per day, i.e., twice during the absorption of a single meal (176). Losses are equal to bile salt synthesis, about 300–600 mg per day, and they are related to body weight (179). Synthesis as a function of body surface area is lower in the neonate and infants but exceeds the fecal loss (180). Maintenance of the pool is dependent on intestinal function and the ability of the liver to compensate for intestinal losses.

Basic Mechanisms Intestinal absorption of bile salts is an active as well as a passive process. In man, passive permeability of the jejunum to bile salts is twice that of the ileum (181, 182) and greater for nonionized than for ionized species by a factor of 4 in the rat (184). Uptake of ionized bile salts is limited by mucosal surface permeability, but the rate-limiting step for nonionized bile acids may be the adjacent unstirred layer (183). Passive permeability decreases with the number of hydroxyl groups and with conjugation. Passive bile salt absorption also occurs in the colon with the same order of permeabilities.

Active intestinal absorption of bile salts takes place in the ileum. Maximal active absorption is greater for tri- than for dihydroxy salts, and K_m values are greater for conjugated species. The number of hydroxy groups either has no effect (184) or decreases affinity (185). Adaptive increase of active ileal absorption occurs after jejunal resection, so that the bile salt pool increases in the rat (186).

Intestinal bacteria produce "secondary bile salts" by deconjugation and 7-dehydroxylation. In man, the principal secondary bile salts are deoxy- and lithocholate. In man, 18% of glycocholate is deconjugated or 7-dehydroxylated per cycle (189). Dehydroxylation generally occurs on free bile salts but seems to be possible in man without deconjugation (188).

Liberated amino acids are catabolized to CO_2, and this is the basis of a breath test which measures the extent of bacterial metabolism of bile acids in diarrheal states. The virtual absence of lithocholic acid in human bile is now explained (187) by sulfation in the liver which prevents further enterohepatic cycling, sulfated lithocholate not being absorbed (189). The presence of bacteria in the intestine increases fecal elimination of bile salts; the bile salt pool size is greater in germ-free rats than in conventional animals (190).

The enzymatic steps of bile acid synthesis from cholesterol and their mechanism of control have been recently reviewed (191, 192). Microsomal 7α-hydroxylation of cholesterol is the initial reaction and is the rate-limiting step, although chenodeoxycholic acid may partly originate from another metabolic pathway (193). Microsomal 7α-hydroxylase activity in vitro has been generally measured by the transformation of exogenous radioactive cholesterol into product (194), though methods measuring the mass of product are probably more reliable (195). It is difficult to achieve substrate saturation of the enzyme in vitro. In vivo bile salt synthesis can be calculated from tracer methods as developed by Lindstedt (196) or from bile salt secretion after biliary drainage. Newly synthesized cholesterol is the preferred substrate for 7α-hydroxylation. This reaction as well as bile salt synthesis is increased by interruption of the enterohepatic circulation (197). In the rat, complete biliary diversion induces an increase that is greater than the basal secretion rate (198). In the rhesus monkey, synthesis is increased 4-fold, and, when enterohepatic circulation decreases by more than 20%, bile salt secretion begins to decrease (199). Supplementation of the diet with bile salts generally depressed 7-hydroxylation of cholesterol and bile salt synthesis (200), with some differences among the salts (200, 202). In man, chenodeoxycholate decreases cholate synthesis by 50% (201).

Influences on Bile Secretion Diurnal variations of hydroxylation of cholesterol occur independently of changes in bile salt enterohepatic circulation; in rat, bile salt synthesis increases during the night (203). However, newly synthesized bile salts usually represent only 2–5% of the daily biliary secretion, thus these changes are probably of minor importance. Diurnal variations of bile salt secretion most probably occur in relation to changes of enterohepatic circulation induced by meals, particularly in species having a gall bladder. Post-prandial influx of bile salt to the liver has been shown by an increase of serum bile salts (204) and this role of meals persists in cholecystectomized man (204). This shows that changes of bile salt secretion depend on bile salts pool and frequency of enterohepatic recycling of this pool. These two factors may play a role in cholelithiasis: a diminished bile acid pool size has been found in gallstone patients (217). It may be related in part to a gall bladder dysfunction since cholecystectomy in animals is associated with a decrease of the bile salt pool (220). Its role in the production of supersaturated bile is not proven. Cholecystectomy decreases (217) or does not modify (218) the cholesterol saturation of bile. The diminished pool size of bile salts in gallstone patients seems related to an increased frequency of enterohepatic cycling (205) and therefore may

(221) or may not (205) be associated to a drop of bile salt secretion. Interruption of enterohepatic cycling due to an overnight fasting increases the cholesterol saturation of bile in normal man and gallstone patients (205). Attempts were made to decrease cholesterol index saturation of bile by administration of bile salts to gallstone patients. Cholate increases or does not influence cholesterol saturation (148). Low doses of decoxycholate, 150 mg per day, increased cholesterol saturation of bile (206); 750 mg per day had no effect (207). Chenodeoxycholate administration decreased cholesterol saturation of bile mainly by decreasing cholesterol secretion relative to bile salt secretion (208). Intraduodenal administration during acute interruption of the enterohepatic circulation in cholecystectomized man showed the same differences: cholate increased bile cholesterol secretion but chenodeoxycholate did not (209). There are marked species differences in response to bile salt administration, but, as chronic and acute administration were rarely compared in the same species, the basis for differences is not clear. In dog, chenodeoxycholate infusion increased cholesterol secretion much more than did taurocholate, and, in perfused dog liver, chenodeoxycholate induced a greater secretion of phospholipid and cholesterol (210). In the cat, taurodeoxycholate and taurochenodeoxycholate increased cholesterol secretion more than did taurocholate (211), but the three bile salts had the same effects on lecithin secretion.

In rat, chronic administration of cholate or chenodeoxycholate had similar effects on saturation of bile with cholesterol (202). Chronic administration of cholate induced cholelithiasis in mice (212); chenodeoxycholate, on the contrary, was not lithogenic. However, in mice, when sitosterol (1% of the diet) is associated with either bile salt, cholate as well as chenodeoxycholate has the same effect on biliary cholesterol secretion relative to bile salt secretion (unpublished results).

These various results are difficult to explain simply on the basis of differences in the lipid binding capacities of various bile salt micelles. It was shown that mixed micelles formed with lecithin and dihydroxycholanates had a cholesterol holding capacity higher than taurocholate mixed micelles (213). However, the differences are not important for concentrations similar to those encountered in bile. Differences in the action of various bile acids on cholesterol metabolism probably play a role which is as yet undefined: (*a*) chenodeoxycholate in rat and hamster is more inhibitory to hepatic cholesterol biosynthesis (3 methyl-3-glutaryl-CoA reductase) than to cholesterol oxidation (7-hydroxylase). Cholate administration, on the other hand, decreased the activity of the two enzymes equally (202, 214). However, the relationships, if any, between these enzymic steps and cholesterol saturation of bile are still unknown. (*b*) It is also possible that cholesterol absorption may depend on the structure of the associated bile salt. It has been reported that in man cholate administration increases cholesterol absorption but chenodeoxycholate probably does not (215). However, compartmental analysis did not confirm this result (216)

Factors other than Bile Salts Administration of lecithin precursors may decrease bile cholesterol saturation, e.g., glycerophosphate ingestion decreases the bile lithogenic index. Dietary choline does not change lecithin secretion or the normal output of bile salts, but it does increase the lecithin secretion plateau level reached during bile salt administration (225). In the same way, the cholesterol secretion for a given secretion of bile salts may vary. As discussed above, the slope of the regression line relating cholesterol and bile salt secretion may vary as a function of the molecular structure of the bile salt. In Pima Indians, who have a high prevalence of gallstones, more cholesterol was secreted in bile per molecule of bile salt than in Caucasians (221). In patients who have a high cholesterol secretion, there is frequently an increased lecithin secretin (221, 222). Obese patients have higher secretion rates of cholesterol in bile than do controls (223), possibly in relation to an increased caloric intake (223a). Clofibrate and ethinyl estradiol (224) increase bile saturation with cholesterol by a selective increase of bile cholesterol secretion.

There are, therefore, probably many ways of modifying the cholesterol saturation index of bile. Theoretically a simultaneous increase of the bile salt pool and bile salt secretion with a decrease of cholesterol secretion may be obtained with bile salts such as chenodeoxycholate. Simultaneous administration of cholate and inhibitors of cholesterol absorption may lead to the same results, with possibly further improvement by adding lecithin precursors.

BILIARY MOTILITY

Recently, Shelhammer (226) found that common bile duct (CBD) contractions elicited an active expulsion of bile in the opossum. However, the CBD did not show any rhythmic activity (227, 228), although isolated CBD was able to contract in response to epinephrine (229), CCK (230, 231), or acetylcholine (232).

The existence of the sphincter of Oddi (SO) was questioned for a long time, and closure of the lower CBD was attributed to duodenal muscle. Persson and Ekman (233) compared, under constant flow, pressures through the cat SO with those through an artificial sphincter composed of vein fragments crossing the duodenal wall. They found that morphine increased while isoprenaline decreased pressure in both models. It was concluded that duodenal muscle is able to occlude the lower part of the CBD. Nevertheless, the existence of an independent SO, previously proven by anatomical investigation (234), is presently recognized by electrophysiological criteria; the SO exhibits an intrinsic electrical activity (235). In the rabbit, bipolar suction electrodes recorded a regular pattern, i.e., bursts of spikes every 3–5 s (236), different from that of the duodenum and closely related to pressure variations in the CBD (237).

The lower part of the CBD exhibits a rhythmic activity of opening and closing movements in agreement with CBD pressure waves demonstrated in man

(238–240), dog (241, 242), and cat (243). Rhythmic contractions of the SO, which moved bile into the duodenum by a pumping mechanism, were also described (230, 231), although simultaneous electromyography and radiocinema tography showed that bile flow stops during each burst of spikes (244). In the fasting state, there is a permanent tonic contraction of the SO, as directly observed through a Thomas cannula (245), which could be essential for gall bladder filling (246).

Actions of Cholecystokinin

Cholecystokinin (CCK) is believed to be the main stimulus for gall bladder contraction in response to a meal (247). A pressure increase inside the gall bladder, after CCK administration, has been observed in the anesthetized cat (248), in man (249), and in the conscious dog (250). CCK-induced contraction of the isolated gall baldder was demonstrated for rabbit (251–254), guinea pig (250–257), opossum (258), and man (259).

The mechanism of CCK action on the gall bladder is not known, though it is clear that it is not mediated by the nervous system. CCK action on the in situ gall bladder is not modified in conscious dogs by vagotomy (260) nor on isolated gall bladders by atropine in dog (230), rabbit (251, 253), and guinea pig (257). Moreover, α- and β-adrenergic blockers have no effect either on the conscious dog gall bladder (261) or on CCK-induced contraction of the isolated gall bladder (230). Finally, tetrodotoxin does not inhibit CCK action (257). It may be concluded that CCK acts directly on gall bladder muscle under both in vivo and in vitro conditions (257, 262).

It is possible to assay the CCK activity of human serum using the isolated rabbit gall bladder in vitro (263). CCK increases phosphodiesterase activity in the gall bladder (264) and decreases its cAMP concentration (255, 256). Prostaglandin E_2, which also decreases cAMP levels, was shown to have a CCK-like activity (251) and could be the mediator of CCK action (266), but this has not been confirmed (267). Alternatively, it has been suggested that CCK acts by increasing cGMP (268). Lin (261, 269) has shown that endogenous CCK, released by introduodenal amino acid infusion, elicited a dose-related contraction of the gall bladder in conscious dogs.

The question of CCK action the CBD is still controversial. In the dog, CCK increased CBD contraction (230) and increased pressure when the CBD was occluded at both ends (270). Recordings of intracellular electrical activity showed an increase of spikes when CCK was added to in vitro guinea pig CBD (271), but CCK does not induce a propulsive motility (272).

Since Sandblom, Voegtlin, and Ivy's work (273), it has been believed that CCK decreases resistance through the SO and increases bile flow as many papers have confirmed in cat (233, 274–277), dog (278–281), and man (280, 282, 283). In man, as studied by a Kehr drain, the decrease in resistance could be preceded by a hypertonic phase (284). In the dog, some workers found that CCK increased sphincter activity (230, 282, 285). In the rabbit, it was found that bolus CCK injection gave a marked rise of sphincter activity, reaching its maximum at about 60 s (286). Increasing doses of CCK elicited a log-dose response curve. There is a simultaneous pressure increase and flow decrease which could be related to species-related anatomical differences in SO structure (288).

As with gall bladder, CCK action on the SO seems to be a direct action on muscle cells, since it is not abolished by vagotomy (288) or atropine (243). Similarly, PGE_2 increases cAMP levels, as does CCK, probably by activating adenylate cyclase (289), and could be the mediator of CCK action.

Action of Other Gastrointestinal Peptides

CCK activity depends upon the COOH-terminal amide heptapeptide (290), a sequence also found in gastrin (291). The cholecystokinetic potency of gastrin is only about 1/15th that of CCK as determined in dog (280, 292), cat (293), and in rabbit isolated gall bladder (252). However, this action is observed only with pharmacological doses (294). A CCK-like action of gastrin was observed also on the CBD and the SO (230, 295). Lin observed that resistance of the SO was decreased by physiological doses, but increased by pharmacological doses, of gastrin (261). In cholecystectomized patients, gastrin has no effect on bile pressure (296).

The synthetic COOH-terminal octapeptide of CCK (OP-CCK) (298) exhibits the same effects as the entire CCK molecule. In the dog OP-CCK-induced gall bladder contraction was not dose related (258, 298). In the healthy man, 20 ng/kg of OP-CCK reduces gall bladder size by 40% (299). On a molar basis OP-CCK is eight times more potent than CCK on guinea pig gall bladder (257).

Secretin exhibits a cholecystokinetic effect on isolated dog gall bladder (230) and increases bile pressure in the conscious dog (250). This cholecystokinetic effect can also be shown on the SO (231), although it is very small. Lin and Spray (282) found that secretin decreased CBD resistance in the conscious dog. On the contrary, secretin increases CCK-induced gall bladder contraction in the cat, guinea pig (293), and dog (294). In human patients with a Kehr drain its action is not significant (296).

Glucagon was found to have no effect on gall bladder (293) and not to decrease bile pressure by relaxing the SO, as CCK does (295), although Lin (269) found that glucagon relaxed both the gall bladder and the SO in dogs (281). Glucagon's action on the biliary tract seems to be a direct one on muscle cells because its relaxant effect was not blocked by a variety of drugs, viz., propanolol, phenoxybenzamine, atropine, and pentolinium (261).

Caerulein, a decapeptide isolated from the skin of a frog (*Hyla caerulea*), has the same structure as the COOH-terminal portion of CCK. Caerulein elicits a contraction of the isolated dog gall bladder, three times that induced by CCK (300), and a gall bladder contraction in the conscious dog, even after a fatty meal (300). Similarly, caerulein relaxes the SO, and more so when the SO is more contracted (300, 301). In the anesthetized rabbit, caerulein, like CCK, increases electromyographic activity and the CBD pressure (286). Caerulein action seems to be a direct one on muscle cells, since it is not abolished by atropine, tetrodotoxin splanchnicectomy, adrenalectomy, or reserpine (302).

Action of Autonomic Nervous System

Any action of the vagus nerve is still controversial. Electrical stimulation of hepatic and celaic branches of the vagus nerve does not modify CBD pressure in the conscious dog (303). On the contrary, stimulation of lesser omentum parasympathetic nerves elicit a pressure increase related to current intensity and abolished by atropine (304). In the dog, electrical stimulation of the cephalic end of the sectioned vagus nerve decreased bile flow through the SO (305). Similarly, electrical stimulation of the vagus nerve induced an increased electromyographic activity of the SO in the rabbit (306). Discrepancies between results are probably related to innervation complexities (307).

Vagotomy induces a gall bladder dilatation in dog (308, 309) and in man (310, 311), without delaying gall bladder emptying (233, 312). Others reported that vagotomy does not modify gall bladder function in the dog (313). Vagotomy was believed to enhance the incidence of biliary lithiasis in animals (310, 314, 315) and in man (316), but this has not been proved (317). Vagotomy was shown to induce a 10% decrease in the nervous fibers of the gall bladder wall (318).

Vagotomy influence on the SO is controversial. Some papers have reported sphincter relaxation (319, 320). In the dog, results are variable (322) or not significant (321), while in the rabbit, vagotomy does not change electromyographic patterns (235, 323). In the conscious dog, metacholine was reported to elicit a spasm of CBD (261).

The presence of sympathetic fibers in the biliary tract was demonstrated by a fluorescence method (324). However, stimulation of the sympathetic system induced variable results. In some cases stimulation of the right splanchnic nerve induces a contraction of the SO, which is abolished by α-blockers, and a gall bladder dilatation, abolished by β-blockers (325). Others have reported that stimulation of the peripheral sympathetic system elicits a gall bladder contraction with closure of the SO (304, 305). Endogenous catecholamines decrease the SO tonus (326), but this effect occurs after vagotomy only (327). On the other hand, sympathetic blockade does not have any effect on the biliary tract (328). If α- and β-receptors are present in the gall bladder, the latter are dominant and β-2 in type (248). At the level of SO, both types of receptors have been demonstrated. Some authors report that α-receptors are dominant (227), which would explain contraction due to sympathetic stimulation. Most papers find a dominance of β-receptors, whose stimulation elicits opening of the SO and a pressure decrease (329, 330). Lesions induced by chronic irritation of the right splanchnic nerve by powdered pumice stone closely mimicked those of human odditis and were found together with a pressure increase and a gall bladder dilatation (331). A normal reflex arc, including splanchnic and vagus nerves, is necessary to observe this phenomenon. On the contrary, there could be a vago-sympathetic reflex arc whose stimulation induces a closure of SO (305).

It is difficult to understand nervous phenomena interacting on the biliary tract. With regard to the SO, it is possible that nervous influences act on muscle fibers through blood flow mechanisms. In turn, these blood phenomena could modify the pressure of the sphincter (332) through the opening of arterio-venous shunts (327).

Other Factors

Intraduodenal administration of 0.1 N HCl regularly induces a spasm of the SO in the conscious Thomas-fistula dog great enough to completely stop the bile flow response to a fatty meal (245).

SO contracture after morphine administration has been known for a long time. Within 10 min of administration, morphine induced a strong and long-lasting spasm, which was not always relieved by atropine. The action of morphine could be prevented by procaine (333) or isoprenaline (233) infused through the CBD, that is, by β-receptor stimulation, contrary to the findings of Crema et al. (334). However, it was shown by Persson and Ekman (233) that morphine effects were the same on the artificial sphincter as on SO. Similarly, Crema et al. (334) found that morphine-induced contraction of the SO was greater in situ than in the in vitro organ, both in the calf and cat, suggesting that the morphine-increased threshold for bile flow through the SO is related to a duodenal mechanism. However, electromyography shows a noteworthy increase of SO action potentials after morphine administration (235). Moreover, Mester et al. (335) induced stenosis by 3 months of morphine administration in the rabbit. It is concluded that, in addition to a prominent action on the duodenal wall, morphine can also act directly on the SO. CCK relaxed the morphine-induced CBD spasm in guinea pig (336) and in conscious dog (261).

The effects of alcohol on the SO are a matter of discussion. It is found that alcohol induces hypertonia of the sphincter in man (337) and a closure of the papilla in the conscious Thomas-fistula dog (245). In the anesthetized rabbit, alcohol greatly increases both sphincter electrical activity and biliary pressure. Since this hypertonic effect disappears in vitro and in vivo after a 2 mg/kg hr^{-1} hexamethonium infusion, it can be assumed that a nervous pathway is required. Vagotomy greatly reduces, but does not abolish, alcohol effects on electromyographic activity (236, 323).

In the cat, infected bile induced a long-lasting spasm of the sphincter (338) related to deconjugation and dehydroxylation of bile salts by bacteria. These same biochemical modifications were related also to the greater toxicity of infected, as against normal, bile for pancreas (339), which could explain some cases of acute pancreatitis.

Antibiotics are reported to reduce SO tonus (340). Neuroleptic drugs, such as Fontanyl and dehydrobenzoperidol, increase SO tonus and biliary pressure (341, 342).

REFERENCES

1. Erlinger, S., and Dhumeaux, D. (1974). Mechanisms and control of secretion of bile water and electrolytes. Gastroenterology 66:281.
2. Bissel, D. M. (1975). Formation and elimination of bilirubin. Gastroenterology 69:519.
3. Elias, H., and Sherrick, J. C. (1969). Morphology of the Liver, p. 1. Academic Press, New York.
4. Steiner, J. C., and Carruthers, J. S. (1961). Studies on the fine structure of the biliary tree. Am. J. Pathol. 38:639.
5. Schaffner, F. (1975). The ultrastructure of bile secretion. *In* W. Gerok and K. Sickinger (eds.), Drugs and the Liver, p. 91. Stuttgart, New York.
6. Farquhar, M. G., and Palade, G. E. (1963). Junctional complex in various epithelia. J. Cell Biol. 17:375.
7. Compagno, J., and Grisham, J. W. (1974). Scanning electron microscopy of extrahepatic biliary obstruction. Arch. Pathol. 97:348.
8. Layden, T. J., Schwartz, J., and Boyer, J. L. (1975). Scanning electron microscopy of the rat liver. Studies of the effect of taurolithocholate and other models of cholestasis. Gastroenterology 69:724.
9. Oschman, J. L., and Berridge, M. J. (1971). The structural basis of fluid secretion. Fed. Proc. 30:49.
10. Beams, H. W., and Kessel, R. G. (1968). The Golgi apparatus: structure and function. Int. Rev. Cytol. 23:209.
11. Schaffner, F., Bacchin, P. G., Hutterer, F., Scharnbeck, H. H., Sarkozi, L. L., Denk, H., and Popper, H. (1971). Mechanisms of cholestasis 4. Structural and biochemical changes in the liver and serum in rats after bile duct ligation. Gastroenterology 60:888.
12. Phillips, M. J., Oda, M., Fisher, M. M., Jeejeebhoy, K. N., and Steiner, J. W. Possible relationship between hepatic microfilaments and bile flow. Second NATA Advanced Study Institute on the Biliary System. In press.
13. Phillips, M. J., Oda, M., Mak, E., Fisher, M. M., and Jeejeebhoy, K. N. (1975). Microfilament dysfunction as a possible cause of intrahepatic cholestasis. Gastroenterology 69:48.
14. Graf, J. (1975). Possible role of a microtubular microfilamentous system in bile formation. Digestion 12:306.
15. Bauduin, H., Stock, C., Vincent, D., and Grenier, J. F. (1975). Microfilamentous system and secretion of enzyme in the exocrine pancreas. Effect of cytochalasin Br. J. Cell Biol. 66:165.
16. Brauer, R. W. (1959). Mechanisms of bile secretion. J. Am. Med. Assoc. 169:1462.
17. Hanzon, V. (1952). Liver cell secretion under normal and pathologic conditions studied by fluorescence on living rats. Acta Physiol. Scand. (Suppl.) 28:101.
18. Nohrwold, D. L., and Shariatzedeh, A. N. (1971). Role of the common bile duct in formation of bile and in gastrin induced cholesis. Surgery 70:147.
19. Chenderovitch, J. (1972). Secretory function of the rabbit common bile duct. Am. J. Physiol. 223:695.
20. Forker, E. L. (1968). Bile formation in guinea pigs: analaysis with inert solutes of graded molecular radius. Am. J. Physiol. 215:56.
21. Forker, E. L. (1967). Two sites of bile formation as determined by mannitol and erythritol clearance in the guinea pig. J. Clin. Invest. 26:1189.
22. Wheeler, H. O., Ross, E. D., and Bradley, S. E. (1968). Canicular bile production in dogs. Am. J. Physiol. 214:866.

23. Sacks, J., and Bakshy, S. (1957). Inulin and tissue distribution of pentose in nephrectomized cats. Am. J. Physiol. 189:339.
24. Schanker, L. S., and Hogben, C. A. M. (1961). Biliary excretion of inulin sucrose and mannitol: Analysis of bile formation. Am. J. Physiol. 200:1087.
25. Glasinovic, J. C., Dumont, M., Duval, M., Erlinger, S., and Benhamou, J. P. (1972). Hepatocellular uptake of erythritol, mannitol and sucrose in the dog. Digestion 6:254.
26. Forker, E. L. (1970). Hepatocellular uptake of inulin, sucrose and mannitol in rats. Am. J. Physiol. 219:1568.
27. Erlinger, S., Dhumeaux, D., Berthelot, P., and Dumont, M. (1970). Effects of inhibitors of sodium transport on bile formation in the rabbit. Am. J. Physiol. 219:416.
28. Wheeler, H. O., and King, K. K. (1972). Biliary excretion of lecithin and cholesterol in the dog. J. Clin. Invest. 51:1337.
29. Forker, E. L., Hicklin, T., and Sornson, H. (1967). The clearance of mannitol and erythritol in rat bile. Proc. Soc. Exp. Biol. Med. 126:115.
30. Boyer, J. L. (1971). Canalicular bile formation in the isolated perfused rat liver. Am. J. Physiol. 221:1156.
31. Erlinger, S., Dhumeaux, D., and Benhamou, J. P. (1969). La secretion biliaire du lapin. Preuves en faveur d'une importante fraction independante des sels biliaires. Rev. Fr. Et. Clin. Biol. 14:144.
32. Berthelot, P., Erlinger, S., Dhumeaux, D., and Preaux, A. M. (1970). Mechanism of phenobarbital-induced hypercholeresis in the rat. Am. J. Physiol. 219:809.
33. Graf, J., and Peterlik, M. Mechanism of transport of inorganic ions into bile. Second NATO Advanced Study Institute on the Biliary System. In press.
34. Graf, J., and Peterlik, M. (1973). Route of sodium transport into bile. Digestion 8:484.
35. Sadig, S., Rao, S. P., and Enquist, I. F. (1972). Hepatic congestion and bile secretion. Arch. Surg. 105:749.
36. Richards, T. G., and Thomson, J. Y. (1961). The secretion of bile against pressure. Gastroenterology 40:705.
37. Strasberg, S. M., Dorn, B. C., Redinger, R. N., Small, D. M., and Egdehl, R. H. (1971). Effects of alteration of biliary pressure on bile composition. A method for study. Primate biliary physiology V. Gastroenterology 61:357.
38. Wheeler, H. O. (1968). Water and electrolytes. *In* C. F. Code (ed.), Handbook of Physiology 6: Alimentary Canal, Vol. 5, p. 2409. American Physiological Society, Washington, D.C.
39. Moore, E. W., and Dietschy, J. M. (1964). Na and K activity coefficients in bile and bile salts determined by glass electrodes. Am. J. Physiol. 206:111.
40. Chenderovitch, J., Phocas, E., and Rautureau, M. (1963). Effects of hypertonic solutions on bile formation. Am. J. Physiol. 205:863.
41. Brauer, R. W., Leong, G. F., and Holloway, R. J. (1954). Mechanics of bile secretion. Effect of perfusion pressure and temperature on bile flow and bile secretion pressure. Am. J. Physiol. 177:103.
42. Brauer, R. W., Pessoti, R. L., and Pizzolato, P. (1951). Isolated rat liver preparation. Bile production and other basic properties. Proc. Soc. Exp. Biol. Med. 78:174.
43. Sperber, I. (1959). Secretion of organic anions in the formation of urine and bile. Pharmacol. Rev. 11:109.
44. Preisig, R., Cooper, H. L., and Wheeler, H. O. (1962). The relationship between taurocholate secretion rate and bile production in the unanesthetized dog during cholinergic blockade and during secretin administration. J. Clin. Invest. 41:1152.
45. Balabaud, C. H., Kronk, and Gumucio, J. J. (1975). The bile salt non-dependent fraction of canalicular bile water in the rat. Gastroenterology 69:805 (Abstr.).
46. Gerolami, A., Crotte, C., Montet, J. C., Vigne, J. C., Grangier, M., and Mule, A. (1972). Métabolisme de l'acide dehydrocholique. Etude chromatographique de ses dérivés dans la bile. Etude *in vitro* de leur capacité à former des micelles mixtes. Biol. Gastroenterol. (Paris) 5:265.

47. Solloway, R. D., Hofmann, A. F., Thomas, P. J., Schoenfield, L. J., and Klein, P. D. (1973). Triketocholanoïc acid: hepatic metabolism and effect in bile flow and biliary lipid secretion in man. J. Clin. Invest. 52:715.
48. Erlinger, S., Dumont, M., Berthelot, Panol, and Dhumeaux, D. (1971). Comparison of the choleretic effects of dehydrocholate and glycodeoxycholate in the rabbit. Digestion 4:144.
49. Hoenig, V., and Preisig, R. (1973). Organic anionic choleresis in the dog: comparative effects of bromsulfalein, ioglycamide and taurocholate. Biomedicine 18:23.
50. Meyer Brunot, H. G., and Keberle, H. (1971). What role do choleretic agents play in bile formation? Digestion 4:166.
51. Boyer, J. L., and Klatskin, G. (1970). Canalicular bile flow and bile secretory pressure: Evidence for a non bile salt-dependent fraction in the isolated perfused rat liver. Gastroenterology 59:853.
52. Dowling, R. H., Mack, E., Picott, J., Berger, T., and Small, D. M. (1968). Experimental model for the study of the enterohepatic circulation of bile in rhesus monkeys. J. Lab. Clin. Med. 72:169.
53. Preisig, R., Buher, H., Stirnemaun, H., and Tauber, J. (1969). Postoperative choleresis following bile duct obstruction in man. Rev. Fs. Et. Clin. Biol. 14:151.
54. Prandi, D., Erlinger, S., Glasinovic, J. C., and Dumont, M. (1975). Canicular bile production in man. Eur. J. Clin. Invest. 5:1.
55. Graf, J., and Peterlik, M. (1972). Inɔrganic ion transport, bile flow and bile secretory pressure in the isolated perfused rat liver: Evidence for sodium dependent bile flow. Digestion 6:255 (Abstr.).
56. Skou, J. C. (1965). Enzymatic basis for active transport of Na^+ and K^+ across cell membranes. Physiol. Rev. 45:596.
57. Graf, J., and Peterlik, M. (1972). Choleretic effects of ouabain and ethacrynic acid in the isolated perfused rat liver. Naunyn Schmidebergs Arch. Pharmacol. 272:230.
58. Shaw, H., Caple, I., and Heath, T. (1972). Effect of ethacrynic acid on bile formation in sheep, dogs, rats, guinea pigs and rabbits. J. Pharmacol. Exp. Ther. 182:27.
59. Allen, J. L., and Scwartz, A. (1969). A possible biochemical explanation for the insensitivity of the rat to cardiac glycosides. J. Pharmacol. Exp. Ther. 168:42.
60. Boyer, J. L., and Reno, D. (1975). Properties of Na^++K^+-activated ATPase in rat liver plasma membranes enriched with bile canaliculi. Biochim. Biophys. Acta 401:59.
61. Boyer, J. L. Second NATO Advanced Study Institute on the Biliary System. In press.
62. Dhumeaux, D., Erlinger, S., Benhamou, J. P., and Fauvert, R. (1970). Effects of rose bengal on bile secretion in the rabbit: inhibition of a bile salt independent fraction. Gut 11:134.
63. Laperche, Y., Launay, A., and Oudea, P. (1972). Effects of phenobarbital and rose bengal on the ATPase of plasma membranes of rat and rabbit liver. Gut 13:920.
64. Morris, T. Q. (1972). Choleretic responses to cyclic AMP and theophylline in the dog. Gastroenterology 62:187.
65. Barnhart, J., Ritt, D., Ware, A., and Combes, B. (1973). A comparison of the effects of taurocholate and theophylline on BSP excretion in dogs. *In* G. Paumgartner and R. Preisig (eds.), The Liver: Quantitative Aspects of Structure and Function, p. 315. Karger, Basel.
66. Erlinger, S., and Dumont, M. (1973). Influence of theophylline on bile formation in the dog. Biomedicine 19:27.
67. Dyck, W. P., and Janowitz, H. D. (1971). Effect of glucagon on hepatic bile secretion in man. Gastroenterology 60:400.
68. Jones, R. S., Geist, R. E., and Hall, A. D. (1971). The choleretic effects of glucagon and secretin in the dog. Gastroenterology 60:64.
69. Klaassen, C. D. (1969). Biliary flow after microsomal enzyme induction. J. Pharmacol. Exp. Ther. 168:218.
70. Gumuccio, J. J., and Gray, R. H. (1974). Phenobarbital increases the rate of synthesis of some proteins of a bile canalicular rich membrane preparation. Gastroenterology 66:886 (Abstr.).
71. Rutizhauser, S. C. B., and Stone, S. L. (1975). Comparative effects of sodium

taurodeoxycholate and sodium taurocholate on bile secretion in the rat, dog and rabbit. J. Physiol. (Lond.) 245:583.
72. Mikai, K., and Fisher, M. M. (1971). The hepatotoxicity of chenodeoxycholic acid. Gastroenterology 60:189.
73. Javitt, N. B., and Emerman, S. (1968). Effect of sodium taurolithocholate on bile flow and bile acid excretion. J. Clin. Invest. 47:1002.
74. King, J. E., and Schoenfield, L. J. (1971). Cholestasis induced by sodium taurolithocholate in isolated hamster liver. J. Clin. Invest. 50:2305.
75. Schaffner, F., and Javitt, N. B. (1966). Morphologic changes in hamster liver during intrahepatic cholestasis induced by taurolithocholate. Lab. Invest. 15:1783.
76. Javitt, N. B. (1973). Excretion of monohydroxy bile acid ester sulfates in the rat. *In* G. Paumgartner and R. Preisig (eds.), The Liver: Quantitative Aspects of Structure and Function, p. 355. Karger, Basel.
77. Liersch, M., Czygan, P., and Stichl, A. (1975). Studies on hepatic uptake and secretion of bile salt sulfates by the isolated perfused rat liver. Digestion 12:326.
78. Scratcherd, T. (1965). Electrolyte composition and control of biliary secretion in the cat and rabbit. *In* W. Taylor (ed.), The Biliary System, p. 515. Blackwell, Oxford.
79. Grossman, M. I., Janovitz, H. D., Ralston, H., and Kim, K. M. (1949). The effect of secretin on bile formation in man. Gastroenterology 12:133.
80. Wheeler, H. O., and Mancusi-Ungaro, P. L. (1966). Role of bile ducts during secretin choleresis in dogs. Am. J. Physiol. 210:1153.
81. Waitman, A. M., Duck, W. P., and Janovitz, H. D. (1969). Effect of secretin and acetazolamide on the volume and electrolyte composition of hepatic bile in man. Gastroenterology 56:286.
82. Nahrwold, D. L., and Shariatzedeh, A. N. (1971). Role of the common bile duct in formation of bile and in gastrin-induced choleresis. Surgery 70:147.
83. Geist, R. E., and Jones, R. S. (1971). Effect of selective and truncal vagotomy on insulin stimulated bile secretion in dogs. Gastroenterology 60:566.
84. Roze, C., and Feldmann, D. (1971). Stimulation par l'insuline d'une fraction de la cholerese independante des sels biliares chez le rat. C.R. Acad. Sci. (Paris) 273:887.
85. Baldvin, J., Heer, F. W., and Albo, R. (1966). Effect of vagus nerve stimulation on hepatic secretion of bile in human subjects. Am. J. Surg. 3:66.
86. Krapup, N., Larsen, J. A., and Munck, A. (1975). Choleretic effect of prostaglandin PGE_1 and PGE_2 in cats. Digestion 12:272 (Abstr.).
87. Goreski, C. A. (1964). Initial distribution and rate of uptake of sulfobromophtalein in the liver. Am. J. Physiol. 207:13.
88. Smyth, D. H. (1972). Intestinal transfer mechanisms measurements and analogy. J. Clin. Pathol. (Suppl.) 24(5):1.
89. Cowen, A. E., Dorman, M. G., Hofmann, A. F., and Thomas, P. F. (1975). Plasma disappearance of radioactivity after intravenous injection of labeled bile acids in man. Gastroenterology 68:1567.
90. Berk, P. D., Howe, R. B., Bloomer, J. R., and Berlin, N. I. (1969). Studies of bilirubin kinetics in normal adults. J. Clin. Invest. 48:2176.
91. Mia, A. S., Gronwall, R. R., and Cornelius, C. E. (1970). [^{14}C]bilirubin turnover studies in normal and mutant Southdown sheep with congenital hyperbilirubinemia. Proc. Soc. Exp. Biol. Med. 133:955.
92. Barber Riley, G., Goetzee, A. E., Richards, T. G., and Thomson, J. Y. (1961). The transfer of bromsulphtalein from the plasma to the bile in man. Clin. Sci. 20:149.
93. Wheeler, H. O., Meltzer, J. I., and Bradley, S. E. (1960). Biliary transport and hepatic storage of sulfobromophtalein sodium in the unanesthetized dog, in normal man and in patients with hepatic disease. J. Clin. Invest. 39:1131.
94. McIntyre, N., Mulligan, R., and Carson, E. (1973). BSP Tm and S. A critical reevaluation. *In* G. Paumgartner and R. Preisig (eds.), The Liver: Quantitative Aspects of Structure and Function, pp. 417–427. Karger, Basel.
95. Arias, I. M. (1972). Transfer of bilirubin from blood to bile. Sem. Hematol. 9:55.
96. Levi, A. J., Gatmaitan, Z., and Arias, I. M. (1969). Two cytoplasmic protein fractions, Y and Z and their possible role in the hepatic uptake of bilirubin, sulfobromophtalein and other anions. J. Clin. Invest. 48:2156.

97. Habig, W., Pabst, M., Fleishner, G., Gatmaitan, Z., Arias, I. M., and Jakoby, W. (1974). The identity of glutathione transferase B with ligandin, a major binding protein of liver. Proc. Natl. Acad. Sci. USA 71:3879.
98. Kaplowitz, N., Pery Robb, I. W., and Javitt, N. B. (1973). Role of hepatic anion binding protein in bromsulphtalein conjugation. J. Exp. Med. 138:483.
99. Arias, I. M. Intracellular molecules in hepatic transport. Second NATO Advanced Study Institute on the Biliary System. In press.
100. Reyes, H., Levi, A. J., and Arias, I. M. (1971). Studies of Y and Z: two hepatic cytoplasmic organic anion binding proteins. Effects of drugs, chemicals, hormones and cholestasis. J. Clin. Invest. 50:2242.
101. Levi, A. J., Gatmaitan, Z., and Arias, I. M. (1969). Deficiency of hepatic organic anion binding protein: a possible cause of physiological jaundice in the newborn. Lancet ii:139.
102. Meuwissen, J. A. T. P. (1975). Binding proteins and hepatic uptake and transport of bilirubin. Digestion 12:276.
103. Kamisaka, K., Listowsky, I., Gatmaitan, Z., and Arias, I. M. (1975). Interactions of bilirubin and other ligands with ligandin. Biochemistry 14:2175.
104. Kenwright, S., and Levi, A. J. (1974). Sites of competition in the selective hepatic uptake of rifamycin SV, flawaspidic acid, bilirubin and bromsulphtalein. Gut 19:220.
105. Scharschmidt, B. F., Waggoner, J. C., and Berk, P. (1975). Hepatic organic anion uptake in the rat. J. Clin. Invest. 56:1280.
106. Kuenzle, C. C. (1970). Bilirubin conjugates of human bile. The excretion of bilirubin as the acyl glycosides of aldobiouronic acid, pseudo aldobiouronic acid and hexuronosyl hexuronic acid, with a branched chain hexuronic acid as one of the components of the hexuronosylhexuronide. Biochem. J. 119:411.
107. Heirwegh, K. P. M., Van Hees, G. P., Blanckaert, N., Fewery, J., and Compernolle, F. Comparative studies on the structures of conjugated bilirubin IX and changes in cholestasis. Second NATO Advanced Study Institute on the Biliary System. In press.
108. Robinson, S., Yannoni, C., and Nagasawa, S. (1971). Bilirubin excretion in rats with normal and impaired bilirubin conjugation. Effect of phenobarbital. J. Clin. Invest. 50:2606.
109. Jones, E. A., Bloomer, J. R., and Berlin, N. I. (1971). The measurement of the synthetic rate of bilirubin from hepatic hemes in patients with acute intermittent porphyria. J. Clin. Invest. 50:2259.
110. Kelman-Sraer, J., Erlinger, S., Peignoux, M., and Benhamou, J. P. (1973). Influence of dehydrocholate on hepatic uptake and biliary excretion of rose bengal in the rabbit. Biomed. Exp. 19:415.
111. Alpert, S., Mosher, M., Shanske, A., and Arias, I. M. (1969). Multiplicity of hepatic excretory mechanisms for organic anions. J. Gen. Physiol. 53:238.
112. Berthelot, P., and Fauvert, R. (1967). L'excrétion de bilirubine non conjuguée dans le bile du rat. Modification de cette excrétion par la novobiocine. Rev. Fr. Et. Clin. Biol. 12:702.
113. Ostrow, J. D. (1971). Photocatabolism of labeled bilirubin in the congenitally jaundiced (Gunn) rat. J. Clin. Invest. 50:707.
114. Paumgartner, G., and Reichen, J. Kinetics of hepatic uptake and excretion of organic anions. Second NATO Advanced Study Institute on the Biliary System. In press.
115. Reichen, J., and Paumgartner, G. (1975). Kinetics of taurocholate uptake by the perfused rat liver. Gastroenterology 68:132.
116. Reichen, J., and Paumgartner, G. (1975). Sodium dependence of hepatocellular uptake. Evidence for carrier mediated transport. Digestion 12:273.
117. Dietmaier, A., Gasser, R., Graf, J., and Peterlik, M. (1975). Sodium-dependent transport of bile acids in the isolated perfused rat liver. Digestion 12:273.
118. Paumgartner, G., and Reichen, J. (1974). Influence of chemical structure on hepatic uptake of bile salts. Gastroenterology 67:810.
119. Glasinovic, J. C., Dumont, M., and Duval, M. (1975). Hepatocellular uptake of taurocholate in the dog. J. Clin. Invest. 55:419.
120. Glasinovic, J. C., Dumont, M., Duval, M., and Erlinger, S. (1975). Hepatocellular uptake of bile acids in the dog. Evidence for a common carrier mediated transport system. An indicator dilution study. Gastroenterology 69:973.

121. Rudman, D., and Kendall, F. E. (1957). Bile acid content of human serum. II. The binding of cholanic acids by human plasma proteins. J. Clin. Invest. 36:538.
122. Schwartz, L. R., Burr, R., Schwenk, M., Pfaff, E., and Greim, H. (1975). Uptake of taurocholic acid into isolated rat liver cells. Eur. J. Biochem. 55:617.
123. Hoffman, N. E., Iser, J. H., and Smallwood, R. A. (1975). Hepatic bile acid transport. Effect of conjugation and position of hydroxyl group. Am. J. Physiol. 229:298.
124. Aharonian, H. S., Balabaud, Ch., and Gumuccio, J. J. (1975). Intrahepatic compartmentation of taurocholate in the rat. Gastroenterology 69:808.
125. Forker, E. L., and Gibson, G. (1973). Interaction between sulfobromophtalein (BSP) and taurocholate. The kinetics of transport from liver cells to bile in rats. *In* G. Paumgartner and R. Preisig (eds.), The Liver: Quantitative Aspects of Structure and Function, pp. 326–336. Karger, Basel.
126. O'Maille, E. R. L., Richards, T. G., and Short, A. H. (1965). Acute taurine depletion and maximal rates of hepatic conjugation and secretion of cholic acid in the dog. J. Physiol. (Lond.) 180:67.
127. O'Maille, E. R. L., Richards, T. G., and Short, A. H. (1967). The influence of conjugation of cholic acid on its uptake and secretion: hepatic extraction of taurocholate and cholate in the dog. J. Physiol. (Lond.) 189:337.
128. Liersch, M., Czygen, P., and Stichl, A. (1975). Studies on hepatic uptake and secretion of bile salt sulfates by the isolated perfused rat liver. Digestion 12:326.
129. Paumgartner, G., and Reichen, J. (1975). Separate carrier mediated transport systems for the hepatocellular uptake of bilirubin and bile acids. Digestion 12:259.
130. Delage, Y., Erlinger, S., Duval, M., and Benhamou, J. P. (1975). Influence of dehydrocholate and taurocholate on bromsulphtalein uptake storage and excretion in the dog. Gut 16:105.
131. Czok, G., Schulze, P. J., and Meyer, H. (1975). Hepatic uptake and biliary excretion of unconjugated and conjugated sulfobromophtalein in the rat under the influence of cholic acid and dehydrocholic acid. Digestion 12:275.
132. O'Maille, E. R. L., Richards, T. G., and Short, A. H. (1966). Factors determining the maximal rate of organic anion secretion by the liver and further evidence on the hepatic site of action of the hormone secretin. J. Physiol. (Lond.) 186:424.
133. Ritt, D. J., and Combes, B. (1967). Enhancement of apparent excretory maximum of sulfobromophtalein sodium (BSP) by taurocholate and dehydrocholate. J. Clin. Invest. 46:1108.
134. Erlinger, S., and Dumont, M. (1973). Influence of canalicular bile flow on sulfobromophtalein transport maximum in bile in the dog. *In* G. Paumgartner and R. Preisig (eds.), The Liver: Quantitative Aspects of Structure and Function, pp. 306–314. Karger, Basel.
135. Goresky, C. A., Haddad, H. H., Kluger, W. S., et al. (1974). The enhancement of maximal bilirubin excretion with taurocholate-induced increments in bile flow. Can. J. Physiol. Pharmacol. 52:389.
136. Macarol, V., Morris, T. Q., Baker, K. J., and Bradley, S. E. (1970). Hydrocortisone choleresis in the dog. J. Clin. Invest. 49:1714.
137. Barnhardt, J., Ritt, D., Ware, A., and Combes, B. (1973). A comparison of the effects of taurocholate and theophylline on BSP excretion in dogs. *In* G. Paumgartner and R. Preisig (eds.), The Liver: Quantitative Aspects of Structure and Function, p. 315. Karger, Basel.
138. Verschure, J. C. M., and Mijnlieff, P. F. (1956). The dominating macromolecular complex of human gallbladder bile. Clin. Chim. Acta 1:154.
139. Bouchier, I. A. D., and Cooperband, S. R. (1967). Sephadex filtration of a macromolecular aggregate associated with bilirubin. Clin. Chim. Acta 15:303.
140. Bloomer, J. R., Boyer, J. L., and Klatskin, G. (1973). Inhibition of bilirubin excretion in man during dehydrocholate choleresis. Gastroenterology 65:929.
141. Vonk, R. J., Jekel, P. A., and Meijer, D. K. F. (1975). Influence of bile salt administration and biliary micelle formation on the biliary elimination organic anions. Digestion 12:258.
142. Spitzer, H. L., Kyriakides, E. C., and Balint, J. A. (1964). Biliary phospholipids in various species. Nature 204:288.
143. Bourges, M., Small, D. M., and Dervichian, D. G. (1967). Biophysics of lipidic

association. III. The quaternary system lecithin, bile salt, cholesterol, water. Biochim. Biophys. Acta 144:189.
144. Small, D. M. (1970). The formation of gallstones. Adv. Med. 16:243.
145. Admirand, W. H., and Small, D. M. (1968). The physicochemical basis of cholesterol gallstone formation in man. J. Clin. Invest. 47:1043.
146. Holzbach, R. T., March, M., Obszewski, M., and Holan, K. (1973). Cholesterol solubility in bile–evidence that supersaturated bile is frequent in healthy man. J. Clin. Invest. 52:1477.
147. Thomas, P. J., and Hofmann, A. F. (1973). A simple calculation of the lithogenic index. Expressing biliary lipid composition on rectangular co-ordinates. Gastroenterology 65:689.
148. Thistle, J. L., and Schoenfield, L. J. (1971). Induced alterations in composition of bile of persons having cholelithiasis. Gastroenterology 61:488.
149. Thistle, J. L., and Hofmann, A. F. (1973). Efficacy and specificity of chenodeoxycholic therapy for dissolving gallstones. N. Engl. J. Med. 289:655.
150. Balint, J. A., Beeler, D. A., Treble, D. H., and Spitzer, H. L. (1967). Studies in the biosynthesis of hepatic and biliary lecithins. J. Lipid Res. 8:486.
151. Balint, J. A., Beeler, D. A., Kyriakides, E. C., and Trebel, D. H. (1971). The effect of bile salts upon lecithin synthesis. J. Lab. Clin. Med. 77:122.
152. Gregory, D. H., Vlahcevic, Z. R., Schatzki, P., and Swell, L. (1975). Mechanism of secretion of biliary lipids. I. Role of bile canalicular and microsomal membranes in the synthesis and transport of biliary lecithin and cholesterol. J. Clin. Invest. 55:105.
153. Schersten, T., Gottfries, A., Nilsson, S., and Samuelsson, B. (1967). Incorporation of plasma free fatty acids into bile lipids in man. Life Sci. 6:1175.
154. Curstedt, T., and Sjovall, J. (1975). Origin of phosphatidylcholins in bile. *In* S. Mater, J. Hackenschmidt, P. Back, and W. Gerok (eds.), p. 209. F. K. Schattauer Verlag, Stuttgart.
155. Curstedt, T., and Sjovall, J. (1974). Biosynthetic pathways and turnover of individual biliary phosphatidyl-cholines during metabolism of $[1,1^2H]$ ethanol in the rat. Biochim. Biophys. Acta 369:173.
156. Swell, L., Bell, C. C., and Entenmuan, C. (1968). Bile acids and lipid metabolism. III. Influence of bile acids on phospholipids in liver and bile of the isolated perfused dog liver. Biochim. Biophys. Acta 164:278.
157. Nilsson, S., and Schersten, T. (1970). Influence of bile acids on the synthesis of biliary phospholipids in man. Eur. J. Clin. Invest. 1:109.
158. Nilsson, S., and Schersten, T. (1970). Importance of bile acids for phospholipid secretion into human hepatic bile. Gastroenterology 57:525.
159. Hardison, W. G., and Apter, J. T. (1972). Micellar theory of biliary cholesterol excretion. Am. J. Physiol. 222:61.
160. Gerolami, A., Crotte, C., Mule, A., and Vigne, J. L. (1972). Experimental gallstones in the mouse. Mechanism of dehydrocholic acid. Rev. Eur. Et. Clin. Biol. 17:500.
161. Montet, J. C., Montet, A. M., Gerolami, A., and Hauton, J. C. (1975). Effects of 3-acetyl fusidate on the biliary secretion of lipids in the rat. Biol. Gastroenterol. (Paris) 8:53.
162. Schersten, T. (1973). Formation of lithogenic bile in man. Digestion 9:469.
163. Beaudoin, A., Carey, M. C., and Small, D. M. (1975). Effects of taurodihydrofusidate, a bile salt analogue on bile formation and biliary lipid secretion in the rhesus monkey. J. Clin. Invest. 56:1431.
164. Gerolami, A., Crotte, C., and Mule, A. (1973). Mécanisme de la sécrétion biliaire des lipides au cours de la lithiase expérimentale de la souris. Biol. Gastroenterol. (Paris) 6:337.
165. Hofmann, A. F., and Small, D. M. (1967). Detergent properties of bile salts. Correlation with physiological functions. Am. Rev. Med. 18:333.
166. Gerolami, A., Crotte, C., Mule, A., Domingo, N., and Durbec, J. P. (1974). Secretion of phospholipids and cholesterol in the bile on rats. Action of phenobarbitone. Biomedicine 20:160.
167. Small, D. M. (1971). The physical chemistry of cholanic acids. *In* P. P. Nair and D. Kritchevsky (eds.), The Bile Acids, Vol. 1, Chemistry, p. 248. Plenum Press, New York.

168. Phillips, M. C., Finer, E. G., and Hauser, H. (1972). Differences between conformations of lecithins and phosphatidyl ethanolamine polar group and their effects on interactions of phospholipid bilayer membranes. Biochim. Biophys. Acta 290:397.
169. Staple, E., and Gurin, S. (1954). The incorporation of radioactive acetate into biliary cholesterol and cholic acid. Biochim. Biophys. Acta 15:372.
170. Mitropoulos, K. A., Myant, N. B., Gibbons, G. F., Balasubramanian, S., and Reeves, B. E. A. (1974). Cholesterol precursor pools for the synthesis of cholic and chenodeoxycholic acid in rats. J. Biol. Chem. 249:6052.
171. Lafont, H., Lairon, D., Domingo, N., Nalbone, G., and Hauton, J. C. (1974). Does a lecithin polypeptide association in bile originates from membrane structural submits? Biochimie 56:465.
172. Nalbone, G., Lafont, H., Domingo, N., Lairon, D., Pautrat, G., and Hauton, J. C. (1973). Ultramicroscopic study of the bile lipoprotein complex. Biochimie 55:1503.
173. Lairon, D., Lafont, H., and Hauton, J. C. (1972). Lack of mixed micelles bile salt lecithin cholesterol in bile and presence of a liporoteic complex. Biochimie 54:529.
174. Hoffman, N. E., and Hofmann, A. F. (1974). Metabolism of steroid and amino acid moieties of conjugated bile acids in man. IV. Description of a multicompartimental model. Gastroenterology 67:887.
175. Vlahcevik, Z. R., Miller, J. R., Farrar, J. T., and Swell, L. (1971). Kinetics and pool size of primary bile acids in man. Gastroenterology 61:85.
176. Dowling, R. H. (1972). The enterohepatic circulation. Gastroenterology 62:122.
177. Small, D. M., Dowling, R. H., and Redinger, R. N. (1972). The enterohepatic circulation of bile salts. Arch. Intern. Med. 130:552.
178. Carey, J. M. (1973). Bile salt metabolism in man. *In* P. P. Nair and D. Kritchevsky (eds.), The Bile Acids, Vol. 2, Physiology and Metabolism, pp. 55–82. Plenum Press, New York.
179. Miettinen, T. A. (1973). Clinical implications of bile acid metabolism in man. *In* P. P. Nair and D. Kritchevsky (eds.), The Bile Acids, Vol. 2, Physiology and Metabolism, pp. 191–247. Plenum Press, New York.
180. Weber, A. M., Chartrand, L., Doyon, C., Gordon, S., and Roy, C. C. (1972). The quantiation of foecal bile acids in children by the enzymatic method. Clin. Chim. Acta 39:524.
181. Lack, L., and Weiner, I. M. (1973). Bile salt transport systems. *In* P. P. Nair and D. Kritchevsky (eds.), The Bile Acids, Vol. 3. Physiology and Metabolism, pp. 33–54. Plenum Press, New York.
182. Krag, E., and Phillips, S. F. (1974). Active and passive bile acid absorption in man perfusion studies of the ileum and jejunum. J. Clin. Invest. 53:1686.
183. Wilson, F. A., and Dietschy, J. M. (1972). Characterization of bile acid absorption across the unstirred water layer and brush border of the rat jejunum. J. Clin. Invest. 51:3015.
184. Schiff, E. R., Small, N. C., and Dietschy, J. M. (1972). Characterization of the kinetics of the passive and active transport mechanisms for bile acid, absorption in the small intestine and colon of the rat. J. Clin. Invest. 51:1351.
185. Firpi, A., Walker, J. T., and Lack, L. (1975). Interactions of cationic bile salt derivatives with the ileal bile salt transport system. J. Lipid Res. 16:379.
186. Mok, H. Y. I., Perry, P. M., and Dowling, R. H. (1974). The control of bile acid pool size: Effect of jejunal resection and phenobarbitone on bile acid metabolism in the rat. Gut 15:247.
187. Cohen, A. F., Korman, M. G., Hofmann, A. F., Cass, O. W., and Coffin, S. B. (1975). Metabolism of lithocholate in healthy man. II. Enterohepatic circulation. Gastroenterology 69:67.
188. Hepner, G. W., Hofmann, A. F., and Thomas, P. J. (1975). Metabolism of steroid and amino acid moieties of conjugated bile acids in man. I. Cholyglycine. J. Clin. Invest. 51:1889.
189. Palmer, R. H. (1971). Bile acid sulfates. II. Formation metabolism and excretion of lithocholic acid sulfates in the rat. J. Lipid Res. 12:680.
190. Kellog, T. F. (1973). Bile acid metabolism in gnotobiotic animals. *In* P. P. Nair and D. Kritchevsky (eds.), The Bile Acids, Vol. 2, Physiology and Metabolism, p. 283. Plenum Press, New York.

191. Mosbach, E. H., and Salen, G. (1974). Bile acid biosynthesis. Pathways and regulation. Amer. J. Dig. Dis. 19:920.
192. Bjorkhem, I., and Danielsson, H. (1974). Hydroxylations in biosynthesis and metabolism of bile acids. Mol. Cell. Biochem. 4:79.
193. Mitropoulos, K. A., and Myant, N. B. (1967). The formation of lithocholic acid, chenodeoxycholic acid and α- and β-muricholic acids from cholesterol incubated with rat liver mitochondria. Biochem. J. 103:472.
194. Balasubramanian, S., Mitropoulos, K. A., and Myant, N. B. (1973). Evidence for compartmentation of cholesterol in rat liver microsomes. Eur. J. Biochem. 34:77.
195. Bjorkhem, I., and Danielsson, H. (1975). 7 α-hydroxylation of exogenous and endogenous cholesterol in rat liver microsomes. Eur. J. Biochem. 53:63.
196. Lindstedt, S. (1953). The turnover of cholic acid in man: bile acids and steroïds. Acta Physiol. Scand. 40:1.
197. Eriksson, S. (1957). Biliary excretion of bile acids and cholesterol in bile fistula rats. Proc. Soc. Exp. Biol. Med. 94:578.
198. Danielsson, H., Einarsson, K., and Johansson, G. (1967). Effect of biliary drainage on individual reactions in the conversion of cholesterol to taurocholic acid. Eur. J. Biochem. 2:44.
199. Small, D. M., Dowling, R. H., and Redinger, R. N. (1972). The enterohepatic circulation of bile salts. Arch. Intern. Med. 130:552.
200. Danielsson, H. (1973). Influence of dietary bile acids on formation of bile acids in rats. Steroïds 21:667.
201. Danzinger, R. G., Hofmann, A. F., Thistle, J. L., and Schenfield, L. J. (1973). Effect of oral chenodeoxycholic acid in bile kinetics and biliary lipid composition in women with cholelithiasis. J. Clin. Invest. 52:2809.
202. Shafer, S., Hauser, S., Lapar, V., and Mosbach, E. H. (1973). Regulatory effects of sterols and bile acids on hepatic 3-hydroxy 3-methyl glutaryl CoA reductase and cholesterol 7-α-hydroxylase in the rat. J. Lipid Res. 14:573.
203. Mitropoulos, K. A., Balasubramanian, S., and Myant, N. B. (1973). The effect of interruption of the enterohepatic circulation of bile acids and of cholesterol feeding on cholesterol 7α-hydroxylase in relation to the diurnal rhythm in its activity. Biochim. Biophys. Acta 326:428.
204. La Russo, N. F., Korman, M. G., Hoffman, N. E., and Hofmann, A. F. (1974). Dynamics of the enterohepatic circulation of bile acids. Post-prandial serum concentrations of conjugates of cholic acid in health, cholecystectomized patients and patients with bile acid malabsorption. N. Engl. J. Med. 291:689.
205. Northfield, T. C., and Hofmann, A. F. (1975). Biliary lipid output during three meals and an overnight fast. I. Relationship to bile acid pool size and cholesterol saturation of bile in gallstone and control subjects. Gut 16:12.
206. Low Beer, T. S. (1975). Influences on the regulation of bile salts synthesis. *In* S. Matern, J. Hackenschmidt, P. Back, and W. Gerok (eds.), Advances in Bile Acid Research. III. Bile Acid Meeting, pp. 201–208. Schattauer Verlag F.K., Stuttgart, New York.
207. La Russo, N. F., and Hofmann, A. F. (1975). Influence of deoxycholic acid ingestion on biliary lipid secretion and bile acid kinetics in man lack of effect on cholesterol saturation. Gastroenterology 69:840.
208. Northfield, T. C., La Russo, M. C., Hofmann, A. F., and Thistel, J. L. (1975). Biliary lipid output during three meals and an overnight fast. II. Effect of chenodeoxycholic acid treatment in gallstone subjects. Gut 16:12.
209. Swell, L., Schwartz, C. C., Halloran, L. G., and Vlahcevic, Z. R. (1975). Rapid feedback inhibition of endogenous cholic and chenodeoxycholic acid synthesis by exogenous chenodeoxycholic acid in man. Biochem. Biophys. Res. Commun. 64:1083.
210. Hoffman, N. E., Donald, D. E., and Hoffmann, AF. (1975). The effect of chenodeoxycholyltaurine and cholyltaurine on bile lipid secretion from the perfused dog liver. Am. J. Physiol. 229:714.
211. Hoffman, N. E., and Smallwood, R. A. (1975). Effect of bile acid structure on bile lipid composition in the cat. Gastroenterology 69:A29/829.

212. Gerolami, A., and Montet, J. C. (1973). Pathogénie de la lithiase biliaire. Sécretion des lipides biliaires et cholélithiase–cholestérolique-lère partie. Biol. Gastroenterol. (Paris) 6:63.
213. Hegardt, F. G., and Dam, H. (1971). The solubility of cholesterol in aqueous solutions of bile salts and lecithin. Z. Ernaehrungswiss. 10:223.
214. Schoenfield, L. J., Bonovris, G. G., and Ganz, P. (1973). Induced alterations in the rate limiting enzymes of hepatic cholesterol and bile acid synthesis in the hamster. J. Lab. Clin. Med. 82:858.
215. Adler, R. D., Bennion, L. D., Duane, W. C., and Grundy, S. M. (1975). Effects of low dose chenodeoxycholic acid feeding on biliary lipid metabolism. Gastroenterology 68:326.
216. Hoffman, N. D., Hofmann, A. F., and Thistle, J. L. (1974). Effect of bile acid feeding on cholesterol metabolism in gallstone patients. Mayo Clin. Proc. 49:236.
217. Vlahcevic, Z. R., Bell, C. C. Jr., Buhac, L., Farrar, J. T., and Swell, L. (1970). Diminished bile acid pool size in patients with gallstones. Gastroenterology 59:165.
218. Shaffer, E. A., Braasch, J. W., and Small, D. M. (1972). Bile composition at and after surgery in normal persons and patients with gallstones. N. Engl. J. Med. 287:1317.
219. Almond, H. R., Vlahcevic, R. A., Bell, C. C. Jr., Gergory, D. H., and Swell, L. (1973). Effect of cholecystectomy on bile acid metabolism. N. Engl. J. Med. 289:1213.
220. Bergman, F., and Van der Linden, W. (1974). Bile acid pool size in hamsters during gallstone formation and after cholecystectomy. Zeitsch. Ernährungswiss 13:37.
221. Grundy, S. M., Metzger, A. L., and Adler, R. D. (1972). Mechanisms of lithogenic bile formation in American Indian women with cholesterol gallstones. J. Clin. Invest. 51:3026.
222. Gerolami, A., Crotte, C., Reynier, M. O., Mule, A., Domingo, N., and Sarles, H. (1975). Diurnal changes of bile lipid concentration in hepatic bile of cholecystectomized man. Digestion 12:209.
223. Bennion, L. J., and Grundy, S. M. (1975). Effects of obesity and calorie intake on biliary lipid metabolism in man. J. Clin. Invest. 56:996.
223a. Sarles, H., Crotte, C., Gerolami, A., Mule, A., Domingo, N., and Hauton, J. C. (1971). The influence of calorie intake and of dietary protein on the bile lipids. Scand. J. Gastroenterol. 6:189.
224. Pertsimilidis, D., Manvelivalda, D., and Arkens, E. H. (1974). Effect of cholifrate and of an estrogen–progestin combination on fasting biliary lipids and cholic acid kinetic in man. Gastroenterology 66:565.
225. Robins, S. J., and Armstrong, M. (1975). Role of dietary choline and lecithin synthesis in biliary lipid secretion. Gastroenterology 69:856 (Abstr.).
226. Shelhammer, J. (1973). Physiology of bile transport: manometric studies of the common bile duct and sphincter of Oddi. Gastroenterology 64:43/686.
227. Stalport, A. (1968). Voie biliaire principale et complexe oddien. Ascia, Bruxelles.
228. Wyatt, A. P. (1968). Dynamics of the common duct. Lancet i:360.
229. Ludwick, J. R. (1969). Observations on the smooth muscle and contractile activity of the common bile duct. Gastroenterology 56:1178.
230. Toouli, J., and Watts, J. McK. (1972). Actions of cholecystokinin–pancreozymin, secretin and gastrin on extra biliary tract motility *in vitro.* Am. J. Surg. 175:439.
231. Watts, J. McK., and Dunphy, J. E. (1966). The role of the common bile duct in biliary dynamics. Surg. Gynecol. Obstet. 122:1207.
232. Mirizzi, P. L. (1942). Functional disturbances of the choledocus and hepatic bile duct. Surg. Gynecol. Obstet. 74:306.
233. Persson, C. G. A., and Ekman, M. (1972). Effect of morphine, cholecystokinin and sympathomimetics on the sphincter of Oddi and intramural pressure in cat duodenum. Scand. J. Gastroenterol. 7:345.
234. Boyden, E. A. (1965). The comparative anatomy of the sphincter of Oddi in mammals with special reference to the choledocoduodenal junction in man. *In* The Biliary System, pp. 15–40. Blackwell Sci. Publ., Oxford.
235. Ishioka, T. (1959). Electromyographic study of the choledocoduodenal junction and duodenal wall muscle. Tohoku J. Exp. Med. 70:73.
236. Sarles, J. C., Midejean, A., and Devaux, M. A. (1975). Electromyography of the

sphincter of Oddi: technic and experimental results in the rabbit. Am. J. Gastroenterol. 63:221.
237. Sarles, J. C., Midejean, A., and Gayne, F. (1974). Etude électromyographique du sphincter d'Oddi. I. Technique et résultats chez le lapin "*in vitro*" et "*in vivo*". Biol. Gastroenterol. (Paris) 7:19.
238. Cushieri, A., Hughes, M., and Cohen, M. (1972). Biliary pressure during cholecystectomy. Br. J. Surg. 59:267.
239. Pinotti, H. W., De Sousa, O. M., Parolari, J. B., Conte, V. P., Raia, A., and Netto, A. C. (1967). The dynamics of the main biliary duct and the mechanism of the Am. J. Dig. Dis. 12:878.
240. Schein, C. J., and Beneventano, J. C. (1968). Choledocal dynamics in man. Surg. Gynecol. Obstet. 126:591.
241. Hauge, C. W., and Mark, J. B. D. (1965). Common bile duct motility and sphincteric mechanism. I. Pressure measurement with multiple lumen catheter in dogs. Am. J. Surg. 162:1028.
242. Menguy, R. B., Hallenbeck, G. A., Bollman, J. L., and Grindlay, J. H. (1958). Intraductal pressures and sphincteric resistance in canine pancreatic and biliary ducts after various stimuli. Surg. Gynecol. Obstet. 106:306.
243. Hedner, P., and Rorsman, G. (1969). On the mechanism of action of the effect of CCK on the choledocoduodenal junction in the cat. Acta Physiol. Scand. 76:248.
244. Ono, N., Watanabe, N., Suzuki, N., Tsuchida, H., Sugiyama, Y., and Abo, M. (1908). Bile flow mechanism in man. Arch. Surg. 96:809.
245. Shore, S. M., Silverman, A., Siegel, M., and Bakal, M. (1971). Direct observation of the canine sphincter of Oddi. Am. J. Surg. 174:264.
246. Tansy, M. F., Innes, D. L., Martin, J. S., and Kendall, F. M. (1974). The role of intraduodenal common bile duct in the filling of the canine gall bladder. Surg. Gynecol. Obstet. 139:585.
247. Jorpes, J. E., and Mutt, V. (1973). Secretin, cholecystokinin, pancreozymin and gastrin. Handbook of Experimental Pharmacology, Vol. 34. Springer Verlag, Berlin.
248. Persson, C. G. A. (1972). Adrenoceptors in the gallbladder. Acta Pharmacol. Toxicol. 31:177.
249. Torsoli, A., Ramorino, M. L., and Alessandrini, A. (1970). Mobility of the biliary tract. Rendic. Gastroenterol. 2:67.
250. Lin, T. M., and Spray, G. F. (1971). Choledochal, hepatic and cholecystokinetic actions of secretin (S); potentiation by cholecystokinin (CCK). Gastroenterology 60:783.
251. Amer, M. S. (1969). Mechanism of action of cholecystokinin. Clin. Res. 17:520.
252. Amer, M. S. (1969). Studies with cholecystokinin. II. Cholecystokinetic potency of porcine gastrins I and II and related peptides in three systems. Endocrinology 84:1277.
253. Amer, M. S. (1972). Studies with cholecystokinin *in vitro*. III. Mechanism of the effect on the isolated rabbit gallbladder strips. J. Pharmacol. Exp. Ther. 183:527.
254. Amer, M. S. (1969). A sensitive *in vitro* method for the assay of cholecystokinin. J. Endocrinol. 43:637.
255. Andersson, K. E., Andersson, R., and Hedner, P. (1972). Cholecystokinetic effect and concentration of cyclic AMP in gallbladder muscle *in vitro*. Acta Physiol. Scand. 85:511.
256. Jung, F. T., and Greengard, H. (1933). Response of the isolated gallbladder to cholecystokinin. Am. J. Physiol. 103:275.
257. Yau, W. M., Makhlouf, G. M., Edwards, L. E., and Rarrar, J. T. (1973). Mode of action of cholecystokinin and related peptides on gall bladder muscle. Gastroenterology 65:451.
258. Ryan, J., and Cohen, S. (1975). Influence of gastrointestinal hormonal stimulation and luminal volume on gallbladder function. Gastroenterology 68:A120/977 (Abstract).
259. Mack, A. J., and Todd, J. K. (1968). A study of human gall bladder muscle *in vitro*. Gut 9:546.
260. Inberg, M. V., Ahonen, P. J., and Scheinin, T. M. (1970). Gallbladder function and

bile composition after selective gastric and truncal vagotomy in the dog. Scand. J. Clin. Lab. Invest. 25:113.
261. Lin, T. M. (1975). Action of gastrointestinal hormones and related peptides on the motor function of the biliary tract. Gastroenterology 69:1006.
262. Harvey, R. F. (1975). Hormonal control of gastrointestinal tract. Am. J. Dig. Dis. 20:523.
263. Berry, H., and Flower, R. J. (1971). The assay of endogenous cholecystokinin and factors influencing its release in the dog and cat. Gastroenterology 60:409.
264. Amer, M. S., and McKinney, G. R. (1970). On the mechanism of action of cholecystokinin (CCK). Effect on phosphodiesterase (PDE). Pharmacologist 12:291.
265. Andersson, K. E., Andersson, R., Hedner, P., and Persson, C. G. A. (1972). Changes in the tissue levels of cAMP in the gallbladder and sphincter of Oddi induced by the C-terminal octapeptide of cholecystokinin. Acta Physiol. Scand. 84:19A.
266. Andersson, K. E., Andersson, R., Hedner, P., and Persson, C. G. A. (1973). Analogous effects of cholecystokinin and prostaglandin E_2 on mechanical activity and tissue levels of cAMP in biliary smooth muscle. Acta Physiol. Scand. 87:41A.
267. Andersson, K. E., Hedner, P., and Hedner, C. G. A. (1974). Differentiation of the contractile effects of prostaglandin E_2 and the C-terminal octapeptide of cholecystokinin in isolated guinea pig gallbladder. Acta Physiol. Scand. 90:657.
268. Amer, M. S. (1974). Cyclic guanosine 3′5′-monophosphate and gallbladder contraction. Gastroenterology 67:333.
269. Lin, T. M. (1974). Action of secretin, glucagon, cholecystokinin and endogenously released secretin and cholecystokinin on gallbladder, choledochus and bile flow in dogs. Fed. Proc. 33:391 (Abstr.).
270. Watts, J. M., and Dunphy, J. E. (1966). The role of the common bile duct in biliary dynamics. Surg. Gynecol. Obstet. 122:1207.
271. Golenhofen, K., Loh, D. V., and Lynen, F. K. (1971). Intracellular recording of electrical activity in smooth muscle of common bile duct. Experientia 27:650.
272. Wakim, K. G. (1971). Passive role of the bile duct system in the delivery of bile into the intestine. Surg. Gynecol. Obstet. 133:826.
273. Sandblom, P., Voegtlin, W. L., and Ivy, A. C. (1935). The effect of cholecystokinin on the choledocoduodenal mechanism. Am. J. Physiol. 113:175.
274. Hedner, P., and Rorsman, G. (1969). On the mechanism of action for the effect of cholecystokinin on the choledocoduodenal junction in the cat. Acta Physiol. Scand. 76:248.
275. Liedberg, G. (1969). The effect of vagotomy on gallbladder and duodenal pressure during rest and stimulation with cholecystokinin. Acta Chir. Scand. 135:695.
276. Liedberg, G., and Halabi, M. (1970). The effect of vagotomy on flow resistance at the choledoco-duodenal junction. Acta Chir. Scand. 136:208.
277. Liedberg, G., and Persson, C. G. A. (1970). Adrenoreceptors in the cat choledochoduodenal junction studied *in situ.* Br. J. Pharmacol. 39:619.
278. Persson, C. G. A., and Ekman, M. (1972). Effect of morphine, cholescystokinin and sympathomimetics on the sphincter of Oddi and intramural pressure in cat duodenum. Scand. J. Gastroenterol. 7:345.
279. Menguy, R. B., Hallenbeck, G. A., Bollman, J. L., and Grindlay, J. H. (1958). Intraductal pressures and sphincteric resistance in canine pancreatic and biliary ducts after various stimuli. Surg. Gynecol. Obstet. 106:306.
280. Sandblom, P. H., Voegtlin, W. L., and Ivy, A. C. (1935). The effect of cholecystokinin on the choledochoduodenal mechanism (sphincter of Oddi). Am. J. Physiol. 113:175.
281. Raih, T. J., Ashmore, C. S., Wilson, S. D., Decosse, J. J., Hogan, W. S., Dodds, W. J., and Stef, J. J. (1973). Effect of enteric hormones on the canine choledochal sphincter. Gastroenterology A104.
282. Lin, T. M., and Spray, G. F. (1969). The effect of pentagastrin, cholecystokinin, caerulein and glucagon on the choledochal resistance and bile flow of conscious dog. Gastroenterology 56:1178.
283. Torsoli, A., Ramorino, M. L., and Carratu, R. (1973). On the use of cholecystokinin in the roentgenological examination of the extrahepatic biliary tract and intestines. *In*

J. E. Jorpes and V. Mutt (eds.), Secretin, Cholecystokinin, Pancreozymin and Gastrin, Handbook Exp. Pharm. XXXIV, pp. 247–257.
284. Plessier, J. (1973). The use of cholecystokinin in the roentgenological examination. Clinical aspects. *In* J. E. Jordes and V. Mutt (eds.), Secretin, Cholecystokinin, Pancreozymin and Gastrin, Handbook Exp. Pharm. XXXIV, pp. 311–338.
285. Salin, J. O., Siegel, C. I., and Mendeloff, A. I. (1973). Biliary duodenal dynamics in man. Radiology 106:1.
286. Toouli, J., and Watts, J. Mck. (1970). The spontaneous motility and the action of cholecystokinin–pancreozymin, secretin and gastrin on the canine extrahepatic biliary tract. Br. J. Surg. 57:858.
287. Sarles, J. C. Bidart, J. M., Devaux, M. A., and Castagnini, A. Action of cholecystokinin–pancreozymin and caerulein on the sphincter of Oddi. Electromyographical, manometrical and flow rate study on the rabbit. Digestion. In press.
288. Zaffagnin, B., and Taccani, C. (1954). Studio morfologico comparativo dell' apparato muscolare del choledoco terminale. Patol. sperimen. 13:201.
289. Liedberg, G. (1969). The effect of vagotomy on gallbladder and duodenal pressure during rest and stimulation with cholecystokinin. Acta Chir. Scand. 136:208.
290. Andersson, K. E., Andersson, R., Hedner, P., and Persson, C. G. A. (1973). Parallelism between mechanical and metabolic response to CCK and prostaglandin E_2 in extrahepatic biliary tract. Acta Physiol. Scand. 89:571.
291. Jorpes, J. E. (1968). The isolation and chemistry of secretin and cholecystokinin. Gastroenterology 55:157.
292. Grossman, M. I. (1972). Gastrointestinal hormones, some thoughts about clinical application. Scand. J. Gastroenterol. 7:97.
293. Vagne, M., and Grossman, M. I. (1968). Cholecystokinetic potency of gastrointestinal hormones and related peptides. Am. J. Physiol. 215:881.
294. Audigier, JC. Vagne, M., Fargier, MC., and Declutieux, A. (1972). Effet cholécystokinétique des hormones gastro-intestinales. Biol. Gastroenterol. (Paris) 5:287.
295. Grossman, M. I. (1970). Gastrin, cholecystokinin and secretin act on one receptor. Lancet i:1088.
296. Nebec, O. T. (1975). Effect of enteric hormones on the human sphincter of Oddi. Gastroenterology 68:A105/962 (Abstr.).
297. Bonfils, S., Gilson, J., and Pointner, H. (1972). Action de la sécrétine et de la gastrine isolément et en association sur les pressions intracholédocienees chez l'homme cholécystectomisé. Arch. Fr. Mal. App. Dig. 61:382.
298. Ondetti, M. A., Plusec, J., and Sabo, F. F. (1970). Synthesis of cholecystokinin–pancreozymin. I. The C-terminal dodecapeptide. J. Am. Soc. 92:195.
299. Amberg, J. R. (1974). Canine cholecystokinesis: effect of octapeptide of cholecystokinin. Work in progress.
300. Shore, J. M., Silverman, A., Siegel, M., and Bakal, M. (1971). Direct observations of the canine sphincter of Oddi. Ann. Surg. 264.
301. Erspamer, V. (1970). Progress report: caerulein. Gut 11:79.
302. Carratu, R., Arcangeli, G., and Pallone, F. (1971). Effects of caerulein on the human biliary tract. Rendic. Gastroenterol. 3:28.
303. Nakamura, N., Koyoma, Y., Kostima, T., and Takahira, H. (1973). Effect of caerulein on intestinal tract and gallbladder. Jpn. J. Pharmacol. 23:107.
304. Hopton, D., and White, T. T. (1972). Effect of hepatic and celiac vagal stimulation on common bile duct pressure. Am. J. Dig. Dis. 16:1095.
305. Satler, J. J., Sakakihara, Y., Nussbaum, M., and Tumen, H. J. (1972). The effect of electrical stimulation of the hepatic periarterial nerve on the dynamics of the biliary tract of the dog. Acta Hepatogastroenterol. 19:234.
306. Tansy, M. F., Mackowiak, R. C., and Chaffee, R. B. (1971). A vagosympathetic pathway capable of influencing common bile duct motility in the dog. Surg. Gynecol. Obstet. 131:225.
307. J. C. Sarles. Unpublished communication.
308. Kyosola, K. (1974). Cholinesterase histochemistry of the innervation of the smooth muscle sphincters around the terminal intramural part of the ductus choledochus in the cat and the dog. Acta Physiol. Scand. 90:278.

309. Williams, R. D., and Huang, T. T. (1969). The effects of vagotomy on biliary pressure. Surgery 66:353.
310. Hopton, D. K. (1973). The influence of the vagus nerves on the biliary system. Br. J. Surg. 60:216.
311. Schein, C. J., and Gliedman, M. L. (1970). The influence of vagotomy on the normal and discarded gallbladder. Digestion 3:243.
312. Parkin, G. T. S., Smith, R. B., and Johnston, D. (1973). Gallbladder, volume and contractibility after truncal, selective and highly selective vagotomy in man. Ann. Surg. 178:581.
313. William, E. J., and Irvine, W. T. (1966). Functional and metabolic effects of total or selective vagotomy. Lancet i:1053.
314. Isaza, J., Jones, D. T., Dragstedt, L. R., and Woodward, E. R. (1972). The effect of vagotomy on motrice function of the gallbladder. Surgery 70:616.
315. Amberg, J. R. (1968). Effect of vagotomy on cholestanol cholelithogenesis in the rabbit. Invest. Radiol. 3:239.
316. Cowie, A. G. A., and Clarck, C. G. (1972). The lithogenic effect of vagotomy. Br. J. Surg. 59:365.
317. Tompkins, R. K., Kraft, A. R., Zimmerman, E., Lichtensten, J. E., and Zollinger, R. M. (1972). Clinical and biochemical evidence of increased gallstone formation after complete vagotomy. Surgery 71:196.
318. Bouchier, I. A. D. (1970). The vagus, the bile and gallstones. Gut 11:799.
319. Dowling, B. L. (1971). The effect of vagotomy and coeliac gangliectomy on the innervation of the canine gallbladder. Br. J. Surg. 58:303 (Abstract).
320. Beneventano, T. C., Rosen, R. C., and Schein, C. J. (1969). The physiological effect of acute vagal section on canine biliary dynamics. J. Surg. Res. 9:331.
321. Amdrup, B. M., and Griffith, C. A. (1970). The effects of vagotomy upon biliary function in dogs. J. Surg. Res. 10:209.
322. Levasseur, J. C., Germain, M., and Martin, E. (1975). Les effets de la vagotomie bitronculaire dans l'activité du sphincter d'Oddi. Etude chez le chien. Ann. Chir. 29:151.
323. Sarles, J. C., Midejean, A., Devaux, M. A., and Castagnini, A. Electromyographie du sphincter d'Oddi chez le lapin. II. Action de la vagotomie, du pentolinium et de l'alcohol. Biol. Gastroenterol. (Paris). In preparation.
324. Baumgarten, H. G., and Lange, W. (1964). Extrinsic adrenergic innervation of the extra-hepatic biliary duct system in guinea pig, cat and rhesus monkey. Z. Zelforsch. Mikrosk. Anat. 100:606.
325. Persson, C. G. A. (1973). Dual effects of the sphincter of Oddi and gallbladder induced by stimulation of the right great splanchnic nerve. Acta Physiol. Scand. 87:334.
326. Tansy, M. F., Innes, D. C., Lawach, D. L., Martin, J. S., and Kendall, F. M. (1974). Influence of intraduodenal folds on the opening pressure of the choledoco-duodenal function. Fed. Proc. 33:391 (Abstr.).
327. Dardick, H., Gliedman, M. L., Christ, R., Koslow, A., and Schein, C. J. (1970). Neuroendocrine influences on the dynamics of the choledocal sphincter. Surg. Gynecol. Obstet. 130:675.
328. Schein, C. J., Tawil, V. E., Dardik, H., and Beneventano, T. C. (1970). Common duct dynamics in man. The influence of the sympathetic block. Am. J. Surg. 119:261.
329. Dardick, H., Schei , C. J., Warren, A., and Gliedman, M. L. (1969). Adrenergic receptors in the canine biliary tract. Surg. Gynecol. Obstet. 128:823.
330. Mori, J., Azuma, H., and Fujiwara, M. (1971). Adrenergic innervation and receptors in the sphincter of Oddi. Eur. J. Pharmacol. 14:365.
331. Sarles, J. C., Sarles, H., and Devaux, M. A. (1975). Experimental Odditis and cholelithiasis in the dog. Role of the autonomic nervous system. Am. J. Gastroenterol. 63:147.
332. Tansy, M. F., Innes, D. L., Martin, J. S., and Kendall, F. M. (1974). An evaluation of neural influences on the sphincter of Oddi in the dog. Am. J. Dig. Dis. 19:423.
333. Lambelin, G., and Roba, J. (1969). Experimental Intersuchung des lokales Wirkung von Procain auf den Sphincter Oddi. Arzneim. Forsch. 19:789.

334. Crema, A., Benzi, G., Frigo, G. M., and Berte, F. (1965). The responses of the terminal bile duct to morphine and morphine-like drugs. J. Pharmacol. Exp. Ther. 149:373.
335. Mester, E., Kantor, E., and Gyenes, G. (1970). Données expérimentales pour la pathogénie du sphincter d'Oddi. Lyon Chir. 66:84.
336. Agosti, A., Mantovani, P., and Muri, L. (1971). Action of caerulein and related substances on the sphincter of Oddi. Arch. Pharmacol. 208:114.
337. Capitaine, Y., and Sarles, H. (1971). Action de l'éthanol sur le tonus du sphincter d'Oddi chez l'homme. Biol. Gastroenterol. (Paris) 3:231.
338. Poncelet, P. R., and Thompson, A. G. (1973). Role of infected bile in spasm of the sphincter of Oddi. Am. J. Surg. 126:387.
339. Hansson, K. (1967). Experimental and clinical studies in etiologic role of bile reflux in acute pancreatitis. Acta Chir. Scand. (Suppl.) 375.
340. Benzi, G. M., Arrigoni, E., Frigo, G. M., Ferrera, A., Panceri, P., Berte, F., and Crema, A. (1972). Antibiotics excreted into the bile. Action on the tone and mobility of the bile duct. Digestion 7:54.
341. Kantor, E., Jakab, T., and Szabo, L. (1969). Secretin, pancreozymin and cholecystokinin. Gastroenterology 18:183.
342. Uray, E., and Kosa, C. S. (1969). Wirkung der bei Neuroleptanagesie verwendeter Medikament auf die Druckwerke der Gallenwege. Anesthetist 18:74.

International Review of Physiology
Gastrointestinal Physiology II, Volume 12
Edited by Robert K. Crane

8
Intestinal Secretion

T. R. HENDRIX AND H. T. PAULK
The Department of Medicine
The Johns Hopkins University
School of Medicine, Baltimore, Maryland

THE INTESTINAL MUCOSA 259

COMPOSITION OF INTESTINAL SECRETION 260

CHOLERA AS A MODEL FOR INTESTINAL SECRETION 261
Cholera Enterotoxin and Epithelial Cell Interaction 261
Site of Production and Composition of Cholera Enterotoxin-induced Intestinal Fluid 262
Mechanism of Production of Cholera Enterotoxin-induced Intestinal Fluid 262
Increased Permeability of Epithelium 262
Impaired Intestinal Absorption 263
Intestinal Secretion 263
Mechanism of Bicarbonate Secretion 266
Role of cAMP 266
Summary 268
Crypts of Lieberkühn as Source of Intestinal Secretion 269

SECRETORY EFFECTS OF OTHER ENTEROTOXINS AND INVASIVE BACTERIA 270
Escherichia coli Enterotoxins 270
Shigella dysenteriae I Enterotoxin 271
Other Bacterial Enterotoxins 271
Invasive Bacteria 271

HORMONES AFFECTING INTESTINAL FLUID MOVEMENT 272
Vasoactive Intestinal Peptide 272
Secretin, Glucagon, and Gastric Inhibitory Peptide 273
Gastrin and Cholecystokinin 273
Thyrocalcitonin 273
Antidiuretic Hormone 273

INHIBITORS OF INTESTINAL SECRETION 274

CONCLUSIONS AND SPECULATIONS 276

Physiologists during the early part of this century generally believed that intestinal secretions, the *succus entericus*, played an important part in digestion because of their content of enzymes. This view seemed to be supported by the fact that the intestine is lined by glands. Howell, in the ninth edition of his textbook, stated, "The small intestine is lined by tubular glands, the crypts of Lieberkühn, which give rise to a liquid secretion. . . . Preyl estimates that as much as three liters may be formed in the whole of the small intestine in the course of a day . . ." (1). In 1941, however, Florey et al. concluded that intestinal secretion played a very minor, if any, direct role in intraluminal digestion (2). Enterokinase, which converts trypsinogen to trypsin and thus leads to the activation of the remainder of the proteolytic enzymes of the pancreas, was the only important digestive function that could be identified. All other enzymes, except amylase, appeared to be derived from shed enterocytes. This led to general acceptance of the view that intestinal fluids were not a "digestive secretion" and interest waned to the point where most physiologists ignored the secretions and even denied their existence. In point of fact, however, Florey et al. had not denied the existence and importance of intestinal secretion. Their summary of the state of understanding is as applicable today as when written 35 years ago: "Our knowledge of histology of the mucosa, which is considerable, contrasts with our lack of knowledge of the function of its several component cells. It has not, for instance, been settled whether the crypts secrete and the villi absorb or whether they have a common function, as the common origin of their epithelial covering might be taken to indicate, in which case both crypts and villi would secrete and absorb." Although there were no experimental data to support the position, Florey and his colleagues obviously favored the notion that crypts and villi had different transport functions, for they concluded their review, stating, "it may be assumed that during digestion a fluid is secreted which has other functions besides that of contributing enzymes. It may be necessary for a constant secretion of fluid to take place from the crypts of Lieberkühn . . . and as the products of digestion are absorbed, water and salts go with them. One may envision a circulation of fluid during active digestion, the secretion passing out from the crypts of Lieberkühn into the lumen and back into the villi" (2). However, with physiological and clinical investigation focused solely on absorption, the questions of secretion raised by Florey et al. remained unresolved.

Current interest in intestinal secretion began with studies of Asiatic cholera prompted by the recent, seventh pandemic of that disease (3–5). The then prevalent hypotheses, e.g., mucosal damage, increased epithelial permeability, or impaired absorption, were found inadequate to explain the massive movement of fluid into the intestinal lumen that characterizes cholera; attention turned to intestinal secretion (6). However, the notion that intestinal secretion was responsible for the intestinal fluid production and consequent fulminant diarrhea of cholera was not new. In 1855 Dr. John Snow, whose epidemiological studies focusing on the Broad Street pump identified fecal contamination of the water supply as the cause of the London cholera epidemic, suggested, "It would seem that the cholera poison when produced in sufficient quantity acts as an irritant on the surface of the stomach and intestine, or what is still more probable, it withdraws fluid from the blood circulating in the capillaries by a power analogous to that by which the epithelial cells of the various organs abstract the different secretions in the healthy body" (6). Twenty years later, Cohnheim, in his Lectures on General Pathology, concluded that "the process of cholera may be interpreted by supposing under the influence of the virus, which probably has entered the intestine from without, there takes place, *an extraordinary profuse secretion from the glands of the small intestine*" (7).

In the sections to follow we will discuss the composition of intestinal secretion, secretory stimuli, inhibitors of secretion, and the source of secretion.

THE INTESTINAL MUCOSA

The intestine is lined by a simple columnar epithelium arranged in tubular glands, the crypts of Lieberkühn, which surround the bases of the villi. The major cell type in the crypts is undifferentiated. These cells divide to produce new cells to move up onto the villi. As the cells move out of the crypts, the rate of protein synthesis decreases and the microvilli of the brush border becomes taller, more closely packed, and develops a variety of enzymes: disaccharidases, alkaline phosphatase, and peptidases. After 4 days or so, the columnar cell has migrated to the tip of the villus and is extruded into the lumen. Maturation is completed about one-half to two-thirds of the way up the villus.

In addition to undifferentiated crypt cells and their progeny, the absorptive cells, there are also goblet cells, found both in crypts and on villi, Paneth cells, located at the bases of the crypts, and enterochromaffin cells, which are most abundant in the crypts. These last are the source of the peptide intestinal hormones. The epithelial cells are joined near their luminal surface by a fusion of their plasma membranes into a "tight" junction. Although this tight junction excludes proteins and molecules as small as glucose, it is believed to be the anatomic location of the paracellular shunt pathway which may account for as much as 85% of fluid traversing the intestinal epithelium. The tight junction also provides the blind end of the space between the absorptive cells, which makes it possible to develop the osmotic gradient necessary for bulk fluid absorption (4, 8, 9).

The arrangement of the blood supply to the intestinal mucosa will determine in part the characteristics of intestinal absorption and secretion in vivo. Reynolds et al. describe in the monkey a rich capillary network surrounding the crypts which anastomoses with the capillary network of the villi (10). Both networks are supplied by the same arteriole, but veins drain both the villous and crypt capillaries so that there appears to be the potential to pass blood past the crypt cells and then into the villi or in other circumstances to decrease flow to the villi by opening up secondary veins draining the capillary plexus of the crypts. This vascular arrangement would serve Florey's crypt-villus fluid circuit well. On the other hand, there is evidence that there may be considerable species variation which needs to be documented. For example, if fluid and electrolytes leave the capillaries as blood passes the crypts, the resultant increased osmotic pressure might facilitate absorption as this blood passed through the capillaries of the villi (see Chapter 1).

COMPOSITION OF INTESTINAL SECRETION

Little has been added to the observations of DeBeer et al. (11) which showed the concentration of sodium, potassium, calcium, and total anions in fasting intestinal fluid to be essentially the same in jejunum and ileum and little different from blood plasma. On the other hand, the concentrations of the principal anions, chloride and bicarbonate, differ in the upper and lower small intestine; the intraluminal concentration of bicarbonate is less than plasma in the jejunum and greater than plasma in the ileum, whereas the opposite is found for chloride. More recent studies have confirmed these findings (12, 13). McGee and Hastings (14) in a study of jejunal fluid in fasting man found that the bicarbonate concentration was approximately one half that of plasma, while the CO_2 tension was twice that of venous blood. They suggested that these findings could be produced by the mixing of an acid and an alkaline secretion in the intestinal lumen. It has been calculated that the addition of 15 mM HCl per liter of plasma ultrafiltrate would be required to produce the bicarbonate concentrations and CO_2 tensions found in the jejunum (15).

It should be made clear that, in these studies, the intestinal contents measured are a composite determined by the volume and composition of several fluids entering and leaving the segment under study: 1) the fluid flowing into the segment from above, 2) the fluid being absorbed, 3) the fluid diffusing across the mucosa to promote electrochemical equilibrium, and 4) the fluid secreted. At the present time, there is no satisfactory method for distinguishing the volume and composition of the fluid secreted from the others. Until recently, there seemed to be no necessity to try to measure the secretion component since in the models studied it was possible to attribute all fluid movement into the intestine to diffusion down activity gradients. This passive movement of water and electrolytes into the intestine does not, however, provide a satisfactory explanation for secretory states such as encountered in cholera. Only Nasset and Ju (16) have described the direct collection of intestinal secretion from the *crypt ostia*, but their studies were concerned with enzyme secretion rather than with the ionic composition of the secreted fluid. There are no published reports of the application of this technique to studies of the ionic composition of intestinal fluid produced in response to secretory stimuli.

CHOLERA AS A MODEL FOR INTESTINAL SECRETION

Interest in intestinal secretion was awakened by the desire to explain the voluminous diarrhea of cholera and to develop a rational effective treatment. Studies were facilitated because all manifestations of cholera are the consequence of fluid and electrolyte losses. The diarrhea of cholera is the consequence of fluid production by the small intestine in response to an enterotoxin produced by the cholera organism which is effective in all animal species studied. It is little wonder therefore that cholera enterotoxin-induced intestinal secretion became the experimental model of intestinal secretion, though how generally applicable the findings are to intestinal secretion is yet to be determined. For example, it is not yet known whether cholera-induced secretion is a unique pathological phenomenon or the turning full on of a normal physiological process generally obscured by absorptive processes, or whether there is a single secretory mechanism or several.

Cholera Enterotoxin and Epithelial Cell Interaction

Cholera enterotoxin is a heat-labile neutral protein with a molecular weight estimated to be between 79,000 and 84,000 (17). It has two major components: one, with a molecular weight of 28,000–29,000, is believed to be responsible for activation of adenylate cyclase and stimulation of intestinal secretion, while the other, with a molecular weight of 50,000–60,000, is responsible for binding to cell membranes (18). Holmgren has proposed that the latter is composed of seven subunits, whereas Gill's calculations indicate five (18, 19). Toxin binding to the intestinal epithelium is rapid and essentially irreversible. Once bound, the effects of the toxin cannot be blocked by specific antitoxin or ganglioside. A 5-min exposure of the intestine to cholera enterotoxin is as effective in eliciting a maximal secretory response as is continuous contact (20). The component of the cell membrane to which the toxin binds appears to be GM_1 ganglioside (18); if cholera enterotoxin is premixed with ganglioside, the toxin does not bind and intestinal secretion is not stimulated (21). Following binding, intestinal secretion is stimulated and adenylate cyclase is activated with consequent elevation of cellular levels of cyclic adenosine 3′:5′-monophosphate (cAMP) (21, 22). Gill has presented evidence, using pigeon erythrocyte membranes, that once the toxin is bound to the plasma membrane by the binding subunit, "B," the active subunit, "A," is inserted into the membrane or even into the cytosol where it participates in reactions, believed to be enzymic and requiring NAD, that lead to the production of a permanently activated adenylate cyclase (19).

Binding of cholera enterotoxin and activation of adenylate cyclase have been seen in all cells tested, viz, hepatocytes, adipocytes, erythrocytes, lymphocytes, adrenal cortical cells, ovarian cells, platelets, etc.

Site of Production and Composition of Cholera Enterotoxin-induced Intestinal Fluid

The massive outpouring of fluid induced by cholera enterotoxin occurs in the small intestine. The volume secreted per unit length is greatest in the duodenum and least in the ileum. The colon does not participate (23–25). The secreted fluids maintain their characteristic ion composition regardless of the rate of fluid production (26). Also, their protein content is very low, in the range of 85 mg/dl, well below levels found in inflammatory bowel disease and congestion of the splanchnic circulation (27).

Mechanism of Production of Cholera Enterotoxin-induced Intestinal Fluid

Movement of water is determined by the activity gradients across the membranes separating the luminal, intracellular, and extracellular compartments. A change in rate or direction of water movement may be due to 1) a change in membrane permeability, or 2) a change in driving force, be it hydrostatic, osmotic, electrochemical, or any combination of these. These have been discussed in other reviews and will only be summarized here (4, 6, 28–30).

Increased Permeability of Epithelium The hypothesis that cholera is a fluid exudate due to gross epithelial damage, although long popular, may be dismissed because exudates are high in protein, whereas cholera secretions are low. Also, the epithelium in cholera is intact both in man (31) and the experimental animal (32).

The hypothesis that cholera enterotoxin, although producing no morphological abnormalities, increases mucosal permeability has also not held up to careful scrutiny. First, a more permeable epithelium should be less selective, i.e., with increasing rates of fluid production, the ionic composition of jejunal and ileal fluids would be expected to approach the concentration in plasma. This is not observed. Second, although an appropriate increase in driving force, osmotic or hydrostatic, could contribute to fluid production in cholera, the required changes have not been found. In fact, the osmotic gradient changes in cholera are in the wrong direction; with the diarrhea and consequent hemoconcentration caused by cholera, plasma protein concentrations may reach values as high as 13.4 g/dl (33). Changes in hydrostatic pressure show little or no effect on the rate of intestinal fluid production, e.g., in experimental cholera, a reduction of mesenteric blood flow to 30% of normal did not alter the rate of cholera-induced fluid production (34). In a study measuring the rate, distribution, and volume of blood flow by analysis of the clearance curves of intra-arterially infused ^{133}xenon no difference was found between control animals and those secreting in response to cholera enterotoxin (35). Also, a 50% reduction of mean arterial blood pressure by phlebotomy did not change the secretion rate in the cholera animals, although in both these and the control animals there were marked changes in volume and distribution of blood flow to the various intestinal compartments. Finally, direct measurement of permeability in human and experimental cholera using isotopically labeled probe molecules shows no evidence of an increase (36–38). In fact, cholera enterotoxin and cAMP increase electrical resistance and decrease conductance of the intestinal mucosa (39, 40).

Impaired Intestinal Absorption It is not surprising that early studies of cholera focused on impaired absorption since most formulations of intestinal function stressed absorption; movement into the lumen was attributed to passive flux down activity gradients (33). Whether or not there is any defect in sodium absorption is an unresolved question. Experimental results are contradictory. Some have reported no change in $J^{Na^+}_{M-S}$ (unidirectional sodium flux from intestinal lumen to blood) (41–44), while others have concluded that sodium absorption is decreased and contributes to the intestinal loss (45, 46).

In vitro studies of rabbit ileal mucosa in Ussing-type chambers have uniformly shown that cholera enterotoxin abolishes the "active" absorption of sodium normally found with all gradients, electrochemical, osmotic, and hydrostatic, eliminated between the luminal and serosal aspects (39, 47, 48). It is not known, however, what contribution, if any, this in vitro electrogenic sodium absorption makes to total in vivo ileal sodium absorption in the presence and absence of cholera enterotoxin, nor is it known what effect cholera enterotoxin has on in vitro jejunal sodium absorption. Since the jejunum makes the largest contribution to the total cholera-induced secretion, information on these points would help clarify the relation between in vitro and in vivo findings.

Regardless of the status of electrogenic sodium absorption, all studies in vitro and in vivo agree that glucose absorption and glucose-associated sodium absorption are unimpaired by cholera enterotoxin (42, 48, 49). Indeed the capacity of this active transport mechanism to entrain sodium, water, and other electrolytes along with the actively transported glucose has been used successfully to provide oral water and electrolyte replacement of fecal losses in cholera (50–52). The absorption of other actively transported solutes that are coupled with sodium absorption, such as amino acids, are also unimpaired in cholera (53, 54). It seems clear that impaired absorption is not a major factor in intestinal fluid production in cholera.

Intestinal Secretion The fact that jejunal and ileal fluids maintain their normal anionic compositions during cholera suggests that there is an active secretory process involved. Support is provided by the demonstration that when intestinal loops in vivo, closed or perfused, are pretreated with cholera toxin the volume of fluid in the loop is greater at the end than at the beginning of the study (24, 25). Unidirectional ion fluxes showing an increase in J_{SM} with J_{MS} unchanged and a J_{net} in the reverse blood → lumen direction also suggest that active intestinal secretion is the driving force when no other driving force is apparent (42). More direct evidence has been provided by measurements of unidirectional fluxes across rabbit ileal mucosa in an Ussing chamber in the absence of any gradient (47). The potential difference (PD) across the mucosa was reduced to zero by imposing an opposite potential. Thus all gradients, electrical, hydrostatic, and osmotic, were reduced to zero. The current flowing under these conditions, the short circuit current (SCC), is related to the algebraic sum of the ion fluxes, and net movement of an ion across the mucosa is the result of an active process. Under these conditions, in control ileal mucosa, there was net movement of Na and Cl from mucosa to serosa and there was a "residual ion flux" in the opposite direction which was attributed to HCO_3 ($J_{net} = SCC - J^{Na}_{net} + J^{Cl}_{net}$) (48, 55). After exposure to cholera toxin, net Na flux was reduced to zero, net Cl flux was reversed, becoming S → M, and there was no change in the "residual ion flux" (Figure 1).

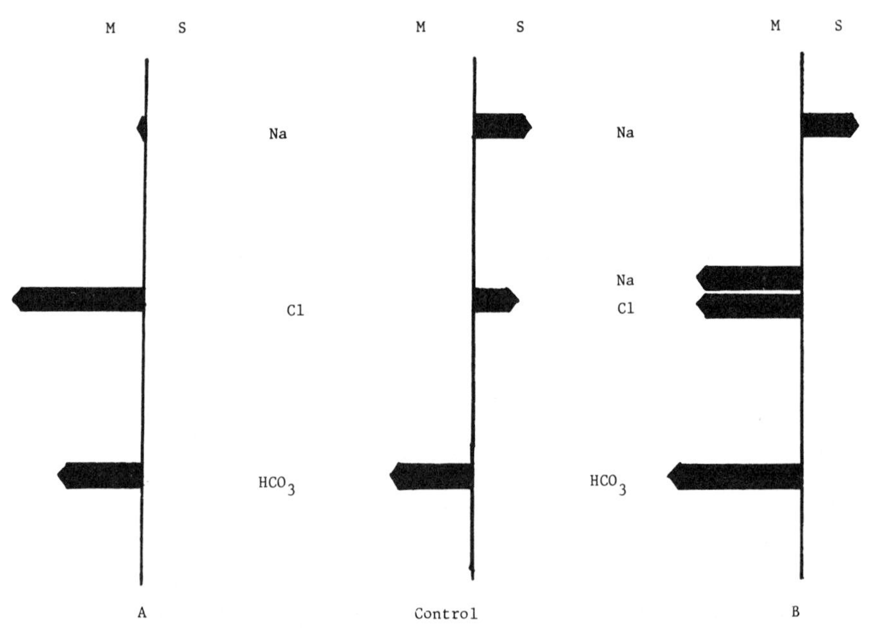

Figure 1. Schematic representation of active ion transport by short circuited rabbit ileal mucosa mounted in an Ussing-type chamber. Under control conditions (*center*), there is net absorption of Na and Cl and secretion of bicarbonate. The secretion of bicarbonate was not measured but deduced from difference between short circuit current and the net fluxes of Na and Cl ($J^R_{Net}(HCO_3) = SCC - (J^{Na}_{Net} - J^{Cl}_{Net})$). *A*, after exposure of the mucosa to cholera enterotoxin or cAMP, Field found that net Na flux was reduced to zero and net Cl flux was reversed without any change in HCO_3 flux (residual ion flux J^R_{Net}). The increase in SCC current observed was accounted for by the striking reversal of Cl movement which more than made up for the reduction in Na absorption. *B*, Powell, on the other hand, attributed the increased short circuit current produced by cholera enterotoxin to maintenance of electrogenic Na absorption and an increase in J^R_{Net}. In addition he interpreted the observation as indicating inhibition of coupled NaCl absorption and induction of coupled NaCl secretion.

Using a similar experimental technique and obtaining similar but not identical results, Powell et al. (40, 56) came to somewhat different conclusions as to the mechanism involved in cholera-induced secretion. Powell et al. and Field (47) agree 1) that cholera enterotoxin was effective only at the mucosal surface, 2) that it increased PD and SCC and decreased tissue conductance, and 3) that theophylline produced a greater increase in SCC when added to control mucosa than when added to enterotoxin-treated mucosa. They differed, however, in their observations and interpretations of the effects of cholera enterotoxin on ion transport. Instead of a reduction of net sodium flux to zero, Powell et al. found a reversal of net sodium flux as well as of net chloride flux. They also found an enterotoxin-induced increase in J^R_{net} (HCO_3) which though sizable did not reach statistical significance. Field found a similar change in J^R_{net} but did not include it in his formulation because of the absence of statistical significance. To summarize, Powell et al. found that cholera enterotoxin 1) caused secretion of both Na and Cl and possibly an increase in J^R_{net} (attributed to HCO_3 secretion) in Ringer solution; 2) stimulated Cl secretion and reduced Na absorption to zero in absence of HCO_3; 3) had no effect on net Na transport or electrical measurements in absence of transportable anions (Cl and HCO_3); and 4) in the absence of Na caused an increase in PD and SCC but had no effect on Cl transport. On this basis, they suggested that the major effect of cholera enterotoxin was to stimulate an electrically neutral S → M transport of NaCl and/or $NaHCO_3$ without affecting electrogenic Na absorption. How much of the difference between these two studies is due to technique (Field had 7.5–10 mM glucose in the serosal solution) and how much to different interpretations of similar data is difficult to say at this point. For example, Field interpreted the increase in SCC as indicating electrogenic secretion of Cl, whereas Powell et al. minimized the increase in SCC, attributing it either to an increase in electrogenic Na absorption or to a minor electrogenic anion secretory mechanism. A strong argument against electrogenic Cl secretion is the observation of Binder et al. that in the absence of Na there was no net Cl flux (57). The Powell formulation is easier to reconcile with in vivo findings, i.e., no defect in Na absorption in the presence of an increased S → M movement of HCO_3^- and a smaller increase in Cl secretion. However, there is no explanation for the difference between these in vitro findings and findings in vivo, showing that the steepest electrical-chemical gradient is for HCO_3 secretion (58). In the in vivo studies the normal canine ileum absorbed fluid from a perfusing solution having a starting composition of Na, 135, K, 5, Cl, 115, and HCO_3, 25 mEq/liter. After exposure to cholera enterotoxin, fluid absorption decreased, and by 4 hr net fluid secretion appeared. After 7 hr the electrolyte composition of the perfusate was Na, 135, K, 13, Cl, 65, and HCO_3, 80 mEq/liter. In these studies, both HCO_3 and Cl were secreted against an electrochemical gradient. Potassium was expected to enter the lumen passively, but the concentration achieved was greater than predicted from only the electrochemical gradient. These differences in vitro and in vivo are probably not due to species variation since the in vivo rabbit ileal contents also have a high HCO_3 concentration (25, 59). More likely, the differences are due, in part at least, to the absence of the capillary circulation and lymph drainage in the in vitro preparation. In addition, in vivo, the two surfaces of the mucosa were not bathed with solutions of identical composition.

Mechanism of Bicarbonate Secretion Understanding of bicarbonate accumulation in the intestine is complicated because the same net effect can be achieved by HCO_3 secretion, OH secretion, or H absorption. On the basis of studies in man, Turnberg et al. suggested a double ion exchange model to account for PD, pH, and net fluxes. In this model Na from the lumen is exchanged for H from the cell, and Cl from the lumen is exchanged for HCO_3; the result is nonelectrogenic absorption (60). In response to cholera enterotoxin both HCO_3 and Cl moved against electrochemical gradients. The explanation suggested for Cl movement being less than HCO_3 was that NaCl absorption was normal. However, this leaves unexplained the increased PD during cholera-induced secretion.

In vitro, no difference between HCO_3 secretion in control and cAMP-treated tissues was found and the ion fluxes were not those predicted by the double ion exchange model (61). In order to better understand this, Hubel (62) compared net water and HCO_3 fluxes, luminal HCO_3, and pCO_2 in luminal fluid and arterial blood. He found that the HCO_3 concentration in ileal loops of control rabbits was 42.6 mM, whereas in cholera enterotoxin-treated loops it was 65.5. If the increased HCO_3 was due to HCO_3 secretion, luminal pCO_2 should rise due to partial dissociation of HCO_3 ($HCO_3^- \rightarrow OH^- + CO_2$); if due to OH^- secretion or H^+ absorption, pCO_2 should fall. In treated loops, pCO_2 fell by 8.4 torr in 45 min, whereas it rose by 4.4 in the controls. At high rates of bicarbonate accumulation, CO_2 is removed from luminal fluid by OH more rapidly than it can diffuse into the lumen.

Role of cAMP There is a great deal of evidence to support a mediator role for cAMP in cholera enterotoxin-induced secretion. First, elevation of ileal cAMP alters in vitro ileal transport in a manner similar to that produced by cholera enterotoxin (39, 48, 55, 56), no matter whether the level of cAMP is increased by addition of cAMP or dibutyryl cAMP, by blocking degradation with theophylline, or by increasing synthesis by activation of adenylate cyclase with prostaglandins or vasoactive intestinal peptide (VIP). Second, in vivo intestinal secretion is produced by cholera enterotoxin, theophylline, and prostaglandins (64). Third, cholera enterotoxin activates intestinal adenylate cyclase and elevates intestinal cAMP (63, 66, 67). Intestinal adenylate cyclase activity is elevated in man during cholera and returns to normal in convalescence (68, 69). Fourth, adenylate cyclase levels are highest in the duodenum and lowest in the ileum (63) parallel to the secretory responses to cholera enterotoxin (24). Finally, the magnitude of the secretory response parallels the level of adenylate cyclase activity induced by cholera enterotoxin (70).

There are, however, a number of discrepancies which need to be resolved. First, although SCC, PD, and undirectional as well as net fluxes are similar when cAMP, theophylline, or cholera enterotoxin are added to rabbit ileal mucosa, in vitro, the magnitude and time course of the changes are considerably different (56). In vitro the changes induced by cholera enterotoxin tend to be less than by theophylline, whereas, in vivo, the secretory response to cholera enterotoxin greatly exceeds that which can be achieved by maximal infusions of theophylline or prostaglandins into the superior mesenteric artery (64). Second, although there is a relationship between the level of adenylate cyclase activity and net water and sodium fluxes (70), an increase in cAMP preceding the appearance of intestinal secretion in vivo has not been demonstrated (71). This failure suggests either that the two phenomena are associated but not by cause and effect or that only a small fraction of the cell population is involved in cholera-induced secretion.

The interaction between cholera enterotoxin and adenylate cyclase differs from that observed with prostaglandins or VIP. With the latter, elevation of cAMP is prompt but rapidly disappears when the hormone is removed. With cholera enterotoxin the delay between time of contact and measurable effect on secretion has been reported as short as 15 min with in vivo rabbit jejunum (72) and on cAMP levels as long as 1 hr with rabbit ileal mucosa in vitro (73). Also, the enterotoxin effect continues long after the toxin is removed; a 5-min exposure stimulates as much secretion as does continuous exposure (20) and the effect lasts at least 24 hr (24, 70). Indeed, it has been suggested that cholera enterotoxin leads to an activation of adenylate cyclase that continues for the life of the stimulated cell.

Several explanations have been advanced to explain the delay in enterotoxin actions. First, it has been proposed that the delay reflects the time required for the synthesis of new adenylate cyclase. However, Kimburg et al. have demonstrated that NaF-stimulated adenylate cyclase activity is no greater after exposure of the mucosa to active than to inactive enterotoxin (63). Second, although binding of the enterotoxin is rapid, penetration to the level of adenylate cyclase is considerably slower: compatible with this, adenylate cyclase activity has been reported to be mostly found in the basolateral membranes of the cell (74), although preliminary observations from our laboratory indicate that with proper assay conditions abundant adenylate cyclase activity is found in the brush border membrane. Nonetheless, the elegant experiments of Gill and King with pigeon erythrocytes strongly support a penetration hypothesis (19). Twenty minutes were required for a detectable rise in adenylate cyclase activity after contact of enterotoxin with intact erythrocytes. With lysed erythrocytes, adenylate cyclase activity rose immediately. Third, it has been suggested that enterotoxin stimulation involves a mediator, the activation, release, or synthesis of which accounts for the delay. The presence of a circulating mediator is suggested by the secretory response of an intestinal loop not itself exposed to enterotoxin but perfused with blood from a loop secreting in response to enterotoxin (75, 76), despite the fact that intravenously administratered enterotoxin has no secretory effect on the intestine (77).

Prostaglandins have been suggested as the mediator (78), and this suggestion is strengthened by the facts that 1) prostaglandins activate intestinal adenylate cyclase (63) and stimulate intestinal fluid secretion (65, 79), and 2) acetylsalicylic acid and indomethacin, which block prostaglandin synthesis, also inhibit the in vivo secretory response to enterotoxin (80, 81).

However, there is compelling evidence for rejecting prostaglandins as simple mediators: 1) prostaglandin-induced in vivo intestinal secretions, at levels unassociated with cellular damage contain four to five times more protein than do secretions in response to enterotoxin or theophylline (65); 2) prostaglandin E_1 and enterotoxin stimulation of adenylate cyclase activity are additive (63), 3) indomethacin in doses sufficient to block prostaglandin synthesis in rabbit ileal mucosa did not alter enterotoxin-induced elevation of intestinal cAMP (73), and 4) no significant elevation of prostaglandin E in rabbit intestinal mucosa was found at the same time after exposure to enterotoxin in vivo when there were significant increases in intestinal fluid volume and cAMP levels (82).

These findings argue strongly against a role for prostaglandins in enterotoxin-induced adenylate cyclase activation but they do not rule out a role for prostaglandins in cholera-induced secretion nor do they explain the inhibitory effects of aspirin and indomethacin. Recently, Wald et al. confirmed that indomethacin inhibits enterotoxin-induced secretion without altering enterotoxin-induced elevation of cAMP levels (71). They suggested the following possible interpretations: 1) indomethacin may inhibit a cAMP-independent, enterotoxin-induced secretory process mediated by prostaglandins, 2) indomethacin may inhibit a prostaglandin-mediated step beyond the generation of cAMP, 3) indomethacin may act on some other biological mechanism involved in secretion, 4) if enterotoxin activation of adenylate cyclase and stimulation of intestinal secretion are not cause and effect, indomethacin may act only on the secretory mechanism 5) if the cells involved in secretion are only a small proportion of the total, effects on cAMP specifically related to secretion may well go undetected in studies that measure the cAMP content of the entire mucosa.

Summary Cyclic AMP and maneuvers that raise cAMP through increasing its production or decreasing its destruction cause intestinal secretion in vivo and produce active chloride secretion in vitro. In addition, elevation of cAMP in vivo increases transmucosal PD, SCC, and conductance. Cholera enterotoxin produces similar changes but the time required to achieve them is much greater and once achieved they are persistent rather than transient. These findings suggest a uniqueness to enterotoxin action and an activation of adenylate cyclase that is qualitatively different from that achieved by prostaglandins or hormones. Alternatively, different cyclase receptors are involved. The nature of the secretory response to elevated intestinal cAMP is unknown. It is not known whether there is one secretory mechanism activated by several types of stimuli or whether there are several independent pathways. The latter seems the more likely since 1) *Shigella* enterotoxin causes intestinal secretion similar to that produced by cholera enterotoxin without an activation of cAMP (83), 2) secretion caused by hypervolemia is associated with anatomic changes, i.e., dilatation of the intercellular spaces between the columnar cells, not seen in the enterotoxin-stimulated intestine and this secretion is not associated with elevation of cAMP (84), 3) secretion can be produced by intestinal distension that is associated neither with histological changes nor an elevation of cAMP (85), and 4) intestinal fluid produced in response to an osmotic gradient has a different electrolyte composition from that produced in response to an enterotoxin (86, 87). In addition, it should be made clear that intestinal fluid accumulation is not synonomous with intestinal secretion since impaired absorption or increased filtration through the mucosa can singly or in combination lead to intestinal fluid accumulation without involving a secretory mechanism.

Crypts of Lieberkühn as Source of Intestinal Secretion Previous generations were more impressed with the relation between structure and function than is general today and never thought to question that glands, even simple tubular ones like the crypts of Lieberkühn, were designed to secrete. Cohnheim's 1881 lecture, quoted in the introduction, introduced the "crypt cell secretion hypothesis." However, as we also noted above, investigations were concerned for a long time only with absorption since movement from blood to lumen could be rationalized by passive movement down activity gradients. The prime stimulus for reviving Cohnheim's "crypt cell secretion hypothesis" was observations that indicated that absorptive and secretory rates could be manipulated independent of one another (4). Since it is generally accepted that absorption is greatest in cells at the villus crest (88), it was suggested that secretion is carried out by other cells. In the "crypt cell secretion hypothesis," glucose and other such solutes could increase absorption of water and electrolytes by cells on the villus irrespective of any secretion produced by cholera toxin in the undifferentiated cells of the crypts of Lieberkühn (4, 42).

Several observations suggest that cholera enterotoxin-induced secretion originates in the crypts: 1) lumina of the long, branch crypts in the dog are dilated in experimental cholera (32), 2) hypertonic Na_2SO_4 produces histological damage at the villous crests and decreases glucose absorption, but enterotoxin-induced secretion is unaffected (89), and 3) cycloheximide in appropriate doses produces mitotic arrest in the crypts and inhibition of enterotoxin-induced secretion without depressing glucose absorption (90). This effect is associated with inhibition of protein synthesis but does not interfere with enterotoxin-induced activation of adenylate cyclase and elevation of cAMP (91). At higher doses cell damage is evident in the crypts and with still higher doses in the entire epithelium, and 4) theophylline decreases intracellular electrical potential in cells adjacent to the crypts, whereas villus cells are unaffected (92).

On the other hand, deJonge has concluded from the results of simple but ingeniously designed experiments that there is no evidence for a specific role for the crypts in enterotoxin-induced secretion; in fact, the evidence was considered incompatible with the crypt cell secretion hypothesis (93). He found that exposure of the intestine to cholera enterotoxin for 2 min or 1 hr resulted in different patterns of adenylate cyclase activation. After the 1-hr exposure, adenylate cyclase activity rose in parallel fashion in both crypts and villi, although activity was greater in the villus fraction and reached a plateau by 5 hr. Net fluid transport changed from absorption to secretion 2 hr after contact. After the 2-min exposure, on the other hand, villus cell adenylate cyclase followed the same course as after the longer exposure, whereas crypt cell adenylate cyclase did not rise. Net fluid absorption was reduced to zero by 3 hr but net secretion was not produced.

However, interpretations other than a functional homogeneity of villus and crypt cells in response to enterotoxin should be considered; increases of cAMP are associated with a variety of changes in intestinal function, active intestinal secretion being but one of them. For example, coupled sodium chloride influx across the brush border is inhibited by theophylline through the mediation of cAMP (94, 95). Thus, it is possible that the effects on net fluid absorption observed by deJonge following the brief exposure were mediated in part by diminished coupled NaCl absorption. On the other hand, the effects of cAMP must be complex since elevation of cAMP enhances neutral amino acid uptake, which is also a sodium coupled process (96).

Using a somewhat different approach, Weiser and Quill (97, 98) obtained data which they also conclude do not support the crypt cell secretion hypothesis. They measured guanylate and adenylate cyclase activities in the upper villus, mid-villus, and crypt regions of rat intestine. Adenylate cyclase was highest in the crypts and lowest in the upper villus, whereas guanylate cyclase had the opposite distribution. They also measured adenylate cyclase activity in these same fractions after exposing the intestine to cholera enterotoxin for 3 hr. Adenylate cyclase increases were 3½ times greater in the upper villus than in the crypt. In addition, ^{125}I-labeled enterotoxin binding after a 15-min exposure, sufficient time to stimulate maximal secretion, was limited to the villi. Schwartz et al. (99) found similar relative adenylate cyclase levels between crypt and villus in rabbit and rat. In rabbit after cholera enterotoxin the enzyme in both regions was increased to the same level. PGE increased the rat crypt level to almost twice that of the villus cells.

This controversy over the role of the crypts in intestinal secretion needs to be resolved, though it will be difficult to do so. Measurements and flux studies of the entire mucosa may only provide an algebraic sum of opposing fluxes generated by two cell populations, further compounded by the paracellular pathway, the conductance of which can be altered by changing levels of intestinal cAMP (100–102). The secretion of one substance, secretory IgA, has been shown to be clearly limited to the crypts of Lieberkühn (103).

SECRETORY EFFECTS OF OTHER ENTEROTOXINS AND INVASIVE BACTERIA

Escherichia coli Enterotoxin

A variety of *Escherichia coli* isolates from man and pigs produce a heat-labile enterotoxin which stimulates the action of cholera enterotoxin (104–107). Both are associated with elevation of adenylate cyclase (108, 109). The secretory response to *E. coli* enterotoxin is shorter than with cholera enterotoxin, which may be attributed in part to differences in binding of the toxins to the mucosa. Cholera enterotoxin binds specifically with G_{MI} ganglioside, whereas no specific binding could be found for *E. coli* enterotoxin (110).

Shigella dysenteriae I Enterotoxin

Shigella enterotoxin induces a secretion similar to cholera enterotoxin-induced secretion in electrolyte and protein content and at similar maximal secretory rates. The delay between contact with the mucosa and onset of secretion is, however, greater, *Shigella* enterotoxin binds less rapidly than does cholera enterotoxin (111). It stimulates secretion without increasing intestinal cAMP (83). In the Ussing chamber, *Shigella* enterotoxin produces different changes than cholera enterotoxin: 1) there is no increase in cAMP; 2) there is no change in SCC nor the theophylline response; and 3) glucose produces less augmentation of SCC in *Shigella* enterotoxin-treated mucosa than in control or cholera enterotoxin-treated mucosa (112). Despite these differences, cholera and *Shigella* enterotoxins are not additive in their effects,suggesting that a common final pathway is involved (83, 111). *Shigella* enterotoxin is cytotoxic (113), but its effect on secretion is achieved without histologic alternation of the mucosa (111).

OtherBacterial Enterotoxins

Enterotoxins have been shown to be produced by staphylococci (114), *Clostridium perfringens* (115, 116), *Pseudomonas aeruginosa* (117), and *Klebsiella pneumoniae* (118), but little is known of their mode of action, whether, for example, they are additive or act through a common mechanism and whether they activate adenylate cyclase or not.

Invasive Bacteria

Salmonella, Shigella, and invasive *E. coli* invade the mucosa of the ileum and colon and produce an acute inflammatory reaction, suggesting that the intestinal secretion and diarrhea associated with these infections is a direct consequence of mucosal damage. However, the pathogenesis of intestinal secretion in this situation proves to be more complex. Fromm et al. (119) found that only one of two invasive strains of *Salmonella typhimurium* caused intestinal secretion in vivo. In vitro rabbit ileum invaded with the nondiarrheagenic strain behaved normally with respect to PD, NaCl absorption, and response to glucose and theophylline. On the other hand, ileal mucosa invaded by the diarrheagenic strain behaved like ileum exposed to cholera toxin. Of even more interest were studies of the *Salmonella*-infected rhesus monkey in which secretion was found in jejunum as well as in ileum and colon even though the numbers of salmonella in the jejunum were few, there were no histological alterations, and the salmonella could not be shown to produce an enterotoxin (120). These differences were extended by finding that the invasive strain that produced secretion also stimulated adenylate cyclase activity similar to the stimulation found with cholera enterotoxin, whereas the inactive strain produced no change. In addition, indomethacin, which decreased cholera enterotoxin-induced secretion without altering the cAMP response, inhibited *Salmonella*-induced intestinal secretion and activation of cAMP without altering the pattern of invasion and inflammation in the ileal mucosa (121). The fluid response to *S. typhimurium* was completely inhibited, whereas cholera enterototoxin- and invasive *Shigella*-induced secretion was inhibited by 40–60% (122). Invasion of the mucosa by the salmonella and shigella organisms was unaffected by indomethacin. Since no enterotoxin production by these invasive organisms has been identified, a humorally mediated stimulant of secretion may be suggested, whether of bacterial origin or produced by the host. Since the secretory response is blocked by indomethacin, it is tempting to suggest that prostaglandins are the mediator, but, as was pointed out above, indomethacin suppression alone is not sufficient evidence to implicate prostaglandins. Altogether, however, these observations should stimulate a search for a humoral mediator which may play a role in toxigenic secretion as well as normal intestinal function.

HORMONES AFFECTING INTESTINAL FLUID MOVEMENT

Based on the above, the question may be asked whether the intestinal secretion provoked by bacterial enterotoxins or mucosal invasion is a protective mechanism evolved, perhaps, to dilute and wash away irritant chemicals and pathogenic micro-organisms or whether it is a "turned-on" physiological process integral to normal intestinal absorptive function. The latter supposition is strengthened by observations that patients with certain specific endocrine tumors have diarrhea, presumably based on increased intestinal secretion, as their primary symptom. If a tumor leads to the production of a hormone in such quantities to produce diarrhea, it is likely that the cell of origin of the tumor produces a hormone involved in the normal control of intestinal secretion (see Chapter 3). In "pancreatic cholera," for example, a refractory watery diarrhea is associated with non-beta islet cell tumors of the pancreas (123–125). Although endocrine cell tumors may secrete any of a variety of hormones, the one most regularly associated with diarrhea is VIP, a 28-amino acid peptide with structural similarities to secretin, glucagon, and gastric inhibitory peptide (126, 127).

Vasoactive Intestinal Peptide

VIP produces net fluid secretion in the canine intestine (128, 129), with the secretory response in the jejunum considerably greater than in the ileum. With rabbit ileum mounted in an Ussing chamber, VIP produces changes in SCC and unidirectional Na and Cl fluxes equivalent to those produced by theophylline and prostaglandins. The time course of the SCC changes are indistinguishable. In addition, VIP, like theophylline and prostaglandins, increases intestinal cAMP (130). Whether VIP and prostaglandin E_1 were additive was not determined. VIP is a "candidate" secretory hormone and may be the secretagogue described by Nasset many years ago and called enterocrinin (131).

Secretin, Glucagon, and Gastric Inhibitory Peptide

Since VIP and these peptide hormones belong to the same hormone structural family (see Chapter 3), it might be expected that they would produce intestinal secretion associated with elevation of intestinal cAMP.

In one in vitro study (everted hamster gut sacs) and two in vivo studies (canine Thiry-Vella loops), secretin failed to alter water and electrolyte transport (129, 132, 133). In man, using a triple-lumen tube perfusion technique in the jejunum, secretin decreased net absorption of Na, K, Cl, and H_2O without altering HCO_3. These studies could not differentiate between decreased absorption or increased secretion or a combination of the two (134). In vitro, secretin produced no increase in intestinal cAMP whereas VIP in parallel experiments caused a striking increase (130).

Gastric inhibitory peptide (GIP) produced intestinal secretion in canine Thiry-Vella loops (128, 129) but failed to alter intestinal cAMP in vitro (130). Glucagon gave the same results.

Gastrin and Cholecystokinin

Intravenous pentagastrin inhibited absorption of fluid and electrolytes from canine ileal and jejunal loops (133). Although effect was attributed to impaired absorption because no increase in volume of empty loops was found after pentagastrin administration, secretion is not excluded because voluminous secretion must be produced in the dog before the absorptive capacity of the intestine is exceeded (72). Using perfused loops with a nonabsorbable marker, Barbezat (129) interpreted his results as evidence that gastrin produced intestinal secretion rather than inhibiting absorption. Gastrin does not alter intestinal cAMP (130).

Intravenous cholecystokinin (CCK) (4 U/kg/hr) inhibited fluid and Na, K, and Cl absorption in both jejunum and ileum of dogs but stimulated HCO_3 secretion in the ileum. This response was believed to be physiological because it was duplicated by infusion of fat into the upper intestine, thus releasing CCK but possibly other gastrointestinal hormones as well (135). It is impossible in these studies to differentiate with certainty whether the effect is on J_{M-S} or J_{S-M} or both.

Thyrocalcitonin

In man, using Fordtran's intestinal perfusion technique of the jejunum, thyrocalcitonin (salmon calcitonin 1 U/kg/hr) produced prompt secretion of water Na, K, and Cl, and decreased HCO_3 absorption without affecting calcium absorption (136).

Antidiuretic Hormone

Using the same technique, Soergel et al. (137) found that antidiuretic hormone (ADH) interfered with sodium and water absorption from the human intestine and occasionally produced an increase in intraluminal volume. No evidence was found for an increase in intestinal permeability.

INHIBITORS OF INTESTINAL SECRETION

Interest in inhibitors of intestinal secretion is 2-fold. First, they provide instruments for dissection of the intestinal secretory mechanism or mechanisms, and second they may lead to rational therapy for diarrheal diseases.

Complexing cholera enterotoxin with its specific antibody prevents binding to the mucosa and no secretion is induced (138). If the epithelial binding sites are occupied by prior exposure to choleragenoid (enterotoxin minus the "A" or active fraction), enterotoxin does not bind and no secretion is produced (139).

$G_{M\,I}$ ganglioside has been shown to be the specific receptor for cholera enterotoxin in the intestinal cell membrane. If cholera enterotoxin is premixed with G_{MI} ganglioside, there is no binding and no stimulation of secretion (140–142).

The highly charged antibiotic, polymixin, binds to cell membrane phospholipids and was tried as an inhibitor of cholera enterotoxin binding. Polymixin did inhibit enterotoxin-induced secretion (cholera as well as *E. coli*) but not by this mechanism. Polymixin pretreatment of loops exposed to cholera enterotoxin for only 5 min did not prevent toxin binding but delayed activation of adenylate cyclase and stimulation of intestinal secretion. The appearance of secretion was correlated with leaching of the polymixin out of the mucosa, suggesting that bound polymixin interferes with transfer of the active toxin subunit to its site of action (143, 144).

The antibiotic cycloheximide reversibly inhibits protein synthesis and prevents cholera enterotoxin-induced secretion, but it does not interefere with enterotoxin activation of adenylate cyclase (90, 145). Presumably it interferes with some protein synthetic step beyond the activation of adenylate cyclase that is required for the secretory process. If cycloheximide is given after cholera enterotoxin secretion is established, fluid production persists for 2.5–3 hr before inhibition is detectable. An animal pretreated with cycloheximide is resistant to the secretory stimulant effect of enterotoxin for 8–12 hr. It becomes responsive again when protein synthesis, as manifest by the reappearance of mitotic figures in the crypts, returns. These findings suggest that the protein(s) required for cholera-induced secretion has a functional existence of 2.5–3 hr. Although cycloheximide depresses both $J_{M \to S}$ and $J_{S \to M}$ in control and cholera-exposed loops, its major effect is on $J_{S \to M}$ and it is only the effect on $J_{S \to M}$ that accounts for the inhibition of cholera-induced secretion (44).

The protein synthesis inhibitor tenuazonic acid has qualitatively similar effects to cycloheximide (146).

Indomethacin is a non-steroid anti-inflammatory agent which blocks prostaglandin synthesis. This action is believed to be the basis of its biological effects. Its effects on cholera enterotoxin-induced secretion are similar to those produced by cycloheximide in that secretion is blocked without decreasing the cAMP response and its effects on unidirectional fluxes and net fluid movement are the same. The indomethacin effect, however, lasts for only 1–2 hr and is not associated with a defect in protein synthesis (71). Its effect on jejunal secretion in salmonellosis supports the notion that there is a mediator between mucosal contact with the stimulant and the secretory process (122). Whether this is unique to this situation or is a more general phenomenon is conjectural.

Methyl prednisolone increases (Na^+-K^+)-ATPase and the absorption of water, sodium, and chloride in rat intestine. Cholera enterotoxin does not alter intestinal (Na^+-K^+)-ATPase. When cholera enterotoxin-exposed rats are treated with methyl prednisolone, induced secretion is prevented and may even be reversed to net absorption without any alteration of enterotoxin activation of adenylate cyclase. Although unidirectional flux studies will have to be done to prove the point, it appears that methyl prednisolone produces its effect by increasing intestinal absorption without affecting enterotoxin-induced secretion (147). The net result is similar to the effect associated with oral fluid therapy of cholera with glucose electrolyte solutions in which absorption is stimulated to match or exceed secretory fluid losses.

The polyene antibiotic amphotericin B binds to membrane sterol groups and increases membrane permeability. It inhibits cholera enterotoxin-induced secretion. It also increases water and sodium absorption in control as well as treated intestinal loops and decreases toxin-induced fluid loss. This effect is due to an increased $J^{Na}_{M \to S}$. It is postulated that Na can move more rapidly through the amphotericin B-treated brush border membrane to become more available to the basolateral membrane pump, which can then produce a larger osmotic gradient in the lateral spaces, which in turn increases fluid absorption (148).

Catecholamines inhibit hormone and prostaglandin augmentation of cAMP levels. Epinephrine and norepinephrine do not alter cAMP levels in control or theophylline-treated rabbit ileal mucosa mounted in an Ussing chamber but both produce a rapid, striking reduction of cAMP levels induced by PGE_1 or cholera enterotoxin. This effect is blocked by phenoxybenzamine but not by propranolol, hence it has characteristics of an α-adrenergic response. Epinephrine augments active Na and Cl absorption in control tissue whereas its effect in cholera enterotoxin-stimulated tissue is minimal; decreases in net Na and Cl secretion were not statistically significant and there was no change in HCO_3 secretion. A decrease in SCC and an increase in conductance were significant (149). This is a paradox, since epinephrine produces a striking decrease in cholera enterotoxin-induced elevation in cAMP which is associated with minimal changes in ion fluxes. Several interpretations can be advanced: 1) cholera enterotoxin-induced secretion is not mediated by cAMP; 2) the relation between cAMP level and secretion is obscured since cholera enterotoxin increases cAMP levels in all cells but only a small proportion participate in secretion; and 3) once the secretory epithelium is primed by elevated cAMP, secretion continues for a time in its absence. It is by now abundantly clear that intestinal secretion is complex and cannot be equated merely with levels of cAMP or adenylate cyclase activity.

Acetazolamide decreases cholera enterotoxin-induced secretion by rabbit ileum in vivo (59). It does not inhibit theophylline-induced elevation of intestinal cAMP in vitro. It inhibits coupled neutral NaCl influx at the brush border and thus abolishes active Cl absorption and reduces active Na absorption, but it does not inhibit spontaneous or theophylline-stimulated active chloride secretion (150).

Ethacrynic acid inhibits in vivo cholera enterotoxin-induced intestinal secretion in the dog (151). It does not interefere with the increased cAMP induced by theophylline or cholera enterotoxin in vitro. It inhibits cholera enterotoxin- and theophylline-induced active chloride secretion but does not inhibit sodium absorption or glucose coupled absorption (152).

CONCLUSIONS AND SPECULATIONS

The small intestine normally secretes fluid and electrolytes, but the absorptive mechanisms are so avid in most species that secretion has gone unrecognized. In pathological states associated with enteric infections and endocrine tumors producing gastrointestinal hormones like VIP, intestinal secretion outdistances the absorptive capacity of the small intestine and colon so that secretion is obvious and difficult to ignore.

Stimulation of the intestine by cholera enterotoxin and VIP has provided convenient, reproducible models for the study of intestinal secretion. It has been assumed that these two stimuli activate the same secretory mechanism since the action of both is associated with activation of adenylate cyclase and consequent elevation of intracellular cAMP. In addition, intestinal secretion is stimulated in association with elevation of intracellular cAMP regardless of whether this elevation is achieved by inhibiting its degradation or by increasing its rate of production. On the basis of fragmentary evidence available at present, however, it is unwarranted to assume that elevation of cyclic AMP is synonymous with the activation of a single secretory mechanism. As detailed above, examples of heterogeneity are many. Among these are 1) the differences between cholera enterotoxin and prostaglandins, 2) the differences between cholera enterotoxin and salmonella invasion, 3) the difference in effects of adrenergic stimulation on mucosal cAMP and ion fluxes, and 4) the differences between the effects of acetazolamide on theophylline- and cholera enterotoxin-treated mucosa.

The differences between in vivo and in vitro studies, particularly the apparent difference in anion transport, remain unexplained, although they are probably in large part attributable to the contribution to the "micro-environment" of the epithelial cells made by capillary perfusion. Both in vivo and in vitro techniques make their unique contributions to our understanding of intestinal secretion, but the differences in findings must be reconciled.

No longer can sodium absorption be considered due to a single mechanism, for there is active neutral sodium absorption coupled with chloride, active electrogenic sodium absorption, sodium absorption coupled with glucose absorption, sodium absorption coupled with neutral amino acid absorption (the latter two are influenced differently by cAMP), and, finally, passive sodium absorption via the shunt pathway driven by an osmotic gradient between the lumen and the lateral intercellular spaces. If sodium absorption can be factored into so many components, it seems possible that the other electrolytes and consequent water flow may have similar complexities. Surely, some of the differences between studies are due to differences in the relative contributions of these several mechanisms.

Finally, all of the differences are increasingly difficult to reconcile with a simple model that treats the intestine as though lined by a homogeneous epithelium or an epithelium made up of mature, actively transporting cells and immature and transport inert cells. Although there may be reversible transport mechanisms that absorb under one circumstance and secrete under another, it is difficult to conceive of any mechanism which can do both at the same time at an increased rate. The epithelium of the intestine is surely heterogeneous and possibly made up of one group of cells that function primarily in absorption and another group that function primarily in secretion, although these may be nominally the same cells which function differently because of location. In the environment of the crypts the membrane carriers of these cells may be oriented in a secretory direction, and, with migration to the villus crests, orientation may reverse to the absorptive direction. Whatever the case, however, acceptance of heterogeneity as a working hypothesis may make it possible to explain some of the anomalies in the relation between cAMP and transport activities, especially if it is kept in mind that the ultimate consequence of elevating intracellular levels of cAMP is determined by the characteristics of the cell in which the elevation occurs. In addition to affecting different sodium transport mechanisms differently, evidence has been obtained which suggests that the level of cAMP also alters conductance through the paracellular pathway. The physiological role of this change may be to maximize the osmotic gradient developed in the lateral intercellular spaces.

In conclusion, our attention has been focused on intestinal secretion in pathological states but it seems likely that, as Florey et al. (2) suggested, intestinal secretion is an integral part of the absorptive function of the intestine (see Chapter 11).

REFERENCES

1. Howell, W. H. (1926). A Text-Book of Physiology for Medical Students and Physicians, p. 802. W.B. Saunders, Philadelphia.
2. Florey, H. W., Wright, R. D., and Jennings, M. A. (1941). The secretions of the intestine. Physiol. Rev. 21:36.
3. Hendrix, T. R., and Banwell, J. G. (1969). Pathogenesis of cholera. Gastroenterology 57:751.
4. Hendrix, T. R., and Bayless, T. M. (1970). Intestinal secretion. Annu. Rev. Physiol. 32:139.
5. Carpenter, C. C. J. (1971). Cholera enterotoxin–recent investigation yield insights into transport processes. Am. J. Med. 50:1.
6. Hendrix, T. R. (1971). Pathophysiology of cholera. Bull. N.Y. Acad. Med. 47:1169.

7. Cohnheim, J. F. (1890). Lectures on general pathology. A handbook for practioners and students, Section III. *In* A. B. McKee (trans.), The Pathology of Digestion, p. 949. New Sydenham Society, London.
8. Trier, J. S. (1968). Morphology of the epithelium of the small intestine. *In* C. F. Code (ed.), Handbook of Physiology, Section 6, Alimentary Canal, III. Intestinal Absorption, pp. 1125–1175. American Physiological Society, Washington, D.C.
9. Fordtran, J. S., and Dietschy, J. M. (1966). Water and electrolyte movement in the intestine. Gastroenterology 50:263.
10. Reynolds, D. G., Brim, J., and Sheehy, T. W. (1967). The Vascular Architecture of the Small Intestinal Mucosa of the Monkey (Macaca Mulatta). Anat. Rec. 159:211.
11. DeBeer, E. J., Johnston, C. G., and Wilson, D. W. (1935). The composition of intestinal secretions. J. Biol. Chem. 108:113.
12. Fordtran, J. S., and Locklear, T. W. (1966). Ionic constituents of gastric and small intestinal fluids after eating. Am. J. Dig. Dis. 11:503.
13. Phillips, S. F., and Summerskill, W. H. J. (1967). Water and electrolyte transport during maintenance of isotonicity in human jejunum and ileum. J. Lab. Clin. Med. 70:686.
14. McGee, L. C., and Hastings, A. B. (1942). The carbon-dioxide tension and acid base balance of jejunal secretions in man. J. Biol. Chem. 142:893.
15. Fordtran, J. S., and Ingelfinger, F. J. (1968). Absorption of water, electrolytes and sugars from the human gut. *In* C. F. Code (ed.), Handbook of Physiology, Section 6, Alimentary Canal. III Intestinal Absorption, p. 1457. American Physiological Society, Washington, D.C.
16. Nasset, E. S., and Ju, J. S. (1973). Micropipet collection of succus entericus at crypt ostia of guinea pig jejunum. Digestion 9:205.
17. Finkelstein, R. A., and LoSpalluto, J. J. (1970). Production of highly purified choleragen and choleragoid. J. Infect. Dis. 12:163.
18. Holmgren, J., Lonnroth, I., and Svennerholm, L. (1973). Tissue receptor for cholera exotoxin: postulated structure from studies with G m ganglioside and related glycolipids. Infect. Immun. 8:208.
19. Gill, D. M., and King, C. A. (1975). The mechanism of action of cholera toxin in pigeon erythrocyte lysates. J. Biol. Chem. 250:6424.
20. Goodgame, J. T., Banwell, J. G., and Hendrix, T. R. (1973). The relationship between duration of exposure to cholera toxin and the secretory response of rabbit jejunal mucosa. Johns Hopkins Med. J. 132:117.
21. Kimberg, D. V., Field, M., Johnson, J., Henderson, A., and Gershow, E. (1971). Stimulation of intestinal adenyl cylase by cholera enterotoxin and prostaglandins. J. Clin. Invest. 50:1218.
22. Shafer, D. E., Lust, W. D., Sircar, B., and Goldberg, N. D. (1970). Elevation of adenosine 3'5'cyclic monophosphate concentration after treatment with cholera toxin. Proc. Natl. Acad. Sci. USA 67:851.
23. Carpenter, C. C. J., and Greenough, W. B. (1968). Response of canine duodenum to intraluminal challenge with cholera exotoxin. J. Clin. Invest. 47:2600.
24. Carpenter, C. C. J., Sack, R. B., Feeley, J. C., and Steinberg, R. W. (1968). Site and characteristics of electrolyte doses and effect of intraluminal glucose in experimental canine cholera. J. Clin. Invest. 47:1210.
25. Leitch, G. J., and Burrows,W. (1968). Experimental cholera in the rabbit ligated intestine: ion and water accumulation in the duodenum, ileum and colon. J. Infect. Dis. 118:349.
26. Banwell, J. G., Pierce, N. F., Mitra, R. C., Brigham, K. L., Caranasos, G. J., Keimowitz, R. I., Fedson, D. S., Thomas, J., Gorbach, S. L., Sack, R. B., and Mondal, A. (1970). Intestinal fluid and electrolyte transport in human cholera. J. Clin. Invest. 49:183.
27. Morris, H. T., and Majno, G. (1968). On the role of the ileal epithelium in the pathogenesis of experimental cholera. Am. J. Pathol. 53:263.
28. Field, M. (1971). Intestinal secretion: effect of cyclic AMP and its role in cholera. N. Engl. J. Med. 284:1137.
29. Schultz, S. G., and Frizzell, R. A. (1972). A review of intestinal absorptive and secretory processes. Gastroenterology 63:161.

30. Field, M. (1974). Intestinal secretion. Gastroenterology 66:1063.
31. Gangarosa, E. J., Beisel, W. R., Benyajati, C., Spring, H., and Piyaratin, P. (1960). The nature of the gastrointestinal lesion in Asiatic cholera and its relation to pathogenesis: a biopsy study. Am. J. Trop. Med. Hyg. 9:125.
32. Elliott, H. L., Carpenter, C. C. J., Sack, R. B., and Yardley, J. H. (1970). Small bowel morphology in experimental canine cholera. A light and electron microscopic study. Lab. Invest. 22:112.
33. Walten, R. H., Morgan, F. M., Songkhla, Y. N., Vanikiati, B., and Phillips, R. A. (1959). Water and electrolyte studies in cholera. J. Clin. Invest. B8:1879.
34. Carpenter, C. C. J., Greenough, W. B., and Sack, R. B. (1969). The relationship between superior mesenteric blood flow to gut electrolyte loss in experimental cholera. J. Infect. Dis. 119:182.
35. Norris, T. H., and Sumner, D. S. (1974). Distribution of blood flow to the layers of the small bowel in experimental cholera. Gastroenterology 66:973.
36. Gordon, R. S., Gardner, J. D., and Kinzie, J. L. (1972). Low mannitol clearance into cholera stool as evidence against filtration as a source of stool fluid. Gastroenterology 63:407.
37. Rhode, J. E., and Chen, L. C. (1972). Permeability and selectivity of canine and human jejunum during Asiatic cholera. Gut 13:191.
38. Scherer, R. W., Harper, D. T., Banwell, J. G., and Hendrix, T. R. (1974). Absence of concurrent permeability changes in intestinal mucosa with secretion. Johns Hopkins Med. J. 134:156.
39. Field, M., Fromm, D., Al-Awqati, Q., and Greenough, W. B. III. (1972). Effect of cholera enterotoxin on ion transport across isolated ileal mucosa. J. Clin. Invest. 51:796.
40. Powell, D. W., Binder, J. H., and Curran, P. F. (1973). Active electrolyte secretion stimulated by choleragen in rabbit ileum *in vitro.* Am. J. Physiol. 225:781.
41. Banwell, J. G., Pierce, N. F., Mitra, R., Caranosos, G. J., Keimowitz, R. I., Mondal, A., and Manji, P. M. (1968). Preliminary results of a study of small intestinal water and solute movement in acute and convalecent human cholera. Indian J. Med. Res. 56:633.
42. Iber, F. L., McGonagle, T. J., Serebro, H. L., Lenblens, E. L., Bayless, T. M., and Hendrix, T. R. (1969). Unidirectional sodium flux in small intestine in experimental canine cholera. Am. J. Med. Sci. 258:340.
43. Love, A. H. G. (1969). Water and sodium absorption by the intestine in cholera. Gut 10:63.
44. Grayer, D. I., Serebro, H. A., Iber, F. L., and Hendrix, T. R. (1970). The effect of cycloheximide on unidirectional sodium fluxes in the jejunum after cholera exotoxin exposure. Gastroenterology 58:815.
45. Swallow, J. H., Code, C. F., and Feter, R. (1968). Effect of cholera toxin on water and ion fluxes in the canine bowel. Gastroenterology 54:35.
46. Phillips, R. H. (1968). Asiatic cholera (with emphasis on the pathophysiological effects of the disease). Annu. Rev. Med. 19:69.
47. Field, M. (1971). Intestinal secretion: effect of cyclic AMP and its role in cholera. N. Engl. J. Med. 284:1137.
48. Field, M. (1971). Ion transport in rabbit ileal mucosa. II. Effects of cyclic 3′5′ AMP. Am. J. Physiol. 221:992.
49. Serebro, H. A., Bayless, T. M., Hendrix, T. R., Iber, F. L., and McGonagle, T. (1968). Absorption of D-glucose by rabbit jejunum during cholera induced diarrhea. Nature 217:1272.
50. Pierce, N. F., Banwell, J. G., Mitra, R. C., Caranosos,G. J., Keimowitz, R. I., Mondal, A., and Manji, P. M. (1968). Effect of intragastric glucose–electrolyte infusion upon water and electrolyte balance in Asiatic cholera. Gastroenterology 55:333.
51. Hirschhorn, N., Kenzie, J. L., Sachar, D. B., Northrup, R. S., Taylor, J. O., Ahmad, Z., and Phillips, R. A. (1968). Decrease in net stool output in cholera during intestinal perfusion with glucose-containing solutions. N. Engl. J. Med. 279:176.
52. Pierce, N. F., Sack, R. B., Mitra, R. C., Banwell, J. G., Brigham, K. L., Fedson, D. S., and Mondal, A. (1969). Replacement of water and electrolyte losses in cholera by an oral glucose–electrolyte solution. Ann. Intern. Med. 70:1173.

53. Nalin, D. R., Cash, R. A., Rahman, M., and Yunus, M. (1970). Effect of glycine and glucose on sodium and water absorption in patients with cholera. Gut 11:768.
54. Rohde, J. E., and Cash, R. E. (1973). Transport of glucose and amino acids in human jejunum during Asiatic cholera. J. Infect. Dis. 127:190.
55. Field, M., Fromm, D., and McColl, I. (1971). Ion transport in rabbit ileal mucosa. I. Na and Cl fluxes and short circuit current. Am. J. Physiol. 222:1388.
56. Powell, D. W., Farris, R. K., and Carbonetto, S. T. (1974). Theophylline, cyclic AMP, choleragen, and electrolyte transport by rabbit ileum. Am. J. Physiol. 227:1428.
57. Binder, H. J., Powell, D. W., Tai, Y. H., and Curran, P. F. (1973). Electrolyte transport in rabbit ileum. Am. J. Physiol. 225:776.
58. Moore, W. L., Jr., Bieberdorf, F. A., Morawski, S. G., Finkelstein, R. A., and Fordtran, J. S. (1971). Ion secretion during cholera-induced ileal secretion in the dog. J. Clin. Invest. 50:312.
59. Norris, H. T., Curran, P. F., and Schultz, S. G. (1969). Modification of intestinal secretion in experimental cholera. J. Infect. Dis. 119:117.
60. Turnberg, L. A., Brebendorf, F. A., Morawski, S. G., and Fordtran, J. S. (1970). Interrelationships of chloride, bicarbonate, sodium and hydrogen transport in human ileum. J. Clin. Invest. 49:557.
61. Dietz, J., and Field, M. (1973). Ion transport in rabbit ileal mucosa. IV. Bicarbonate secretion. Am. J. Physiol. 225:858.
62. Hubel, K. A. (1973). The mechanism of bicarbonate secretion in rabbit ileum exposed to choleragen. J. Clin. Invest. 53:1964.
63. Kimburg, D. V., Field, M., Johnson, J., Henderson, A., and Gershon, E. (1971). Stimulation of intestinal mucosal adenyl cyclase by cholera enterotoxin and prostaglandins. J. Clin. Invest. 50:1218.
64. Schwartz, C. J., Kimberg, D. V., Sheerin, H. E., Field, M., and Said, S. I. (1974). Vasoactive intestinal peptide stimulation of adenylate cyclase and active electrolyte secretion in intestinal mucosa. J. Clin. Invest. 54:536.
65. Pierce, N. F., Carpenter, C. C. J., Elliott, H. L., and Greenough, W. B., III. (1971). Effects of prostaglandins, theophylline, and cholera exotoxin upon transmucosal water and electrolyte movement in the canine jejunum. Gastroenterology 60:22.
66. Shafer, D. E., Lunt, W. D., Sircar, B., and Goldberg, N. D. (1970). Elevation of adenosine 3′5′ cyclic monophosphate concentration in intestinal mucosa after treatment with cholera toxin. Proc. Natl. Acad. Sci. USA 67:851.
67. Sharp, G. W. G., and Huprie, S. (1971). Stimulation of intestinal adenyl cyclase by cholera toxin. Nature 229:266.
68. Chen, L. C., Rohde, J. E., and Sharp, W. G. W. (1971). Intestinal adenyl-cyclase activity in human cholera. Lancet i:939.
69. Chen, L. C., Rohde, J. E., and Sharp, W. G. W. (1972). Properties of adenyl-cyclase from human jejunal mucosa during naturally acquired cholera and convalescence. J. Clin. Invest. 51:731.
70. Guerrant, R. L., Chen, L. C., and Sharp, G. W. G. (1972). Intestinal adenyl-cyclase activity in canine cholera: correlation in fluid accumulation. J. Infect. Dis. 125:377.
71. Wald, A., Gotterer, G. S., Rajendra, G. R., and Hendrix, T. R. (1975). Effect of indomethacin on cholera induced intestinal AMP and fluid movement. Clin. Res. 23:396.
72. McGonagle, T. J., Serebro, H. A., Iber, F. I., Bayless, T. M., and Hendrix, T. R. (1969). Time of onset of action of cholera toxin in dog and rabbit. Gastroenterology 57:5.
73. Kimberg, D. V., Field, M., Gershon, E., and Henderson, A. (1974). Effects of prostaglandins and cholera enterotoxin on intestinal mucosal cyclic AMP accumulation. Evidence against an essential role for prostaglandins in the action of toxin. J. Clin. Invest. 53:941.
74. Parkinson, D., Ebel, H., DiBona, D. R., and Sharp, G. W. G. (1972). Localization of the action of cholera toxin on adenyl cyclase in mucosal epithelial cells of rabbit intestine. J. Clin. Invest. 51:2292.
75. Serebro, H. A., McGonagle, T., Iber, F. L., Royall, R., and Hendrix, T. R. (1968). An effect of cholera toxin on the small intestine without mucosal contact. Johns Hopkins Med. J. 123:229.

76. Vaughan-Williams, E. M., and Dohadwalla, A. N. (1969). The appearance of a choleragenic agent in the blood of infant rabbits infected intestinally with *Vibrio cholerae*, demonstrated by cross circulation. J. Infect. Dis. 120:658.
77. Pierce, N. F., Graybill, J. R., Kaplan, M. M., and Bouwman, D. L. (1972). Systemic effects of parenteral cholera enterotoxin in dogs. J. Lab. Clin. Med. 79:145.
78. Bennett, A. (1971). Cholera and prostaglandins. Nature 231:536.
79. Matrichansky, C., and Bernier, J. J. (1973). Effect of prostaglandin E_1 on glucose, water and electrolyte absorption in the human jejunum. Gastroenterology 64:1111.
80. Jacoby, H. I., and Marshall, C. H. (1972). Antagonism of cholera enterotoxin by anti-inflammatory agents in the rat. Nature 235:163.
81. Finck, A. D., and Katz, R. L. (1972). Prevention of cholera-induced intestinal secretion in the cat by aspirin. Nature 238:273.
82. Hudson, N., ElHindi, S., Wilson, D. E., and Poppe, L. (1975). Prostaglandin E in cholera toxin induced intestinal secretion. Lack of intermediary role. Am. J. Dig. Dis. 20:1035.
83. Flores, J., Grady, G. F., McIver, J., Witkum, P., Beckman, B., and Sharp, G. W. G. (1974). Comparison of the effects of enterotoxins of *Shigella dysenteriae* and *Vibrio cholerae* on the adenylate cyclase system of the rabbit intestine. J. Infect. Dis. 130:374.
84. DiBona, D. R., Chen, L. C., and Sharp, G. W. G. (1974). A study of intracellular spaces in the rabbit jejunum during acute volume expansion and after treatment with cholera toxin. J. Clin. Invest. 53:1300.
85. Turjaman, N. A., and Gotterer, G. S. The effect of intestinal distension on levels of cAMP and fluid secretion. Submitted for publication.
86. Halsted, C. H., Luebbers, E. H., Bayless, T. M., and Hendrix, T. R. (1971). A comparison of intestinal fluid production in response to cholera exotoxin and hypertonic mannitol. Johns Hopkins Med. J. 129:179.
87. Halsted, C. H. Grayer, D. I., Luebbers, E. H., Yardley, J. H., and Hendrix, T. R. (1971). Effect of cycloheximide on intestinal secretion induced by hypertonic glucose. Gut 12:262.
88. Kinter, W. G., and Wilson, T. H. (1965). Autoradiographic studies of sugar and amino acid absorption by everted sacs of hamster intestine. J. Cell Biol. 25:19.
89. Roggin, G. M., Banwell, J. G., Yardley, J. H., and Hendrix, T. R. (1972). Unimpaired response of rabbit jejunum to cholera toxin after selective damage to villus epithelium. Gastroenterology 63:981.
90. Serebro, H. A., Iber, F. L., Yardley, J. H., and Hendrix, T. R. (1969). Inhibition of cholera toxin action in the rabbit by cycloheximide. Gastroenterology 56:506.
91. Kimberg, D. V., Field, M., Gershon, E., Schooley, R. T., and Henderson, A. (1973). Effects of cycloheximide on the response of intestinal mucosa to cholera enterotoxin. J. Clin. Invest. 52:1376.
92. Hershhorn, N., and Frazier, H. S. (1973). Electrical profile of stripped isolated rabbit ileum. Johns Hopkins Med. J. 132:271.
93. deJonge, H. R. (1975). The response of small intestinal villous and crypt epithelium to choleragen in the rat and guinea pig. Evidence against a specific role of the crypt cells in choleragen-induced secretion. Biochim. Biophys. Acta 381:128.
94. Nellans, H. N., Frizzell, R. A., and Schultz, S. G. (1973). Coupled sodium chloride influx across brush border of rabbit ileum. Am. J. Physiol. 225:467.
95. Frizzell, R. A., Nellans, H. N., Rose, R. C., Markscheid-Kaspi, L., and Schultz, S. G. (1973). Intracellular Cl concentrations and influxes across the brush border of rabbit ileum. Am. J. Physiol. 224:328.
96. Kinzie, J. L., Ferrendelli, J. A., and Alpers, D. H. (1973). Adenosine cyclic 3′5′-monophosphate-mediated transport of neutral and dibasic amino acids in jejunal mucosa. J. Biol. Chem. 248:7018.
97. Weiser, M. M., and Quill, H. (1975). Intestinal villus and crypt cell response to cholera toxin. Gastroenterology 69:479.
98. Quill, H., and Weiser, M. M. (1975). Adenylate and guanylate cyclase activities and cellular differentiation in rat small intestine. Gastroenterology 69:470.
99. Schwartz, C. J., Kimberg, D. V., and Ware, P. (1975). Adenylate cyclase in intestinal

crypt and villus cells: stimulation by cholera enterotoxin and prostaglandin E. Gastroenterology 68:94.
100. Frizzell, R. A., and Schultz, S. G. (1972). Ionic conductances of extracellular shunt pathway in rabbit ileum: influence of shunt on transmural sodium transport and electrical potential differences. J. Gen. Physiol. 59:318.
101. Schultz, S. G., Frizzell, R. A., and Nellans, H. N. (1974). Ion transport by mammalian small intestine. Annu. Rev. Physiol. 36:51.
102. Powell, D. W. (1974). Intestinal conductance and permselectivity changes with theophylline and choleragen. Am. J. Physiol. 227:1436.
103. O'Daly, J. A., Craig, S. W., and Cebra, J. J. (1971). Localization of b markers, a stain and SC of SI_gA in epithelial cells lining Lieberkühn crypts. J. Immunol. 106:286.
104. Sack, R. B., Gorbach, S. L., Banwell, J. G., Jacobs, B., Chetterjee, B. D., and Nutra, R. C. (1971). Enterotoxigenic *Escherichia coli* isolated from patients with severe cholera-like illness. J. Infect. Dis. 123:378.
105. Gyles, C. L., and Barnum, D. A. (1969). A heat-labile enterotoxin from strains of *Escherichia coli* enteropathic for pigs. J. Infect. Dis. 120:419.
106. Banwell, J. G., and Sherr, H. (1973). Effect of bacterial enterotoxin on the gastrointestinal tract. Gastroenterology 65:467.
107. Sherr, H. P., Banwell, J. G., Rothfeld, A., and Hendrix, T. R. (1973). Pathophysiological response of rabbit jejunum to *Escherichia coli* enterotoxin. Gastroenterology 65:895.
108. Al-Awqati, Q., Wallace, C. K., and Greenough, W. B. III. (1972). Stimulation of intestinal secretion *in vitro* by culture filtrates of *Escherichia coli*. J. Infect. Dis. 125:300.
109. Kantor, H. S., Tao, P., and Gorbach, S. L. (1974). Stimulation of intestinal adenyl cyclase by *Escherichia coli* enterotoxin: comparison of strains from an infant and an adult with diarrhea. J. Infect. Dis. 129:1.
110. Holmgren, J. (1973). Comparison of the tissue receptors for *Vibrio cholerae* and *Escherichia coli* enterotoxins by means of gangliosides and natural cholera toxoid. Infect. Immun. 8:851.
111. Steinberg, S. E., Banwell, J. G., Yardley, J. H., Keusch, G. T., and Hendrix, T. R. (1975). Comparison of secretory and histological effects of shigella and cholera enterotoxins in rabbit jejunum. Gastroenterology 68:309.
112. Donowitz, M., Keusch, G. T., and Binder, H. J. (1975). Effect of shigella enterotoxin on electrolyte transport in rabbit ileum. Gastroenterology 69:123.
113. Keusch, G. T., Grady, G. F., Takeuchi, A., and Sprinz, H. (1972). The pathogenesis of shigella diarrhea. II. Enterotoxin induced acute enteritis in the rabbit ileum. J. Infect. Dis. 126:92.
114. Sullivan, R., and Asano, T. (1971). Effects of staphylococcal enterotoxin B on intestinal transport in the rat. Am. J. Physiol. 220:1973.
115. Duncan, C. L., and Strong, D. H. (1969). Ileal loop fluid accumulation by cell-free products of *Clostridium perfringens*. J. Bacteriol. 100:86.
116. Grady, G. F., and Keusch, G. T. (1971). Pathogens of bacterial diarrheas. N. Engl. J. Med. 285:831 and 891.
117. Kubota, Y., and Liu, P. V. (1971). An enterotoxin of *Pseudomonas aeruginosa*. J. Infect. Dis. 123:97.
118. Klipstein, F. A., Horowitz, I. R., Angert, R. F., and Schenk, E. A. (1975). Effect of *Klebsiella pneumoniae* enterotoxin on intestinal transport in the rat. J. Clin. Invest. 56:799.
119. Fromm, D., Gianella, R. A., Formal, S. B., Quyano, R., and Collins, H. (1974). Ion transport across isolated ileal mucosa invaded by salmonella. Gastroenterology 66:215.
120. Rout, W. R., Formal, S. B., Damin, G. J., and Gianella, R. A. (1974). Pathophysiology of salmonella diarrhea in the rhesus monkey: Intestinal transport, morphological and bacteriological studies. Gastroenterology 67:59.
121. Giannella, R. A., Gots, R. E., Charney, A. N., Greenough, W. B., III., and Formal, S. B. (1975). Pathogenesis of *Salmonella*-mediated intestinal fluid secretin. Activation of adenylate cyclase and inhibition by indomethacin. Gastroenterology 69:1238.

122. Gots, R. E., Formal, S. B., and Gianella, R. A. (1974). Indomethacin inhibition of *Salmonella typhimurium, Shigella flexneri* and cholera-mediated rabbit ileal secretion. J. Infect. Dis. 130:280.
123. Verner, J. V., and Morrinon, A. B. (1958). Islet cell tumor and a syndrome of refractory watery diarrhea and hypokalemia. Am. J. Med. 25:374.
124. Marks, I. N., Banks, S., and Louw, J. H. (1967). Islet cell tumor of the pancreas with reversible watery diarrhea and achlorhydria. Gastroenterology 52:695.
125. Kraft, A. R., Tompkins, R. K., and Zollinger, R. M. (1970). Recognition and management of the diarrheal syndrome caused by non-beta cell tumors of the pancreas. Am. J. Surg. 119:163.
126. Bloom, S. R., Polak, J. M., and Pearse, A. G. E. (1973). Vasoactive intestinal peptide and watery diarrhea syndrome. Lancet ii:14.
127. Said, S. I., and Faloona, G. R. (1975). Elevated plasma and tissue levels of vasoactive intestinal polypeptide in the watery diarrhea syndrome due to pancreatic, eroncho-genic and other tumors. N. Engl. J. Med. 293:155.
128. Barbezat, G. O., and Grossman, M. I. (1971). Intestinal secretion: stimulation by peptides. Science 174:422.
129. Barbezat, G. O. (1973). Stimulation of intestinal secretion by polypeptide hormones. Scand. J. Gastroenterol. 8 (Suppl. 22):1.
130. Schwartz, C. J., Kimberg, D. V., Sheerin, H. E., Field, M., and Said, S. I. (1974). Vasoactive intestinal peptide stimulation of adenylate cyclase and active secretion in intestinal mucosa. J. Clin. Invest. 54:536.
131. Fink, R. M., and Nasset, E. S. (1942). The physiological response to enterocrinin considered quantitatively. Am. J. Physiol. 139:626.
132. Gardner, J. D., Peskin, G. W., Cerda, J. J., and Brooks, F. P. (1967). Alterations of *in vitro* fluid and electrolyte absorption by gastrointestinal hormones. Am. J. Surg. 113:57.
133. Bynum, T. E., Jacobson, E. D., and Johnson, L. R. (1971). Gastrin inhibition of intestinal absorption in dogs. Gastroenterology 61:858.
134. Hicks, T., and Turnberg, L. A. (1973). The influence of secretin on ion transport in human jejunum. Gut 14:485.
135. Bussjaeger, L. J., and Johnson, L. R. (1973). Evidence for hormonal regulation of intestinal absorption by cholecystokinin. Am. J. Physiol. 224:1276.
136. Gray, T. K., Bieberdosf, F. A., and Fordtran, J. S. (1973). Thyrocalcitonin and the jejunal absorption of calcium, water and electrolytes in normal subjects. J. Clin. Invest. 52:3084.
137. Soergel, K. H., Whalen, G. E., Harris, J. A., and Gennen, J. E. (1968). Effect of antidiuretic hormone on human small intestinal water and solute transport. J. Clin. Invest. 47:1968.
138. Kasai, G. J., and Burrows, W. (1966). The titration of cholera toxin and antitoxin in the rabbit ileal loop. J. Infect. Dis. 116:606.
139. Cuatrecasas, P. (1973). *Vibrio cholerae* choleragenoid. Mechanism of inhibition of cholera toxin action. Biochemistry 12:3577.
140. Pierce, N. F. (1973). Differential inhibitory effects of cholera toxoid and ganglioside on the enterotoxins of *Vibrio cholerae* and *Escherichia coli.* J. Exp. Med. 137:1008.
141. Cuatrecasas, P. (1973). Interaction of *Vibrio cholerae* enterotoxin with cell membranes. Biochemistry 12:3547.
142. Holmgren, J., Lonnroth, I., Mansson, J. E., and Svennerholm. (1975). Interaction of cholera toxin and membrane G_{MI} ganglioside of small intestine. Proc. Natl. Acad. Sci. USA 72:2520.
143. Maimon, H. N., Banwell, J. G., and Hendrix, T. R. (1972). Inhibitions of cholera toxin induced secretion. Clin. Res. 20:459.
144. Maimon, H. N., Mitch, W. E., Banwell, J. G., and Hendrix, T. R. (1976). Inhibition of enterotoxin-induced intestinal secretion by the polypeptide antibiotic polymyxin. Johns Hopkins Med. J. 138:82.
145. Kimberg, D. V., Field, M., Gershon, E., Schooley, R. T., and Henderson, A. (1973). Effects of cycloheximide on response of intestinal mucosa to cholera enterotoxin. J. Clin. Invest. 52:1376.

146. Moritz, M., and Wornelsdorl, A. H. (1973). Rabbit cholera: inhibitory effect of tenuazonic acid on cholera-induced secretion of water and electrolytes. Gastroenterology 65:259.
147. Charney, A. N., and Donowitz, M. (1975). Prevention and reversal of cholera toxin-induced intestinal secretion by methylprednisolone. Presented at the 11th Joint Conference on Cholera, November 4–6, 1975, New Orleans.
148. Chen, L. C., Guerrant, R. L., Rohde, J. E., and Casper, A. G. T. (1973). Effect of amphotericin B on sodium and water movement across normal and cholera-toxin challenged canine jejunum. Gastroenterology 65:252.
149. Field, M., Sheerin, H. E., Henderson, A., and Smith, P. L. (1975). Catecholamine effects on cyclic AMP levels and ion secretion in rabbit ileal mucosa. Am. J. Physiol. 229:86.
150. Nellans, H. N., Frizzell, R. A., and Schultz, S. G. (1975). Effect of acetazolamide on sodium and chloride transport by *in vitro* rabbit ileum. Am. J. Physiol. 228:1808.
151. Carpenter, C. C. J., Curlin, G. T., and Greenough, W. B. (1969). Response of canine Thiry–Vella jejunal loops to cholera exotoxin and its modification by ethacrynic acid. J. Infect. Dis. 120:332.
152. Al-Awqati, Q., Field, M., and Greenough, W. B., III. (1974). Reversal of cyclic AMP-mediated intestinal secretion by ethacrynic acid. J. Clin. Invest. 53:687.

International Review of Physiology
Gastrointestinal Physiology II, Volume 12
Edited by Robert K. Crane

9
Mechanisms Underlying the Absorption of Water and Ions

H. J. BINDER

Department of Internal Medicine
Yale University, New Haven, Connecticut

ELECTROLYTE TRANSPORT IN THE ILEUM 286
Contrasting in Vivo versus in Vitro Models 286
Ion Secretion and Paracellular Pathways 287
In Vitro Studies with Guinea Pig Ileum 288
In Vitro Studies with Rabbit Ileum 289
"Consensus" Model of Electrolyte Transport 289
Electrogenic Na Absorption 290
Neutral NaCl Absorption 290
Neutral NaCl Secretion 292
Role of pH and Bicarbonate 293
Summary 293

ELECTROLYTE TRANSPORT IN JEJUNUM 294
In Vivo Studies 294
In Vitro Studies 294
Summary 294

HORMONE CONTROL OF ION TRANSPORT 295
In Vivo Studies 295
Glucagon 295
Secretin and CCK 295
Thyrocalcitonin 296
Vasoactive Intestinal Peptide 296
In Vitro Studies 297
Summary 297

GLUCOSE STIMULATION OF SODIUM ABSORPTION 297
In Vitro Model 297
In Vivo Model 298

Reconciliation of the Controversy 299
Summary 301

This review will provide a summary of recent observations of electrolyte transport across mammalian small intestine. In the past few years conflicting models of ion transport have been proposed and recent experiments may provide possible resolution of these varied hypotheses. The subject will be divided into three separate topics: "normal" ion transport, glucose stimulation of sodium absorption, and hormonal control of ion movement.

ELECTROLYTE TRANSPORT IN THE ILEUM

Review of the many investigations of electrolyte transport across small intestinal epithelium does not permit a clear and precise understanding of the mechanism by which the intestinal mucosa transports sodium, chloride, and bicarbonate from the lumen to the plasma in vivo or from mucosa to serosa in vitro. Some, though certainly not all, of the confusion results from the heterogeneity of the experimental methods used. During the past few years, however, detailed study of ion transport across isolated rabbit ileal mucosa has clarified and amplified our understanding of electrolyte movement and has resulted in partial delineation of the processes responsible for ion transport across this tissue. Although at this time it is not certain whether the model of transport that will eventually evolve from these studies of rabbit ileum will be relevant to the understanding of transport processes in other species, especially man, the differences that have been noted in different species may be more apparent than real. Since significant differences between in vivo jejunal and ileal transport (1, 2) have been described, we will discuss models of jejunal transport separate from those of ileal transport, although this separation may be arbitrary.

Contrasting in Vivo versus in Vitro Models

Two contrasting and conflicting sets of observations can be identified in the large number of studies reported during the past 20 years: 1) Schultz and Zalusky a decade ago adopted the Ussing chamber technique to the study of rabbit ileum in vitro, and reported studies which suggested a model of electrogenic sodium transport (3); and 2) Fordtran and associates, studying fluid and electrolyte movement in man with triple lumen perfusion techniques, have proposed that ileal electrolyte movement is neutral and can be explained by double ion exchange pumps (2).

In 1964 Schultz and Zalusky demonstrated in the rabbit ileum that the short circuit (Isc) was totally accounted for by net sodium absorption (J_{net}^{Na}) (3).

Moreover, Isc was a linear function of the mucosal sodium concentration, and both Isc and J_{net}^{Na} could be completely inhibited by ouabain. In these studies no evidence of either active chloride absorption or active chloride secretion was observed (4). The model that evolved from these studies was relatively straightforward and was similar to previous studies of sodium transport across several other epithelia. These experiments indicated that 1) sodium transport was electrogenic; 2) there was no evidence of neutral or coupled sodium transport; 3) sodium transport was probably driven by (Na-K)-ATPase at the basolateral border; and 4) chloride movement was not active (3, 4).

In contrast to this model of electrogenic sodium transport across rabbit ileal mucosa, a model of neutral sodium chloride absorption was proposed by Turnberg et al. which was based on ileal perfusion studies in normal human subjects (2). In these experiments sodium, chloride, and bicarbonate movement occurred against an electrochemical gradient and both absorption and secretion of fluid were observed. During perfusion with a plasma-like solution a negligible electrical potential difference (PD) was recorded; replacement of chloride by sulfate did not change the PD but resulted in luminal acidification. Furthermore, chloride movement was sensitive to luminal concentration of bicarbonate, and acetazolamide inhibited both Na and Cl absorption. These investigators felt that their results could best be explained by double ion exchanges: a cation Na:H exchange and an anion Cl:HCO_3 exchange.

Thus, two contrasting models had been proposed for electrolyte movement in the ileum: one in vitro demonstrating electrogenic sodium absorption, and the other in vivo indicating a neutral sodium chloride process.

Ion Secretion and Paracellular Pathways

Before reviewing recent experimental results which indicate that the differences between these two models may not be as great as initial review suggests, it is important to consider briefly some recently emphasized phenomena of intestinal ion transport.

First, electrolyte transport cannot be discussed solely as an absorptive phenomenon but must be evaluated in terms of the interaction of the several processes that may be occurring simultaneously. Net fluid and electrolyte movement will represent the difference between absorption and secretion. Although intestinal electrolyte secretion was discussed frequently before 1940, the subject "disappeared" during the interval from 1940 to 1968 (5, 6). Recent interest in the study of cholera and other diarrheal illnesses has provided impetus for the renewed investigation of intestinal electrolyte secretion. Although intestinal secretion is discussed in another chapter of this volume, it is impossible to separate completely absorption and secretion (7).

Second, transmural ion movement occurs via two different pathways: 1) a transcellular route and 2) a paracellular or shunt pathway. Although movements through these two pathways vary, recent studies indicate that across low resistance epithelia like the small intestine, the shunt pathway is the route for most ion movement (8). The proportions of ion flux via these two pathways may vary with different methods and species, and may account for some of the differences in the models of ion transport proposed to date.

In Vitro Studies with Guinea Pig Ileum

Recent studies of electrolyte transport in the guinea pig ileum have provided a model of ion transport that may help in understanding the results of studies of rabbit ileum (9, 10). Powell et al. observed that during in vivo perfusion of the guinea pig ileum, net absorption of sodium and water did not occur, but rather there was a net secretion of water, sodium and chloride (11). Detailed study of the guinea pig ileum thus afforded an opportunity to evaluate a naturally occurring secretory process—most other secretory phenomena require a secretagogue and involve several effects of the secretagogue in addition to that of secretion. It is not known why the small intestine of the guinea pig secretes or what physiological advantage is provided by this secretion. Spontaneous secretion is not, however, unique; an occasional normal human subject may secrete during ileal perfusion studies, and electrolyte secretion occurs regularly in the rabbit appendix. In the guinea pig studies, the ileal mucosa was studied in a modified Ussing chamber, and Na and Cl fluxes were determined by methods perfected by Schultz and Zalusky for study of rabbit ileum (9, 10). Under short circuit conditions net sodium absorption was present but was consistently less than Isc; net Cl secretion was also demonstrated.

Ion replacement studies resulted in the development of a model in which overall net Na transport resulted from two oppositely directed transport processes: an electrogenic Na absorptive system and an electrically neutral secretion of Na and an anion (Cl and/or HCO_3). The Na absorption process is electrogenic and completely accounts for the short circuit current. The secretory process does not contribute to the short circuit current, and it has a greater affinity for bicarbonate than for chloride. Measured sodium transport represents the difference between electrogenic sodium absorption and neutral sodium anion secretion. Chloride secretion increases following removal of bicarbonate from the media. These studies have not provided a locus for the secretory process. It is still uncertain whether the secretory and absorptive processes are located in the same or different cells, whether the secretory process is located along the basolateral membrane or along the luminal surface of the epithelial cell, or whether the secretory process is located in a crypt or a villous epithelial cell.

Additional studies demonstrated that these two oppositely directed transport processes were under separate control, which permitted their further characterization. Removal of both bicarbonate and chloride resulted in inhibition of the secretory process but had no effect on electrogenic Na absorption. Under these conditions, Isc was almost completely accounted for by J^{Na}_{net}. Na removal abolished both absorptive and secretory processes. Neutral sodium and anion secretion was increased by theophylline but was insensitive to ouabain. In contrast, Na absorption was inhibited by ouabain. 3-*O*-Methylglucose increased only electrogenic Na absorption, i.e., equal increases in both Isc and J^{Na}_{net} were noted. Glucose and galactose, however, increased both electrogenic sodium absorption and neutral sodium and anion secretion, i.e., the presence of either hexose resulted in a greater increase in Isc than J^{Na}_{net}. Finally, fructose appeared to stimulate only neutral sodium chloride absorption. These studies emphasize that under so-called "normal" conditions, net Na and Cl movement represents the interplay of a number of different events, and different substrates alter these events differently. It is anticipated that these studies of guinea pig ileum will assist in understanding the processes involved in ion movement across human or rabbit ileum.

In Vitro Studies with Rabbit Ileum

Interest in ion transport across rabbit ileum is related to the understanding of normal transport phenomena, as well as the effect of several secretagogues on ion movement (e.g., cholera enterotoxin, *Escherichia coli* enterotoxin, prostaglandin E_1, vasoactive intestinal polypeptide (12–16)). Despite many studies of ion transport across rabbit ileal mucosa, no model of ion transport is consistent with all experimental observations to date. Conflicting observations have been reported by both the same and different investigators. Varying experimental conditions may in part be responsible since both stripped and unstripped intestine, high and low potassium concentration, high and low bicarbonate concentration with resulting different solution pH levels, have been employed in many of these studies (3, 17, 18). Furthermore, there is evidence of differences between different "normal" populations of rabbits (19). Whether these differences between population groups are related to inapparent infection with coccidia is not known (20). Additional considerations include the problems associated with the presence of oppositely directed transport processes, so that a decrease in net absorption, for example, could represent either a decrease in the absorptive process or an increase in the secretory process. Across low resistance epithelia like that of the intestine the small net movement which is present is determined by the measurement of relatively large unidirectional fluxes. Bicarbonate movement has been largely determined by calculation of the so-called residual flux, which depends on independent determination of five different observations, each with its own experimental errors (21). It should not be surprising that different results have been obtained when experimental conditions have been somewhat different.

"Consensus" Model of Electrolyte Transport

The following discussion will attempt to provide a model based on a "consensus" of the various studies reported to date.

The model presented in Figure 1 does not employ the typical epithelial cell, since there is little evidence to show whether different transport processes are located in all epithelial cells or only in certain cells, or whether villous and crypt cells possess identical or different ion transport mechanisms. Although influx studies of ion transport provide evidence of brush border processes, it has been difficult to localize other transport processes to brush border or basolateral membranes. Finally, it is well appreciated that this model (Figure 1) has omitted the possibility of $Cl:HCO_3$ exchanges at the brush border membrane or of Na:H exchanges at the basolateral membrane, although there is evidence to indicate their presence.

Electrogenic Na Absorption In this model there is both neutral and electrogenic Na absorptive and neutral Na and anion (Cl and HCO_3) secretory processes. Recent studies indicate that electrongenic Na absorption may represent between 50 and 80% of total Na absorption (i.e., Isc $\neq J_{net}^{Na}$) (22). In contrast, the initial studies of Schultz and Zalusky proposed that Na absorption was entirely electrogenic (i.e., Isc $= J_{net}^{Na}$) (3). There is to date no adequate explanation for these differences. Electrogenic Na absorption (process a) is increased by actively transported hexoses but is not affected by changes in mucosal cyclic adenosine 3′:5′-monophosphate (cyclic AMP) (23, 24). Removal of Cl or HCO_3 from the incubation media appears to decrease electrogenic Na absorption (17, 25), although the mechanism of this phenomenon is ill defined. The short circuit current is accounted for under most situations by the electrogenic Na absorption process, but measured Na transport is the result of oppositely directed absorptive and secretory processes.

Neutral NaCl Absorption Recent studies from Schultz's laboratory have provided compelling evidence for a neutral NaCl process (process b) across brush

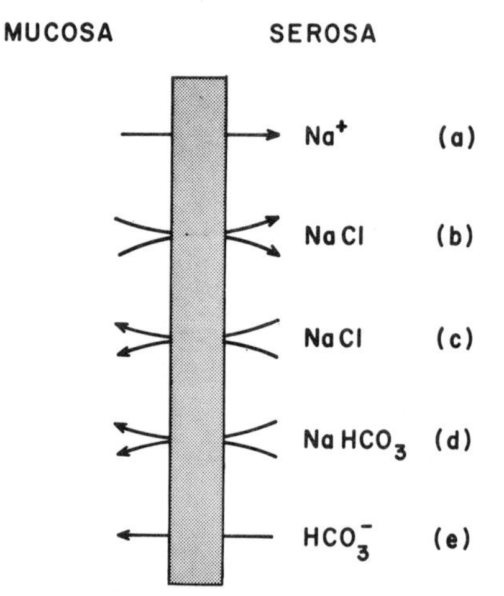

Figure 1. A model of ion transport across rabbit ileal mucosa. As noted in text, no attempt has been made to distinguish mucosal brush border processes from basolateral membrane processes. In addition, a chloride-bicarbonate exchange is probably also present though not shown. (*a*) Electrogenic Na absorption is increased by actively transported hexoses (e.g., 3-*O*-methylglucose) and inhibited by ouabain; (*b*) neutral or coupled NaCl absorption is increased by catecholamines and metabolizable hexoses (e.g., fructose) and is inhibited by cyclic AMP and acetazolamide; (*c*) neutral NaCl secretion is increased by cyclic AMP and high serosal pCO_2; (*d*) neutral $NaHCO_3$ secretion is increased by high serosal pCO_2; (*e*) electrogenic HCO_3 secretion is inhibited by catecholamines.

border cells which contributes to transmural transport of both Na and Cl (22, 25, 26). Figure 2 summarizes the experimental data that support this neutral or coupled system (22). In these particular studies, removal of Cl resulted in an approximate 20% decrease in Na influx, and removal of Na resulted in a 20% reduction of Cl influx. Furthermore, both Na and Cl influxes are mediated by cyclic AMP. Theophylline, which inhibits cyclic nucleotide phosphodiesterase, inhibited 20% of both Na and Cl influx but only when Cl or Na, respectively, were present. Theophylline in either Na-free or Cl-free solutions had no inhibitory effect on Cl or Na influx. These studies suggested that this neutral NaCl absorptive process was located at the brush border membrane, and additional studies have indicated that acetazolamide also inhibits this neutral NaCl process. Other studies have suggested that a similar neutral NaCl absorptive process operates in rabbit gall bladder (27) and possibly in the rat colon (28).

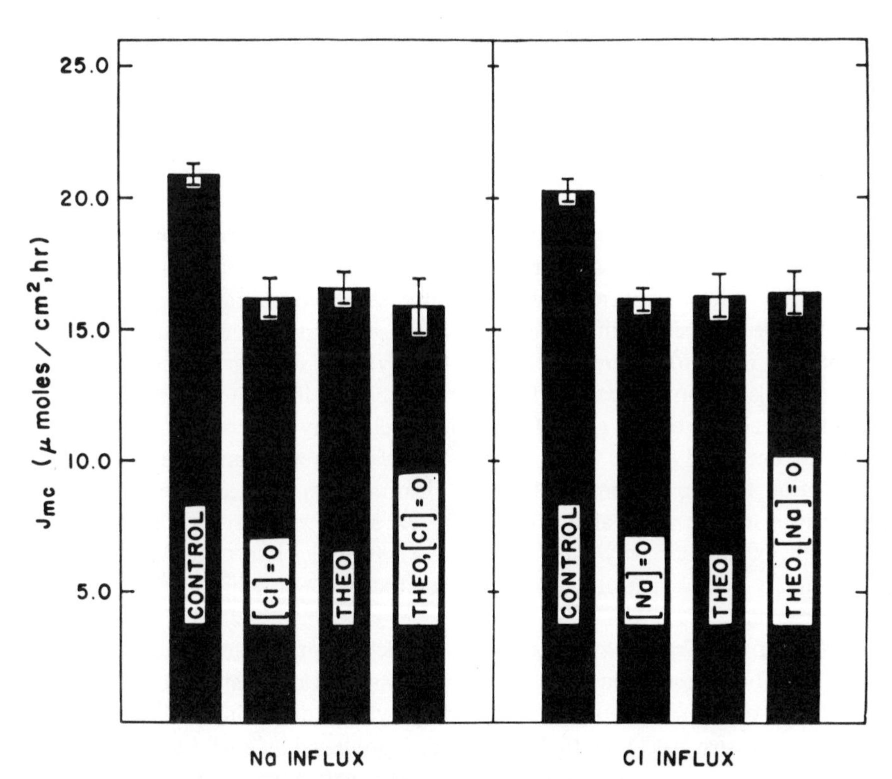

Figure 2. Summary of results of sodium and chloride influx determinations across brush border of rabbit ileum (Nellans et al. (22)). Sodium influx is partially inhibited by either theophylline or chloride removal and chloride influx by either theophylline or sodium removal. In the absence of sodium or chloride theophylline does not inhibit chloride or sodium influx, respectively. These results suggest that approximately 20% of sodium and chloride movement across the brush border is mediated by a coupled NaCl process that is inhibited by cyclic AMP. (Reproduced from Nellans et al. (22).)

The sensitivity of the neutral NaCl process to cylic AMP is consistent with the many observations that choleragen, cyclic AMP, and prostaglandin E_1 inhibit substrate-independent Na and Cl transport (12, 24, 29). In many of these studies in which transmural movement of both Na and Cl were determined, the primary effect of cyclic AMP was to decrease J_{ms}^{Na} and J_{ms}^{Cl} without altering J_{sm}^{Na} and J_{sm}^{Cl}. Although recent studies by Powell et al. (29, 30) and Desjeux et al. (31) indicate that cyclic AMP also stimulates active secretion of neutral NaCl (process c), there is little doubt that cyclic AMP inhibits neutral NaCl absorption. Other effects of cyclic AMP on ileal ion transport include a decrease of tissue conductance and a decrease in cation selectivity of the shunt pathway (30–32).

In addition to the inhibition of coupled NaCl influx by cyclic AMP, stimulation of this process may also occur. Catecholamines and certain sugars will increase neutral NaCl absorption (33, 34). The addition of epinephrine and norepinephrine results in stimulation of the neutral NaCl absorptive mechanism (33). In addition, epinephrine abolishes bicarbonate secretion (process e). These effects of catecholamines are blocked by α-adrenergic blockers but not by β-adrenergic blockers.

Additional studies suggest that metabolizable hexoses also may increase neutral NaCl absorption (34). Fructose and mannose, sugars which are metabolized but not actively transported, increase J_{net}^{Na} and J_{net}^{Cl} without altering Isc. In addition, serosal glucose produces qualitatively similar results. It is important to emphasize that in several recent studies glucose has been present in the serosal solution. Previous in vitro studies by Smyth and collaborators using the rat jejunum also suggested that nonmetabolized sugars stimulated neutral sodium absorption (35, 36). Therefore, Na and Cl movement is under hormonal control and may also be influenced by the type of substrate in the incubation medium.

As already noted, cyclic AMP results in electrolyte secretion. Choleragen, theophylline, and cyclic AMP produce net Cl secretion and reduce net Na absorption to zero but minimally alter bicarbonate movement (determined by calculation of the residual flux) (12, 24, 29, 30). First, there has not been an adequate explanation why cyclic AMP in in vitro studies results in Cl secretion as the primary anion change, while in in vivo studies the principal anion alteration is bicarbonate.

Neutral NaCl Secretion Recent studies indicate that changes in mucosal cyclic AMP have been associated with net Na secretion and not just a decrease in net sodium absorption. Powell et al. (30) have provided compelling evidence that an increase in active NaCl secretion (process c) is produced by cyclic AMP. Furthermore, Desjeux et al. (31) have provided data which indicate that serosa to mucosa Na movement consists of two pathways—a diffusional flux through a paracellular shunt pathway, and a flux that is independent of applied potential and presumably involves a transcellular pathway. The transcellular transport process is stimulated by theophylline and inhibited by both metabolic inhibitors and the removal of chloride and bicarbonate (31). These results are consistent with neutral secretory transport of sodium coupled to either bicarbonate or chloride. Nellans et al. (25) by similar techniques could not identify a non-diffusional component of serosa to mucosa transmural Na movement. It is uncertain whether differences in the bicarbonate or potassium concentration or in pH are responsible for these divergent results. The recent results of Sheerin and Field (37) would indicate that the concentration of bicarbonate in the serosal solution might be an important factor. In conclusion, there is support for a neutral Na anion secretory process which is affected by pH and bicarbonate concentraton in addition to mucosal cyclic AMP concentration.

Role of pH and Bicarbonate The role of pH and bicarbonate concentration in ion transport has recently been explored (37). Increasing the concentration of bicarbonate in the serosal solution from 10 to 50 mM results in an increase in the serosa to mucosa movement of Na, Cl, and bicarbonate. More importantly, there appear to be equal increases in Na and anion (Cl plus bicarbonate) (processes c and d) movement. A change from net Cl absorption to net Cl secretion was also observed when pCO_2 was decreased (or pH increased). Therefore, it is apparent that changes in both serosal bicarbonate concentration and pH will alter ileal ion transport and that some of the differences between several investigations may possibly be explained by the different pH and bicarbonate concentrations employed. In contrast to the stimulation of NaCl secretion by cyclic AMP, increasing the serosal bicarbonate concentration increases both NaCl and $NaHCO_3$ secretion. These observations may be relevant to the unexplained phenomenon that cyclic AMP increases Cl secretion in vitro but bicarbonate secretion in vivo.

Summary

The model presented in Figure 1 is complicated, yet probably incomplete. There are significant experimental data to support the existence of all five processes. Furthermore, as already noted, a chloride-bicarbonate exchange is probably also present. Variations in experimental conditions may explain differing results. This model of ion transport in the rabbit ileum, which includes neutral NaCl absorptive and secretory processes, more closely resembles that proposed for the human ileum by Turnberg et al. (2) than the original model of Schultz and Zalusky (3) in rabbit ileum. The inhibition of Na and Cl absorption by acetazolamide was believed to represent evidence supporting the model of double ion exchanges by Turnberg et al. since acetazolamide was presumed to act by inhibition of carbonic anhydrase activity (2). Nellans et al. have demonstrated that acetazolamide inhibits mucosal influx of Na and Cl, but these investigators do not believe that this inhibition is secondary to carbonic anhydrase inhibition (19). It is quite reasonable to accept the concept that ion transport in both rabbit and human ileum is similar but that under different experimental conditions different processes may be dominant. Additional experiments in the next few years will probably indicate whether a single model of ion transport is present in both these species and whether the differences which are present represent qualitative or quantitative differences.

ELECTROLYTE TRANSPORT IN JEJUNUM

As with results of ileal ion transport, studies of jejunal transport also are conflicting. However, there are far fewer jejunal studies to reconcile.

In Vivo Studies

In vivo perfusion studies of jejunum in volunteer subjects have indicated that in the absence of bicarbonate net sodium movement does not occur against the electrochemical concentration gradient (1, 39). Furthermore, sodium absorption which is observed when bicarbonate is added to the perfusion solution most likely represents a Na:H exchange. In these studies a finite PD was not observed (1, 39), although in subsequent studies by Fordtran a PD of between 1.5 and 3.5 mV has been calculated by the Nernst equation when using steady state luminal and plasma potassium concentrations (40). This model of Na:H exchange is in contrast to that proposed in the ileum in which two double ion exchanges have been proposed.

In Vitro Studies

Few in vitro studies of jejunal ion transport have been reported (41–44). Probably most relevant to this discussion are the two studies of human tissue obtained by peroral biopsy or at operation (41, 42). In both studies, bidirectional Na fluxes were determined under short circuited conditions in a bicarbonate-free solution. Rohde and Anderson (41) studied jejunal mucosa from patients with acute or convalescence stage cholera. In the recovery group J_{net}^{Na} was essentially equivalent to Isc. In the studies of jejunal mucosa obtained at operation, Binder (42) employed a HCO_3-free, Cl-free solution and observed a finite PD, and also demonstrated that J_{net}^{Na} was equal to Isc.

Summary

The in vitro studies are not incompatible with the in vivo studies of Turnberg et al. (1). The PD observed in the in vitro studies (−1.4 mV) is less than that which can be reliably determined in vivo. Furthermore, although both the in vivo and in vitro studies employed a bicarbonate-free solution, there were significant differences in methods, since in the in vivo experiments it was not possible to alter the composition of the serosal medium. This may be one explanation for the inability to demonstrate in vivo phenomena in vitro. Since the zero net sodium movement observed in vivo may represent the result of opposing absorptive and secretory processes, similar to those present in the guinea pig ileum (9), an additional explanation for the discrepancies between in vivo and in vitro studies may be that in vitro the secretory system is either not operating or is not detectable. The in vitro studies exclude the possibility that species differences alone are responsible for the differences in results obtained in experimental animals and in intact human jejunum. Most likely, both electrogenic sodium transport and a neutral Na:H exchange process are present in human jejunum, but it depends on the experimental conditions whether one will observe either one or both of these two processes.

HORMONE CONTROL OF ION TRANSPORT

Considerable study has been directed toward the effect of various hormones on intestinal electrolyte transport. These studies have involved both descriptive observations of the effect of the hormone on fluid and electrolyte movement in vivo, and detailed study of the effect of vasoactive intestinal polypeptide (VIP) and catecholamines on ion transport in vitro. Many of these studies were initiated to evaluate the role of these several hormones as "potential mediators" of the diarrhea associated with pancreatic islet cell adenomas and other tumors (45–48). Although these studies indicate that several hormones alter electrolyte transport, it is uncertain whether many of the observed effects are physiological and not just pharmacological phenomena.

In Vivo Studies

The effects of glucagon, secretin, pentagastrin, cholecystokinin (CCK), gastric inhibitory polypeptide (GIP), and vasoactive intestinal polypeptide (VIP) have been studied in vivo, either with a triple lumen perfusion technique in man or in Thiry-Vella loops in dogs (49–55). All of these compounds have been found either to decrease net absorption or to cause net secretion of fluid and electrolytes. It is important to re-emphasize that under these experimental conditions it is impossible to distinguish whether a decrease in *net* absorption is secondary to a stimulation of the secretory process or to a decrease in the absorptive process.

Glucagon Hicks and Turnberg evaluated the effect of intravenous porcine pancreatic glucagon on jejunal absorption of water, Na, and Cl in healthy volunteers (52). At an infusion rate of 0.6 and 1.2 μg/kg/hr, glucagon significantly decreased water, Na, and Cl absorption to zero, while mean transit time and mean volume of fluid in the test segment were increased. Whether this effect of glucagon on fluid movement is physiological and whether the glucagon effects on transport are related to those on transit will require additional studies. Of significance, however, are recent studies of Whalen et al. (56) which were designed to determine whether the glucagon alters fluid movement under physiological conditions. Glucagon released by arginine infusion failed to alter water absorption despite prolongation of mean transit time.

Secretin and CCK The effect of secretin and CCK was also evaluated in man with the triple lumen perfusion techniques by Moritz et al. (54) In these studies, administration of natural porcine secretin (2 U/kg/hr) resulted in a reversal of the normal absorptive process in human subjects; net secretion of water, sodium, and chloride was observed. In parallel studies Moritz et al. also observed that infusion of CCK resulted in net fluid and electrolyte secretion. Hicks and Turnberg (53) also evaluated the effect of secretin and reported that secretin significantly reduced water, Na, and Cl absorption. These effects were observed only in the proximal jejunum, not in the mid-jejunum. This latter observation is in contrast to a recent report in which the effect of cholera enterotoxin was studied in both the duodenum and jejunum of the rabbit. Fromm (57) observed that there was spontaneous secretion in the duodenum but that cholera enterotoxin induced fluid secretion only in the mid-jejunum and not in the duodenum.

Matuchansky et al. (55) also studied the effect of CCK on jejunal fluid and electrolyte movement and transit time in man. In these studies, intravenous administration of CCK resulted in net water, Na, and Cl secretion. Determination of unidirectional sodium movement with ^{22}Na revealed that a decrease in lumen to plasma Na movement was induced by CCK without significant change in plasma to lumen sodium movement. Simultaneous measurement of mean transit time revealed that CCK reduced mean transit time but also decreased the volume of fluid on the intestinal segment. These investigators indicated that according to their results, the net secretion observed was secondary to primary changes in motility. This author is not as confident that this casual relationship has been proved unequivocally. Although rapid movement of fluid through an isolated segment of intestine (i.e., a decrease in mean transit time) might result in a decrease in net absorption, it is not clear how net secretion could be produced by such a decrease in transit time. At present, a definite association between alterations in motility and fluid movement remains to be established. Additional studies are required to determine the relationship between intestinal motility, intestinal blood flow, and ion transport.

Thyrocalcitonin In another investigation of hormone control of jejunal fluid and electrolyte movement, Gray, Bieberdorf, and Fordtran (58) evaluated the effect of thyrocalcitonin on intestinal absorption. These studies were prompted by the suggestion that the diarrhea associated with medullary carcinoma of the thyroid was related to tumor production of thyrocalcitonin. Infusion of synthetic salmon thyrocalcitonin resulted in net secretion of water, sodium, chloride, and potassium during perfusion with either a bicarbonate-free or a bicarbonate-containing solution. The mechanism by which thyrocalcitonin alters fluid and electrolyte movement is not understood, especially since thyrocalcitonin may induce secretion in vivo in the rabbit but has no effect on isolated rabbit ileal mucosa (59, 60).

Vasoactive Intestinal Peptide Clinical interest in the effect of various hormones on fluid and electrolyte movement relates to the problem of identifying a hormonal agent responsible for the significant diarrhea present in the syndrome of pancreatic cholera. In this syndrome, an islet cell tumor which does not secrete gastrin has been associated with profuse watery diarrhea, hypokalemia, and achlorhydria (WDHA syndrome) (45, 46). All five hormones that have at one time been suggested as the responsible agent have now been demonstrated to cause intestinal fluid secretion: VIP, GIP, secretin, prostaglandin E_1, and thyrocalcitonin (49–51, 58, 61). Current opinion favors the concept that although more than one hormone may be produced by these pancreatic adenomas, VIP is most likely to be responsible for the diarrhea in many of these patients, and that VIP may be produced by tumors other than pancreatic islet cell adenomas, for example, bronchogenic carcinoma (46–48).

In Vitro Studies

In addition to the studies of the effect of intravenous administration of hormones on in vivo movement of fluid and electrolytes, additional studies of the effect of at least two hormones on ion transport in vitro have been reported. Schwartz et al. (16) have studied the effect of VIP on ion transport across isolated rabbit and human ileum. In these studies, VIP increased Isc, decreased J_{net}^{Na}, and produced net Cl secretion. These changes qualitatively resembled those produced by cholera enterotoxin and theophylline, both of which produced increased mucosal cyclic AMP levels. VIP also increased mucosal cyclic AMP levels in parallel studies. These investigators suggested that VIP was the hormone mediator of the diarrhea of WDHA syndrome and that the diarrhea was mediated by VIP stimulation of cyclic AMP-induced ion secretion. In the same report, the effect on mucosal cyclic AMP of several other hormones known to alter fluid and electrolyte absorption was also studied. GIP, glucagon, gastrin, CCK, and vasopressin did not alter mucosal cyclic AMP content. If these observations are confirmed and if these compounds produce secretion in rabbit in vivo, then one must suggest that either these hormones produce secretion by a noncyclic AMP secretory process or they stimulate an intermediary which in turn stimulates adenylate cyclase and results in an increase in cyclic AMP content.

Field and McColl have evaluated the effect of catecholamines on ion transport (33). Epinephrine and norepinephrine decreased Isc. Flux measurements revealed that epinephrine and norepinephrine increased Na and Cl absorption and also decreased the so-called residual flux. These effects of epinephrine on Isc were blocked by α-adrenergic but not by β-adrenergic blockers.

Summary

Intestinal electrolyte movement is controlled by multiple hormone factors which may alter ion movement in different directions. Complete understanding of normal ion movement will require delineation of its hormone control.

GLUCOSE STIMULATION OF SODIUM ABSORPTION

In Vitro Model

Glucose stimulation of sodium and water absorption is a well accepted phenomenon in the mammalian small intestine (62, 63). Despite the constancy of this observation and its application in clinical medicine as oral therapy in the treatment of cholera (64), there has been considerable controversy regarding the mechanism by which glucose augments the absorption of sodium. Studies over several years have repeatedly confirmed that glucose increases sodium absorption from the small intestine. Until Schultz and Zalusky demonstrated that 3-*O*-methylglucose, an actively transported but nonmetabolized sugar, also increased sodium absorption, it was generally thought that this augmentation was secondary to an increase in cellular metabolism of glucose (23). Since both the absence of sodium and the inhibition of sodium transport also inhibited glucose transport, the observation, at least in rabbit ileum, the 3-*O*-methylglucose stimulated sodium transport indicated that the transport of glucose was directly related to sodium transport. Additional in vitro studies have suggested that glucose entry into the mucosal cell is coupled to sodium entry (65). The presence of a sodium gradient between mucosal solution and cell is required for glucose accumulation, and glucose augments sodium transport by increasing the rate of sodium entry into the mucosal cell, thereby resulting in an increase in sodium extrusion from the cell. This sodium gradient hypothesis indirectly links the uphill movement of glucose into the epithelial cell to the metabolic energy requirement for the maintenance of the sodium gradient. Alternative mechanisms have been proposed by Csaky and Kimmich, who have also favored glucose stimulation of active sodium transport (66, 67). Schultz and Curran (63) recently discussed the relative merits of these various proposals and stressed their belief that the available evidence best supports the sodium gradient hypothesis of Crane (62) rather than the proposals of either Csaky or Kimmich.

In Vivo Model

In contrast to these in vitro studies which have provided the basis and support for the sodium gradient hypothesis, Fordtran has proposed that glucose stimulation of sodium absorption is linked to solvent drag and that bulk water flow is the major determinant of sodium movement in the jejunum (39). This hypothesis has been based on studies of fluid and electrolyte movement in man which have employed triple lumen perfusion techniques. These studies indicate that glucose absorption results in an increase in water flow which is the driving force for the increased sodium absorption. In these in vivo studies, glucose does not increase the electrical PD. Therefore, in the jejunum, glucose stimulation of sodium movement is secondary to solvent drag: sodium movement is sensitive to the magnitude and direction of bulk water flow. In contrast, in ileal studies, glucose increases water absorption but sodium movement is not significantly altered (39). The electrical potential difference in the ileum is increased when glucose is added to the perfusate. These observations can best be explained by an increase in lumen to plasma movement of sodium secondary to the absorption of glucose, balanced by an equal increase in plasma to lumen movement of sodium driven by the increased PD. In either model, water movement across the cell from the lumen is driven by the increased concentration of solute (either glucose or glucose and sodium, respectively) as originally proposed by Curran and

MacIntosh's double membrane theory, or as modified by Diamond and Bossert's standing gradient hypothesis (68, 69).

In part, this controversy exists because most of the studies that support the sodium gradient hypothesis have been performed in vitro with experimental animals, whereas Fordtran's experiments that indicate that jejunal sodium movement is markedly sensitive to bulk water flow and not directly coupled to glucose transport have been performed exclusively in he human jejunum in vivo. Recent reports have provided important experimental observations indicating that both glucose-coupled sodium movement and sodium movement linked to glucose-stimulated water flow (solvent drag) may operate in the human jejunum (42, 70).

Reconciliation of the Controversy

In in vitro experiments which employed the methods and apparatus of Schultz and Zalusky, Binder studied normal jejunum obtained at operation (42). The addition of either 3-*O*-methylglucose or glucose resulted in an increase in PD and Isc. Furthermore, the increase in net sodium movement produced by glucose was equal to the increase in Isc, and the increase in net sodium movement was equivalent to the increase in mucosa to serosa sodium movement. Although these observations might be explained by the development of a diffusion potential or a streaming potential secondary to solvent drag, analysis of the unidirectional Na fluxes would make this interpretation less likely. Therefore, these studies suggest that under certain in vitro conditions, glucose stimulation of active sodium transport can be demonstrated across human jejunum. Species differences cannot be the explanation for the contrast between these views of glucose-sodium interaction in the jejunum.

Modigliani and Bernier, in triple lumen perfusion studies of the jejunum, and ileum, determined unidirectional sodium movement (71). The investigators concluded that in the jejunum, glucose augmentation of Na movement was secondary to solvent drag, but a "direct link between the transport of the sugar and sodium" was also operating. In the ileum, they felt, only coupled glucose-sodium transport was present.

Fordtran, in an elegant series of studies, has recently provided additional evidence that under some but certainly not all in vivo conditions, glucose stimulates active sodium absorption in the human jejunum (70). In these studies the absorption of water, sodium, potassium, and urea from isosmolar sodium chloride solutions containing mannitol, fructose, or glucose were compared. These studies were designed to determine the relative contribution of active and passive sugar stimulation of sodium absorption. The difference in absorption between the fructose and mannitol solutions was compared with that observed between the glucose and mannitol solutions. If fructose absorption is entirely passive, then the increase in water and sodium absorption produced by fructose compared to that observed in the presence of mannitol should represent increased water and sodium absorption secondary to passive sugar absorption. At equal rates of sugar absorption, any increase in water and sodium absorption produced by glucose that is greater than that produced by fructose should then represent active sugar stimulation of water and sodium absorption. Perfusion with fructose at 130 mM and with glucose at 65 mM resulted in equal rates of sugar absorption, but water and sodium absorption was unequal. Although urea absorption was greater in the glucose solution than in the fructose solution, potassium movement demonstrated a different pattern. Glucose stimulated potassium secretion but fructose stimulated potassium absorption. Assuming that all potassium movement is passive and driven by the existing electrochemical concentration gradients, calculation by the Nernst equation indicated that glucose significantly increased PD (lumen negative) but that fructose did not alter the PD. These results are presented in part in Figure 3.

Fordtran concluded that glucose stimulates both active and passive sodium transport. Since glucose absorption also increases the PD which causes sodium (and potassium) secretion, the overall effect is that that portion of net sodium

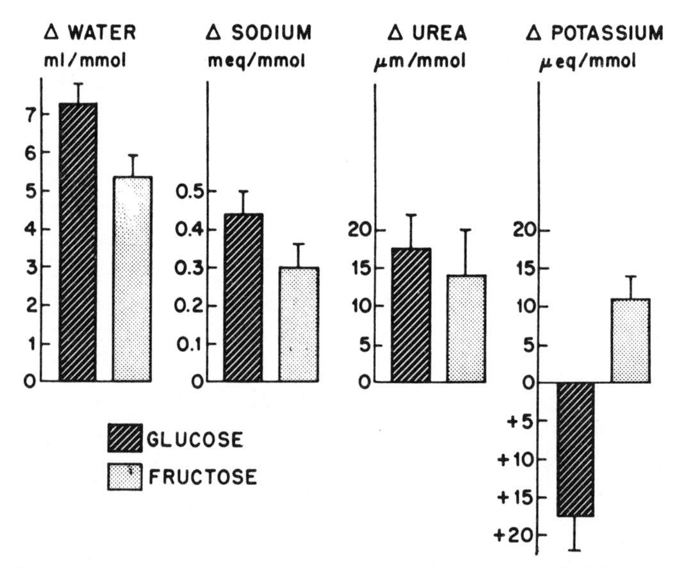

Figure 3. Results of water, sodium, urea, and potassium movement in jejunum of control subjects during perfusion with mannitol, glucose, and fructose solutions. These results represent the difference between that observed with either glucose or fructose and mannitol. The plus signs indicate net secretion of potassium. In these experiments 130 mM fructose and 65 mM glucose solutions resulted in equal absorption of the hexose but water, sodium, and urea absorption was greater with the glucose solution than with the fructose solution. In contrast, fructose increased potassium absorption but glucose stimulated potassium secretion. Nernst equation calculations indicated that glucose increased the electrical potential difference (PD) but that fructose did not alter the PD. These results are compatible with the concept that glucose stimulates both active and passive sodium absorption but that the passive component is the predominant one. (Reproduced from Fordtran (70).)

movement secondary to active sugar absorption may not increase, and only the passive component of glucose-stimulated water and sodium absorption is observed. Finally, Fordtran estimated that 66–100% of "net sodium absorption stimulated by glucose is mainly the passive consequence of solvent drag" (70).

Summary

It is now probable that all previous observations and hypotheses concerning the mechanism of glucose stimulation of sodium absorption in the human jejunum have been correct. Under in vitro conditions glucose stimulation of active sodium absorption can easily be demonstrated, and depending on the experimental conditions may also be observed in vivo. It would appear, however, that the major fraction of glucose-stimulated sodium absorption in vivo is driven by solvent drag, since Fordtran's recent studies demonstrated that even under conditions when active sodium absorption can be demonstrated, most glucose-stimulated sodium absorption is still secondary to bulk water movement (70).

It is important to re-emphasize that experimental results supporting the presence of one mechanism rarely can be used to provide evidence against alternate or additional proposals. The two mechanisms by which glucose increases sodium absorption are probably both present, but their relative contribution may depend on such variables as species (man, rat, etc.), segment of intestine (jejunum or ileum), methods (in vivo or in vitro), and other factors not as yet identified.

REFERENCES

1. Turnberg, L. A., Fordtran, J. S., Carter, N. W., and Rector, F. C. (1970). Mechanism of bicarbonate absorption and its relationship to sodium transport in the human jejunum. J. Clin. Invest. 49:548.
2. Turnberg, L. A., Bieberdorf, F. A., Morawski, S. G., and Fordtran, J. S. (1970). Interrelationships of chloride, bicarbonate, sodium and hydrogen transport in the human ileum. J. Clin. Invest., 49:557.
3. Schultz, S. G., and Zalusky, R. (1964). Ion transport in isolated rabbit ileum. I. Short-circuit current and Na fluxes. J. Gen. Physiol. 47:567.
4. Schultz, S. G., Zalusky, R., and Gass, A. E. (1964). Ion transport in isolated rabbit ileum. III. Chloride fluxes. J. Gen. Physiol. 48:375.
5. Hendrix, T. R., and Bayless, T. M. (1970). Digestion: intestinal secretion. Annu. Rev. Physiol. 32:139.
6. Field, M. (1974). Intestinal secretion. Gastroenterology 66:1063.
7. Hendrix, T. R., and Paulk, H. T. (1976). Intestinal secretion. This volume.
8. Frizzell, R. A., and Schultz, S. G. (1972). Ionic conductances of extracellular shunt pathway in rabbit ileum. Influence of shunt on transmural sodium transport and electrical potential differences. J. Gen. Physiol. 59:318.
9. Powell, D. W., Binder, H. J. and Curran, P. F. (1972). Electrolyte secretion by the guinea pig ileum *in vitro*. Am. J. Physiol. 223:531.
10. Binder, H. J., Powell, D. W., and Curran, P. F. (1972). Effect of hexoses on ion transport in guinea pig ileum. Am. J. Physiol. 223:538.
11. Powell, D. W., Malawer, S. J., and Plotkin, G. R. (1968). Secretion of electrolytes and water by the guinea pig small intestine *in vivo*. Am. J. Physiol. 215:1226.
12. Field, M., Fromm, D., Al-Awqati, Q., and Greenough, W. B. (1972). Effect of cholera enterotoxin on ion transport across isolated ileal mucosa. J. Clin. Invest. 51:796.

13. Al-Awqati, Q., Wallace, C. K., and Greenough, W. B. (1972). Stimulation of intestinal secretion *in vitro* by culture filtrates of *Escherichia coli.* J. Infect. Dis. 125:300.
14. Al-Awqati, Q., and Greenough, W. B. (1972). Prostaglandins inhibit intestinal sodium transport. Nature 238:26.
15. Kimberg, D. V., Field, M., Gershon, E., and Henderson, A. (1974). Effects of prostaglandins and cholera enterotoxin on intestinal mucosal cyclic AMP accumulation. Evidence against an essential role for prostaglandins in the action of toxin. J. Clin. Invest. 53:941.
16. Schwartz, C. J., Kimberg, D. V., Sheerin, H. E., Field, M., and Said, S. I. (1974). Vasoactive intestinal polypeptide stimulation of adenylate cyclase and active electrolyte secretion in intestinal mucosa. J. Clin. Invest. 54:536.
17. Binder, H. J., Powell, D. W., Tai, Y. H. and Curran, P. F. (1973). Electrolyte transport in rabbit ileum. Am. J. Physiol. 225:776.
18. Field, M., Fromm, D., and McColl, I. (1971). Ion transport in rabbit ileal mucosa. I. Na and Cl Fluxes and short circuit current. Am. J. Physiol. 220:1388.
19. Nellans, H. N., Frizzell, R. A., and Schultz, S. G. (1975). Effect of acetazolamide on sodium and chloride transport by *in vitro* rabbit ileum. Am. J. Physiol. 228:1808.
20. Al-Awqati, Q., Garcia-Bunuel, R., Field, M., and Greenough, W. B. (1970). The influence of structural abnormality on ion transport in rabbit ileum. Proc. Soc. Exp. Biol. Med. 135:598.
21. Dietz, J., and Field, M. (1973). Ion transport in rabbit ileal mucosa. IV. Bicarbonate secretion. Am. J. Physiol. 225:858.
22. Nellans, H. N., Frizzell, R. A., and Schultz, S. G. (1973). Coupled sodium-chloride influx across the brush border of rabbit ileum. Am. J. Physiol. 225:467.
23. Schultz, S. G., and Zalusky, R. (1964). Ion transport in isolated rabbit ileum. II. The interaction between active sodium and active sugar transport. J. Gen. Physiol. 47:1043.
24. Field, M. (1971). Ion transport in rabbit ileal mucosa. II. Effect of cyclic 3′,5′-AMP. Am. J. Physiol. 221:992.
25. Nellans, H. N., Frizzell, R. A., and Schultz, S. G. (1974). Brush-border processes and transepithelial Na and Cl transport of rabbit ileum. Am. J. Physiol. 226:1131.
26. Frizzell, R. A., Nellans, H. N., Rose, R. C., Markscheid-Kaspi, L., and Schultz, S. G. (1973). Intracellular Cl concentrations and influxes across the brush border of rabbit ileum. Am. J. Physiol. 224:328.
27. Binder, H. J., and Rawlins, C. L. (1973). Electrolyte transport across isolated large intestinal mucosa. Am. J. Physiol. 225:1232.
28. Frizzell, R. A., Dugas, M. C., and Schultz, S. G. (1975). Sodium chloride transport by rabbit gallbladder. Direct evidence for a coupled NaCl influx process. J. Gen. Physiol. 65:769.
29. Powell, D. W., Binder, H. J., and Curran, P. F. (1973). Active electrolyte secretion stimulated by choleragen in rabbit ileum *in vitro.* Am. J. Physiol. 225:781.
30. Powell, D. W., Farris, R. K., and Carbonetto, S. T. (1974). Theophylline, cyclic AMP, choleragen, and electrolyte transport by rabbit ileum. Am. J. Physiol. 227:1428.
31. Desjeux, J. F., Tai, Y. H., and Curran, P. F. (1974). Characteristics of sodium flux from serosa to mucosa in rabbit ileum. J. Gen. Physiol. 64:274.
32. Powell, D. W. (1974). Intestinal conductance and permselectivity changes with theophylline and choleragen. Am. J. Physiol. 227:1436.
33. Field, M., and McColl, I. (1973). Ion transport in rabbit ileal mucosa. III. Effects of catecholamines. Am. J. Physiol. 225:852.
34. McCall, W., Binder, H. J., Powell, D. W., and Curran, P. F. (1975). Effect of hexoses on ion transport in rabbit ileum. Unpublished observations.
35. Barry, R. J. C., Smyth, D. H., and Wright, E. M. (1965). Short-circuit current and solute transfer by rat jejunum. J. Physiol. (Lond.) 181:410.
36. Barry, R. J. C., Eggenton, J., and Smyth, D. H. (1969). Sodium pumps in rat small intestine in relation to hexose transfer and metabolism. J. Physiol. (Lond.) 204:299.
37. Sheerin, H. E., and Field, M. (1975). Ileal HCO_3 secretion: relationship to Na and Cl transport and effect of theophylline. Am. J. Physiol. 228:1065.
38. Frizzell, R. A., Markscheid-Kaspi, L., and Schultz, S. G. (1974). Oxidative metabolism of rabbit ileal mucosa. Am. J. Physiol. 226:1142.

39. Fordtran, J. S., Rector, F. C., and Carter, N. W. (1968). The mechanisms of sodium absorption in the human small intestine. J. Clin. Invest. 47:884.
40. Ireland, P., and Fordtran, J. S. (1973). Effect of dietary calcium and age on jejunal calcium absorption in humans studied by intestinal perfusion. J. Clin. Invest. 52:2672.
41. Rohde, J. E., and Anderson, B. (1973). *In vitro* measurement of ion fluxes across biopsies of human jejunal mucosa during cholera. J. Appl. Physiol. 35:557.
42. Binder, H. J. (1974). Sodium transport across isolated human jejunum. Gastroenterology 67:231.
43. Fromm, D., (1973). Na and Cl transport across isolated proximal small intestine of the rabbit. Am. J. Physiol. 224:110.
44. Volpe, B. T., and Binder, H. J. (1975). Bile salt alterations of ion transport across jejunal mucosa. Biochim. Biophys. Acta 394:597.
45. Verner, J. B., and Morrison, A. B. (1958). Islet cell tumor and a syndrome of refractory watery diarrhea and hypokalemia. Am. J. Med. 25:374.
46. Said, S. I., and Faloona, G. R. (1975). Elevated plasma and tissue levels of vasoactive intestinal polypeptide in the watery diarrhea syndrome due to pancreatic, bronchogenic and other tumors. N. Engl. J. Med. 293:155.
47. Schmitt, M. G., Soergel, K. H., Hensley, G. T., and Chey, W. T. (1975). Watery diarrhea associated with pancreatic islet cell carcinoma. Gastroenterology 69:206.
48. Rambaud, J. C., Modigliani, R., Matuchansky, C., Bloom, S., Said, S. I., Pessayre, D., and Bernier, J. J. (1975). Pancreatic cholera. Studies on tumoral secretions and pathophysiology of diarrhea. Gastroenterology 69:110.
49. Barbezat, G. O., and Grossman, M. I. (1971). Glucagon stimulates intestinal secretion. Lancet 1:918.
50. Barbezat, G. O., and Grossman, M. I. (1971). Cholera-like diarrhea induced by glucagon plus gastrin. Lancet 1:1025.
51. Barbezat, G. O., and Grossman, M. I. (1971). Intestinal secretion: stimulation by peptides. Science 174:442.
52. Hicks, T., and Turnberg, L. A. (1974). Influence of glucagon on the human jejunum. Gastroenterology 67:1114.
53. Hicks, T., and Turnberg, L. A. (1973). The influence of secretin on ion transport in the human jejunum. Gut 14:485.
54. Moritz, M., Finkelstein, G., Meshkinpour, H., Fingergut, J., and Lorber, S. H. (1973). Effect of secretin and cholecystokinin on the transport of electrolyte and water in human jejunum. Gastroenterology 64:76.
55. Matuchansky, C., Huet, P. M., Mary, J. Y., Rambaud, J. C., and Bernier, J. J. (1972). Effects of cholecystokinin and metoclopramide on jejunal movements of water and electrolytes and on transit time of luminal fluid in man. Eur. J. Clin. Invest. 2:169.
56. Whalen, G. E., Wu, W. C., Ganeshappa, K. P., Wall, M. J., Kalkhoff, R. K., and Soergel, K. H. (1973). The effect of endogenous glucagon on human small bowel function. Gastroenterology 64:A-139, 822 (Abstr.).
57. Fromm, D., Dibala, R. P., and Sullivan, H. W. (1975). Ion transport by rabbit jejunum *in vivo.* Am. J. Physiol. 228:160.
58. Gray, T. K., Bieberdorf, F. A., and Fordtran, J. S. (1973). Thyrocalcitonin and the jejunal absorption of calcium, water and electrolytes in normal subjects. J. Clin. Invest. 52:3084.
59. Cheung, L. Y., Jubiz, W., Moore, J. G., and Frailey, J. (1975). Gastric prostaglandin E (PGE) output during basal and stimulated acid secretion in normal subjects and patients with peptic ulcer Gastroenterology 68:A-16,873 (Abstr..).
60. Gray, T. K., Juan, D., and Powell, D. W. (1975). Salmon calcitonin and water and electrolyte transport in rabbit ileum. Proc. Soc. Exp. Biol. Med. 150:151.
61. Cummings, J. H., Newman, A., Misiewicz, J. J., Milton-Thompson, G. J., and Billings, J. A. (1973). Effect of intravenous prostaglandin $F_{2\alpha}$ on small intestinal junction in man. Nature 243:169.
62. Crane, R. K. (1962). Hypothesis of mechanism of intestinal active transport of sugars. Fed. Proc. 21:891.
63. Schultz, S. G., and Curran, P. F. (1970). Coupled transport of sodium and organic solutes. Physiol. Rev. 50:637.

64. Hirschhorn, N., Kinzie, J. L., Sachar, D. B., Northrup, R. S., Taylor, J. O., Ahmad, S. Z., and Phillips, R. A. (1968). Decrease in net stool output in cholera during intestinal perfusion with glucose-containing solutions. N. Engl. J. Med. 279:176.
65. Goldner, A. M., Schultz, S. G., and Curran, P. F. (1969). Sodium and sugar fluxes across the mucosal border of rabbit ileum. J. Gen. Physiol. 53:362.
66. Csaky, T. Z. (1963). A possible link between the active transport of electrolytes and non-electrolytes. Fed. Proc. 22:3.
67. Kimmich, G. A. (1970). Active sugar accumulation by isolated epithelial cells. A new model for sodium-dependent metabolite transport. Biochemistry 9:3669.
68. Curran, P. F., and MacIntosh, J. R. (1962). A model system for biological water transport. Nature 193:347.
69. Diamond, J. M., and Bossert, W. H. (1967). Standing-gradient osmotic flow. A mechanism for coupling of water and solute transport in epithelia. J. Gen. Physiol. 50:2061.
70. Fordtran, J. S. (1975). Stimulation of active and passive sodium absorption by sugars in the human jejunum. J. Clin. Invest. 55:728.
71. Modigliani, R., and Bernier, J. J. (1972). Effects of glucose on net and unidirectional movements of water and electrolytes in the human small intestine. Biol. Gastroenterol. V:165.

International Review of Physiology
Gastrointestinal Physiology II, Volume 12
Edited by Robert K. Crane

10
Digestion and Absorption of Lipids

B. BORGSTRÖM
Department of Physiological Chemistry,
University of Lund, Lund, Sweden

DIGESTION OF LIPIDS 306
Lipolytic Enzymes 306
Pregastric Lipases 306
Pancreatic Lipases and Colipase 306
Classic Pancreatic Lipase 307
Carboxylic Ester Hydrolase-Sterol Ester Hydrolase 309
Phospholipases 309
Colipase 310
Lipases from Other Sources 310
Intestinal Lipase 310
Bile Salt-stimulated Milk Lipase 311

INTRALUMINAL PHASE OF FAT DIGESTION 311

UPTAKE OF LIPIDS INTO THE ENTEROCYTE 311

INTRACELLULAR METABOLISM OF ABSORBED LIPIDS 313

ABSORPTION OF SPECIFIC LIPIDS 314
Phospholipids 314
Sterols 315
Various Lipids 317

CONCLUSIONS 317

When the field of fat digestion and absorption was reviewed 2 years ago in this series (1) special emphasis was focused on the absorption process as a physicochemical problem. The rather spectacular progress in this field made during the last few years against this background was summed up. The progress during the last 2–3 years has not been so spectacular and the reader is therefore referred to the last review for a discussion of this topic. The present review will naturally build upon the previous one and will emphasize the development in some of the fields which have been of special interest to this reviewer rather than being a complete coverage of the literature in all the different aspects of digestion and absorption of lipids. The subject of fat digestion and absorption has been reviewed in the recent literature (2–5).

DIGESTION OF LIPIDS

Lipolytic Enzymes

Pregastric Lipases It was noted as long ago as 1957 that a considerable lipolysis of fed triglyceride occurred in the stomach content of man (6). It was not clear if this lipolysis was an effect of regurgitated intestinal content or was attributable to a gastric lipase. The presence of a lipase in gastric aspirates of man different from pancreatic lipase was later indicated (7) and it was assumed that this lipase contributed to the digestion of much triglyceride in infants as well as to hydrolysis of medium chain triglycerides (MCT) in adults. In the rat Hamosh and Scow (8) found a lipase secreted by the serous glands of the tongue which they named lingual lipase. This enzyme, which has not yet been obtained in pure form, was characterized by a low pH range (optimum around 5) and the preferential formation of diglyceride plus fatty acid. The enzyme was active against esters with short as well as long chain fatty acids and evidence was obtained to show that this enzyme, acting in the stomach, is important as a first step in the digestion of dietary lipids. In man the so-called gastric lipase is most probably secreted by glands in or near the pharynx which have not been identified and has been renamed the pharyngeal lipase (9). It has a pH optimum at 5.4, it is not affected by bile salt, and its major effect is to hydrolyze triglyceride to diglyceride and fatty acid. It is proposed that this enzyme catalyzes the first step in digestion of dietary fat and that the formation of fatty acids facilitates the further emulsification of triglyceride fat in the stomach and duodenum.

Pancreatic Lipases and Colipase Lipase by definition is an enzyme that preferentially acts on water-insoluble substrates (4), i.e., it is active at the substrate-water interface and has only a rather low activity against water-soluble substrates. It also has certain requirements as to the structure of the ester bond to be split and the rate of hydrolysis is affected by neighboring groups. Within the limits of this definition at least two different lipase activites have been recognized in pancreatic juice of higher animals, one of which is the classic pancreatic lipase. Bradshaw and Rutter (10) detected two lipases, designated A and B, in pancreatic extracts and juice of the rat. Both of these enzyme activities hydrolyzed emulsified triolein; lipase B was inhibited by bile salt while lipase A was stimulated. Reasoning from these properties and the substrate specificities it can be concluded that lipase B is identical with the classic pancreatic lipase while lipase. Bradshaw and Rutter (10) detected two lipases, designated A and B, in carboxylic ester hydrolase (11, 14), nonspecific lipase (12, 13), or sterol ester hydrolase (14). The designation of lipase A and B (10) may be unfortunate as it can easily be confused with pancreatic lipase L_A and L_B (15) used for two molecular species of the classic pancreatic lipase which differ in their glycopeptide composition (16).

Classic Pancreatic Lipase This enzyme has been purified from different sources, namely, pig (15), rat (17), man (18, 19), cattle (20), and sheep (21). The molecular weights of all these lipases are in a close range of 46,000–49,000 and the amino acid compositions are very similar. The isoelectric point differs for some of these enzymes as well as the NH_2-terminal amino acid, which for the porcine enzyme is a serine (20) and for the ovine, lysine (21). Bovine lipase has been found to contain two free sulfhydryl groups of different reactivities (20), while only one reactive thiol group was reported for porcine lipase (22). The primary structure of porcine lipase has been partly worked out (23), the results indicating that the enzyme is composed of a single peptide chain with an NH_2-terminal serine and a COOH-terminal half-cystine. It is of interest that the relative amount of hydrophobic amino acid residues in the pancreatic lipases from different sources does not differ from that found for other pancreatic enzymes.

A monomer with a molecular weight of 15,000 has been reported for lipase purified from human pancreatic glands obtained at postmortem (19). The preparation started from a 100,000 $\times g$ supernatant of a water homogenate of the gland and included salt fractionation, butanol extraction, precipitation with acetone, and finally electrofocusing. Straube et al. (19) report different aggregation sizes of lipase, ranging from 15,000 to more than 240,000, all of which contain varying amounts of lipids. However, the different sizes of proteins were not characterized as proteins. Consequently, it is difficult to evaluate these results at the present time. It is known that pancreatic gland extracts contain lipase in macromolecular form, most probably in association with lipids corresponding to "lipase rapide" (24, 25) and lipase F (26). Hence, it is questionable whether this monomeric form of lipase has any physiological significance. It should probably be regarded as an artifact formed when lipase comes into contact with cellular membrane fragments or other lipids.

The action of lipase on water-soluble substrates has been reinvestigated. The addition of water-miscible organic compounds such as dioxane increases the activity of lipase severalfold against monomeric tripropionin (27). This suggests that the organic compounds change the conformation of lipase or the structure of the water in the vicinity of the substrate. The rate of hydrolysis by lipase is still low relative to the rate with insoluble esters, but not substantially different from hydrolysis rates found for several ordinary esterases. It was previously suggested that lipase acts on substrates in micellar form (28). However, there is no good evidence for this view. Such interpretations as to the importance of the physical state of the substrate for the action of pancreatic lipase are difficult to support in the light of the finding that adsorption of the substrate to any interface greatly increases the surface concentration that is available to the enzyme (27, 29, 30). Siliconized glass beads increase the rate of hydrolysis of the water-soluble compounds, tripropionin (29) and *p*-nitrophenylacetate (300-fold), most probably because of the binding of the enzyme to the interface. Kinetic studies of the action of lipase on substrate monolayers have been of continued interest and a method has been described for isobaric titration of the activity of lipolytic enzymes (30). In this method measurement is made of the amount of insoluble substrate that must be added to keep the surface pressure constant.

The effect of bile salts on the lipase-catalyzed hydrolysis of insoluble triglycerides has in the past mostly been described as an activation. However, results with purified or pure lipase have shown that bile salts inhibit lipase activity and that this inhibition occurs with bile salt concentration in the range of their critical micellar concentration (18, 31, 32). The interpretation of these findings is not clear. On the one hand, this may be a general detergent effect (31) inasmuch as it has been shown that bile salts will displace lipase from the substrate interface in the range of the micellar concentration and above pH 6.5 (33). On the other hand, this may be a charge effect, i.e., the substrate surface may become negatively charged by the adsorption of bile salt anions which repulse the negatively charged lipase. However, other interpretations are also open and need to be investigated before a definite explanation of the bile salt effect can be given. For example, it has also been shown that bile salts bind to pancreatic lipase in the range of their micellar concentration although this binding is weak and only 2–4 mol of bile salt bind per mol of lipase (34).

Bile salts have also been shown to protect lipase from irreversible inactivation at an interface including that of its own substrate (18, 33, 35). This effect may be the result of the binding of the bile salt to the enzyme, stabilization of its conformation, or to the clearing of lipase from interfaces (33). Pancreatic lipase has an affinity for hydrophobic interfaces and it appears that a prerequisite for its activity is binding to the substrate interface, which in turn makes possible the binding of the substrate ester bond to the active site.

With respect to the lipase-catalyzed reaction, a change in the absorption spectrum of a lipase substrate riboflavin 4′,5′-dibuturate when mixed with lipase has been taken to indicate the formation of an enzyme-substrate complex (36). Also, a transient acetyl-lipase intermediate formed during hydrolysis of *p*-nitrophenylacetate by lipase was isolated by Sémériva et al. (37). The acyl enzyme derivative was stable at pH 5.0 and could be isolated by Sephadex filtration, but it was deacylated at pH 7.4. The earlier finding that pancreatic lipase has no specificity towards the primary positions in the triglyceride has been confirmed. The enzyme, however, preferentially attacks unsaturated fatty acids in these positions (38).

As to the preparation of lipase, affinity chromatography of lipase to hydrophobic Sepharose (aliphatic amine-Sepharose 4B) may be of preparative as well as theoretical interest (39). The chain length of the aliphatic amine used is of importance and the lipase can be desorbed by bile salts.

New methods for lipase assay appear steadily. The tributyrine method (40) has appeared in an automated version for the determination of pancreatic lipase in duodenal juice by continuous single time point titration (41). The advantage with the tributyrine method is that no emulsifier or other additions have to be made to obtain good linearity between enzyme added and rate of titration. This is especially important in kinetic experiments to study lipase function but it is also useful for methods used for the clinical determination of lipase in pancreatic juice and duodenal content. With long chain triglycerides as substrates the addition of bile salts to stabilize the emulsion may lead to erroneous results. In this situation the amount of colipase (see below) may well be the limiting factor (42). The use of a commercially available olive oil emulsion (stabilized with gum arabic) for the determination of lipase in human duodenal content has been validated (43).

Carboxylic Ester Hydrolase-Sterol Ester Hydrolase As mentioned earlier, pancreatic juice contains enzyme activities against several different substrates which have been given different names: nonspecific lipase, lipase A, carboxylic ester hydrolase, sterol ester hydrolase. The current indication is that all these activities are the effects of one enzyme, as has recently been reported in two publications (14, 44). The enzyme from rat has a monomer weight of 70,000 and aggregates in the presence of cholic acid to a complex which on gel filtration moves corresponding to a molecular weight of $\cong$ 400,000. The complex is believed to be the "active enzyme" (44), and is claimed to contain six subunits of enzyme. However, no experimental evidence for these assumptions is given. With other substrates, such as *p*-nitrophenylacetate, the enzyme is active in its monomeric form although its activity is greatly increased by the presence of bile salts (12). According to the results published by Erlanson (14) the specificity for tri-OH bile salt is not absolute, and the same effect can be achieved by di-OH bile salts at higher concentrations. Although the micellar form seems to be the preferred physical form for the substrate it is apparent that there is a wide specificity for the chemical form of the substrate. The enzyme hydrolyzes, for example, long chain monoglycerides as well as retinol esters in the presence of bile salts and thus is in a way complementary to the classic lipase, which hydrolyzes these esters in the absence of bile salts (45). A corresponding enzyme is also found in human pancreatic juice (25, 46), but it has not been characterized beyond gel filtration experiments which indicate a molecular weight around 300,000. Whether this represents a polymeric form or a monomer is not known.

Phospholipases Two prophospholipases A_2 have been isolated from the porcine pancreas (47). The form previously described (called form II) differs from the one now isolated (form I) in the length of the activation peptide. The two forms are most probably present in porcine pancreatic juice. The kinetic parameters of the two prophospholipases have been reported (48). Ox and sheep pancreas also contain two forms of prophospholipase A_2 (49). The ox proteins on activation release an identical heptapeptidase but the two isoenzymes have different isoelectric points. The two sheep pro-enzymes differ only in the composition of the activation peptide. The ruminant phospholipases A_2 differ from the porcine enzyme in several respects such as pH optimum and specific activity. Porcine pancreatic phospholipase A_2 as well as its zymogen bind calcium ions in a 1:1 molar ratio. The spectral differences occurring on calcium binding have been studied and interpreted to indicate a conformational change caused by this binding (50). A phospholipase A_1 activity has been isolated from beef pancreas (51). The enzyme, which is not identical with lipase, is not secreted in the pancreatic juice. It has both phospholipase A_1 and lysophospholipase activity; deoxycholate stimulates the former and inhibits the latter activity. The effect of porcine phospholipase A_2 on the substrate present as a monolayer film has been studied and a special kinetic model constructed (52).

Colipase Colipase, a polypeptide cofactor for pancreatic lipase present in rat (53, 54), human (55), and pig (57) pancreatic juice and isolated from the porcine pancreatic gland (56), has been further characterized (32, 57). A family of biologically active peptides with from 68 to 106 amino acid residues have been isolated (32, 58) but they are most probably proteolytic degradation products of the two peptides isolated from the porcine gland which differ only by one free Glu/Asp (58). The free carboxyls of glutamic and aspartic acid present in the peptide are essential for its biological activity (58). The amino acid sequence and the position of the disulfide bridges of porcine colipase has been worked out (59, 60).

Bile salts inhibit lipase (33) by preventing the binding of the enzyme to its substrate; colipase effects the binding of lipase to the substrate in the presence of bile salts probably by the formation of a lipase-colipase complex (33). This is the biological importance of colipase. Although bile salts bind to colipase as micelles (34, 61), this binding may not be of importance for the physiological function of colipase. The rate-limiting step of lipolysis at alkaline pH is indicated to be the slow formation of a ternary lipase-colipase-bile salt complex and its adsorption to the substrate interface (62). The binding between colipase and lipase at the substrate interface is strong; the K_m for colipase measured as the concentration of colipase at which half-maximal activity of lipase is observed with tributyrine as substrate is in the order of 10^{-8} M (33). Colipase also has a stimulating effect on lipase in the absence of bile salt. The increase is about 1.4-fold (18, 31) and the binding between lipase and colipase under these conditions appears to be rather weak (observations in the author's laboratory). Lipase and colipase also seem to bind to mixed bile salt/lecithin micelles (63) although the importance of such binding is not clear.

Lipases from Other Sources

Intestinal Lipase A lipase has been partially purified from homogenates of the small intestinal mucosa of various species including man that is different from pregastric and pancreatic lipase and is not of bacterial origin (64). The enzyme is mainly active against short and medium chain tryglycerides but will also hydrolyze triolein as well as ethyl acetate, 1-mono-olein but not 2-mono-olein. The physiological importance of the enzyme has not been defined.

Bile Salt-stimulated Milk Lipase This lipase is not produced in the intestinal tract but is secreted by the mammary gland of man (65) and gorillas (66). It is probably physiologically important for the digestion of milk triglycerides in the intestine (66). It is resistant to the acidity of the gastric content and is greatly stimulated by the primary bile salts but not by deoxycholate.

INTRALUMINAL PHASE OF FAT DIGESTION

Little progress has been made since the last review in defining small intestinal content from a physicochemical point of view. From gel filtration experiments with intestinal content from rats it is suggested that association of lipolysis products with the lipoprotein fraction of bile represents a major state of organization in intestinal content (67). The results do not exclude the possibility, however, that the mixed bile salt/lecithin/lipolytic product micelles are excluded to a larger extent due to bigger size than the bile salt/lipolysis product micelles. The importance of the biliary proteins for the physical state of the intestinal lumen lipids has so far been neglected and needs to be further studied.

The monomer concentrations of lipids in intestinal content is of great interest as it represents a third phase (emulsion, micellar, monomer). No methods have so far been available for its determination although an interesting approach has been made using partitioning into polyethylene disks (68).

The concentration of conjugated and free bile salts in the content from different levels of the human small intestine after ingestion of a normal diet has been defined (69). The concentration of total bile salts in the upper jejunum was reported in the range of 4–6 mM. Free bile salts were found in appreciable quantities only at the level of the upper ileum with increases distally where they are most probably normal components. The formation of free bile salts may be an important factor for the efficiency of normal bile salt reabsorption.

The detergents present in the intestinal tract (stomach) of the Crustacea has been further studied (70). The general structure of these compounds found in crayfish and lobster is that of fatty acyl-taurine or fatty acyl-dipeptides. It is of interest that this type of detergent inhibits pancreatic lipase from mammals and that the inhibition is reversed by colipase (31).

The effect of the nonabsorbable anion exchange resin cholestyramine on intestinal absorption in long term experiments in children has been reported (71).

UPTAKE OF LIPIDS INTO THE ENTEROCYTE

In principle two different types of experiments to study tissue uptake of lipids have traditionally been utilized: perfusion in vivo or uptake into an isolated tissue preparation in the form of everted sacs, or intestinal rings and sheets with the intestinal surface exposed by other means. Isolated enterocytes have not yet been used for this purpose, possibly because of difficulties with the exposure of the basolateral membrane to detergent solutions. Isolated enterocytes (72, 73) have, however, recently been used to study the effect of bile diversion on the metabolism of these cells. Perfusion studies have been performed in man to compare the rate of absorption of micellar and emulsified oleic acid (74). A word of caution regarding the interpretation of jejunal perfusion studies has been given. Segmental perfusion studies may involve a varying and much longer length of the intestine than appreciated as a result of a gathering (concertina effect) of the intestine over the tube (75).

Work has continued to appear showing the importance of the unstirred layer as a diffusion barrier for uptake of highly insoluble long chain fatty acids and the importance of bile salt to overcome this resistance (76, 77). An apparatus has been described to make possible a reproducible and systematic rate of stirring for measurements of unstirred layer thickness and tissue uptake rates (78). Uptake of different lipids by in vitro intestinal preparations has been reported for α-tocopherol (79), phylloquinone (80), and cholesterol (81). As has been found for most other lipids studied in in vitro preparation, α-tocopherol uptake appears to be a passive diffusion process. In contrast intestinal absorption of phylloquinone (vitamin K_1) has been reported to show characteristics indicating an energy-mediated process (80).

Cholesterol absorption has long been known to have an absolute requirement for bile salts. This question was reinvestigated using intestinal sheets of rat jejunal mucosa (81). Uptake of cholesterol from micellar solutions of different detergents including bile salts was found to be insignificant. On the addition of mono-olein or fatty acids, uptake took place from bile salt solutions but not to any significant extent from solutions of other detergents as dodecyl sulfate, Tween 80, or cetyltrimethyl ammonium bromide. It is concluded that specific interactions of bile salts and fatty acid or monoglyceride with the plasma membrane have to be postulated. Cholesterol and other sterols may be taken up in the monomer form and the simultaneous presence of bile salt in the enterocyte membrane may be necessary for their uptake and further transport into the cell.

In an important paper El-Gorab et al. (82) have investigated the effect of different detergents on the uptake of retinol and β-carotene by everted gut sacs. They found similarities but also striking differences in behavior between the two lipids. Binding to the mucosal surface took place for both from micellar solutions of bile salts as well as of Tween 20 and hexadecyltrimethyl ammonium bromide. Cleavage of the β-carotene to retinol, however, occurred only when bile salt solutions were used. The results do not indicate whether this is an effect of bile salts on membrane permeability or on intracellular handling. Retinol, on the other hand, was taken up and esterified from solutions of bile salts as well as the other nonbiological detergents. The uptake from bile salt micelles was more rapid, which may reflect the size of the micelles as previously discussed. The requirement of β-carotene uptake for bile salts is thus similar to that for sterols. Both lipids can be classified as water insoluble and nonpolar (2), and it may be that the requirement for bile salts is a physicochemical property of this class of lipids rather than a specific chemical effect.

INTRACELLULAR METABOLISM OF ABSORBED LIPIDS

A very rapid diffusion of long chain fatty acids in bile solution from the lumen of the small intestine of cats, in vivo, into the enterocyte has been reported (83). The fatty acids were found diffusely distributed throughout the whole epithelial cell after a 1-min incubation. The authors explain the fact that the free fatty acids are not transported rapidly to the bloodstream by a hindrance due to a "short circuiting" in the mucosal counter-current exchanger.

These results are of interest in relation to the previous demonstration of a fatty acid binding protein in the mucosa cell cytosol, which has been further characterized (84). A 12,000 molecular weight protein has been isolated from the intestinal mucosa and other tissues capable of binding long chain fatty acids. The concentration of this protein was found to respond to the fat content of the diet.

It has previously been indicated that fatty acid uptake is most probably rate limiting for esterification in the intestinal mucosa cell (85). An increase in mucosal esterification rate has, however, been shown to result in an increase in oleic acid flux. These results would indicate that the intracellular fatty acid pool may regulate uptake (86). On the other hand, during maximal rate of infusion of [1-^{14}C]triolein into the small intestine of the unanesthetized rat luminal lipolysis seemed to regulate absorption rate. The lipid concentration in the cells, predominantly triglyceride, reached a steady state in the upper part of the small intestine (87).

The question of dietary adaptation of the triglyceride synthesizing enzyme of the mucosal cell has been studied. A lack of adaptation of the acyl-CoA: monoglyceride acyltransferase in the rat and hamster to rat feeding was reported (88). During starvation this enzyme activity remained unchanged when expressed per mg of mucosal protein or weight (89). Three other enzymes engaged in triglyceride synthesis in the enterocyte (diglyceride acyltransferase, cholinephosphotransferase, and lysolecithin acyltransferase) showed a complex picture in response to lipid feeding (90). In bile fistula rats the level of acyl-CoA:monoglyceride acyltransferase was decreased and the level of acyl-CoA synthetase increased in distal ileum possibly because of an adaptation to the composition of the lipid mixture taken up (91). The localization of the acyl-CoA:monoglyceride and diglyceride acyltransferase activities of the mucosa cell to the endoplasmic reticulum has been confirmed (92). It is localized in the villus tip cells but it is not a brush border enzyme. In the essential fatty acid-deficient rat the incorporation of fatty acid into triglyceride was reduced and the microsomal esterifying enzyme activities were lowered (93). An accumulation of triglyceride in the mucosa of such animals indicated a deficient synthesis or release of chylomicron lipid.

Changes in the enzymatic activities as revealed by histochemical methods were reported in biopsies of human intestinal mucosa after infusion of fat (94). Dietary sterols and bile acids have been shown to be important in regulating small intestinal hydroxymethylglutaryl-CoA (HMG-CoA) reductase (95).

ABSORPTION OF SPECIFIC LIPIDS

Phospholipids

The finding of O'Doherty et al. (96) that the lecithin of bile is important for the production and release of chylomicrons has stimulated renewed interest in the mechanism of phospholipid absorption. Rodgers et al. (97) have shown that in the rat the lysolecithin derived from luminal lecithin is absorbed intact and utilized for lecithin resynthesis, in confirmation of earlier observations (98, 99). The newly synthesized lecithin rapidly appears in lipoprotein membranes of the enterocyte which may be needed for the normal transport of lipids out of the cell as chylomicrons (97). A part of the lysolecithin is further hydrolyzed to glycerophosphorylcholine, glycerophosphate, glycerol, and P_i (100). The glycerophosphate is partly reutilized to form triacylglycerol (99). A further attempt to determine the importance of luminal lecithin for maintaining normal lipid absorption showed slight decreases of lipid re-esterifying enzymes in vitro but no indication that biliary or dietary lecithin is essential for a normal lipid absorption (101), nor could the same author demonstrate any block in the transport of triglyceride fat out of the mucosal cell in bile fistula animals supplied with bile salts but no lecithin, in contrast to a previous report (96). Several differences in the experimental design as an explanation of the difference in results are discussed. Significant amounts of lecithin were recovered from the intestinal lumen of the bile fistula rats used by Rodgers (101).

Working with isolated intestinal mucosa cells from normal and bile-diverted rats O'Doherty et al. (72, 73) have been able to demonstrate that cells obtained from the bile-diverted rats show a marked impairment in their ability to release chylomicrons. Addition of lysophosphatidylcholine or choline to the cells in vitro resulted in a release of chylomicrons. The addition of these compounds to cells from bile fistula rats greatly stimulated incorporation into phosphatidylcholine and probably its de novo synthesis. Protein synthesis is also decreased in these cells, a fact which may be related to the deficiency of phosphatidyl precursors and membrane synthesis. Biliary diversion under the conditions used (72) thus results in an acute choline deficiency of the enterocytes with important consequences for lipoprotein synthesis. An important difference between these experiments and those of Rodgers (101) may possibly be the fact that the latter incorporated methionine in the diet. The source of the lecithin in the gut lumen in these experiments must be investigated. The total biliary output of lecithin in man has been quantitated (102) and is in the order of ≅ 11 g/24 hr.

The question concerning the existence of an enterohepatic circulation of lecithin as previously suggested (103) has been further investigated (104). Uniformly ^{14}C-labeled lecithin was administered in the intestine of pigs. Radioactivity was recovered in the bile lecithin fraction indicating that hydrolytic products of the bile lecithin were precursors of biliary lecithin synthesis. Also, the recovery of radioactivity in the bile was the same whether the thoracic duct lymph was diverted or not indicating that the lecithin, or its precursors, was transported via the portal blood. However, the recovery of labeled lecithin in the bile in these experiments was so low that it would hardly seem to justify the conclusion that there is an enterohepatic circulation of lecithin.

Sterols

The usefulness of the dual isotope plasma ratio technique to measure cholesterol absorption, as originally introduced by Zilversmit (105), has obtained further support (106) and has been extended to the primates (107). In the primate study absorption was also measured by direct determination of isotope excreted in the feces. The results indicate a poor fecal recovery of isotope in monkeys compared to that previously found in rats (106), which needs to be further investigated. The reabsorption of endogeneous cholesterol in the intestine was found to be in the range of 30–80% in both species.

Other methods for determination of cholesterol absorption include the measurement of the ratio in the fecal sterol fraction of cholesterol and sitosterol fed in labeled form. This method is başed on the assumption that cholesterol and sitosterol are degraded to the same extent by bacterial action (if any) and gives figures similar to previously used methods which include total fecal collection (108). For a discussion of the pros and cons of different methods for calculating sterol absorption the reader is also referred to the work of Quintao et al. (109).

Cholesterol introduced into the colon of guinea pigs has been shown to undergo bacterial transformation to estradiol which is absorbed and excreted in the urine (110). The magnitude of the transformation was quantitatively unimportant, being some 2% of the administered dose, but this is the first time any degradation of the sterol side chain has been proven to take place in the intestinal tract, although some indications have previously been obtained for bacterial degradation of sterols in the intestinal tract of man (111). The occurrence of such a degradation may be rare (112) but it is important to take it into consideration as it will invalidate the results of isotope kinetic calculations on cholesterol metabolism.

The specific requirements for bile salts in cholesterol uptake in vitro (81) have been discussed above. Natural bile and lecithin caused reduced uptake of cholesterol from bile salt micelles in vitro (113, 114), an effect which was also apparent with tissues cooled to 0°C indicating that these components interefered with a nonenergy-requiring step in absorption. It is suggested that lecithin affects the physical properties of the micelles, changing the partition of cholesterol between the micellar and monomer phases.

The importance of the small intestine in the regulation of cholesterol metabolism has been further emphasized (115, 116). In the intestinal lumen exogenous and endogenous cholesterol are mixed and absorbed into the mucosal cell. The cholesterol present in the mucosal cell is either synthesized de novo or arises from absorption from the lumen. Isotope kinetic data in the rat indicate that cholesterol derived from the two sources exists in different pools with different turnover rates. The cholesterol present in the mucosa cell is either structural or in transit and the experiments indicate that transport of absorbed cholesterol is more efficient than transport of synthesized cholesterol, the appearance in lymph of which may be dependent on random collision (117). The source of cholesterol transported from the intestine in the lymph have been identified in the rat (118). On a cholesterol-free diet approximately 80% of the chyle cholesterol leaving the small intestine was derived from the rapidly exchangeable body pool and the rest was synthesized de novo in the intestine. The latter accounted for approximately one-fifth of the total daily turnover of cholesterol in the rat. Of the cholesterol newly synthesized in the mucosa an appreciable fraction is transported to the intestinal lumen making up as much as four-fifths of the fecal sterols in animals on a sterol-free diet (119). Dietary cholesterol did not affect the mucosal cholesterol content or transport to the lumen which was considered to be mainly by active secretion. In bile-diverted rats the mucosal pool was expanded as was the luminal content. The results of these papers also indicate that absorbed dietary cholesterol is selectively and rapidly transported to the chyle and that cholesterol feeding in the rat greatly reduces de novo cholesterol synthesis in the proximal intestine. The important secretion of cholesterol into the gut lumen of the rat must lead to differences in results obtained for cholesterol metabolism when studied by the balance technique and by isotopic kinetic studies. The situation in the rat may not be directly applicable to man.

In the rat it has been reported that addition of taurocholate (0.5%) to the diet resulted in a significant increase of cholesterol absorption (120). Addition of the same amount of tauro-cheno-deoxycholate had no such effect. Absorption of dietary cholesterol was calculated as the difference between cholesterol intake and excretion in feces of cholesterol of dietary sources (calculated from specific activity of fecal sterols after labeling of the exchangeable pool by [^{14}C]mevalonate). No explanation for this effect is offered. The same isotope method was used for measuring cholesterol absorption in man (121). The intestinal absorption of cholesterol was found to be proportional to dietary intake with a factor of 0.4 up to even high dietary levels. These experiments were done on subjects obviously in a steady state as regards sterol metabolism and the data would also indicate that the total amounts of cholesterol absorbed from the gastrointestinal tract are directly related to the amounts available for absorption (dietary + endogeneous) (121). No differences were found in the intestinal absorption of cholesterol in normal man or hypercholesterolemic subjects (122). In these experiments no difference in the extent of absorption of cholesterol was found depending on the physical form of the cholesterol fed (egg yolk, crystalline or incorporated into a breakfast).

Feeding plant sterols, essentially sitosterol, inhibits cholesterol absorption of both dietary and endogeneous origin in the rat and prevents the feedback inhibition of hepatic cholesterolgenesis (123–125). The HMG-CoA reductase of the liver was double that in control animals. An interesting observation was that sitosterol feeding resulted in the excretion of a considerable quantity of fecal cholesterol which was most probably synthesized in the intestine (125). The intestinal absorption of sitosterol in the rat was found to be around 5% confirming previous estimates (125). The metabolism of sitosterol and cholesterol, when injected intravenously, is different but no evidence was obtained for a mutual interference in each other's metabolism (126).

Various Lipids

Evidence has been reported for an enterohepatic circulation of 25-hydroxy-vitamin D_3 in man (127).

An analysis of the importance of different factors for the absorption of vitamin E in children with malabsorption has been reported. Chylomicron formation and an adequate intraluminal concentration of bile salts were found the most important factors for a normal absorption (128).

DDT (dichlordiphenylchloroethane) was found to be transported from the intestine largely in the triglyceride phase of chylomicrons (129).

CONCLUSIONS

Progress in our knowledge of the process of fat digestion and absorption during the last 2–3 years has been more obvious in certain fields than in others. It has perhaps been most marked in our understanding of the properties of the lipolytic enzymes, their chemistry and physical chemistry as well as their interaction with bile salts, cofactors, and substrate.

Some attention has also continued to be given to the process of lipid uptake by the enterocyte and the importance of the physical state of the lipids for absorption. Our understanding of the special properties and importance of the main biological detergents of the intestinal content, the bile salts, is steadily increasing (130).

This piece by piece knowledge of the digestion and absorption process is of course very much needed and important. In the future, however, it would be desirable to have more information on the function of the complete system as it occurs in the intestine. Enzymologists generally work with pure systems at very low enzyme concentrations and initial rates of reaction. However, in the intestine enzyme concentrations are very much higher than those the enzymologists use, by the order of several micrograms per ml. Also, intestinal content is not a pure bile salt solution nor a diluted bile; it contains high concentrations of proteins from dietary and endogenous sources. Consequently, at this point in time, we do not know much about the enzymology and physical chemistry of the very complicated system as it actually occurrs in the intestinal content. Such information is needed and will be vital for an understanding of the function of the intestinal tract in health as well as in disease.

REFERENCES

1. International Review of Physiology, Gastrointestinal Physiology (1974).
2. Borgström, B. (1974). Fat digestion and absorption. *In* D. H. Smyth (ed.), Biomembranes, Vol. 4 B, pp. 555–620. Plenum Press, New York.
3. Brindly, P. N. (1974). Intracellular phase of fat absorption. *In* D. H. Smyth (ed.), Biomembranes, Vol. 4 B, pp. 621–627. Plenum Press, New York.
4. Brockerhoff, H., and Jensen, R. G. (1974). Lipolytic Enzymes, p. 330. Academic Press, New York.
5. Ockner, R. K., and Isselbacher, K. J. (1974). Recent concept of intestinal fat absorption. Rev. Physiol. Biochem. Pharmacol. 71:107.
6. Borgström, B. Dahlquist, A., Lundh, G., and Sjövall, J. (1957). Studies of intestinal digestion and absorption in the human. J. Clin. Invest. 36:1521.
7. Cohen, M., Morgan, R. G. H., and Hofmann, A. F. (1974). Lipolytic activity of human gastric and duodenal juice against medium and long chain triglycerides. Gastroenterology 60:1.
8. Hamosh, M., and Scow, R. O. (1973). Lingual lipase and its role in the digestion of dietary lipid. J. Clin. Invest. 52:88.
9. Hamosh, M., Klaeveman, H. L., Wolf, R. O., and Scow, R. O. (1975). Pharyngeal lipase and digestion of dietary triglyceride in man. J. Clin. Invest. 55:908.
10. Bradshaw, W. S., and Rutter, W. J. (1972). Multiple pancreatic lipases. Tissue distribution and pattern of accumulation during embryological development. Biochemistry 11:1517.
11. Morgan, R. G. H., Barrowman, J., Filipek-Wender, H., and Borgström, B. (1968). The lipolytic enzymes of rat pancreatic juice. Biochim. Biophys. Acta 167:355.
12. Mattson, F. H., and Volpenhein, R. A. (1966). Carboxylic ester hydrolases of rat pancreatic juice. J. Lipid Res. 7:536.
13. Albro, P. W., and Latimer, A. D. (1974). Pancreatic nonspecific lipase, an enzyme highly specific for micelles. Biochemistry 13:1431.
14. Erlanson, C. (1975). Purification, properties and substrate specificity of a carboxyl esterase in pancreatic juice. Scand. J. Gastroenterol. 10:401.
15. Verger, R., de Haas, G. H., Sarda, L., and Desnuelle, P. (1969). Purification from porcine pancreas of two molecular species with pancreatic lipase activity. Biochim. Biophys. Acta 188:272.
16. Plummer, T. H., and Sarda, L. (1973). Isolation and characterization of the glycopeptides of porcine pancreatic lipases L_A and L_B. J. Biol. Chem. 248:7865.
17. Vandermeers, A., and Christoph, J. (1968). α-Amylase et lipase du pancreas de rat. Purification chromatographique, recherche du poids molecularie et composition en acides amines. Biochim. Biophys. Acta 154:110.
18. Vandermeers, A., Vandermeers-Piret, M. C., Rathé, J., and Christophe, J. (1974). On human pancreatic triacylglycerol lipase isolation and some properties. Biochim. Biophys. Acta 370:257.
19. Straube, E., Schütt, C., Brock, J., and Mücke, D. (1974). Isolierung und charakterisierung der menschlichen pankreas lipase. Acta Biol. Med. Germanica 33:263.
20. Julien, R., Rathelot, J., Canioni, P., Sarda, L. , and Plummer, T. H. (1975). Further characterization of bovine pancreatic lipase. Biochim. Biophys. Acta 379:157.

21. Canioni, P., Benajiba, A., Julien, R., Rathelot, J., Benabdeijlil, A., and Sarda, L. (1975). Ovine pancreatic lipase: purification and some properties. Biochimie 57:35.
22. Verger, R., Sarda, L., and Desnuelle, P. (1972). On the sulfhydryl groups of porcine pancreatic lipase and their possible role in the activity of the enzyme. Biochim. Biophys. Acta 242:580.
23. Rovery, M., Bianchetta, J., and Guidoni, A. (1973). Chemical structure of porcine pancreatic lipase, end group determination and cyanogen bromide peptides. Biochim. Biophys. Acta 328:391.
24. Sarda, L., Maylie, M. F., Roger, J., and Desnuelle, P. (1964). Compartment de la lipase pancreatique sur Sephadex. Biochim. Biophys. Acta 89:183.
25. Erlanson, C., and Borgström, B. (1970). Carboxylic ester hydrolase and lipase of human pancreatic juice and intestinal content. Scand. J. Gastroenterol. 5:395.
26. Kimura, H., Kitamura, T., and Tsuji, M. (1972). Studies in human pancreatic lipase. 1. Interconversion between low and high molecular forms of human pancreatic lipase. Biochim. Biophys. Acta 270:307.
27. Entressangles, B., and Desnuelle, P. (1974). Action of pancreatic lipase on monomeric tripropionin in the presence of water-miscible organic compounds. Biochim. Biophys. Acta 341:437.
28. Entressangles, B., and Desnuelle, P. (1968). Action of pancreatic lipase on aggregated glyceride molecules in an isotropic system. Biochim. Biophys. Acta 159:285.
29. Brockman, H. L., Law, J. H., and Kézdy, F. J. (1973). Catalysis by adsorbed enzymes. The hydrolysis of tripropionin by pancreatic lipase adsorbed in siliconized glass beads. J. Biol. Chem. 248:4965.
30. Brockman, H. L., Kezdy, F. J., and Law, J. H. (1975). Isobaric titration of reacting monolayers: kinetics of hydrolysis of glyceride by pancreatic lipase B. J. Lipid Res. 16:67.
31. Borgström, B., and Erlanson, C. (1973). Pancreatic lipase and co-lipase. Interactions and effects of bile salts and other detergents. Eur. J. Biochem. 37:60.
32. Maylié, M. F., Charles, M., Astier, M., and Desnuelle, P. (1973). On porcine pancreatic colipase: large scale purification and some properties. Biochem. Biophys. Res. Commun. 52:291.
33. Borgström, B. (1975). On the interactions between pancreatic lipase and co-lipase and the substrate and the importance of bile salts. J. Lipid Res. 16:411–417.
34. Borgström, B., and Donnér, J. (1975). Binding of bile salts to pancreatic co-lipase and lipase. J. Lipid Res. 16:287.
35. Brockerhoff, H. (1971). On the function of bile salts and proteins as cofactor for lipase. J. Biol. Chem. 246:5828.
36. Yagi, K., Osamura, M., and Ohishi, N. (1973). Enzyme substrate complex of pancreatic lipase and its artificial substrate. Riboflavin di- and tributyrate. J. Biochem. 73:455.
37. Sémériva, M., Chapus, C., Bovier-Lapierre, C., and Desnuelle, P. (1974). On the transient formation of an acetyl enzyme intermediate during the hydrolysis of *p*-nitrophenyl acetate by pancreatic lipase. Biochim. Biophys. Res. Commun. 58:808.
38. Morely, N. H., Kuksis, A., and Buchnea, D. (1974). Hydrolysis of synthetic triacylglycerols by pancreatic and lipoportein lipase. Lipids 9:481.
39. Kosugi, Y., and Suzuki, H. (1974). Affinity chromatographic studies on the adsorption of lipase on aliphatic amine-Sepharose 4B. J. Ferment. Technol. 52:577.
40. Erlanson, C., and Borgström, B. (1970). Tributyrine as a substrate for determination of lipase activity of pancreatic juice and small intestinal content. Scand. J. Gastroenterol. 5:293.
41. Vandermeers, A., Vandermeers-Priet, M.-C., Rathé, J., and Christophe, J. (1974). A simple automated method for the assay of pancreatic lipase. Clin. Chim. Acta 52:271.
42. Borgström, B., and Hildebrand, H. (1975). Lipase and co-lipase activities of human small intestinal contents after a liquid test meal. Scand. J. Gastroenterol. 10:585–591.
43. Honegger, J., and Hadorn, B. (1973). The determination of lipase activity in human duodenal juice. Biol. Gastroenterol. 6:217.

44. Clame, K. B., Gallo, L., Cheriathundam, E., Vahouny, G. U., and Treadwell, C. R. (1975). Purification and properties of subunits of sterol ester hydrolase from rat pancreas. Arch. Biochem. Biophys. 168:57.
45. Erlanson, C., and Borgström, B. (1968). The identity of vitamin A esterase activity of rat pancreatic juice. Biochim. Biophys. Acta 151:629.
46. Figarella, C. (1973). Les proteines du suc pancreatique humain. Arch. Fr. Mal. App. Dig. 62:337.
47. Nieuwenhuizen, W., Steenbergh, S. P., and de Haas, G. H. (1973). The isolation and properties of two prephospholipases A_2 from porcine pancreas. Eur. J. Biochem. 40:1.
48. Nieuwenhuizen, W., Oomens, A., and de Haas, G. H. (1973). The fluorimetric determination of the kinetic parameters K_m and K_{cat} of the tryptic activation of prephospholipase A_2 and derivatives. Eur. J. Biochem. 40:9.
49. Dutilh, C. L., van Doren, P. J., Verheul, F. E. A. M., and de Haas, G. H. (1975). Isolation and properties of prophospholipase A_2 from ox and sheep pancreas. Eur. J. Biochem. 53:91.
50. Pieterson, W. A., Volwerk, J. J., and de Haas, G. H. (1974). Interaction of phospholipase A_2 and its zymogen with divalent metal ions. Biochemistry 13:1439.
51. Van den Bosch, H., Aarsmar, A. J., and van Deenen, L. L. M. (1974). Isolation and properties of a phospholipase A_1 activity from beef pancreas. Biochim. Biophys. Acta 348:197.
52. Verger, R., Mieras, M. C. E., and de Haas, G. H. (1973). Action of phospholipase A at interfaces. J. Biol. Chem. 248:4023.
53. Morgan, R. G. H., Barrowman, J., and Borgström, B. (1969). The effect of sodium taurodeoxycholate and pH on the gel filtration behaviour of rat pancreatic protein and lipases. Biochim. Biophys. Acta 175:65.
54. Borgström, B., and Erlanson, C. (1971). Pancreatic juice co-lipase. Physiological importance. Biochim. Biophys. Acta 242:509.
55. Morgan, R. G. H., and Hoffman, N. E. (1971). The interaction of lipase, lipase co-factor and bile salts in triglyceride hydrolysis. Biochim. Biophys. Acta 248:143.
56. Maylie, M. F., Gache, C., and Desnuelle, P. (1971). Isolation and partial identification of a pancreatic co-lipase. Biochim. Biophys. Acta 229:286.
57. Erlanson, C., Fernlund, P., and Borgström B. (1973). Purification and characterization of two proteins with co-lipase activity from porcine pancreas. Biochim. Biophys. Acta 310:437.
58. Borgström, B., Erlanson, C., and Sternby, B. (1974). Further characterization of two colipases from porcine pancreas. Biochim. Biophys. Res. Commun. 59:902.
59. Charles, M., Erlanson, C., Biancetta, J., Joffre, J., Guiboni, A., and Rovery, M. (1974). The primary structure of porcine co-lipase. I. The amino acid sequence. Biochim. Biophys. Acta 359:186.
60. Erlanson, C., Charles, M., Astier, M., and Desnuelle, P. (1974). The primary structure of porcine co-lipase. II. The disulfide bridges. Biochim. Biophys. Acta 359:198.
61. Charles, M., Sari, H., Entressangles, B., and Desnuelle, P. (1975). Interaction of pancreatic co-lipase with a bile salt micelle. Biochim. Biophys. Res. Commun. 65:740.
62. Vandermeers, A., Vandermeers-Piret, M. C., Rathé, J., and Christophe, J. (1975). Effect of co-lipase on adsorption and activity of rat pancreatic lipase on emulsified tributyrin in the presence of bile salt. FEBS Lett. 49:334.
63. Lairon, D., Nalbone, G., Domingo, N., Lafont, H., Hauton, J., Julién, R., Rathelot, J., Lanioni, P., and Sarda, L. (1975). In vitro studies on interactions of rat pancreatic lipase and co-lipase with biliary lipids. Lipids 10:262.
64. Serrero, G., Négrel, R., and Ailhaud, G. (1975). Characterization and partial purification of an intestinal lipase. Biochem. Biophys. Res. Commun. 65:89.
65. Hernell, D., and Olivecrona, T. (1974). Human milk lipases. II. Bile salt stimulated lipase. Biochim. Biophys. Acta 396:234.
66. Hernell, O. (1975). Human milk lipases. III. Physiological implications of the bile salt stimulated lipase. Eur. J. Clin. Invest. 5:267.
67. Lairon, D., Lafont, H., Domingo, N., and Hauton, J. (1973). On the state of lipids in the rat small intestine after a fatty meal. Biochemie 55:1165.

68. Sallee, V. L. (1974). Apparent monomer activity of saturated fatty acids in micellar bile salt solutions measured by a polyethylene partiioning system. J. Lipid Res. 15:56.
69. Northfield, T. C., and McColl, I. (1973). Postprandial concentration of free and conjugated bile acids down the length of the normal human small intestine. Gut 14:513.
70. Holwerda, D. A., and Vonk, H. J. (1973). Emulsifiers in the intestinal juice of crustacea. Isolation and nature of surface-active substances from *Astacus leptodactylus* and *Homarus vulgaris.* Comp. Biochem. Physiol. 45 B:51.
71. West, R. J., and Lloyd, J. K. (1975). The effect of cholestyramine on intestinal absorption. Gut 16:93.
72. O'Doherty, P. J. A., Yousef, I. M., Kakis, G., and Kuksis, A. (1975). Protein and glycerolipid biosynthesis in isolated intestinal epithelial cells of normal and bile fistula rats. Arch. Biochem. Biophys. 169:252.
73. O'Doherty, P. J. A., Yousef, I., and Kuksis, A. (1973). Effect of puromycin on protein and glycerolipid biosynthesis in isolated mucosa cells. Arch. Biochem. Biophys. 156:586.
74. Hoffman, N. E., and Hofmann, A. F. (1973). A comparison of the rate of absorption of micellar and nonmicellar oleic acid. A jejunal perfusion study in man. Dig. Dis. 18:489.
75. Cook, G. C., and Carruthers, R. H. (1974). Reaction of human small intestine to an intraluminal tube and its importance in jejunal perfusion studies. Gut 15:545.
76. Sallee, V. L., and Dietschy, J. M. (1973). Determinants of intestinal mucosal uptake of short and medium chain fatty acids and alcohols. J. Lipid Res. 14:475.
77. Wilson, F. A., and Dietschy, J. M. (1974). The intestinal unstirred layer: its surface area and effect on active transport kinetics. Biochim. Biophys. Acta 363:112.
78. Lukie, B. F., Westergaard, H., and Dietschy, J. M. (1974). Validation of a chamber that allows measurements of both tissue uptake rates and unstirred layer thickness in the intestine under controlled stirring conditions. Gastroenterology 67:652.
79. Hollander, D., Rim, E., and Muralidhara, K. S. (1975). Mechanism and site of small intestinal absorption of α-tocopherol in the rat. Gastroenterology 68:1492.
80. Hollander, D. (1973). Vitamin K_1 absorption by everted sacs of the rat. Am. J. Physiol. 225:360.
81. Feldman, E. B., and Cheng, C. Y. (1975). Mucosal uptake in vitro of cholesterol from bile salt and surfactant solutions. Am. J. Clin. Nutr. 28:692.
82. El-Gorab, M. I., Underwood, B. A., and Loerch, J. D. (1975). The roles of bile salts in the uptake of β-carotene and retinol by rat everted gut sacs. Biochim. Biophys. Acta 401:265.
83. Haglund, V., Jodal, M., and Lundgren, O. (1973). An autoradiographic study of the intestinal absorption of palmitic and oleic acid. Acta Physiol. Scand. 89:306.
84. Ockner, R. T., and Manning, J. A. (1974). Fatty acid-binding protein in small intestine. Clin. Invest. 54:326.
85. Bennet Clark, S. (1971). The uptake of oleic acid by rat small intestine: a comparison of methodologies. J. Lipid Res. 12:43.
86. Marubbio, A. T., Morris, J. A., Bennet Clark, S., and Holt, P. R. (1974). Monoglyceride modification of jejunal absorption of fatty acid in the rat. J. Lipid Res. 15:165.
87. Clark, S. B., Lawergren, B., and Martin, J. V. (1973). Regional intestinal absorptive capacities for triolein: an alternative to markers. Am. J. Physiol. 225:574.
88. Rao, G. A., and Abraham, S. (1974). Lack of adaptation of intestinal 2-monoglyceride acyl transferase. Lipids 9:940.
89. Powell, G. K., and McElveen, M. A. (1974). Effect of prolonged fasting on fatty acid re-esterification in rat intestinal mucosa. Biochim. Biophys. Acta 369:8.
90. Mansbach, C. M. (1975). Effect of fat feeding on complex lipid synthesis in hamster intestine. Gastroenterology 68:708.
91. Brand, S. J., and Morgan, R. G. H. (1974). Fatty acid uptake and esterification by proximal and distal intestine in bile fistula rats. Biochim. Biophys. Acta 309:1.
92. Négrel, R., and Ailhaud, G. (1975). Localization of the monoglyceride pathway

enzymes in the villus tip of intestinal cells and their absence from the brush border. FEBS Lett. 54:183.
93. Bennet Clark, S., Ekkers, T. E., Singh, A., Balint, J. A., Holt, P. H., and Rodgers, J. B. (1973). Fat absorption in essential fatty acid deficiency: a model experimental approach to studies of the mechanism of fat malabsorption of unknown etiology. J. Lipid Res. 14:581.
94. Monges, H., Chamlian, A., Cougard, A., and Mathieu, B. (1974). Enzyme histochemical study of fat absorption in human duodenal mucosa. Gut 15:777.
95. Shefer, S., Hauser, S., Lapar, V., and Mosbach, E. H. (1973). Regulatory effects of dietary sterols and bile acids on rat intestine HMG CoA-reductase. J. Lipid Res. 14:400.
96. O'Doherty, O., Kakis, P. J. A., and Kuksis, G. (1973). Role of luminal lecithin in intestinal fat absorption. Lipids 8:249.
97. Rodgers, J. D., O'Brien, R. J., and Balint, J. A. (1975). The absorption and subsequent utilization of lecithin by the rat jejunum. Am. J. Dig. Dis. 20:208.
98. Scow, R. O., Stein, Y., and Stein, O. (1967). Incorporation of dietary lecithin and lysolecithin into lymph chylomicrons in the rat. J. Biol. Chem. 242:4919.
99. Nilsson, Å. (1968). Intestinal absorption of lecithin and lysolecithin by lymph fistula rats. Biochim. Biophys. Acta 152:379.
100. Parthasarathy, S., Subbaian, P. V., and Ganguly, J. (1974). The mechanism of intestinal absorption of phosphatidylcholine in rats. Biochem. J. 140:503.
101. Rodgers, J. B. (1975). Lipid absorption in bile fistula rats. Lack of a requirement of biliary lecithin. Biochim. Biophys. Acta 398:92.
102. Northfield, T. C., and Hofmann, A. F. (1975). Biliary lipid output during three meals and an overnight fast. Gut 16:1.
103. Boucrot, P. (1972). Is there an entero-hepatic circulation of bile phospholipids? Lipids 7:282.
104. Tompkins, R. K., and King, W. (1974). Investigation of the enterobiliary metabolism of lecithin. Surgery 75:243.
105. Zilversmit, D. B. (1972). A single blood sample dual isotope method for the measurement of cholesterol absorption in the rat. Proc. Soc. Exp. Biol. Med. 140:862.
106. Zilversmit, D. B., and Hughes, L. B. (1974). Validation of a dual-isotope plasma ratio method for measurement of cholesterol absorption in rats. J. Lipid Res. 15:465.
107. Corey, J. E., and Hayes, K. C. (1975). Validation of the dual-isotope plasma ratio technique as a measure of cholesterol absorption in old and new world monkeys. Proc. Soc. Exp. Biol. Med. 148:842.
108. Sodhi, H. S., Kudchodkar, B. J., Varughese, P., and Duncan, D. (1974). Validation of the ratio method for calculating absorption of dietary cholesterol in man. Proc. Soc. Exp. Biol. Med. 145:107.
109. Quintao, E., Grundy, S. M., and Ahrens, E. H. (1971). An evaluation of four methods for measuring cholesterol absorption by the intestine in man. J. Lipid Res. 12:221.
110. Goddard, P., and Hill, M. J. (1974). The *in vivo* metabolism of cholesterol by gut bacteria in the rat and guinea pig. J. Steroid Biochem. 5:569.
111. Grundy, S. M., Ahrens, E. H., and Salen, G. (1968). Dietary β-sitosterol as an internal standard to correct for cholesterol losses in sterol balance studies. J. Lipid Res. 9:374.
112. Kudchodkar, B. J., Sodhi, H. S., and Horlich, I. (1972). Lack of degradation of dietary and endogeneous sterols in gastrointestinal tract of man. Metabolism 21:343.
113. Rampone, A. J. (1973). The effect of lecithin on intestinal cholesterol uptake by rat intestine in vitro. J. Physiol. (Lond.) 229:505.
114. Rampone, A. J. (1973). Studies on micellar fatty acid uptake by rat intestine *in vitro* with reference to the role of bile. J. Physiol. (Lond.) 229:495.
115. Chevallier, F., and Lutton, C. (1973). The intestine is the major site of cholesterol synthesis in the rat. Nature (New Biol.) 242:61.
116. McIntyre, N., and Isselbacher, K. J. (1973). Role of the small intestine in cholesterol metabolism. Am. J. Clin. Nutr. 26:647.
117. Sodhi, H. S., Orchard, R. C., Agnish, N. D., Varughese, P. V., and Kudshodkar, B. J. (1973). Separate pools for endogeneous and exogeneous cholesterol in rat intestines. Atherosclerosis 17:197.

118. Ho, K. J., and Taylor, C. B. (1974). Sources of cholesterol in the intestinal lymph in rats fed a cholesterol-free diet. Proc. Soc. Exp. Biol. Med. 147:826.
119. Peng, S. K., Ho, K. J., and Taylor, C. B. (1974). The role of the intestinal mucosa in cholesterol metabolism: its relation to plasma and luminal cholesterol. Exp. Mol. Pathol. 21:138.
120. Raicht, R. F., Cohen, B. I., and Mosbach, E. H. (1974). Effects of sodium taurochenodeoxycholate and sodium taurocholate on cholesterol absorption in the rat. Gastroenterology 67:1155.
121. Kudchodkar, B. J., Sodhi, H. S., and Horlich, L. (1973). Absorption of a dietary cholesterol in man. Metabolism 22:155.
122. Connor, W. E., and Lind, D. S. (1974). The intestinal absorption of dietary cholesterol by hypercholesterolemic and normocholesterolemic humans. J. Clin. Invest. 53:1062.
123. Fishler-Mates, Z., Budowski, P., and Pinsky, A. (1974). Effect of soy sterols on cholesterol metabolism in the rat. Int. J. Vitam. Nutr. Res. 44:497.
124. Fishler-Mates, Z., Budowski, P., and Pinsky, A. (1973). Effect of soy sterols on cholesterol synthesis in the rat. Lipids 8:41.
125. Raicht, R. F., Cohen, B. I., Shefer, S., and Mosbach, E. H. (1975). Sterol balance studies in the rat. Effects of dietary cholesterol and β-sitosterol on balance and rate limiting enzymes of sterol metabolism. Biochim. Biophys. Acta 388:374.
126. Subbiah, M. T. R., and Kuksis, A. (1973). Differences in metabolism of cholesterol and sitosterol following intravenous injection in rats. Biochim. Biophys. Acta 306:95.
127. Arnaud, S. B., Goldsmith, R. S., Lambert, P. W., and Go, V. L. W. (1975). 25-Hydroxyvitamin D_3: evidence of an entero-hepatic circulation in man. Proc. Soc. Exp. Biol. Med. 149:570.
128. Muller, D. P. R., Harris, J. T., and Lloyd, J. K. (1974). The relative importance of the factors involved in the absorption of vitamin E in children. Gut 15:966.
129. Pocock, D. M. E., and Vost, A. (1974). DDT absorption and chylomicron transport in the rat. Lipids 9:374.
130. Borgström, B. (1974). Bile salts–their physiological functions in the gastrointestinal tract. Acta Med. Scand. 196:1.

International Review of Physiology
Gastrointestinal Physiology II, Volume 12
Edited by Robert K. Crane
Copyright 1977 University Park Press Baltimore

11
Digestion and Absorption: Water-soluble Organics

R. K. CRANE

Department of Physiology,
College of Medicine and Dentistry of New Jersey,
Rutgers Medical School,
Piscataway, New Jersey

DIGESTIVE-ABSORPTIVE SURFACE 326

ROLE OF ADSORBED PANCREATIC ENZYMES 326

COMPONENTS OF THE BRUSH BORDER 328

MECHANISMS OF ABSORPTION 334
Brush Border Membrane 334
Mechanism of Energizing Absorption 334
External Fluid Circuit 337
Basolateral Membrane 339

ORGANIZATION OF BRUSH BORDER MEMBRANE FUNCTIONS 340
Interaction between Amino Acid and Sugar Transport 340
Kinetic Advantage 340
Hydrolase-related Transport 341

DEVELOPMENT AND CONTROL OF BRUSH BORDER MEMBRANE FUNCTIONS 342

DELETIONS OF BRUSH BORDER MEMBRANE FUNCTIONS 346

This review is focused primarily on cell membranes and membrane-associated events as they may provide for and control parts of the digestive-absorptive process. No specific cut-off date was chosen for citations to the literature especially as interesting and relevant papers continued to appear during the final stages of manuscript preparation in late 1975 and early 1976.

Absorption begins at the membrane on the luminal surface of the intestinal epithelial cells. This luminal surface is a "brush border" made up of closely packed, parallel cylindrical processes called microvilli. The limiting plasma membrane of the cell follows the contours of the microvilli. Just beneath the brush border along the sides of the cells are specialized junctional structures by means of which the absorptive cells are held together into a more or less continuous sheet. The membrane encasing the cell inward of the junctional complex is called the basolateral membrane.

DIGESTIVE-ABSORPTIVE SURFACE

In 1960, McDougal, Little, and Crane (268) identified the brush border as the location of the active transport processes responsible for the efficient absorption of water-soluble compounds. Miller and Crane (275) isolated brush borders from hamster small intestine as intact, subcellular organelles and showed that they possessed hydrolytic enzymes responsible for terminal phases of carbohydrate digestion. Crane et al. (89) proposed that the transport carrier for sugar also co-transported Na^+ thereby providing a means for coupling sugar accumulation to the trans-membrane electrochemical potential gradient. From these observations developed a concept of the brush border as a "digestive-absorptive surface" as depicted in Figure 1. Also in 1960, the claim appeared (407, 408) that pancreatic amylase contributed its major digestive role by being adsorbed to the mucosal surface of the gut and by being most active in that location.

At the current writing, the concepts as represented in Figure 1 are substantially confirmed. Kinter (223) and Stirling and Kinter (395) confirmed the findings of McDougal et al. Stirling (393) confirmed the findings of Alvarado and Crane (10) and others earlier (73) that phlorizin binds at the brush border presumably, according to kinetic evidence, to the active site of sugar transport carriers. Most recently, Stirling et al. (396) showed that the brush borders of a patient with glucose-galactose malabsorption have a greatly reduced capacity to bind phlorizin.

At the outset it was known (274) that the hydrolases of the brush border were "at a locus external to a diffusion barrier sensitive to phlorizin and external to the active transport process for sugars" (274). This locus was confirmed when a way was found to prepare a pure microvillus membrane fraction from disrupted brush borders (120, 309) and it was shown that the brush border enzymes were present at increased specific activity.

ROLE OF ADSORBED PANCREATIC ENZYMES

As mentioned above, Ugolev (407, 408) proposed that pancreatic amylase is active in digestion principally while adsorbed to the muscosal surface of the

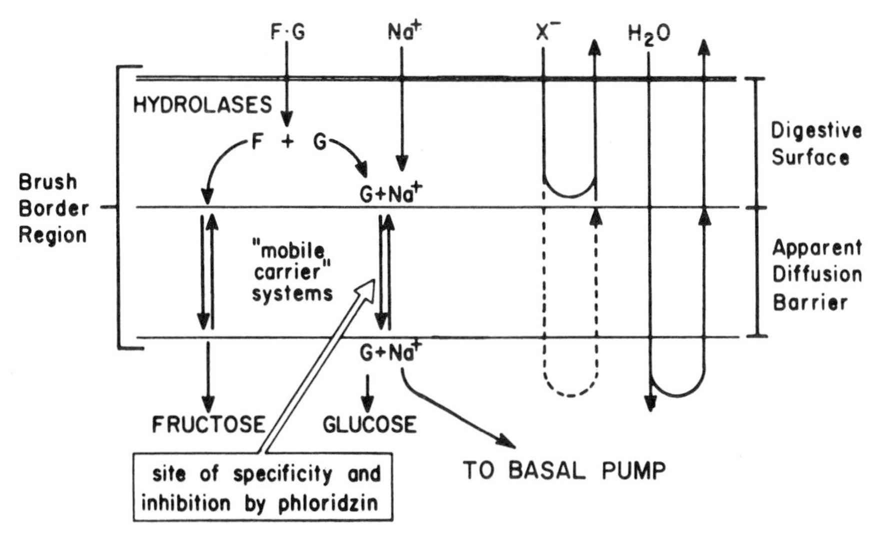

Figure 1. Representation of the digestive-absorptive contributions of intestinal epithelial cell brush border membrane (modified from Crane, Miller, and Bihler (89)).

intestine, and developed this proposal into a concept of membrane digestion (409). In recent years, a number of difficulties have developed. First, the assumed location for adsorbed pancreatic enzymes was described in such terms (102, 410) that the presence of substantial hydrophobic binding regions in the pancreatic enzyme structures would have to be assumed; they are not in evidence. Lately, however, Ugolev (411, 412) has adopted Crane's view that any adsorption of these enzymes is to the fuzzy coat (203) and he has, at the same time, given prominence to the digestive role of the intrinsic brush border enzymes described by Miller and Crane (275). The difference in location is an important point owing to the different kinds of organization between enzymic and transport function which the different locations would allow. The brush border enzymes in the outer layer of the membrane may very well be and there is some evidence that they are actually closely organized with transport systems (78). Pancreatic enzymes adsorbed to the fuzzy coat are too far removed from the membrane surface for this to occur. Second, the question has been seriously raised whether adsorbed pancreatic enzymes are an essential or even an important part of the digestive-absorptive process. The level of unadsorbed α-amylase in the lumen of the gut of man after a meal of starch is high enough (96) to provide some 10 times the capacity needed to account for the observed rapid rates of hydrolysis (167). Also, so far as can be found in the published literature, confirmation of Ugolev's experiments has generally not been forthcoming (5, 95) and, more specifically, a direct attempt (199) to find enhancement of amylase activity by jejunal mucosa as claimed by Jesuitova et al. (206) has failed to show "any effect which could be a significant factor in the digestion of a meal containing starch." Some amounts of pancreatic enzymes, protease (162) as well as amylase (5), bind to the mucosal surface of the gut, but there is as yet no hard evidence to support the idea that this binding is an essential feature of the process of digestion. In fact, the recent finding (376) that goblet cell mucus stimulates proteolytic digestion provides a basis for a completely alternative explanation of any stimulation of pancreatic enzyme activity that Ugolev and his colleagues may have observed. It is in this light that a recent paper (263) on the hydrolysis of starch in various diseases should be read.

Current understanding places the pancreatic enzymes and the brush border enzymes in a sequential relationship for the digestive process. In the adult, pancreatic amylase splits amylose only as far as maltotriose and maltose (271) and amylopectin to maltotriose, maltose, and α-dextrins (418). Brush border saccharidases then take over to cleave free glucose from these products. Brush border enzymes also contribute directly to the digestive capacity of the small intestine for dietary disaccharides. Pancreatic proteases release only a relatively small amount of free amino acid from proteins (213, 292, 293). Oligopeptides and peptides released are split by brush border enzymes but some dipeptides appear to enter the cell rapidly by specific membrane transport systems (266) and to be split intracellularly.

Hydrolysis and absorption of lipid is not discussed because there is as yet no clearly demonstrated contribution of the enzymes or carriers of the brush border membrane to the processes involved in fat digestion and absorption. Long chain free fatty acids and monoglycerides are quite lipid soluble and appear to be absorbed by passive diffusion through the lipoidal matrix of the membrane although short chain fatty acids may use a facilitated diffusion carrier system identified with the transfer of pyruvate (233). The highly water-soluble product, glycerol, being small enough, diffuses for the most part through aqueous channels although there may be a nonenergized carrier system of limited capacity (344).

COMPONENTS OF THE BRUSH BORDER

Brush borders were first isolated in 1960 as intact, subcellular organelles from hamster small intestine (273, 275). A purified plasma membrane fraction was reported in 1965 (120, 309). Revised procedures as they have been adapted to the rat (142), mouse (38), guinea pig (11), and human (198, 355, 425) have recently been collected (121). Very similar procedures provide a means for isolating brush borders from the kidneys of several species (31, 402). The use of exfoliated cells as a starting material to avoid peroral intestinal biopsy in human infants has been described (404).

From the outside inward the morphologically identifiable components of the brush border (for details see Ref. 406) are: 1) a filamentous surface coat attached to 2) the plasma membrane covering the microvilli, each of which contains 3) a rigid bundle of filaments which are rooted in 4) an otherwise rather structureless subjacent zone called the terminal web which "endows the apex with mechanical stability and anchors it to the body of the cell" (48).

The fuzzy coat is not a static component of the surface. The glycoprotein of which it is composed appears to be synthesized in the Golgi region and transported to the cell surface where it is incorporated (27) and contributes to the surface charge of the membrane (421). This glycoprotein is "a relatively complex mixture of glycoproteins with different functions" (140). The fuzzy coat probably accounts for the adsorption of environmental proteins (25) including pancreatic enzymes as discussed above. Ito has recently reviewed its form and function (204).

Johnson (207) proposed that the enzymes of the brush border, which are readily removed by papain (119), are a part of the glycocalyx and appear as particles on the surface of the membrane in electron microscopy with negative staining techniques in hamster (309), rabbit (291), and man (356). However, if the major portions of the enzymes (for example, sucrase) are so superficial in location, the antigenic portion of the molecule is nonetheless very close to the nominal surface of the membrane (161), its active center is in close proximity to the glucose transport system (78), and the enzyme is not a part of the especially extensive fuzzy coat in such animals as the cat, bat, or man (161). Benson, Sacktor, and Greenawalt (28) have reported that the disaccharidases cannot be correlated quantitatively with the number of "knobs" on the brush border membrane. However, there are (in addition to disaccharidases) a sizeable number of brush border enzymes which Benson et al. did not assay and the question is still open.

The core clearly serves a structural function and may serve others as well. Pure preparations of core material are not yet available. Hence, assumptions (129) that the filamentous fraction of Tris-disrupted brush borders represents core material must be treated with due caution. Faust et al. (129) report the binding of D-glucose to the filamentous fraction and assume this binding to represent a step in the pathway of energized glucose absorption. This assumption does not take into account the concerns which others have expressed about the specificity and other characteristics of similar binding (122) and about the possibility of bacterial contamination (158).

Tilney and Mooseker (403) have reported the presence of actin in the brush borders of chicken intestine and identified it with the filaments of the core. In further work, the brush border has been characterized as a Ca^{2+}-regulated contractile apparatus (278) in which Mooseker and Tilney (279) have delineated the organization of an actin filament-membrane complex. The normal development of the microvilli has been described for several species (308, 310, 414). It has been observed recently that cytochalasin B alters the morphology of developing microvilli in embryonic chick intestine in organ culture (49).

The microvillus membrane is about 50% protein (118) and much of this is contributed by the enzymes listed in Table 1.

The exact location of the brush border enzymes, attached to or buried within the outer surface of the microvillus membrane, has not been completely decided and it may be different for different enzymes. The list of enzymes in Table 1 is arbitrary in that entering an activity in the list involves a decision that

Table 1. Brush border enzyme activites (1975)[a]

Enzyme
Dipeptidase (428)
Oligopeptidase (191, 316, 335)
γ-Glutamyltransferase (64)
Enterokinase (189, 295, 356)
Maltase (243, 275)
Oligosaccharidase (glucoamylase) (5, 258)
Lactase (75, 112)
Phlorizin hydrolase (glycosylceramidase) (107, 235, 257)
Sucrase (243, 275)
Isomaltase (α-dextrinase) (167, 208)
Trehalase (208, 256)
Guanylate cyclase (101)
Alkaline phosphatase (99, 191)

[a]The references listed are not necessarily the earliest indications of the brush border localization.

the evidence for location is sufficient. The rule of thumb used for this purpose is that the enzyme be present in purified microvillus membranes at a specific activity increase over the tissue homogenate about the same as that for sucrase.

Some enzymes thought to be brush border membrane enzymes from studies on isolated brush borders (e.g., cholesterol esterase (255)) were probably merely adsorbed inasmuch as they were removed during membrane repurification. Others, for example the monoglyceride pathway enzymes (141), have been shown to belong to the endoplasmic reticulum (289). The (Na^+K^+)-ATPase appears to be entirely contributed by bits of basolateral membrane adhering to the junctional complexes of isolated intact brush borders.

(Na^+K^+)-ATPase was early postulated to be present in the brush border (89). However, this location did not fit with the later findings of Csaky and Hara (91) with respect to ouabain inhibition at the serosal surface of epithelial cells. Also it was not in accord with the later ideas of Schultz and Zalusky (366) and Schultz and Curran (361) that the driving force for salt and water absorption is the same "pump" ATPase. Early findings by Taylor (399), Berg and Chapman (29), and Rosenberg and Rosenberg (340) which appeared to confirm the presence of an ATP-utilizing cation pump in the brush border seem now to have lost out to more recent cell fractionation techniques which indicate that most, if not all, pump ATPase is associated with the basolateral membrane (152, 326).

Most of the enzymes listed in Table 1 have been isolated in purified form and show the common characteristic of having a high carbohydrate content. Sucrase and isomaltase occur together in a complex (370) as do lactase and phlorizin hydrolase (41, 65, 330). The sucrase-isomaltase complex contains 15% carbohydrate (62) and the lactase-phlorizin hydrolase complex contains 17% (41).

Alkaline phosphatase (144) contains a similar low percentage of carbohydrate (12%), but glucoamylase (214), enterokinase (260), and γ-glutamyltransferase (from kidney membrane (306, 307)), all contain over 30% carbohydrate. At least some of the carbohydrate of glucoamylase and sucrase shows blood group antigenicity of great potency (215).

A precursor protein to fully active sucrase has been proposed on the basis of immunofluorescence in human jejunal crypt cells (382) and the presence of a catalytically inactive antigen in rabbit (114). Sucrase is activated by alkali metal ions in a qualitatively similar manner to the nonelectrolyte transport systems of the brush border (370) but the kinetic constants of the two systems, at least in the guinea pig, are markedly different (254). Tris, which is a well known competitive inhibitor of isolated sucrase (370), appears also to be active in vivo (323). The sucrase-isomaltase complex has been separated into its subunit enzymes (18, 45, 67, 325). Most studies of these enzymes have been carried out on papain-released enzyme in which the hydrophobic "anchor" is presumed to be left behind in the membrane (261). However, the isolation of a hydrophobic form of the sucrase-isomaltase complex released by means of detergent rather than peptidase has now been reported (378). Recent studies on the mechanism of the sucrase- and isomaltase-catalyzed reactions (63, 324) are consistent with the intermediate formation of an oxocarbonium ion following fission between the C-1 of the glucose ring and the glucosidic oxygen (389, 432).

The presence of phlorizin hydrolase activity in the outer membrane surface prompted Diedrich (107) to propose, in spite of the clearly competitive kinetics of phlorizin inhibition of sugar transport (10), that the true inhibitory species was the aglycone, phloretin, released by enzyme action. The idea was prompted by the fact that phloretin is the inhibitory species for red cell sugar transport (236) and was briefly supported by Colombo and Semenza (66). Newer measurements of phlorizin and phloretin inhibition of galactose influx in rabbit intestine (127) and a study of lactase-deficient (and phlorizin-hydrolase-deficient) humans (43) fail to support Diedrich.

Trehalase has not been studied in such detail because it is more resistant to removal from the membrane by papain (94, 119). However, movement may now be expected as a result of the description of methods for the purification of this enzyme from kidney (230) and from intestine (351).

Brush border membrane peptidases are currently receiving some attention which appears to further clarify their role in the digestive-absorptive process (247). There are several brush border enzymes showing peptidase activity (119, 153) and at least one of these is more active against tri- and higher peptides than against dipeptides (316, 335, 428). Also the brush border and cytoplasmic peptidase activities are generally considered to be quite different moieties (179, 217) although a close similarity of a brush border peptidase with one intracellular hydrolase has been reported (301). The activity of brush border oligopeptidase is quite high enough to support the concept of a digestive function. However, most dipeptidase activity is intracellular (302, 318). A soluble dipeptidase of broad specificity (298) and a soluble prolidase (385) have both been purified from pig intestine. The mode of binding of brush border aminopeptidase is currently under active study (261).

The story on enterokinase is somewhat complicated. Enterokinase is isolated and purified with brush borders and microvillus membranes and is released by papain action similarly to the other enzymes (246, 294, 296, 356) but its activity does not decrease in celiac disease in parallel with the other enzymes as the membrane surface is reduced (429). This anomalous behavior may be explained by the finding (322, 356) that enterokinase in man occurs also in another location and in this other location is activated during assay. Enterokinase also shows a different cellular distribution than sucrase along the length of the villus (359). The specificity site of the porcine enzyme can be selectively inactivated by heat leaving the catalytic site intact (24).

The digestive function of most of the brush border enzymes is fairly obvious but for three of them the situation is less clear. The enzyme, γ-glutamyltransferase, has been made the basis of an hypothesis of amino acid active transport carried on by a cycle of enzymic activities (269, 270), although this hypothesis would now appear to be moot inasmuch as no deficit of amino acid transport has been found in an individual with nearly complete γ-glutamyltransferase deficiency and exhibiting glutathionemia and glutathionuria (360). Binkley (42) suggests that γ-glutamyltransferase is a secretory enzyme. Rosenberg (339) suggests that it may be involved in folate release prior to absorption. The enzyme, alkaline phosphatase, has been related to Ca^{2+}-ATPase activity (345) which is vitamin D sensitive and may be the basis for energizing the absorption of Ca^{2+} (262) and Zn^{2+} (228). In this regard, though, it may be noted that in the proximal tubules of rat kidney Ca^{2+}-ATPase, like (Na^+-K^+)-ATPase in both tissues, is a basolateral membrane enzyme (222). Alkaline phosphatase has been purified from calf intestine by a method employing affinity chromatography (47). The function of brush border guanylate cyclase is not known but there is a villus-crypt gradient of its activity (327) which suggests a specialized role in the differentiated villus cell.

In addition to the above mentioned enzyme proteins, the external surface of the plasma membrane contains proteins specific for binding Ca^{2+} (46, 423), Zn^{2+} (227), folate (237), and IF$\cdot B_{12}$ complex (264, 405). There are also binding sites for γ-globulins in the newborn and possibly in the adult (420). Working with human microvillus membranes Maestracci et al. (252) have visualized at least 22 membrane-associated proteins and polypeptides by electrophoresis on polyacrylamide gels following solubilization of the membranes with sodium dodecyl sulfate. All of the carbohydrate splitting enzymes plus γ-glutamyltransferase and alkaline phosphatase have been identified with specific bands (251) but peptidases and proteins of molecular weight less than 80,000 are not identified. Similar experience is reported in laboratory animals (3). Based on the means needed to remove them from membrane association, all of the enzyme proteins and the binding protein for folate would appear to be integral membrane proteins in the usage of Singer (383, 384) whereas the binding proteins for Ca^{2+} (46), Zn^{2+} (227), and IF·B_{12} (211, 358) are probably peripheral. It may be anticipated that additional integral proteins will be identified with the specific transport activities of the brush border membrane.

Weiser (424) has made extensive studies of the glycosyltransferase acceptor activities of rat intestinal cell surface membranes. However, these studies do not support the proposal (202) that the glycosyltransferases responsible for the elaboration of the surface carbohydrate coat are brush border membrane enzymes. Kim and Perdomo (218) conclude that glycosyltransferase activity in brush border preparations is due to their "inevitable trace contamination with smooth (microsomal) membranes." It has, in fact, been pointed out (212) that there is no valid experimental basis anywhere for the conclusion that glycosyltransferases involved in mammalian membrane biogenesis are localized on the cell surface. Most work which has been interpreted as supporting the conclusion has revealed the product of the enzyme activity in the membrane rather than the enzyme itself (320). Shur and Roth (377), on the other hand, point out that studies interpreted as indicating a Golgi localization for glycosyltransferases rest on no firmer ground.

Pteroyl-γ-oligoglutamyl endopeptidase (347) would appear to be a candidate brush border membrane enzyme. The starting material for its isolation was butanol extracted in the manner of Morton's original method for the isolation of membrane alkaline phosphatase (281). However, definitive studies on the issue have not been reported (339).

Carbohydrates associated with the protein or lipid of the surface coat differ with location along the intestine (245). They act as binding sites for a variety of lectins (e.g., Ref. 128) as well as for cholera toxin (92, 190, 419).

The lipids of the rat microvillus membrane are cholesterol, phospholipid, and glycosphingolipid in the respective molar ratios of 2:1:1 (143). Although the role, if any, of these components in the various functions of the membrane is not known, it should be noted that the high glycosphingolipid content is extraordinary for an extraneural plasma membrane. Altogether, the lipids of this membrane, no differently than other plasma membranes, appear to form a bilayer lipoidal matrix composed of the fatty chains of phospholipids and glycosphingolipids interspersed with cholesterol through which lipid-soluble materials (including products of lipid digestion) may pass readily by simple diffusion (105). The membrane is, however, a substantial barrier to the rapid diffusion of large, highly water-soluble compounds (like hexoses) because these do not readily enter the lipoidal matrix and are too large for any occasional aqueous channels in the membrane as the dimensional properties of the latter are equivalent to pores of no more than 4–5 Å in radius (138, 239). There are aqueous channels between the cells because the junctional complexes are not tight (151). These paracellular channels provide a shunt pathway for ions and small highly water-soluble organic compounds (150) and may under some circumstances provide an alternate route of generally unappreciated magnitude for molecules even of the size of hexoses (336); the Na^+-independent mucosal to serosal fluxes of 3-*O*-methylglucose and mannitol are equal and approximately one-third the Na^+-dependent flux of 3-*O*-methylglucose in in vitro rabbit ileum (see also Ref. 44).

MECHANISMS OF ABSORPTION

Brush Border Membrane

In Figure 1, it is indicated that fructose and glucose enter the cell by different pathways; the one for glucose being Na^+ dependent, the one for fructose not. Both pathways are carrier dependent. In recent years the single route indicated for glucose has been shown to include at least two routes (193, 194) which differ, however, only in specificity and location along the length of the small intestine but not in mechanism. Although Gracey, Burke, and Oshin (165) have reported Na^+-dependent fructose uptake in rat gut, application of this finding to other species, or to the normal functions of the digestive-absorptive process is disputed by the results of others (171, 193, 364, 380). There is no solid reason to change the facilitated diffusion carrier mechanism indicated for fructose. Insofar as carrier transport is concerned these two kinds of mechanism are the only ones known in the gut. Vectorial enzyme transport is another matter to be discussed later.

Mechanism of Energizing Absorption The absorption of foodstuffs from the gut is efficient from the point of view of conservation of mass; very little in the way of usable foodstuff that enters the alimentary canal and is digested escapes being absorbed by the small intestine. Efficiency of this degree in biological forms is generally assumed to depend upon the expenditure of energy and most ad hoc hypotheses assume that the energy is expended, directly or indirectly, in the form of ATP. Also, it is generally conceded that the energy input into the process occurs at a plasma membrane. So it is also in the gut (73).

This is not necessarily the case for all compounds. As mentioned above, the absorption of fructose, though carrier mediated, appears to occur by a facilitated diffusion rather than an energized transport mechanism such as used by glucose. The difference between the two transport mechanisms would seem to match the energy demands of the respective absorptive problems.

Fructose levels in the blood during its absorption are low, being only one-tenth those of glucose during its absorption (10–15 mg % as against 150–200 mg %), and fructose is rapidly metabolized, reducing the late or postabsorptive blood fructose to very low levels (103). Consequently, there is no large, stable blood to intestinal gradient of fructose concentration and there may simply be no need for fructose to be absorbed by an energized transport system. The case is different for glucose. The blood in health always contains appreciable glucose and, as first demonstrated by Barany and Sperber (23), a quantitatively important part of the absorption of a load of glucose occurs "uphill" from the lumen to the blood.

As a basis for uphill transport, the energy-coupled translocation process diagrammed in Figure 1 appears, at the current time, to be generally accepted in all three of its important aspects.

First, Bihler et al. (36) demonstrated the "existence of a specific, Na^+-dependent, and energy-independent process mediating the rapid equilibration of certain sugars between the tissue and the medium." This was later confirmed by numerous in vitro studies (for review see Schultz and Curran (362)) with intact tissue carried out by more discriminating test methods. Final confirmation has come from Hopfer et al. (197), who have succeeded in preparing isolated rat brush border membranes in a vesicular form. These vesicles, devoid of cytoplasm and out of contact with energy-yielding reactions, nonetheless displayed the characteristics anticipated from theory. Moreover, these properties are not shared by basolateral membranes (34, 286).

The conceptual problems which have arisen from sugar absorption studies in vivo also appear to be resolved. Early studies with in vivo loops of rat intestine appeared to confirm the anticipated Na^+ dependency of glucose absorption (319), but later ones did not (139). Early peroral perfusion studies in man (303, 349) were also negative although an effect of Na^+ on glucose uptake by human intestine in vitro was readily demonstrated (123).

Substantially ignored in the early in vivo studies was the possible contribution of secretory processes (181, 363); as concluded by Fleshler and Nelson (135) from their studies of L-alanine absorption from Thiry-Vella loops in dogs. These workers believe that the necessary Na^+ ions for absorption are contributed by the mucosal wall, perhaps as discussed earlier (77).

Also ignored were the contributions of the unstirred layer as suggested by Semenza (372) and surface ion binding. Unstirred layers can be demonstrated in vitro (109, 115, 426) and, in vivo, this unstirred layer is, according to Semenza and based on comparative kinetic data (312), even more substantial than the one observed in vitro.

Careful attention to these aspects has recently been paid in Fordtran's laboratory by using replacement substances for Na^+ in the perfusion solutions other than mannitol. With the use of Mg^{2+}, it has been finally demonstrated that as much as 50% of the active absorption of sugars in vivo is clearly Na^+ dependent (33) and it may be concluded that the discrepancy between in vitro and in vivo studies is more apparent than real and has largely a methodological basis.

Second, it was possible, early on, to imagine that Na^+ merely activated transport without participating in it and the idea has been recently revived (267). Nonetheless, for various reasons Crane et al. (89) postulated co-transport of Na^+ and sugar as a basis for the perceived need to couple energy with the carrier. Supportive evidence for co-transport has come from the influence of a reversed Na^+ gradient on sugar flux (76) and in the form of an effect of sugar on the in vitro transmural potential and Na^+ flux (366, 367). Substantial proof of co-transport has come from direct measurements of the simultaneous flux of Na^+ and sugar in rabbit ileum (163).

Kinetically, the handling of glucose by hamster small intestine is different (85) than in rabbit ileum although it is much the same as amino acid transport in rabbit ileum (93). The difference in sugar transport may have to do with the fact that hamster gut is almost entirely jejunum (253). However, comparative studies of rabbit jejunum are lacking. Studies by means of transmural potential (248) have indicated that the rat is like the hamster in this respect.

In all cases studied, however, there is, concomitant with an increase of sugar permeability of the brush border membrane by the presence of Na^+, an increase in Na^+ permeability by the presence of sugar which is demonstrable even at the level of the isolated membrane vesicle (285). Sugar-dependent Na^+ transport occurs in vitro (148). It also occurs in vivo and can be measured in the intact human intestine (346). As measured in rat jejunum (104) the stoichiometry between glucose and Na^+ absorption is 1/1, matching that predicted from in vitro studies and suggesting that Na^+ influx is affected directly by glucose flux and not indirectly by osmotic gradients. Moreover, it would appear that sugar-dependent Na^+ transport forms the basis of a reliable treatment for the salt and water loss of cholera (186, 288) using glucose-containing or amino acid-containing solutions or both; see also Ref. 180.

Altogether these studies support the existence of a carrier for the co-transport of glucose and Na^+ in the brush border membrane which establishes the capacity for energy to be coupled to glucose transport by means of an imposed electrochemical potential of Na^+ across the membrane.

Third, on the bases thus established, the only remaining question is whether the energy in the electrochemical gradient is sufficient to drive substrate accumulation to the observed levels and whether gradient coupling is the sole or principal means by which metabolic energy is transduced.

Gradient coupling between Na^+ and sugar in the gut cell has been demonstrated (76). Vidaver (415) and Terry and Vidaver (400) have demonstrated gradient coupling between Na^+ and amino acid in the pigeon red cell. Eddy (117) and Morville, Reid, and Eddy (282) have demonstrated the same for the Ehrlich mouse ascites tumor cell. In fact, such demonstrations within a more general framework have become commonplace. Mitchell (277) incorporated Crane's ideas on gradient coupling into a slightly broader chemi-osmotic hypothesis in which co-transport was called "symport" and H^+ rather than Na^+ provided the motive force. Kashket and Wilson (210) have shown proton-coupled accumulation of galactoside in a streptococcus. Komor (225) and Komor and Tanner (226) have shown proton-coupled hexose transport in *Chlorella.*

Armstrong et al. (13) have found that the energy available in the Na^+ gradient of the bull frog epithelial cells is theoretically adequate. Others (353) have calculated that there is not enough energy in the cation gradients to account for observed solute gradients in Ehrlich ascites tumor cells. The problems inherent in such calculations are well demonstrated by the recent revision by Heinz and Geck (178) of the earlier calculations to conclude that the energy in the electrochemical potential gradient of Na^+ is "approximately adequate to account for the highest accumulation ratios" reported for α-aminoisobutyric acid. Also the energy available in the membrane potential has not been included (177). In making these calculations inhomogeneity of ion distribution throughout the cell has received some attention (317) but the possible contribution of amino acid accumulation by nuclei (2) has not.

Kimmich (220) has recently reviewed the question of coupling between Na^+ and sugar transport in small intestine. From his own work with isolated chicken cells (219), he has visualized a mechanism in which the energized intermediate of the (Na^+-K^+)-ATPase-driven ion pump is alternatively utilized as the direct driving force for sugar and amino acid accumulation. As noted earlier, the amount of (Na^+-K^+)-ATPase in the brush border membrane is very small and it appears to be almost entirely a basolateral membrane enzyme. If it is, Kimmich's hypothesis cannot possibly be correct. Others working with isolated cells (157) find results "consistent with the ion gradient model" and "not compatible with the direct energy-coupling model" of Kimmich. In related work, Geck, Heinz, and Pfeiffer (159) have provided evidence (for the Ehrlich cell) against a direct coupling between amino acid transport and ATP hydrolysis.

Currently, it may be said that the gradient-coupled theory is the only hypothesis about the way the brush border membrane is involved in Na^+-dependent absorption which is supported by experiments on the membrane itself (14, 50, 197, 379, 381). The hypothesis of convective-diffusive processes through membrane channels (188), for example, is not. Moreover, a review of the literature indicates that energized (i.e., against the gradient) carrier-mediated absorption is more or less uniformly reported as being Na^+ dependent (Table 2) although the twin caveats must be entered that 1) not in all instances has it been adequately shown that the dependence on Na^+ is based in gradient-coupled transport and 2) not in all instances has the work been sufficiently reproduced in laboratories other than that of its origination as to have the findings certified.

External Fluid Circuit

When the concentration of an absorbed substance is higher in the lumen than in the bloodstream the question of energized absorption is largely moot. The value of an energized absorption pathway is apparent only when the concentration of the absorbed substance is lower in the lumen of the gut than in the bloodstream. The gradient-coupled mechanism seems to satisfy the need for the pathways listed in Table 2. However, the need is clearly not satisfied for substances which are absorbed by pathways which are not energized. In such cases complete absorption would require some degree of absorption against the gradient and an additional mechanism to nonenergized facilitated diffusion transport would be required.

Crane (77) proposed such a mechanism based on the fluid circuit[1] ideas of Florey et al. (136) and an original concept (73) of how Na^+ dependency might

[1] The term fluid circuit, as used here, does not have the same meaning as the same term used by Fordtran et al. (137).

Table 2. Pathways of energized absorption for water-soluble organic substrates

Na^+-dependent:	Na^+ dependency not tested:
Amino acids (acidic)[a] (365)	L-dopa[f] (417)
Amino acids (basic)[b] (238, 333)	Folate[g] (386)
Amino acids (neutral) (362)	Nicotinic acid[h] (145)
Ascorbic acid[c] (390)	L-Penicillamine[f] (422)
Bile salts[c] (231)	
Bilirubin (373)	
Biotin (30)	
Choline[d] (185)	
Glucose, galactose and analogs (73)	
Imino acids (284)	
Myoinositol (53)	
Peptides (1, 266)	
Riboflavin (337)	
Thiamin (131)	
Uracil[e] (90)	

[a]Accumulation may not always be seen owing to rapid transamination (315).

[b]The Σ-NH_3^+ of lysine may occupy the Na^+ site when Na^+ is absent, thus masking Na^+ dependency.

[c]In the ileum. In the jejunum uptake is passive (229, 427).

[d]In the chicken. In the rat, uptake is apparently not energized when metabolism is taken into account (350).

[e]In the frog. Although Na^+ dependency was not tested, pyrimidines were also shown to be actively transported in the rat (374) but purines appear not to be actively transported in mammalian intestine (216).

[f]Probably absorbed as amino acid analogs.

[g]This applied to concentrations of folate below 10^{-6} M. At higher concentrations the major mechanism for absorption appears to be passive diffusion (182) aided by solvent drag (386). Active folate transport occurs in the choroid plexus but it does not require Na^+ (59).

[h]In the frog. Nicotinic acid has been reported not to be actively transported in mammalian intestine (388).

serve glucose active transport by a Na^+ pump-driven fluid circuit in the microvillus projections of the brush border. The mechanism is based upon the presumptions 1) that Na^+, Cl^-, water, and the organic foodstuff all enter at (or towards) the tip of the villus and 2) that Na^+, Cl^-, and water (but not the organic foodstuff) are leaked or pumped from the base of the villus. The result would be an apparent active transport without cellular accumulation of the foodstuff (see also chapter 8).

The attractiveness of this hypothesis would seem to be enhanced by the recent work on secretion which strongly indicates that Na^+ efflux from a region near the base of the villus of the kind postulated takes place under normal conditions as well as in disease (132, 181) though the anion is probably bicarbonate rather than chloride (39, 110). Also as would be required, the pathway taken by this secretion is substantially denied to the glucose analog

L-glucose (238). However, the situation is not as clear as one would like. Studies of exsorption of various sugars strongly indicate that actively transported sugars which may be exsorbed would be recaptured by the brush border transport systems before they can reach the lumen to be measured (22). Hence, their exclusion from this pathway cannot be proved. The contribution of solvent drag to the absorption of drugs (299, 300) could involve an external fluid circuit.

Basolateral Membrane

Bihler and Cybulsky (34) studied isolated cells from the mouse treated at the brush border surface before isolation with Hg^{2+} which rapidly reduces the number of active carriers for the entry of sugars and amino acids (394) by interacting with sensitive thiol groups (387). They found a carrier-mediated process for basolateral entry of sugars which does not require Na^+ and to which phlorizin is not inhibitory. Also the specificity of the basolateral process is not as exclusive as the brush border process which requires a C-2 hydroxyl in the glucose position for rapid transport (79). 2-Deoxyglucose appeared to be an excellent substrate for the process. Kimmich and Randles (221) using isolated chick cells have substantially confirmed these findings and have added the observation that the aglycone, phloretin, is inhibitory at the basolateral membrane. The differences in Na^+ dependence and phlorizin sensitivity of the brush border and basolateral membranes have been confirmed with vesicle preparations of each (286). These findings accord with the effects of unilateral Na^+ replacement on the absorption of 3-*O*-methylglucose by in vitro rabbit ileum (336) and would seem to confirm earlier impressions that movement of accumulated sugars out of the cell into the lamina propria is a carrier-mediated, but downhill, diffusion process. However, there are continued reports of weak inward (toward the cell) active transport of galactose (188) and of important outward (toward the lamina propria) active transport, at least of 3-*O*-methylglucose (126), across the basolateral membrane. At the present time observations to the effect that blood sugar concentration is higher than the cellular sugar concentration during in vivo absorption (126) cannot otherwise be easily rationalized. However, an active membrane transport system is not the only possibility. The functional structure of the epithelial cell layer and associated underlying structures, including the vascular bed as well as the external fluid circuit, may be much more important to absorption that has been appreciated. First, Boyd et al. (44) and Fisher and Gardner (133) have demonstrated that the presence of Na^+ in the vascular bed is necessary for maximal rates of trans-epithelial transport (as contrasted to brush border membrane accumulation) and Boyd et al. (44) have shown that Na^+ has important effects on solute transport between mucosal compartments. Second, sieving of substrate accumulated in the lamina propria would be expected as a consequence of the operation of the external fluid circuit (see discussion above). It is well known that cells toward the tip of the villus have the highest specific activity of brush border membrane enzymes (297) as well as of the basolateral (Na^+-K^+)-ATPase (58) and are most active in absorption (223) whereas secretion of organic solute-free saline takes place near the base or from the crypts (181). One consequence of these spatial factors could be an additional accumulation of solute after leaving the base of the cell. Third, water flux through the paracellular pathways may restrict the approach of serosally accumulated substrate without affecting the availability to the substrate of the membrane from the cell side. In this case, a situation would be created in which an apparent active transport, cell to lamina propria, could have its basis in a difference in the area of membrane available for equilibrative diffusion without the intervention of specific energy transduction mechanisms.

ORGANIZATION OF BRUSH BORDER MEMBRANE FUNCTIONS

Interaction between Amino Acid and Sugar Transport

The possibility that amino acids and sugars share some portion of the absorptive process in common was raised 50 years ago by Cori (72), who found mutual inhibition of absorption between amino acids and sugars administered to rats by stomach tube. This mutual inhibition was later confirmed (60, 290, 352). Bingham et al. (40) proposed that the inhibitory interactions reflected a competition for a limited supply of the energy needed to support cellular accumulation of sugars and amino acids. Alvarado (7, 9), on the other hand, argued that the inhibitions resulted from the use by these two kinds of compounds of the same multifunctional, Na^+-dependent carrier.

Recent studies by Bihler and Sawh (37) showed that providing substrates for energy metabolism relieves inhibition of transport and supports the hypothesis of competition for limited energy supplies. Also, according to the gradient-coupling hypothesis discussed above, the energy supplies of importance would be those that maintain the trans-brush border membrane electrochemical gradient of Na^+. Read (332) recognized this and proposed that galactose inhibits cycloleucine transport by dog fish intestinal mucosa because it partially collapses the Na^+ gradient on which the accumulation of both would depend. Read's interpretation, as extended by Semenza (371) and Frizzell and Schultz (149), appears to be confirmed by studies with isolated membrane vesicles. Having demonstrated that Na^+-dependent D-glucose transport in such vesicles is electrogenic (285), Murer and his colleagues (287) now find that Na^+-dependent sugar and amino acid fluxes across the membrane are electrically coupled.

Kinetic Advantage

In Figure 1 the brush border enzymes are shown as external to the carriers responsible for the absorption of their products. However, there is no specific representation of a superficial region which Miller and Crane (274) were obliged to postulate in order to account for their observations which indicated "that hydrolysis occurs in a restricted portion of the cell to give a zone of relatively high concentration from which diffusion occurs into the tissue as a whole and into the medium." Their observations seemed out of keeping with the sense that the interface between hydrolysis (enzyme activity) and transport (carrier activity) was in immediate and complete contact with the bulk medium. In experiments using glucose oxidase to trap released sugar, Miller and Crane (274) obtained evidence to indicate that the brush border membrane enzyme, sucrase, is more intimately associated with glucose transport than the brush border enzyme, alkaline phosphatase.

All in all, the phenomena seemed not to be "extracellular," in the usual sense of that word. The term "periplasmic" (276) was also inappropriately descriptive. "Periplasmic" has a special meaning for bacteria with their clearly defined cell wall outside of a plasma membrane and this kind of organization is not replicated at the brush border pole of the mammalian gut cell. Consequently, the concept of a superficial region was dropped in favor of a concept of a close spatial organization between the enzyme, sucrase, and the glucose transport carriers which would confer a "kinetic advantage" for the absorption of the enzyme products (78). This concept of kinetic advantage was supported by the results obtained by Parsons and Prichard in a series of studies with perfused amphibian intestine begun in 1965 (311). The results (312–314) suggested a very close relationship, perhaps even an identity, between brush border disaccharidases and the glucose transport carrier. There was a highly efficient capture of glucose units released by enzyme activity.

Hydrolase-related Transport

What is now believed to be the correct explanation for the observations of Miller and Crane has been provided by the finding (81, 87) that the disaccharidases of the brush border membrane act as vectorial enzymes to transfer their sugar products directly across the membrane quite independently of the glucose or fructose transport carriers. This represents a new way for molecules to get across membranes and has been termed hydrolase-related transport (83).

The full documentation of hydrolase-related transport (HRT) is given by Malathi et al. (259) and by Ramaswamy et al. (328). The work is substantially confirmed by Hanke and Diedrich (176) and Diedrich and Hanke (108). All disaccharides plus the glycoside, phlorizin, having a counterpart brush border enzyme can utilize this pathway, independently of one another; yet, entirely consistent with the data of Miller and Crane, glucose 1-phosphate which is a substrate for membrane alkaline phosphatase can not. Hydrolase-related transport does not require Na^+ and the glucose released does not mix with the pool of luminal glucose before it crosses the membrane into the cell.

The functional basis of HRT is not yet entirely clear. An early proposal (87) was a variant on the idea of kinetic advantage and of the efficient capture mechanism. It would not account well for the simultaneous transfer of both glucose and fructose. The possibility that intact disaccharides cross the diffusion barrier and are split intracellularly seems highly unlikely since the diffusion barrier would have to be able to restrict the movement of mannitol more than the movement of a two times larger disaccharide. What remains is the possibility that the disaccharidases actually function as vectorial enzymes. This possibility is supported by the findings in Semenza's laboratory (397) that purified sucrase-isomaltase complex when incorporated into black lipid membranes which are normally impermeable to sucrose, glucose, fructose, and mannitol increases the permeability to products of sucrase activity but not to free glucose and fructose. Also, phlorizin does not penetrate the brush border membrane (8, 73, 396) yet it does contribute its glucose moiety in a direct transfer (108). Finally, the phenomenon of HRT has been demonstrated with isolated membrane vesicles (329).

The functional significance of HRT escapes understanding at the present time. On the one hand, its demonstration provides complete corroboration to the very early results of Miller and Crane (274). On the other hand, this new transport system is not obviously related to a major physiological event (82). For example, HRT exists but its activity for sugars has not been clearly demonstrated in vivo. Chain et al. (57) studied sugar absorption with the in vitro technique of Fisher and Parsons (134) and concluded that fructose is better absorbed when given as sucrose and that oligosaccharides are better than glucose. This has also been claimed to be the case in vivo in man (249). However, the most recent comparison of the absorption of sucrose and its monosaccharides (68) shows no difference between them.

It should theoretically be possible for HRT to function as a bypass mechanism for the absorption of some amounts of hexose when carrier function is deleted as in glucose-galactose malabsorption. However, as a practical matter, it would not be expected to be effective, at least at normal dietary loads of di- and oligosaccharides, because diarrhea would still be expected from the large amount of luminal hexose also produced. However, there has been found to be a progressive developing tolerance to carbohydrates in children with glucose-galactose malabsorption (124) and it may be the HRT mechanism which contributes to this tolerance. Elsas and Lambe (124) attempted to test this idea but the published data are inconclusive.

One point should be made clear. The efficient capture mechanism (ECM) described by Parsons and Prichard (see above) is not the same as hydrolase-related transport (HRT). Crane has compared the properties of each system as described in the relevant publications and found that they differ completely in five of the six important parameters; the sixth could not be tested in the frog (314).

DEVELOPMENT AND CONTROL OF BRUSH BORDER MEMBRANE FUNCTIONS

The development of function in the gut may be followed within two time frames. The one is the time frame of the life of the individual from conception to death. The other is the time frame of the epithelial cell from its formation in a crypt of Lieberkühn to its loss from the villus when it has migrated to the tip (61, 242). In the human, the former time frame includes 9 months in utero, and the years after birth. The latter time frame is a matter only of hours and days. Hence it is often difficult to decide whether an agency which influences a function is altering development or exerting control. As is well known, a variety of agencies (chemical, nutritional, and hormonal) do cause changes in the enzymic and transport activities of the brush border membrane. However, the changes are generally not dramatically large unless the agencies cause gross morphological damage or induce or delay development within one or the other of the time frames.

So far as is known, a reasonably full complement of brush border digestive and absorptive functions appears to be achieved by the full term human infant during intrauterine life (224), though not in all cases at the full activity to be seen later. Thus the human is very different from such other mammalian species as rat, rabbit, and pig in which the newborn has high lactase activity but little or none of the other disaccharidases. In the rat, sucrase activity appears at about 16 days or it may be precociously elicited by glucocorticoids (111, 184). Lactase in the rat begins a decline to the low levels of the adult during the 2nd week after birth in a process which appears to depend upon a normal level of circulating thyroid hormone (430). Lactase in the human declines during childhood to low adult levels in most of the world's population (69). The reason why it does not generally decline in people of northern European ancestry is not known although the ability to digest lactose in adulthood appears to be transmitted as an autosomal dominant and it has been concluded that it represents a mutated gene or a polymorphism (331).

There is considerable turnover of the brush border protein during the active life of the cell (86, 205) and, as seems generally to be the case (100, 106), the larger proteins turn over faster (3), possibly, in the gut, as a consequence in part of pancreatic protease action (6, 55). Moreover, as the cell moves outward along the villus, the brush border enzymes increase in specific activity until the cell enters the extrusion zone (98). Thus whatever affects epithelial cell turnover or protein synthesis may be expected to affect brush border enzyme and transport activities (for review, see Ref. 4). Premature induction of digestive enzymes by cortisol appears to involve the addition of new proteins rather than activation of proteins already present (156).

Herbst et al. (183) found that colchicine depressed enzyme levels and interpreted this as an effect directly on the villous cells. However, in these studies, the rate of epithelial turnover was increased 2- to 3-fold. Hence, an equally plausible explanation would be that the increased rate of epithelial cell turnover reduced the time period during which the normally expected increase in enzyme specific activity could be achieved. There are, however, clear examples of direct cellular effects. Frizzell et al. (147) have shown that in as little as 3 hr cycloheximide inhibits influxes requiring the intervention of transport proteins. In 12 hr, amino acid influx is inhibited by tetracycline (241). Both of these effects take place in a shorter time than is required for cell renewal.

The milieu of the lumen is important to the digestive-absorptive functions of the brush border but not in a way that can be clearly seen to be part of a system of precise regulation. For a contrary view, see Ugolev et al. (413). Bile salts are normally released into the gut lumen during the digestion and absorption of a meal and these bile salts, including the conjugated variety which are a necessary part of the digestive secretions as well as the deconjugated variety which are formed by bacterial action, have various effects in addition to their role in lipid digestion and absorption. For example, conjugated bile salts are innocuous in the jejunum but are inhibitory to energized transport in the ileum. This seems to be because conjugated bile salts are actively accumulated in the ileum (but not the jejunum) and intracellular concentrations are achieved in the ileum which inhibit the Na^+ pump (52). Deoxycholate, on the other hand, is also inhibitory in the jejunum (170, 175). Conjugated bile salts are said to be important in the activation of trypsinogen by enterokinase (172). They also seem to have an accelerating effect on the turnover of some, though not all, brush border enzymes (71). Long term (3 days) administration of the deconjugated bile salt, deoxycholate, produces extensive ultrastructural damage (166) which probably underlies some of the inhibitory effects of this compound on transport.

Daily rhythms in enzyme and transport activities of the brush border in the rat (155, 348) and the pig (154) have been reported. Also, detailed study of the effect of a restricted feeding schedule has shown that a rise in activity level may anticipate as well as follow the daily onset of the synchronizer (391, 392). These findings underscore the difficulty of meaningful experimentation in this area. Unless changes in activity levels are followed on a 24-hr basis it may well be impossible to assign changes seen to an effect of the experimental maneuver used on enzyme or transport levels; the effect may be on the rhythm. The reason for this is that the changes in activity level in the work on rhythms as well as in the studies of the effect of diet, such as, for example, the rise in sucrase by specific feeding of fructose-containing carbohydrate diets (343), are not large. They are generally no more than 2-fold and it is hard to believe that a 2-fold change can be part of a physiologically important regulatory mechanism in the normal intestine. Regulation by a 2-fold change in capacity does not seem to square with the fact that the small intestine is generously oversupplied with digestive-absorptive capacity, particularly for carbohydrates.

Holdsworth and Dawson (187), for example, measured the absorptive capacity of a nominal 30-cm segment of intestine in normal human beings using a peroral perfusion tube technique. From their measured values, it may be calculated that the small intestine has the capacity to absorb over 5 kg of sugar and more than 25,000 calaries in a 24-hr period. In making this calculation the total jejuno-ileal length of 421 cm (70) was corrected for the "concertinaing" effect of the insertion of a perfusion tube into the gut, i.e., $\times$ 1/3. This is a large capacity, but studies of the length of small intestine needed to be left in continuity after jejuno-ileal shunt operations for refractory obesity agree. Removal of approximately 90% of the total length (leaving 34 cm of proximal jejunum anastomosed to 13 cm of distal ileum) is required to achieve a satisfactory degree of weight loss (209). Altogether such studies indicate that the absorptive capacity of the gut is anywhere from 5 to 10 times greater than any normal need that might be served. The digestive capacity of the gut as measured by the brush border enzymes, sucrase and lactase, is even greater since free sugar accumulates in the lumen when disaccharides are used in a perfusion experiment like that of Holdsworth and Dawson (187).

Control of digestion and absorption—in the short term such as during a meal and the immediate postprandial period—clearly is not applied at the level of the brush border digestive enzymes and absorptive systems. The work of Fenton (130) and of Reynell and Spray (334) revealed this fact early on and other studies have since confirmed it. Control has been identified with a negative feedback mechanism involving receptors in the upper intestine and the motility of the stomach. Foodstuffs in general, especially fats but also proteins and carbohydrates, when they enter the duodenum through the pyloric valve elicit responses which slow gastric emptying. Under this control, the stomach delivers its contents in a predictable and exponential fashion until it is very nearly empty (200). The actual rate appears to be determined by and to be roughly proportional to the "nutritive density" (kcal/ml) of the mixed foodstuffs delivered (201).

Although short term control of digestive-absorptive capacity seems actually not to be exerted at the level of enzymes and transport proteins, there may be noticeable effects in abnormal states and there is certainly much to be learned about control mechanisms in general by studies of the effects of various agencies on the functioning levels of brush border membrane components.

Several agencies have already been mentioned. The diabetic state is another. Modern interest in this field began with the early and unequivocal demonstration that the activity of brush border glucose transport is enhanced in the diabetic rat (74). Olsen and Rosenberg (305) extended this observation to include enhancement of amino acid transport in the rat and Caspary (51) has now found that the maximal rate of conjugated bile salt transport in the ileum is increased almost 2-fold. Caspary concludes that the effect of diabetes is a general one on Na^+-dependent transport processes because of stimulation of the Na^+ pump at the basolateral membrane; as would probably be agreed by Schedl and Wilson (354). However, in early diabetes enhancement of amino acid absorption is selective (232) and Hopfer records changes in the transport properties of isolated microvillus membranes from diabetic rats (196).

Vinnick et al. (416) found enhanced sugar absorption in human subjects with juvenile-onset diabetes; thus the enhancement of transport was indicated to be a general mammalian phenomenon. Genel et al. (160) have since disputed the findings of Vinnick et al. because they did not observe such an effect. However, the results of Genel et al. seem to be in some doubt because of the short time interval during which their patients were off insulin and a higher level, if it were to occur, could be achieved. Bihler and Freund (35) suggest that diabetes and starvation affect the same controlling factor.

More recently, Olsen and Rogers (304) found brush border sucrase to be enhanced in the diabetic rat. These findings have been confirmed and extended to other rat brush border saccharidases (54, 338, 431). The increased brush border enzyme and transport activities in the rat appear to go along with similar increases in intracellular enzymic machinery (12). Calcium-binding protein, however, is decreased (357).

Cerda et al. (56) reported elevated disaccharidases in human subjects with chronic pancreatic insufficiency and diabetes mellitus. However, Caspary et al. (55) found no difference between controls, juvenile diabetics, and maturity-onset diabetics and the results of Arvanatakis and Olsen (15) with diabetics were similar. Both groups found an elevation of sucrase and maltase in nondiabetic patients with chronic pancreatitis. On the other hand, a report by Tandon et al. (398) agrees with Cerda and colleagues as possibly does that also by DuBois et al. (113).

DELETIONS OF BRUSH BORDER MEMBRANE FUNCTIONS

Many studies have shown that enzymes and transport proteins alike follow, in general, the kinetic predictions of Michaelis and Menten (272) and if relatively simple rather than complex kinetics may be assumed, a comparison between enzymes and carriers suggests that spatial translocation catalyzed by a carrier is formally equivalent to bond transfer catalyzed by an enzyme. Biochemical diseases for a membrane then in which a protein function is deleted may be viewed as one group irrespective of whether the protein functions as an enzyme or in transport. Listed together in Table 3 are the various deficiencies or deletions of brush border membrane function which have been identified to date.

These conditions are of special interest to the membrane physiologist owing to the possibility of studying membranes with an altered component of specifically functional proteins. Methods for isolation of purified membranes from fresh or frozen human surgical and peroral biopsy specimens are now available (355, 425). Also the proteins can be visualized after solubilization of the membranes with sodium dodecyl sulfate and electrophoretic separation on polyacrylamide gels and the position on the gel of the disaccharidases is known (252). Application of these methods to individuals with varying degrees of sucrase-isomaltase deficiency (88, 321) has revealed a progressive reduction in the sucrase-isomaltase band with reduction in enzyme activity to the complete absence of sucrase-isomaltase protein at zero enzyme levels. The absence of recognizable protein not only from the membrane but also from the entire cell has been elegantly confirmed by radioimmunoassay (169).

Table 3. Known deletions or deficiences of brush border membrane functions

Enzymes	Transport proteins
Sucrase-isomaltase (16, 20, 26)	Glucose-galactose malaborption (240, 341)
Congenital lactase (116, 192)	Neutral amino acid malabsorption (195, 368, 375)
Adult lactase[a] (21, 97, 174)	Folate malabsorption (125, 234)
Trehalase (32, 250)	
Enterokinase (173)	Basic amino acid-cystine malabsorption[c] (280, 342, 401)
γ-Glutamyltransferase[b] (360)	Imino acid glycine malabsorption[c] (164)

[a]Phlorizin hydrolase (glycosylceramidase) activity goes together with lactase in the same way as sucrase and isomaltase go together and is absent in adult lactase deficiency (244).

[b]The absence of the enzyme activity has not been demonstrated in brush border membranes by direct study. It is inferred from the absence of activity in skin fibroblast membranes.

[c]These are primarily kidney reabsorptive conditions which represent a complex of genetic lesions. In some cases an intestinal absorptive deficit can be demonstrated but not in others (369).

A small protein band remains in lactase deficiency (88, 146). However, cases of complete deletion of brush border lactase activity such as reported by Asp et al. (17–19) where complete protein deletion might be expected have not been studied.

REFERENCES

1. Addison, J. M., Burston, D., Dalrymple, J. A., Matthews, D. M., Payne, J. W., Sleisenger, M. H., and Wilkinson, S. (1975). A common mechanism for transport of di- and tri-peptides by hamster jejunum *in vitro.* Clin. Sci. Mol. Med., 49:313.
2. Allfrey, V. G., Meudt, R., Hopkins, J. W., and Mirsky, A. E. (1961). Sodium-dependent "transport" reactions in the cell nucleus and their role in protein and nucleic acid synthesis. Proc. Natl. Acad. Sci. USA 47:907.
3. Alpers, D. H. (1972). The relation of size to the relative rats of degradation of intestinal brush border proteins. J. Clin. Invest. 51:2621.
4. Alpers, D. H., and Kinzie, J. L. (1973). Regulation of small intestinal protein metabolism. Gastroenterology 64:471.
5. Alpers, D. H., and Solin, M. (1970). The characterization of rat intestinal amylase. Gastroenterology 58:833.
6. Alpers, D. H., and Tedesco, F. J. (1975). The possible role of pancreatic proteases in the turnover of intestinal brush border proteins. Biochim. Biophys. Acta 401:28.
7. Alvarado, F. (1966). Transport of sugars and amino acids in the intestine: evidence of a common carrier. Science 151:1010.
8. Alvarado, F. (1970). Effect of phloretin and phlorizin on sugar and amino acid transport systems in small intestine. Fed. Eur. Biochem. Symp. 20:131.
9. Alvarado, F. (1970. Intestinal transport of sugars and amino acids: independence or federalism? Am. J. Clin. Nutr. 23:824.
10. Alvarado, F., and Crane, R. K. (1962). Phlorizin as a competitive inhibitor of the active transport of sugars by hamster small intestine, *in vitro.* Biochim. Biophys. Acta 56:170.

11. Andersen, Knut-Jan, Von der Lippe, G., Morkrid, L. and Schjonsby, H. (1975). Purification and characterization of guinea-pig intestinal brush border. Biochem. J. 152:157.
12. Anderson, J. W. (1974). Glucose metabolism in jejunal mucosa of fed, fasted and streptozotocin-diabetic rats. Am. J. Physiol. 226:226.
13. Armstrong, W. McD., Byrd, B. J., and Hamang, P. M. (1973). The Na^+ gradient and D-galactose accumulation in epithelial cells of bullfrog small intestine. Biochim. Biophys. Acta 330:237.
14. Aronson, P. S., and Sacktor, B. (1975). The Na^+ gradient-dependent transport of D-glucose in renal brush border membranes. J. Biol. Chem. 250:6032.
15. Arvanatakis, C., and Olsen, W. A. (1974). Intestinal mucosal disaccharidases in chronic pancreatitis. Am. J. Dig. Dis. 19:417.
16. Asp, N.-G., Berg, N.-O., Dahlqvist, A., Gudman-Höyer, E., Jarnum, S., and McNair, A. (1975). Intestinal disaccharides in Greenland Eskimos. Scand. J. Gastroenterol. 10:513.
17. Asp, N.-G, Dahlqvist, A., Kuitunen, P., Launiala, K., and Visakorpi, J. K. (1973). Complete deficiency of brush border lactase in congenital lactose malabsorption. Lancet ii:329.
18. Asp, N.-G., and Dahlqvist, A. (1973). Separation of human small-intestinal sucrase from isomaltase. FEBS Lett. 35:303.
19. Asp, N.-G., and Dahlqvist, A. (1974). Intestinal β-galactosidases in adult low lactase activity and in congenital lactase deficiency. Enzyme 18:84.
20. Auricchio, S., Prader, A., Mürset, G., and Witt, G. (1961). Saccharose intoleranz: Durch fall infolge hereditären Mangels an intestinaler Saccharase aktivität. Helv. Paediatr. Acta 16:483.
21. Auricchio, S., Rubino, A., Landolt, M., Semenza, G., and Prader, A. (1963). Isolated intestinal lactose deficiency in the adult. Lancet ii:324.
22. Axon, A. T. R., and Creamer, B. (1975). The exsorption characteristics of various sugars. Gut 16:99.
23. Barany, E., and Sperber, E. (1939). Absorption of glucose against a concentration gradient by the small intestine of the rabbit. Scand. Arch. Physiol. 81:290.
24. Barns, R. J., and Elmslie, R. G. (1974). The active site of porcine enteropeptidase selective inactivation of the peptidase activity. Biochim. Biophys. Acta 350:495.
25. Bell, L. G. E. (1962). Polysaccharide and cell membranes. J. Theor. Biol. 3:132.
26. Bell, R. R., Draper, H. H., and Bergan, J. G. (1973). Sucrose, lactose and glucose tolerance in northern Alaskan Eskimos. Am. J. Clin. Nutr. 26:1185.
27. Bennett, G., and LeBlond, C. P. (1970). Formation of cell coat material for the whole surface of columnar cells in the rat small intestine, as visualized by radioautography with L-fucose-^{3}H. J. Cell Biol. 46:409.
28. Benson, R. L., Sacktor, B., and Greenawalt, J. W. (1971). Studies on the ultrastructural localization of intestinal disaccharidases. J. Cell Biol. 48:711.
29. Berg, G. G., and Chapman, B. (1965). The sodium and potassium activated ATPase of intestinal epithelium. I. Location of enzymatic activity in the cell. J. Cell. Comp. Physiol. 65:361.
30. Berger, E., Long, E., and Semenza, G. (1972). The sodium activation of biotin absorption in hamster small intestine *in vitro*. Biochim. Biophys. Acta 255:873.
31. Berger, S. J., and Sacktor, B. (1970). Isolation and biochemical characterization of brush borders from rabbit kidney. J. Cell Biol. 47:637.
32. Bergoz, R. (1971). Trehalose malabsorption causing intolerance to mushrooms; report of a probable cure. Gastroenterology 60:909.
33. Bieberdorf, F. A., Morawski, S., and Fordtran, J. S. (1975). Effect of sodium, mannitol, and magnesium on glucose galactose, 3-*O*-methylglucose, and fructose absorption in the human ileum. Gastroenterology 68:58.
34. Bihler, I., and Cybulsky, R. (1973). Sugar transport at the basal and lateral aspect of the small intestinal cell. Biochim. Biophys. Acta 298:429.
35. Bihler, I., and Freund, N. (1975). Sugar transport in the small intestine of obese hyperglycemic, fed and fasted mice. Diabetologia 11:387.

36. Bihler, I., Hawkins, K. A., and Crane, R. K. (1962). Studies on the mechanism of intestinal absorption of sugars. VI. The specificity and other properties of Na^+-dependent entrance of sugars into intestinal tissue under anaerobic conditions, *in vitro.* Biochim. Biophys. Acta 59:94.
37. Bihler, I., and Sawh, P. C. (1973). The role of energy metabolism in the interaction between amino acid and sugar transport in the small intestine. Can. J. Physiol. Pharmacol. 51:378.
38. Billington, T., and Nayudu, P. R. V. (1975). Studies on the brush border membrane of the mouse duodenum. I. Membrane isolation and analysis of protein components. J. Membr. Biol. 21:49.
39. Binder, H. J., Powell, D. W., Tai, Y.-H, and Curran, P. F. (1973). Electrolyte transport in rabbit ileum. Am. J. Physiol. 225:776.
40. Bingham, J. K., Newey, H., and Smyth, D. H. (1966). Specificity of the inhibitory effects of sugars on intestinal amino acid transfer. Biochim. Biophys. Acta 120:314.
41. Birkenmeier, E., and Alpers, D. H. (1974). Enzymatic properties of rat lactase–phlorizin hydrolase. Biochim. Biophys. Acta 350:100.
42. Binkley, F., Wiesemann, M. L., Groth, D. P., and Powell, R. W. (1975). γ-Glutamyl transferase: a secretory enzyme. FEBS Lett. 51:168.
43. Blum, A. L., Haemmerli, U. P., and Lorenz-Meyer, H. (1975). Is phlorizin or its aglycon the inhibitor of intestinal glucose transport? A study in normal and lactase deficient man. Eur. J. Clin. Invest. 5:285.
44. Boyd, C. A. R., Cheeseman, C. I., and Parsons, D. S. (1975). Effects of sodium on solute transport between compartments in intestinal mucosal epithelium. Nature 256:747.
45. Braun, H., Cogoli, A., and Semenza, G. (1975). Dissociation of small-intestinal sucrase–isomaltase complex into enzymatically active subunits. Eur. J. Biochem. 52:475.
46. Bredderman, P. J., and Wasserman, R. H. (1974). Chemical composition, affinity for calcium, and some related properties of the vitamin D dependent calcium-binding protein. Biochemistry 13:1687.
47. Brenna, O., Perrella, M., Pace, M., and Pietta, P. G. (1975). Affinity-chromatography purification of alkaline phosphatase from calf intestine. Biochem. J. 151:291.
48. Brunser, O., and Luft, J. H. (1970). Fine structure of the apex of absorptive cells from rat small intestine. J. Ultrastruct. Res. 31:291.
49. Burgess, D. R., and Grey, R. D. (1974). Alterations in morphology of developing microvilli elicited by cytochalasin B. J. Cell Biol. 62:566.
50. Busse, D., Jahn, A., and Steinmaier, G. (1975). Carrier-mediated transfer of D-glucose in brush border vesicles derived from rabbit renal tubules Na^+-dependent versus Na^+-independent transfer. Biochim. Biophys. Acta 401:231.
51. Caspary, W. F. (1973). Increase of active transport of conjugated bile salts in streptozotocin-diabetic rat small intestine. Gut 14:949.
52. Caspary, W. F. (1974). Inhibition of active hexose and amino acid transport by conjugated bile salts in rat ileum. Eur. J. Clin. Invest. 4:17.
53. Caspary, W. F., and Crane, R. K. (1970). Active transport of myo-inositol and its relation to the sugar transport system in hamster small intestine. Biochim. Biophys. Acta 230:308.
54. Caspary, W. F., Rhein, A., and Creutzfeldt, W. (1972). Increase of intestinal brush border hydrolases in mucosa of streptoxotocin-diabetic rats. Diabetologia 8:412.
55. Caspary, W. F., Winckler, K., Lankisch, P. G., and Creutzfeldt, W. (1975). Influence of exocrine and endocrine pancreatic function on intestinal brush border enzymatic activities. Gut 16:89.
56. Cerda, J., Preiser, H., and Crane, R. K. (1972). Brush border enzymes and malabsorption: elevated disaccharidases in chronic pancreatic insufficiency with diabetes mellitus. Gastroenterology 62:841.
57. Chain, E. B., Mansford, K. R. L., and Pocchiari, F. (1960). The absorption of sucrose, maltose and higher oligosaccharides from the isolated rat small intestine. J. Physiol. (Lond.) 154:39.

58. Charney, A. N., Gots, R. E., and Giannella, R. A. (1974). (Na^+-K^+)-stimulated adenosine triphosphatase in isolated intestinal villus tip and crypt cells. Biochim. Biophys. Acta 367:265.
59. Chen, Chi-Po, and Wagner, C. (1975). Folate transport in the choroid plexus. Life Sci. 16:1571.
60. Chez, R. A., Schultz, S. G., and Curran, P. F. (1966). Effect of sugars on transport of alanine in intestine. Science 153:1012.
61. Clarke, R. M. (1973). Progress in measuring epithelial turnover in the villus of the small intestine. Digestion 8:161.
62. Cogoli, A., Mosimann, H., Vock, C., von Balthazar, A.-K., and Semenza, G. (1972). A simplified procedure for the isolation of the sucrase–isomaltase complex from rabbit intestine: its amino acid and sugar composition. Eur. J. Biochem. 30:7.
63. Cogoli, A., and Semenza, G. (1975). A probable oxocarbonium ion in the reaction mechanism of small intestinal sucrase and isomaltase. J. Biol. Chem. 250:7802.
64. Cohen, M. I., Gartner, L. M., Blumenfeld, O. O., and Arias, I. M. (1969). Gamma glutamyl transpeptidase: measurement and development in guinea pig small intestine. Pediatr. Res. 3:5.
65. Colombo, V., Lorenz-Meyer, H., and Semenza, G. (1973). Small intestinal phlorizin hydrolase: The "β-glycosidase complex". Biochim. Biophys. Acta 327:412.
66. Colombo, V. E., and Semenza, G. (1972). An example of mutual competition between transport inhibitors of different kinetic type: the inhibition of intestinal transport of glucalogues by phloretin and phlorizin. Biochim. Biophys. Acta 288:145.
67. Conklin, K. A., Yamashiro, K. M., and Gray, G. M. (1975). Human intestinal sucrase–isomaltase: identification of free sucrase and isomaltase and cleavage of the hybrid into active distinct subunits. J. Biol. Chem. 250:5735.
68. Cook, G. C. (1970.). Comparison of the absorption and metabolic products of sucrose and its monosaccharides in man. Clin. Sci. 38:687.
69. Cook, G. C. (1973). Incidence and clinical features of specific hypolactasia in adult man. *In* B. Borgstrom, A. Dahlqvist, and L. Hambraeus (eds.), Intestinal Enzyme Deficiencies and their Nutritional Implications, pp. 52–73. Symposia of the Swedish Nutrition Foundation XI. (Stockholm).
70. Cook, G. C., and Carruthers, R. H. (1974). Reaction of human small intestine to an intraluminal tube and its importance in jejunal perfusion studies. Gut 15:545.
71. Cook, R. M., Powell, P. M., and Moog, F. (1973). Influence of biliary stasis on the activity and distribution of maltase, sucrase, alkaline phosphatase and leucylnaphthylamidase in the small intestine of the mouse. Gastroenterology 64:411.
72. Cori, C. F. (1926). Absorption of glycine and d,l-alanine. Proc. Soc. Exp. Biol. Med. 24:125.
73. Crane, R. K. (1960). Intestinal absorption of sugars. Physiol. Rev. 40:789.
74. Crane, R. K. (1961). An influence of alloxan diabetes on the active transport of sugars by rat small intestine *in vitro*. Biochem. Biophys. Res. Commun. 4:436.
75. Crane, R. K. (1962). Hypothesis for mechanism of intestinal active transport of sugars. Fed. Proc. 21:891.
76. Crane, R. K. (1964). Uphill outflow of sugar from intestinal epithelial cells induced by reversal of the Na^+ gradient. Its significance for the mechanism of Na^+-dependent active transport. Biochem. Biophys. Res. Commun. 17:481.
77. Crane, R. K. (1965). Na^+-dependent transport in the intestine and other animal tissues. Fed. Proc. 24:1965.
78. Crane, R. K. (1966). Structural and functional organization of an epithelial cell brush border. *In* K. B. Warren (ed.), Symposia of the International Society for Cell Biology, Vol. 5, pp. 71–102. Academic Press, New York.
79. Crane, R. K. (1968*a*). Absorption of sugars. *In* C. F. Code, (ed.), Handbook of Physiology, Sect. 6, Alimentary Canal, Vol. III, Intestinal Absorption, pp. 1323–1351. American Physiological Society, Washington, D.C.
80. Crane, R. K. (1968*b*). A concept of the digestive–absorptive surface of the small intestine. *In* C. F. Code (ed.), Handbook of Physiology, Sect. 6, Alimentary Canal, Vol. 5, Digestion, pp. 2535–2542. American Physiological Society, Washington, D.C.

81. Crane, R. K. (1970). Functional organization at the brush border memberane. *In* Digestion and Intestinal Absorption, 7th Int. Congr. Clin. Chem., Geneva, Evian, Vol. 4, pp. 23–30. Karger, Basel, Munchen, New York.
82. Crane, R. K. (1973). Digestive–absorptive organization in the brush border membrane. *In* B. Borgstrom, A. Dahlqvist, and L. Hambraeus (eds.), Intestinal Enzyme Deficiencies and their Nutritional Implications, Symposia of the Swedish Nutrition Foundation XI, pp. 15–19. Stockholm.
83. Crane, R. K. (1975*a*). Fifteen years of struggle with the brush border. *In* T. Csaky (ed.), Intestinal Absorption and Malabsorption, pp. 127–142. Raven Press, New York.
84. Crane, R. K. (1976*b*). The physiology of the intestinal absorption of sugars. *In* A. Jeanes and J. Hodge (eds.), Physiological Effects of Food Carbohydrates, pp. 1–9. American Chemical Society Series, Washington, D.C.
85. Crane, R. K., Forstner, G., and Eichholz, A. (1965). Studies on the mechanism of intestinal absorption of sugars. X. An effect of Na^+ concentration on the apparent Michaelis constants for intestinal sugar transport *in vitro*. Biochim. Biophys. Acta 109:467.
86. Crane, R. K., and Holmes, T. (1968). Protein turnover in the digestive–absorptive surface (brush border membrane) of the rat small intestine. J. Physiol. (Lond.) 197:79P.
87. Crane, R. K., Malathi, P., Caspary, W. F., and Ramaswamy, K. (1970). Evidence for a second glucose transport system in hamster small intestine specific for glucose release by brush border digestive enzymes. Fed. Proc. 29:1052.
88. Crane, R. K., Menard, D., Preiser, H., and Cerda, J. J. (1976). The molecular basis of brush border membrane diseases in membranes and diseases. *In* L. Bolis, J. Hoffman, and A. Leaf (eds.), Membranes and Diseases, pp. 229–242. Raven Press, New York.
89. Crane, R. K., Miller, D., and Bihler, I. (1961). The restrictions on the possible mechanisms of intestinal active transport of sugars. *In* A. Kleinzeller and A. Kotyk (eds.), Membrane Transport and Metabolism, pp. 439–450. Academic Press, New York.
90. Csaky, T. Z. (1961). Significance of sodium ions in active intestinal transport of nonelectrolytes. Am. J. Physiol. 201:999.
91. Csaky, T. Z., and Hara, Y. (1965). Inhibition of active intestinal sugar transport by digitalis. Am. J. Physiol. 209:467.
92. Cuatrecasas, P. (1973). Interaction of *Vibrio cholerae* enterotoxin with cell membranes. Biochemistry 12:35.
93. Curran, P. F., Schultz, S. G., Chez, R. A., and Fuisz, R. E. (1967). Kinetic relations of the Na–amino acid interaction at the mucosal border of the intestine. J. Gen. Physiol. 50:1261.
94. Dahlqvist, A. (1960). Characteristization of hog intestinal trehalase. Acta Chem. Scand. 14:9.
95. Dahlqvist, A. (1973). General survey on the digestion and absorption of carbohydrates. *In* B. Borgstrom, A. Dahlqvist, and L. Hambraeus (eds.), Intestinal Enzyme Deficiencies and their Nutritional Implications, pp. 9–14. Almqvist and Wiksell, Sweden.
96. Dahlqvist, A., and Borgstrom, B. (1961). Digestion and absorption of disaccharides in man. Biochem. J. 81:411.
97. Dahlqvist, A., Hammond, J. B., Crane, R. K., Dunphy, J., and Littman, A. (1963). Intestinal lactase deficiency and lactose intolerance in adults. Gastroenterology 45:488.
98. Dahlqvist, A., and Nordström, C. (1966). The distribution of disaccharidase activities in the villi and crypts of the small intestinal mucosa. Biochim. Biophys. Acta 113:624.
99. Deane, H. W., and Dempsey, E. W. (1945). The localization of phosphatase in the Golgi region of intestinal and other epithelial cells. Anat. Rec. 93:401.
100. Dehlinger, P. J., and Schimke, R. T. (1970). Effect of size on the relative rate of degradation of rat liver soluble proteins. Biochem. Biophys. Res. Commun. 40:1473.
101. DeJonge, H. R. (1975). The localization of guanylate cyclase in rat small intestinal epithelium. FEBS Lett. 53:237.

102. DeLaey, P. (1967). Die Membramverdauung der Starke. 6. Mitt. Die bindung der Amylase auf der Intestinal-Mucosa. Die Nahrung 11:17.
103. Dencker, H., Meeuwisse, G., Norryd, C., Olin, T., and Tranberg, K.-G (1973). Intestinal transport of carbohydrates as measured by portal catheterization in man. Digestion 9:514.
104. Dennhardt, R., and Haberich, F. J. (1973). Die Wirkung Aktiv Transportierter Zucker auf dens Natrium-, Kaluim.-und Volumentransport am Jejunum und Ileum der Ratte. Pflügers Arch. 345:221.
105. Diamond, J. M., and Wright, E. M. (1969). Molecular forces governing non-electrolyte permeation through cell membranes. Proc. R. Soc. Lond. (Biol.) 172:273.
106. Dice, J. F., Dehlinger, P. J., and Schimke, R. T. (1973). Studies on the correlation between size and relative degradation rate of soluble proteins. J. Biol. Chem. 248:4220.
107. Diedrich, D. F. (1968). Is phloritin the sugar transport inhibitor in intestine. Arch. Biochem. Biophys. 127:803.
108. Diedrich, D. F., and Hanke, D. W. (1975). Relationship between glycosidase activity and sugar transport in the intestine. *In* T. Csaky (ed.), Intestinal Transport and Malabsorption, pp. 143–154. Raven Press, New York.
109. Dietschy, J. M., Sallee, V. L., and Wilson, F. A. (1971). Unstirred water layers and absorption across the intestinal mucosa. Gastroenterology 61:932.
110. Dietz, J., and Field, M. (1973). Ion transport in rabbit ileal mucosa. V. Bicarbonate secretion. Am. J. Physiol. 225:858.
111. Doell, R. G., and Kretchmer, N. (1964). Intestinal invertase: precocious development of activity after injection of hydrocortisone. Science 143:42.
112. Doell, R. G., Rosen, G., and Kretchmer, N. (1965). Immunochemical studies of intestinal disaccharidases during normal and precocious development. Proc. Natl. Acad. Sci. USA 54:1268.
113. Dubois, R. S., Gotlin, R. W., and Rodgerson, D. O. (1975). Lack of dietary regulation of jejunal glycolytic enzymes and disaccharidases in obesity: the role of insulin. Gastroenterology 68:461.
114. Dubs, R., Gitzelmann, R., Steinmann, B., and Lindenmann, J. (1975). Catalytically inactive sucrase antigen of rabbit small intestine: the enzyme precursor. Helv. Paediatr. Acta 30:89.
115. Dugas, M. C., Ramaswamy, K., and Crane, R. K. (1975). An analysis of the D-glucose influx kinetics of *in vitro* hamster jejunum based on considerations of the mass transfer coefficient. Biochim. Biophys. Acta 382:576.
116. Durand, P. (1958). Lattosuria idiopatica in una paziente con diarrea cronica ed acidosi. Minerva Pediatr. 10:706.
117. Eddy, A. A. (1968). The effects of varying the cellular and extracellular concentrations of sodium and potassium ions on the uptake of glycine by mouse ascites tumor cells in the presence and absence of sodium cyanide. Biochem. J. 108:489.
118. Eichholz, A. (1967). Structural and functional organization of the brush border of intestinal epithelial cells. III. Enzymic activities and chemical composition of various fractions of tris-disrupted brush borders. Biochim. Biophys. Acta 135:475.
119. Eichholz, A. (1968). Studies on the organization of the brush border in intestinal epithelial cells. V. Sub-fractionation of enzymatic activities of the microvillus membrane. Biochim. Biophys. Acta 163:101.
120. Eichholz, A., and Crane, R. K. (1965). Studies of organization of the brush border in intestinal epithelial cells. I. Tris-disruption of isolated hamster brush borders and density gradient separation of fractions. J. Cell Biol. 26:687.
121. Eichholz, A., and Crane, R. K. (1974). Isolation of plasma membranes from intestinal brush borders. *In* S. Fleischer and L. Packer (eds.), Methods in Enzymology, Biomembranes, Part A. Vol. 31, pp. 123–134. Academic Press Inc., New York.
122. Eichholz, [illegible], Howell, K., and Crane, R. K. (1969). Studies on the organization of the brush bor[illegible] in intestinal epithelial cells. VI. Glucose binding to isolated intestinal brush bord[illegible] and their subfractions. Biochim. Biophys. Acta 193:179.
123. Elsas, L. J.[illegible] [illegible]illman, R. E., Patterson, J. H., and Rosenberg, L. E. (1970). Renal and

intestinal hexose transport in familial glucose–galactose malabsorption. J. Clin. Invest. 49:576.
124. Elsas, L. J., and Lambe, D. W., Jr. (1973). Familial glucose–galactose malabsorption. Remission of glucose intolerance. J. Pediatr. 83:226.
125. Erbe, R. W. (1975). Inborn errors of folate metabolism. N. Engl. J. Med. 293:807.
126. Esposito, G., Faelli, A., and Capraro, V. (1973). Sugar and electrolyte absorption in the rat intestine perfused *in vivo*. Pflügers Arch. 340:335.
127. Estep, J. A., and Goldner, A. M. (1974). Effect of phlorizin on galactose influx in rabbit intestine. Biochim. Biophys. Acta 367:371.
128. Etzler, M. E., and Branstrator, M. L. (1974). Differential localization of cell surface and secretory components in rat intestinal epithelium by use of lectins. J. Cell Biol. 62:329.
129. Faust, R. G., Shearin, S. J., and Misch, D. W. (1972). Sodium-dependent binding of D-glucose to a filamentous fraction of Tris-disrupted brush border from hamster jejunum. Biochim. Biophys. Acta 255:685.
130. Fenton, P. F. (1945). Response of the gastrointestinal tract to ingested glucose solutions. Am. J. Physiol. 144:609.
131. Ferrari, G., Ventura, U., and Rindi, G. (1971). The Na^+-dependence of thiamine intestinal transport *in vitro*. Life Sci. 10:67.
132. Field, M. (1974). Intestinal secretion. Gastroenterology 66:1063.
133. Fisher, R. B., and Gardner, M. L. G. (1974). Dependence of intestinal glucose absorption on sodium, studied with a new arterial infusion technique. J. Physiol. (Lond.) 241:235.
134. Fisher, R. B., and Parsons, D. S. (1949). A preparation of surviving rat small intestine for the study of absorption. J. Physiol. (Lond.) 110:36.
135. Fleshler, B., and Nelson, R. A. (1970). Sodium dependency of L-alanine absorption in canine Thiry–Vella loops. Gut 11:240.
136. Florey, H. W., Wright, R. D., and Jennings, M. A. (1941). The secretions of the intestine. Physiol. Rev. 21:36.
137. Fordtran, J. S., Rector, F. C., Jr., and Carter, N. W. (1968). The mechanisms of sodium absorption in the human small intestine. J. Clin. Invest. 47:884.
138. Fordtran, J. S., Rector, F. C., Jr., Ewton, M. F., Soter, N., and Kinney, J. (1965). Permeability characteristics of the human small intestine. J. Clin. Invest. 44:1935.
139. Forster, H., and Matthaus, M. (1973). Some comments on the coupling between intestinal absorption of glucose and of sodium. FEBS Lett. 31:75.
140. Forstner, G. G. (1971). Release of intestinal surface–membrane glycoproteins associated with enzyme activity by brief digestion with papain. Biochem. J. 121:781.
141. Forstner, G. G., Riley, E. M., Daniels, S. J., and Isselbacher, K. J. (1965). Demonstration of glyceride synthesis by brush borders of intestinal epithelial cells. Biochem. Biophys. Res. Commun. 21:83.
142. Forstner, G. G., Sabesin, S. M., and Isselbacher, K. J. (1968). Rat intestinal microvillus membranes: purification and biochemical characterization. Biochem. J. 106:381.
143. Forstner, G., and Wherrett, J. R. (1973). Plasma membrane and mucosal glycosphingolipids in the rat intestine. Biochim. Biophys. Acta 306:446.
144. Fosset, M., Chappelet-Tordo, D., and Lazdunski, M. (1973). Intestinal alkaline phosphatase: physical properties and quaternary structure. Biochemistry 13:1783.
145. Fox, K. R., and Hogben, C. A. M. (1974). Nicotinic acid active transport by *in vitro* bullfrog small intestine. Biochim. Biophys. Acta 332:336.
146. Freiburghaus, A. U., Schmitz, J., Schindler, M., Rotthauwe, Hans-W., Kuitunen, P., Launiala, K., and Hadorn, B. (1976). Personal communication.
147. Frizzell, R. A., Nellans, H. N., Acheson, L. S., and Schultz, S. G. (1973). Effects of cycloheximide on influx across the brush border of rabbit small intestine. Biochim. Biophys. Acta 291:302.
148. Frizzell, R. A., Nellans, H. N., and Schultz, S. G. (1973). Effects of sugars and amino acids on sodium and potassium influx in rabbit ileum. J. Clin. Invest. 52:215.
149. Frizzell, R. A., and Schultz, S. G. (1971). Distinction between galactose and phenylalanine effects on alanine transport in rabbit ileum. Biochim. Biophys. Acta 233:485.

150. Frizzell, R. A., and Schultz, S. G. (1972). Ionic conductances of extracellular shunt pathway in rabbit ileum. J. Gen. Physiol. 59:318.
151. Frömter, E., and Diamond, J. (1972). Route of passive ion permeation in epithelia. Nature (New Biol.) 235:9.
152. Fujita, M., Ohta, H., Kawai, K., Matsui, H., and Nakao, M. (1972a). Differential isolation of microvillous and basolateral plasma membranes from intestinal mucosa: mutually exclusive distribution of digestive enzymes and ouabain-sensitive ATPase. Biochim. Biophys. Acta 274:336.
153. Fujita, M., Parsons, D. S., and Wojnarowska, F. (1972b). Oligopeptidases of brush border membranes of rat small intestinal mucosal cells. J. Physiol. (Lond.) 277:377.
154. Furuya, S., and Takahashi, S. (1975). Absorption of L-histidine and glucose from the jejunum segment of the pig and its diurnal fluctuation. Br. J. Nutr. 34:267.
155. Furuya, S., and Yugari, Y. (1974). Daily rhythmic change of L-histidine and glucose absorptions in rat small intestine *in vivo.* Biochim. Biophys. Acta 343:558.
156. Galand, G., and Forstner, G. G. (1974). Isolation of microvillus plasma membranes from suckling-rat intestine: the influence of premature induction of digestive enzymes by injection of cortisol acetate. Biochem. J. 144:293.
157. Gall, D. Grant, Butler, D. G., Tepperman, F., and Hamilton, J. R. (1974). Sodium ion transport in isolated intestinal epithelial cells: the effect of actively transported sugars on sodium ion efflux. Biochim. Biophys. Acta 339:291.
158. Garcia-Castineiras, S., Torres-Pinedo, R., and Alvarado, F. (1973). Bacterial contamination as a source of error in D-glucose-binding studies using intestinal brush border membrane preparation. FEBS Lett. 30:115.
159. Geck, P., Heinz, E. and Pfeiffer, B. (1974). Evidence against direct coupling between amino acid transport and ATP hydrolysis. Biochim. Biophys. Acta 339:419.
160. Genel, M., London, D., Holtzapple, P. G., and Segal, S. (1971). Uptake of alpha-methylglucoside by normal and diabetic human jejunal mucosa. J. Lab. Clin. Med. 77:743.
161. Gitzelman, R., Bachi, Th., Binz, H., Lindenmann, J., and Semenza, G. (1970). Localization of rabbit intestinal sucrase with ferritin antibody conjugates. Biochim. Biophys. Acta 196:20.
162. Goldberg, D. M., Campbell, R., and Roy, A. D. (1971). The interaction of trypsin and chymotrypsin with intestinal cells in man and several animal species. Comp. Biochem. Physiol. 38B:697.
163. Goldner, A. M., Schultz, S. G., and Curran, P. F. (1969). Sodium and sugar fluxes across the mucosal border of rabbit ileum. J. Gen. Physiol. 53:362.
164. Goodman, S. I., McIntyre, C. A., and O'Brien, D. (1967). Impaired intestinal transport of proline in a patient with familial iminoaciduria. J. Pediatr. 71:246.
165. Gracey, M., Burke, V., and Oshin, A. (1972). Active intestinal transport of D-fructose. Biochim. Biophys. Acta 266:397.
166. Gracey, M., Papadimitron, J., Burke, V., Thomas, J., and Bower, G. (1973). Effects of small-intestinal function and structure induced by feeding a deconjugated bile salt. Gut 14:519.
167. Gray, G. (1970). Carbohydrate digestion and absorption. Gastroenterology 58:96.
168. Gray, G. M., and Santiago, N. A. (1966). Disaccharide absorption in normal and diseased human intestine. Gastroenterology 51:489.
169. Gray, G., Conklin, M., and Townley, R. R. W. (1976). Sucrase-isomaltase deficiency: absence of an inactive enzyme variant. N. Engl. J. Med. 294:750.
170. Guiraldes, E., Lamabadusuriya, S. P., Oyesiku, J. E. J., Whitfield, A. E., and Harries J. T. (1975). A comparative study on the effects of different bile salts on mucosal ATPase and transport in the rat jejunum *in vivo.* Biochim. Biophys. Acta 389:495.
171. Guy, M. J., and Deren, J. J. (1971). Selective permeability of the small intestine for fructose. Am. J. Physiol. 221:1051.
172. Hadorn, B., Herr, J., Troesch. V., Verhaage, W., Götze, H., and Bender, S.W. (1974). Role of bile salts in the activation of trypsinogen by enterokinase: disturbance of trypsinogen activation in patients with intrahepatic biliary atresia. Gastroenterology 66:548.

173. Hadorn, B., Tarlow, M. J., Lloyd, J. K , and Wolff, O. H. (1969). Intestinal enterokinase deficiency. Lancet i:812.
174. Haemmerli, V. P., Kistler, H. J., Ammann, R., Auricchio, S., and Prader, A. (1963). Lactasemangel der Dünndarmmucosa als Ursache gewisser Formen Erworbener Milchintoleranz beimerwachsenen. Helv. Med. Acta 30:693.
175. Hajjar, J. J., Khuri, R. N., and Bikhazi, A. B. (1975). Effect of bile salts on amino acid transport by rabbit intestine. Am. J. Physiol. 229:518.
176. Hanke, D. W., and Diedrich, D. F. (1974). Fate of the hydrolyzed glucose moiety from phlorizin in hamster jejunum. Fed. Proc. 33:271.
177. Heinz, E. Personal communication.
178. Heinz, E., and Geck, P. (1974). The efficiency of energetic coupling between Na^+ flow and amino acid transport in Ehrlich cells: a revised assessment. Biochim. Biophys. Acta 339:426.
179. Heizer, W. D., Kerley, R. L., and Isselbacher, K. J. (1972). Intestinal peptide hydrolases differences between brush border and cytoplasmic enzymes. Biochim. Biophys. Acta 264:450.
180. Hellier, M. D., Thirumalai, C., and Holdsworth, C. D. (1973). The effect of amino acids and dipeptides on sodium and water absorption in man. Gut 14:41.
181. Hendrix, T. R., and Bayless, T. M. (1970). Digestion: intestinal secretion. Annu. Rev. Physiol. 32:139.
182. Herbert, V. (1967). Biochemical and hematologic lesions in folic acid deficiency. Am. J. Clin. Nutr. 20:562.
183. Herbst, J. J., Hurwitz, R., Sunshine, P., and Kretchmer, N. (1970). Effect of colchicine on intestinal disaccharidases: correlation with biochemical aspects of cell renewal. J. Clin. Invest. 49:530.
184. Herbst, J. J., and Koldovsky, O. (1972). Cell migration and cortisone induction of sucrase activity in jejunum and ileum. Biochem. J. 126:471.
185. Herzberg, G. R., and Lerner, J. (1973). Intestinal absorption of choline in the chick. Biochim. Biophys. Acta 307:234.
186. Hirschorn, N., Kinzie, J. L., Sachar, D. B., Northrop, R. S., Taylor, J. O., Ahmad, S. Z., and Phillips, R. A. (1968). Decrease in net stool output in cholera during intestinal perfusion with glucose-containing solutions. N. Engl. J. Med. 279:176.
187. Holdsworth, C. D., and Dawson, A. M. (1964). The absorption of monosaccharides in man. Clin. Sci. 27:371.
188. Holman, G. D., and Naftalin, R. J. (1975). Galactose transport across the serosal border of rabbit ileum and its role in intracellular accumulation. Biochim. Biophys. Acta 382:230.
189. Holmes, R., and Lobley, R. W. (1970). The localization of enterokinase to the brush border membrane of the guinea-pig small intestine. J. Physiol. (Lond.) 211:50.
190. Holmgren, J., Lönnroth, I, Mannson, J-E., and Svennerholm, L. (1975). Interaction of cholera toxin and membrane G_{M1} ganglioside of small intestine. Proc. Natl. Acad. Sci. USA 72:2520.
191. Holt, J. H., and Miller, D. (1962). The localization of phosphomonoesterase and aminopeptidase in brush borders isolated from intestinal epithelial cells. Biochim. Biophys. Acta 58, 239.
192. Holzel, A., Schwarz, V., and Sutcliffe, K. W. (1959). Defective lactose absorption causing malnutrition in infancy. Lancet i:1126.
193. Honegger, P., and Gershon, E. (1974). Further evidence for the multiplicity of carriers for free glucalogues in hamster small intestine. Biochim. Biophys. Acta 352:127.
194. Honegger, P., and Semenza, G. (1973). Multiplicity of carriers for free glucalogues in hamster small intestine. Biochim. Biophys. Acta 318:390.
195. Hooft, C., Carton, D., Snoeck, J., et al. (1968). Further investigations in the methionine malabsorption syndrome. Helv. Paediatr. Acta 23:334.
196. Hopfer, U. (1975). Diabetes mellitus: changes in the transport properties of isolated intestinal microvillous membranes. Proc. Natl. Acad. Sci. USA 72:2027.
197. Hopfer, U., Nelson, K., Perrotto, J., and Isselbacher, K. J. (1973). Glucose transport in isolated brush border membrane from rat small intestine. J. Biol. Chem. 248:25.

198. Houghton, S. E., and McCarthy, C. F. (1973). The isolation, partial characterization and subfractionation of human intestinal brush border. Gut 14:529.
199. Hubel, K. A., and Parsons, D. S. (1971). Membrane digestion of starch. Am. J. Physiol. 221:1827.
200. Hunt, J. N., and Knox, M. T. (1968). Regulation of gastric emptying. *In* C. F. Code (ed.), Handbook of Physiology, Vol. 4 Motility, Sect. 6, Alimentary Canal, pp. 1917–1935. American Physiological Society, Washington, D.C.
201. Hunt, J. N., and Stubbs, D. F. (1975). The volume and energy content of meals as determinants of gastric emptying. J. Physiol. (Lond.) 245:209.
202. Isselbacher, K. J. (1974). The intestinal cell surface: some properties of normal, undifferentiated and malignant cells. Ann. Intern. Med. 81:681.
203. Ito, S. (1969). Structure and function of the glycocalyx. Fed. Proc. 28:12.
204. Ito. S. (1974). Form and function of the glycocalyx on free cell surfaces. Phil. Trans. R. Soc. Lond. B. 268:55.
205. James, W. P. T., Alpers, D. H., Gerber, J. E., and Isselbacher, K. J. (1971). The turnover of disaccharidases and brush border proteins in rat intestine. Biochim. Biophys. Acta 230:194.
206. Jesuitova, N. N., DeLacy, P., and Ugolev, A. M. (1964). Digestion of starch *in vivo* and *in vitro* in a rat intestine. Biochim. Biophys. Acta 86:205.
207. Johnson, C. F. (1967). Disaccharidase. Localization in hamster intestine brush borders. Science 155:1670.
208. Jos, J., Frezal, J., Rey, J., Lamy, M., and Wegmann, R. (1967). La localisation hisochimique des disaccharidases intestinales par un nouveau procede. Ann. Histochim. 12:53.
209. Juhl, E., Quaade, F., and Baden, H. (1974). Weight loss in relation to the length of small intestine left in continuity after jejunoileal shunt operation for obesity. Scand. J. Gastroenterol. 9:219.
210. Kashket, E. R., and Wilson, T. H. (1973). Proton-coupled accumulation of galactoside in *Streptococcus lactis* 7962. Proc. Natl. Acad. Sci. USA 70:2866.
211. Katz, M., and Cooper, B. A. (1974). Solubilized receptor for intrinsic factor–vitamin B_{12} complex from guinea pig intestinal mucosa. J. Clin. Invest. 54:733.
212. Keenan, T. W., and Morre, D. James. (1975). Glycosyltransferases: do they exist on the surface membrane of mammalian cells? FEBS Lett. 55:8.
213. Keller, P. J. (1968). Pancreatic proteolytic enzymes. *In* C. F. Code (ed.), Handbook of Physiology, Alimentary Canal, Bile, Digestion, Ruminal Physiology. Vol. V, Sect. 6, pp. 2605–2636. American Physiological Society, Washington, D.C.
214. Kelly, J. J., and Alpers, D. H. (1973*a*). Properties of human intestinal glucoamylase. Biochim. Biophys. Acta 315:113.
215. Kelly, J. J., and Alpers, D. H. (1973*b*). Blood group antigenicity of purified human intestinal disaccharidases. J. Biol. Chem. 248:8216.
216. Khan, A. H., Wilson, S., and Crawhall, J. C. (1975). The influx of uric acid and other purines into everted jejunal sacs of the rat and hamster. Can. J. Physiol. Pharmacol. 53:113.
217. Kim, Y. S., Kim, Y. W., and Sleisenger, M. H. (1974). Studies on the properties of peptide hydrolases in the brush border and soluble fractions of small intestinal mucosa of rat and man. Biochim. Biophys. Acta 370:283.
218. Kim, Y. S., and Perdomo, J. M. (1974). Biosynthesis of membrane glycoproteins in the rat small intestine. FEBS Lett. 44:309.
219. Kimmich, G. A. (1970). Active sugar accumulation by isolated intestinal epithelial cells, a new model for sodium dependent metabolite transport. Biochemistry 9:3669.
220. Kimmich, G. A. (1973). Coupling between Na^+ and sugar transport in small intestine. Biochim. Biophys. Acta 300:31.
221. Kimmich, G. A., and Randles, J. (1975). A Na^+-independent phloretin-sensitive monosaccharide transport system in isolated intestinal epithelial cells. J. Membr. Biol. 23:57.
222. Kinne-Saffran, E., and Kinne, R. (1974). Localization of a calcium-stimulated ATPase in the basal-lateral plasma membranes of the proximal tubule of rat kidney cortex. J. Membr. Biol. 17:263.

223. Kinter, W. B. (1961). Autoradiographic study of intestinal transport. *In* J. Metcoff (ed.), Proc. 12th Ann. Conf. Nephrotic Syndrome, pp. 59–68. National Kidney Disease Foundation, New York.
224. Koldovsky, O. (1969). *In* S. Karger (ed.), Development of the Functions of the Small Intestine in Mammals and Man, pp. 1–204. White Plains, New York.
225. Komor, E. (1973). Proton-coupled hexose transport in *Chlorella Vulgaris.* FEBS Lett. 38:16.
226. Komor, E., and Tanner, W. (1974). The hexose-proton symport system of *Chlorella vulgaris:* specificity, stoichiometry and energetics of sugar-induced proton uptake. Eur. J. Biochem. 44:219.
227. Kowarski, S., Blair-Stanek, C. S., and Schachter, D. (1974). Active transport of zinc and identification of zinc binding protein in rat jejunal mucosa. Am. J. Physiol. 226:401.
228. Kowarski, S., and Schachter, D. (1973). Vitamin D and adenosine triphosphatase dependent on divalent cations in rat intestinal mucosa. J. Clin. Invest. 52:2765.
229. Krag, E., and Phillips, S. F. (1974). Active and passive bile acid absorption in man: perfusion studies of the ileum and jejunum. J. Clin. Invest. 53:1686.
230. Labat, J., Baumann, F., and Courtois, J.-E. (1974). La trehalase renale du pore et de l'homme: purification et quelques proprietes. Biochimie 56:805.
231. Lack, L., and Weiner, I. M. (1966). Intestinal bile salt transport: structure–activity relationships and other properties. Am. J. Physiol. 210:1142.
232. Lal, D., and Schedl, H. P. (1974). Intestinal adaptation in diabetes: amino acid absorption. Am. J. Physiol. 227:827.
233. Lamers, J. M. J., and Holsmann, W. C. (1975). Inhibition of pyruvate transport by fatty acids in isolated cells from rat small intestine. Biochim. Biophys. Acta 394:31.
234. Lanzkowsky, P. (1970). Congenital malabsorption of folate. Am. J. Med. 48:580.
235. Leese, H. J., and Semenza, G. (1973). On the identity between the small intestinal enzymes phlorizin hydrolase and glycosylceramidase. J. Biol. Chem. 248:8170.
236. LeFevre, P. G. (1959). Molecular structural factors in competitive inhibition of sugar transport. Science 130:104.
237. Leslie, G. I., and Rowe, P. B. (1972). Folate binding by the brush border membrane proteins of small intestinal epithelial cells. Biochemistry 11:1696.
238. Lifson, N., Hakin, A. A. and Lender, E. J. (1972). Effects of cholera toxin on intestinal permeability and transport interactions. Am. J. Physiol. 222:1479.
239. Lindemann, B., and Solomon, A. K. (1962). Permeability of luminal surface of intestinal mucosal cells. J. Gen. Physiol. 45:801.
240. Lindquist, B., and Meeuwisse, G. W. (1962). Chronic diarrhoea caused by monosaccharide malabsorption. Acta Paediatr. 51:674.
241. Ling, V., and Morin, C. L. (1971). Inhibition of amino acid transport in rat intestinal rings by tetracycline. Biochim. Biophys. Acta 249:252.
242. Lipkin, M. (1973). Proliferation and differentiation of gastrointestinal cells. Physiol. Rev. 53:891.
243. Lojda, Z. (1965). Some remarks concerning the histochemical detection of disaccharidases and glucosidases. Histochemie 5:339.
244. Lorenz-Meyer, H., Blum, A. L., Haemmerli, H. P., and Semenza, G. (1972). A second enzyme deficit in acquired lactase deficiency: lack of small intestinal phlorizin hydrolase. Eur. J. Clin. Invest. 2:326.
245. Louvard, D., Maroux, S., Baratti, J., Desnuelle, P., and Mutaftschiev, S. (1973). On the preparation and some properties of closed membrane vesicles from hog duodenal and jejunal brush border. Biochim. Biophys. Acta 291:747.
246. Louvard, D., Maroux, S., Baratti, J., and Desnuelle, P. (1973). On the distribution of enterokinase in porcine intestine and on its subcellular localization. Biochim. Biophys. Acta 309:127.
247. Louvard, D., Maroux, S., and Desnuelle, P. (1975). Topological studies on the hydrolases bound to the intestinal brush border membrane: II. Interactions of free and bound aminopeptidase with a specific antibody. Biochim. Biophys. Acta 389:389.
248. Lyon, I., and Crane, R. K. (1966). Studies on *in vitro* transmural potentials in relation

to intestinal absorption. I. Apparent Michaelis constants for Na^+-dependent sugar transport. Biochim. Biophys. Acta 112:278.
249. MacDonald, I., and Turner, C. J. (1968). Serum fructose levels after sucrose or its constituent monosaccharides. Lancet i:841.
250. Madzarovova-Nohejlova, J. (1973). Trehalase deficiency in a family. Gastroenterology 65:130.
251. Maestracci, D., Preiser, H., Hedges, T., Schmitz, J., and Crane, R. K. (1975). Enzymes of the human intestinal brush border membrane: identification after gel electrophoretic separation. Biochim. Biophys. Act 382:147.
252. Maestracci, D., Schmitz, J., Preiser, H., and Crane, R. K. (1973). Proteins and glycoproteins of the human intestinal brush border membrane. Biochim. Biophys. Acta 323:113.
253. Magalhaes, H. (1968). Gross anatomy. *In* R. A. Hoffman, P. F. Robinson, and H. Magalhaes (eds.), The Golden Hamster, pp. 91–109. Iowa State Univ. Press, Ames.
254. Mahmood, A., and Alvarado, F. (1975). The activation of intestinal brush border sucrase by alkali metal ions: an allosteric mechanism similar to that for the Na^+-activation of non-electrolyte transport systems in intestine. Arch. Biochem. Biophys. 168:585.
255. Malathi, P. (1967). Localization of cholesteryl and retinyl ester hydrolases in the microvillus membrane of brush borders isolated from intestinal epithelial cells. Gastroenterology 52:1106.
256. Malathi, P., and Crane, R. K. (1968). Spatial relationship between intestinal disaccharidases and the active transport system for sugars. Biochim. Biophys. Acta 163:275.
257. Malathi, P., and Crane, R. K. (1969*a*). Phlorizin hydrolase, a β-glucosidase of hamster intestinal brush border membrane. Biochim. Biophys. Acta 173:245.
258. Malathi, P., and Crane, R. K. (1969*b*). Glucamylase activity in hamster small intestine. Gastroenterology 56:1182.
259. Malathi, P., Ramaswamy, K., Caspary, W. F., and Crane, R. K. (1973). Studies on transport of glucose from disaccharides by hamster small intestine *in vitro*. I. Evidence for a disaccharidase related transport system. Biochim. Biophys. Acta 307:613.
260. Maroux, S., Baratti, J., and Desnuelle, P. (1971). Purification and specificity of porcine enterokinase. J. Biol. Chem. 246:5031.
261. Maroux, S., Louvard, D., and Desnuelle, P. The intestinal brush border aminopeptidase (β-naphthyl amidase) as a model of enzyme bound to the surface of a membrane. International FEBS Congress. In press.
262. Martin, D. L., Melancon, M. J., Jr., and DeLuca, H. F. (1969). Vitamin D stimulated, calcium-dependent adenosine triphosphatase from brush borders of rat small intestine. Biochim. Biophys. Res. Commun. 35:819.
263. Masevitch, C. H., Ugolev. A.M., Zabelinskii, E. K., and Kisily, N. P. (1975). Lumenal and membrane hydrolysis of starch in some diseases of the small intestine and pancreas. Am. J. Gastroenterol. 63:299.
264. Mathan, V. I., Babior, B. M., and Donaldson, R. M. (1974).Kinetics of the attachment of intrinsic factor*n*bound cobamides to ideal receptors. J. Clin. Invest. 54:598.
265. Matthews, D. M. (1974). Absorption of amino acids and peptides from the intestine. Clin. Endocrin. Metabol. 3:3.
266. Matthews, D. M. (1975). Intestinal absorption of peptides. Physiol. Rev. 55:537.
267. Matthews, R. H., Rice, J., and Och, F. F. (1971). An alternate hypothesis regarding the effect of ions on sugar and neutral amino acid transport. Physiol. Chem. Phys. 3:235.
268. McDougal, D. G., Jr., Little K. D., and Crane, R. K. (1960). Studies on the mechanism of intestinal absorption of sugars. IV. Localization of galactose concentrations within the intestinal wall during active transport *in vitro*. Biochim. Biophys. Acta 45:483.
269. Meister, A. (1973). On the enzymology of amino acid transport. Transport in kidney and probably other tissues is mediated by a cycle of enzymic reaction involving glutathione. Science 180:33.
270. Meister, A. (1974). Glutathione: metabolism and function via the γ-glutamyl cycle. Life Sci. 15:177.

271. Messer, M., and Kerry, K. R. (1967). Intestinal digestion of maltotriose in man. Biochim. Biophys. Acta 132:432.
272. Michaelis, L., and Menten, M. L. (1973). Die Kinetik der Invertinwirkung. Biochem. Z. 49:333.
273. Miller, D., and Crane, R. K. (1960). The concept of a digestive surface in the small intestine: cellular nature of disaccharide and phosphate ester hydrolysis. J. Lab. Clin. Med. 56:928.
274. Miller, D., and Crane, R. K. (1961a). The digestive function of the epithelium of the small intestine. I. An intracellular locus of disaccharide and sugar phosphate ester hydrolysis. Biochim. Biophys. Acta 52:281.
275. Miller, D., and Crane, R. K. (1961b). The digestive function of the epithelium of the small intestine. II. Localization of disaccharide hydrolysis in the isolated brush border portion of intestinal epithelial cells. Biochim. Biophys. Acta 52:293.
276. Mitchell, P. (1961). Approaches to the analysis of specific membrane transport. *In* T. W. Goodwin and O. Lindberg (eds.), Biological Structure and Function, Vol. 2, pp. 581–603. Academic Press, London.
277. Mitchell, P. (1963). Molecule, group and electron translocation through natural membranes. Biochem. Soc. Symposium 22:142.
278. Mooseker, M. S. (1974). Brush border motility: microvillar contraction in isolated brush border membrane. J. Cell Biol. 63:231a.
279. Mooseker, M. S., and Tilney, L. G. (1975). Organization of an actin filament–membrane complex: filament polarity and membrane attachment in the microvilli of intestinal epithelial cells. J. Cell Biol. 67:725.
280. Morin, C. L., Thompson, M. W., Jackson, S. H., and Sass-Kortsak, A. (1971). Biochemical and genetic studies in cystinuria: observations on double heterozygotes of genotype 1/11. J. Clin. Invest. 50:1961.
281. Morton, R. K. (1954). The purification of alkaline phosphatase of animal tissues. Biochem. J. 57:595.
282. Morville, M., Reid, M., and Eddy, A. A. (1973). Amino acid absorption by mouse ascites tumour cells depleted of both endogenous amino acids and adenosine triphosphate. Biochem. J. 134:11.
283. Munck, B. G. (1970). Interactions between lysine, Na^+ and Cl^- transport in rat jejunum. Biochim. Biophys. Acta 203:424.
284. Munck, B. G., and Schultz, S. G. (1972). Effects of sugar and amino acid transport on transepithelial fluxes of sodium and chloride of short circuited rat jejunum. J. Physiol. (Lond.) 223:699.
285. Murer, H., and Hopfer, U. (1974). Demonstration of electrogenic Na^+-dependent D-glucose transport in intestinal brush border membranes. Proc. Natl. Acad. Sci. USA 71:484.
286. Murer, H., Hopfer, U., Kinne-Saffran, E., and Kinne, R. (1974). Glucose transport in isolated brush border and lateral plasma membrane vesicles from intestinal epithelial cells. Biochim. Biophys. Acta 345:170.
287. Murer, H., Sigrist-Nelson, K., and Hopfer, U. (1975). On the mechanism of sugar and amino acid interaction in intestinal transport. J. Biol. Chem. 250:7392.
288. Nalin, D. R., Cash, R. A., Rahman, M., and Yumus, M. D. (1970). Effect of glycine and glucose on sodium and water absorption in patients with cholera. Gut 11:768.
289. Negrel, R., and Ailhaud, G. (1975). Localization of the monoglyceride pathway enzymes in the villus tips of intestinal cells and their absence from the brush border. FEBS Lett. 54:183.
290. Newey, H., and Smyth, D. H. (1964). Effects of sugars on intestinal transfer of amino acids. Nature (Lond.) 202:400.
291. Nishi, Y., Yoshida, T. O., and Takesue, Y. (1968). Electron microscope studies on the structure of rabbit intestinal sucrase. J. Mol. Biol. 37:441.
292. Nixon, S. E., and Mawer, G. E. (1970*a*). The digestion and absorption of proteins in man. I. The site of absorption. Br. J. Nutr. 24:227.
293. Nixon, S. E., and Mawer, G. E. (1970*b*). The digestion and absorption or proteins in man. II. The form in which digested protein is absorbed. Br. J. Nutr. 24:241.

294. Nordstrom, C. (1972). Enzymic release of enteropeptidase from isolated rat duodenal brush borders. Biochim. Biophys. Acta 268:711.
295. Nordstrom, C., and Dahlqvist, A. (1970). The cellular localization of enterokinase. Biochim. Biophys. Acta 198:621.
296. Nordstrom, C., and Dahlqvist, A. (1972). Localization of human enterokinase. Lancet 10:933.
297. Nordstrom, C., and Dahlqvist, A. (1973). Quantitative distribution of some enzymes along the villi and crypts of human small intestine. Scand. J. Gastroenterol. 8:407.
298. Noren, O., Sjostrom, H., and Josefsson, L. (1973). Studies on a soluble dipeptidase from pig intestinal mucosa. I. Purification and specificity. Biochim. Biophys. Acta 327:446.
299. Ochsenfahrt, H., and Winne, D. (1974*a*). The contribution of solvent drag to the intestinal absorption of the acidic drugs benzoic acid and salicylic acid from the jejunum of the rat. Arch. Pharmacol. 281:197.
300. Ochsenfahrt, H., and Winne, D. (1974*b*). The contribution of solvent drag to the intestinal absorption of the basic drugs amidopyrine and antipyrine from the jejunum of the rat. Arch. Pharmacol. 281:175.
301. O'Cuinn, G., Donlon, J., and Fottrell, P. F. (1974). Similarities between one of the multiple forms of peptide hydrolase purified from brush border and cytosol fractions of guinea pig intestinal mucosa. FEBS Lett. 39:225.
302. O'Cuinn, G., and Fottrell, P. F. (1975). Purification and characterization of an aminoacyl proline hydrolase from guinea-pig intestinal mucosa. Biochim. Biophys. Acta 391:388.
303. Olsen, W. A., and Ingelfinger, F. J. (1968). The role of sodium in intestinal glucose absorption in man. J. Clin. Invest. 47:1133.
304. Olsen, W. A., and Rogers, L. (1971). Jejunal sucrase activity in diabetic rats. J. Lab. Clin. Med. 77:838.
305. Olsen, W. A., and Rosenberg, I. H. (1970). Intestinal transport of sugars and amino acids in diabetic rats. J. Clin. Invest. 49:96.
306. Orlowski, M., and A. Meister. (1963). γ-Glutamyl-*p*-nitroanilide: a new convenient substrate for determination and study of C- and D-γ-glutamyl-transpeptidase activities. Biochim. Biophys. Acta 73:679.
307. Orlowski, M., and A. Meister. (1965). Isolation of γ-glutamyl transpepitase from hog kidney. J. Biol. Chem. 240:338.
308. Overton, J. (1965). Fine structure of the free cell surface in developing mouse intestinal mucosa. J. Exp. Zool. 159:195.
309. Overton, J., Eichholz, A., and Crane, R. K. (1965). Studies of the organization of the brush border in intestinal epithelial cells. II. Structure of hamster microvilli as revealed by electron microscopic examination of the fractions of Tris-disrupted brush borders. J. Cell Biol. 26:693.
310. Overton, J., and Shoup, J. (1964). Fine structure of cell surface specialization in the maturing duodenal mucosa of the chick. J. Cell Biol. 21:75.
311. Parsons, D. S., and Prichard, J. S. (1965). Hydrolysis of disaccharides during absorption by the perfused small intestine of amphibia. Nature 208:1097.
312. Parsons, D. S., and Prichard, J. S. (1966). Properties of some model systems for transcellular active transport. Biochim. Biophys. Acta 126:471.
313. Parsons, D. S., and Prichard, J. S. (1968). A preparation of perfused small intestine for the study of absorption in amphibia. J. Physiol. (Lond.) 198:405.
314. Parsons, D. S., and Prichard, J. S. (1971). Relationships between disaccharide hydrolysis and sugar transport in amphibian small intestine. J. Physiol. (Lond.) 212:299.
315. Parsons, D. S., and Volman-Mitchell, H. (1974). The transamination of glutamate and aspartate during absorption *in vitro* by small intestine of chicken, guinea-pig and rat. J. Physiol. (Lond.) 239:677.
316. Peters, T. J. (1970). The subcellular localization of di- and tri-peptide hydrolase activity in guinea pig small intestine. Biochem. J. 120:195.
317. Pietrzyk, C., and Heinz, E. (1972). Some observations on the non-homogeneous distribution inside the Ehrlich cell. *In* E. Heinz (ed.), Na^+-linked Transport of Organic Solutes, pp. 84–90. Springer Verlag, Berlin, Heidelberg, New York.

318. Piggott, C. D., and Fottrell, P. F. (1975). Purification and characterization from guinea-pig intestinal mucosa of two peptide hydrolases which preferentially hydrolyse dipeptides. Biochim. Biophys. Acta 391:403.
319. Ponz, F., and Lluch, M. (1964). Influence of the Na^+ concentration on the *in vivo* intestinal absorption of sugars. Rev. Span. Fisiol. 20:179.
320. Porter, C. W., and Bernacki, R. J. (1975). Ultrastructural evidence for ectoglycosyltransferase systems. Nature 256:648.
321. Preiser, H., Menard, D., Crane, R. K., and Cerda, J. J. (1974). Deletion of enzyme protein from the brush border membrane in sucrase–isomaltase deficiency. Biochim. Biophys. Acta 363:279.
322. Preiser, H., Schmitz, J., Maestracci, D., and Crane, R. K. (1975). Modification of an assay for trypsin and its application for the estimation of enteropeptidase. Clin. Chim. Acta 59:169.
323. Puls, W., and Keup, U. (1975). Inhibition of sucrase by Tris in rat and man, demonstrated by oral loading tests with sucrose. Metabolism 24:93.
324. Quaroni, A., Gershon, E., and Semenza, G. (1974). Affinity labeling of the active sites in the sucrase–isomaltase complex from small intestine. J. Biol. Chem. 249:6424.
325. Quaroni, A., Gershon-Quaroni, E., and Semenza, G. (1975). Tryptic digestion of native small-intestinal sucrase–isomaltase complex: isolation of the sucrase submit. Eur. J. Biochem. 52:481.
326. Quigley, J. P., and Gotterer, G. S. (1972). A comparison of the (Na^+-K^+)-ATPase activities found in isolated brush border and plamsa membrane of the rat intestinal mucosa. Biochim. Biophys. Acta 255:107.
327. Quill, H. and Weiser, M. M. (1975). Adenylate and guanylate cyclase activities and cellular differentiation in rat small intestine. Gastroenterology 69:470.
328. Ramaswamy, K., Malathi, P., Caspary, W. F., and Crane, R. K. (1974). Studies on the transport of glucose from disaccharides by hamster small intestine *in vitro*. II. Characteristics of the disaccharidase-related transport system. Biochim. Biophys. Acta 345:39.
329. Ramaswamy, K., Malathi, P., and Crane, R. K. (1976). Demonstration of hydrolase-related glucose transport in brush border membrane vesicles prepared from guinea pig small intestine. Biochem. Biophys. Res. Commun. 68:162.
330. Ramaswamy, S., and Radhakrishnan, A. N. (1975). Lactase–phlorizin hydrolase complex from monkey small intestine: purification, properties and evidence for two catalytic sites. Biochim. Biophys. Acta 403:446.
331. Ransome-Kuit, O., Kretchmer, N., Johnson, J. O., and Gribble, J. T. (1975). A genetic study of lactose digestion in Nigerian families. Gastroenterology 68:431.
332. Read, C. P. (1967). Studies on membrane transport. I. A. common transport system for sugars. and amino acids? Biol. Bull. 133:630.
333. Reiser, S., and Christiansen, P. A. (1973). The properties of Na^+-dependent and Na^+-independent lysine uptake by isolated intestinal epithelial cells. Biochim. Biophys. Acta 307:212.
334. Reynell, P., and Spray, G. H. (1956). The simultaneous measurement of absorption and transit in the gastrointestinal tract of the rat. J. Physiol. (Lond.) 131:452.
335. Rhodes, J. B., Eichholz, A., and Crane, R. K. (1967). Studies on the organization of the brush border in intestinal epithelial cells. IV. Aminopeptidase activity in microvillus membranes of hamster intestinal brush borders. Biochim. Biophys. Acta 135:959.
336. Rinaldo, J. E., Jennings, B. L., Frizzell, R. A., and Schultz, S. G. (1975). Effects of unilateral sodium replacement on sugar transport across *in vitro* rabbit ileum. Am. J. Physiol. 228:854.
337. Rivier, D. A. (1973). Kinetics and Na^+-dependence of riboflavin absorption by intestine *in vivo*. Experientia 29:756.
338. Rommel, K., Böhmer, R., Goberna, R., and Fussgänger, R. (1972). Influence of insulin and glibenclamid on the intestinal disaccharidases of subtotally pancreatectomized rats. Digestion 6:146.
339. Rosenberg, I. H. (1975). Folate absorption and malabsorption. N. Engl. J. Med. 293:1303.

340. Rosenberg, I. H., and Rosenberg, L. E. (1968). Localization and characterization of adenosine triphosphatase in guinea pig intestinal epithelium. Comp. Biochem. Physiol. 24:975.
341. Rosenberg, L. E. (1976). Intestinal hexose transport in familial glucose–galactose malabsorption. *In* L. Bolis, J. Hoffman, and A. Leaf (eds.), Membranes and Diseases, p. 253. Raven Press, New York.
342. Rosenberg, L. E., Downing, S., Durant, J. L., and Segal, S. (1966). Cystinuria: biochemical evidence for three genetically distinct diseases. J. Clin. Invest. 45:365.
343. Rosenzweig, N. S. (1973). The influence of dietary carbohydrates on intestinal disaccharidase activity in man. *In* B. Borgstrom, A. Dahlqvist, and L. Hambraeus (eds.), Intestinal Enzyme Deficiencies and Their Nutritional Implications, Symposia of the Swedish Nutrition Foundation SI, pp. 52–73. Stockholm.
344. Rubin, A. W., and Deren, J. J. (1974). Studies of glycerol transport across the rabbit brush border. Gastroenterology 66:378.
345. Russell, R. G. G., Monod, A., Bonjour, J.-P., and Fleisch, H. (1972). Relation between alkaline phosphatase and Ca^{2+}-ATPase in calcium transport. Nature 240:126.
346. Sachar, D. B., Taylor, J. P., Saha, J. R., and Phillips, R. A. (1969). Intestinal transmural electric potential and its response to glucose in acute and convalescent cholera. Gastroenterology 56:512.
347. Saini, P. K., and Rosenberg, I. H. (1974). Isolation of pteroyl-γ-oligoglutamyl endopeptidase from chicken intestine with the aid of affinity chromatography. J. Biol. Chem. 249:5131.
348. Saito, M. (1972). Daily rhythmic changes in brush border enzymes of the small intestine and kidney in rats. Biochim. Biophys. Acta 286:212.
349. Saltzman, D. A., Rector, F. C., Jr., and Fordtran, J. S. (1972). The role of intraluminal sodium in glucose absorption *in vivo*. J. Clin. Invest. 51:876.
350. Sanford, P. A., and Symth, D. H. (1972). Intestinal transfer of choline in rat and hamster. J. Physiol. (Lond.) 215:769.
351. Sasajima, K., Kawachi, T., Sato, S., and Sugimura, T. (1975). Purification and properties of α,α-trehalase from the mucosa of rat small intestine. Biochim. Biophys. Acta 403:139.
352. Saunders, S. J., and Isselbacher, K. J. (1965). Inhibition of intestinal amino acid transport by hexoses. Biochim. Biophys. Acta 102:397.
353. Schafer, J. A., and Heinz, E. (1971). The effect of reversal of Na^+ and K^+ electrochemical potential gradients on the active transport of amino acids in Ehrlich ascites tumor cells. Biochim. Biophys. Acta 249:15.
354. Schedl, H. P., and Wilson, H. D. (1974). Jejunal sodium transport in the rat: effects of alloxan diabetes. Biochim. Biophys. Acta 367:225.
355. Schmitz, J., Preiser, H., Maestracci, D., Ghosh, B. K., Cerda, J. J., and Crane, R. K. (1973). Purification of the human intestinal brush border membrane. Biochim. Biophys. Acta 323:98.
356. Schmitz, J., Presider, H., Maestracci, D., Crane, R. K., Troesch, V., and Hadorn, B. (1974). Subcellular localization of enterokinase in human small intestine. Biochim. Biophys. Acta 343:435.
357. Schneider, L. E., Wilson, H. D., and Schedl, H. P. (1973). Intestinal calcium binding protein in the diabetic rat. Nature 245:327.
358. Schneider, R. P., Donaldson, R. M., Jr., and Babior, B. M. (1974). Evidence for the solubilization of the intestinal intrinsic factor receptor by sonication of ileal brush borders. Biochim. Biophys. Acta 373:58.
359. Schneider, R., Troesch, V., and Hadorn, B. (1975). On the cellular distribution of sucrase and enterokinase in different populations of rat intestinal epithelial cells isolated by a vibration method. Biol. Gastroenterol. 8:11.
360. Schulman, J. D., Goodman, S. I., Mace, J. W., Patrick, A. D., Tietze, F., and Butler, E. J. (1975). Glutathionuria: inborn error of metabolism due to tissue deficiency of gamma-glutamyl transpeptidase. Biochem. Biophys. Res. Commun. 65:68.
361. Schultz, S. G., and Curran, P. F. (1968). Intestinal absorption of sodium chloride and water. *In* Handbook of Physiology, Alimentary Canal, Sect. 6, Vol. III, pp. 1245–1275. American Physiological Society, Washington, D.C.

362. Schultz, S. G., and Curran, P. F. (1970). Coupled transport of sodium and organic solutes. Physiol. Rev. 50:637.
363. Schultz, S. G., and Frizzell, R. A. (1972). An overview of intestinal absorptive and secretory processes. Gastroenterology 63:161.
364. Schultz, S. G., and Strecker, C. K. (1970). Fructose influx across the brush border of rabbit ileum. Biochim. Biophys. Acta 211:586.
365. Schultz, S. G., Yu-Tu, L., Alvarez, G. O., and Curran, P. F. (1970). Dicarboxylic amino acid influx across brush border of rabbit ileum. J. Gen. Physiol. 56:621.
366. Schultz, S. G., and Zalusky, R. (1964). The interaction between active sodium and active sugar transport. J. Gen. Physiol. 47:1043.
367. Schultz, S. G., and Zalusky, R. (1965). Interactions between active sodium transport and active amino acid transport in isolated rabbit ileum. Nature 204:292.
368. Scriver, C. R. (1965). Hartnup disease, a genetic modification of intestinal and renal transport of certain neutral alpha-amino acids. N. Engl. J. Med. 273:530.
369. Scriver, C. R., and Bergeron, M. (1974). Amino acid transport in kidney. The use of mutation to dissect membrane and transepithelial transport. *In* W. L. Nyhan (ed.), Heritable Disorders of Amino Acid Metabolism, p. 515. John Wiley & Sons, Inc., New York.
370. Semenza, G. (1968). Intestinal oligosaccharidases and disaccharidases. *In* C. Code (ed.), Handbook of Physiology, Alimentary Canal, Bile, Digestion, Ruminal Physiology, Vol. V. Sect. 6, pp.2543–2566. American Physiological Society, Washington, D.C.
371. Semenza, G. (1971). On the mechanism of mutual inhibition among sodium-dependent transport systems in the small intestine. A hypothesis. Biochim. Biophys. Acta 241:637.
372. Semenza, G. (1972). Some aspects of intestinal sugar transport. *In* W. L. Burland and P. D. Samuel (eds.), Transport Across the Intestine, A Glaxo Symposium, pp. 78–92. Churchill, Livingstone, Edinburgh, London.
373. Serrani, R. E., Corchs, J. L., and E. A. R. Garay. (1973). Sodium effect on bilirubin uptake by the rat intestinal mucosa. Biochim. Biophys. Acta 330:186.
374. Shanker, L. S., and Tocco, D. J. (1960). Active transport of some pyrimidines across the rat intestinal epithelium. J. Pharmacol. Exp. Ther. 128:115.
375. Shih, V. E., Bixby, E. M., Alpers, D. H., Bartsocas, C. S., and Thier, S. O. (1971). Studies of intestinal transport defect in Hartnup disease. Gastroenterology 61:445.
376. Shora, W., Forstner, G. G., and Forstner, J. F. (1975). Stimulation of proteolytic digestion by intestinal goblet cell mucus. Gastroenterology 68:470.
377. Shur, B. D., and Roth, S. (1975). Cell surface glycosyl-trasnferases. Biochim. Biophys. Acta 415:473.
378. Sigrist, H., Ronner, P., and Semenza, G. (1975). A hydrophobic form of the small-intestinal sucrase–isomaltase complex. Biochim. Biophys. Acta 406:433.
379. Sigrist-Nelson, K. (1975). Dipeptide transport in isolated intestinal brush border membrane. Biochim. Biophys. Acta 394:220.
380. Sigrist-Nelson, K., and Hopfer, U. (1974). A distinct D-fructose transport system in isolated brush border membrane. Biochim. Biophys. Acta 367:347.
381. Sigrist-Nelson, K., Murer, H., and Hopfer, U. (1975). Active alanine transport in isolated brush border membranes. J. Biol. Chem. 250:5674.
382. Silverblatt, R., Conklin, K., and Gray, G. M. (1974). Sucrase precursor in human jejunal crypts. J. Clin. Invest. 53:76a.
383. Singer, S. J. (1972). A fluid lipid–globular protein mosaic model of membrane. Ann. N.Y. Acad. Sci. 195:16.
384. Singer, S. J. (1974). The molecular organization of membrane. Annu. Rev. Biochem. 43:805.
385. Sjöström, H., Noren, O., and Josefsson, L. (1973). Purification and specificity of pig intestinal prolidase. Biochim. Biophys. Acta 327:457.
386. Smith, M. E. (1973). The uptake of pteroylglutamic acid by the rat jejunum. Biochim. Biophys. Acta 298:124.
387. Smith, M. W., Ferguson, D. R., and Burton, K. A. (1975). Glucose and phlorhizin-

protected thiol groups in pig intestinal brush border membranes. Biochem. J. 147:617.
388. Spencer, R. P., and Bow, T. M. (1964). *In vitro* transport of radiolabeled vitamins by the small intestine. J. Nucl. Med. 5:251.
389. Stefani, A., Janett, M., and Semenza, G. (1975). Small intestinal sucrase and isomaltase split the bond between glycosyl-C_1 and the glycosyl oxygen. J. Biol. Chem. 250:7810.
390. Stevenson, N. R., and Brush, M. K. (1969). Existence and characteristics of Na^+-dependent active transport of ascorbic acid in guinea pig. Am. J. Clin. Nutr. 22:318.
391. Stevenson, N. R., and Fierstein, J. S. (1976). Circadian rhythms of intestinal sucrase and glucose transport: cued by time of feeding. Am. J. Physiol. 230:731.
392. Stevenson, N. R., Ferrigini, F., Parnicky, K., Day, S., and Fierstein, J. S. (1975). Effect of changes in feeding schedule on the diurnal rhythms and daily activity levels of intestinal brush border enzymes and transport systems. Biochim. Biophys. Acta 406:131.
393. Stirling, C. E. (1967). High-resolution radioautography of phlorizin-^3H in rings of hamster intestine. J. Cell Biol. 35:605.
394. Stirling, C. E. (1975). Mercurial perturbation of brush border membrane permeability in rabbit ileum. J. Membr. Biol. 23:33.
395. Stirling, C. E., and Kinter, W. B. (1967). High resolution radioautography of galactose-^3H accumulation in rings of hamster intestine. J. Cell Biol. 35:585.
396. Stirling, C. E., Schneider, A. J., Wong, M. D., and Kinter, W. B. (1972). Quantitative radioautography of sugar transport in intestinal biopsies from normal humans and a patient with glucose–galactose malabsorption. J. Clin. Invest. 51:438.
397. Storelli, C., Vogeli, H., and Semenza, G. (1972). Reconstitution of a sucrase-mediated sugar transport system in lipid membranes. FEBS Lett. 24:287.
398. Tandon, R. K., Srivastava, I. M., and Pandey, S. C. (1975). Increased disaccharidase activity in human diabetics. Am. J. Clin. Nutr. 28:621.
399. Taylor, C. B. (1962). Cation-stimulation of an ATPase system from the intestinal mucosa of the guinea pig. Biochim. Biophys. Acta 60:437.
400. Terry, P. M., and Vidaver, G. A. (1973). The effect of gramicidin on sodium-dependent accumulation of glycine by pigeon red cells: a test on the cation gradient hypothesis. Biochim. Biophys. Acta 323:441.
401. Thier, S. O., Segal, S., Fox, M., Blair, A., and Rosenberg, L. E. (1965). Cystinuria: defective intestinal transport of dibasic amino acids and cystine. J. Clin. Invest. 44:442.
402. Thuneberg, L., and Rostgaard, J. (1968). Isolation of brush border fragments from homogenates of rat and rabbit kidney cortex. Exp. Cell Res. 51:123.
403. Tilney, L. G., and Mooseker, M. (1971). Actin in the brush border of epithelial cells of the chicken intestine. Proc. Natl. Acad. Sci. USA 68:2611.
404. Torres-Pinedo, R., Rivera, C., and Garcis-Castineiras, S. (1974). Intestinal exfoliated cells in infants: a system for study of microvillous particles. Gastroenterology 66:1154.
405. Toskes, P. P., and Deren, J. J. (1973). Vitamin B_{12} absorption and malabsorption. Gastroenterology 65:662.
406. Trier, J. (1968). Morphology of the epithelium of the small intestine. *In* C. F. Code (ed.), Handbook of Physiology Alimentary Canal, Intestinal Absorption, Vol. III, Sect. 6, pp. 1125–1175. American Physiological Society, Washington, D.C.
407. Ugolev, A. (1960*a*). Parietal (contact) digestion. Bull. Exp. Biol. Med. 40:12.
408. Ugolev, A. (1960*b*). Influence of the surface of the small intestine on enzymatic hydrolysis of starch by enzymes. Nature 188:589.
409. Ugolev, A. M. (1965). Membrane (contact) digestion. Physiol. Rev. 45:555.
410. Ugolev, A. M. (1966). Die Membrane-(Kontakt) verdanungin der Physiologie und Pathologie des magen-darm Traktes. Die Nahrung 10:483.
411. Ugolev, A. M. (1972*a*). Progress report: membrane digestion. Gut 13:735.
412. Ugolev, A. M., and DeLaey, P. (1973). Membrane digestion: a concept of enzymic hydrolysis on cell membranes. Biochim. Biophys. Acta 300:105.

413. Ugolev, A. M., Gruzdkov, A. A., DeLaey, P., Egorova, V. V., Hezuitova, N. N. et al. (1975). Substrate interactions on the intestinal mucosa: a concept for the regulation of intestinal digestion. Br. J. Nutr. 34:205.
414. Van der Stark-van der Molen, L. G., and dePriester, W. (1972). Brush border formation in the midgut of an insect, *Calliphora erythrocephata* Meigen, the formation of microvilli in the midgut during embryonic development. Z. Zellforsch. Mikrosk. Anat. 125:295.
415. Vidaver, G. A. (1964). Glycine transport by hemolyzed and restored pigeon red cells. Biochemistry 3:795.
416. Vinnick, I. E., Kern, F., and Sussman, J. E. (1965). Malabsorption and the diarrhea of diabetes mellitus. J. Lab. Clin. Med. 66:131.
417. Wade, D. N., Mearrick, P. T., and Morris, J. L. (1973). Active transport of L-dopa in the intestine. Nature 243:463.
418. Walker, G. J., and Whelan, W. J. (1960). The mechanism of carbohydrates action. VII. Stages in the salivary α-amylolysis of maltose, amylopectin and glycogen. Biochem. J. 76:257.
419. Walker, W. A., Field, M., and Isselbacher, K. J. (1974). Specific binding of cholera toxin to isolated intestinal microvillous membranes. Proc. Natl. Acad. Sci. USA 71:320.
420. Walker, W. A., and Isselbacher, K. J. (1974). Uptake and transport of macromolecules in the intestine. Gastroenterology 67:531.
421. Wallach, D. F. H., and Kamat, V. B. (1966). The contribution of sialic acid to the surface charge of fragments of plasma membrane endoplasmic reticulum. J. Cell Biol. 30:660.
422. Wass, M., and Evered, D. F. (1970). Transport of penicillamine across mucosa of the rat small intestine *in vitro*. Biochem. Pharmacol. 19:1287.
423. Wasserman, R. H., Taylor, A. N., and Kallfelz, F. A. (1966). Vitamin D and transfer of plasma calcium to intestinal lumen in chicks and rats. Am. J. Physiol. 211:419.
424. Weiser, M. M. (1976). Alterations in the cell surface membrane of the intestinal epithelial cell during mitosis, differentiation, and after neoplastic transformation. *In* L. Bolis, J. Hoffman, and A. Leaf (eds.), Membranes and Diseases, pp. 281–290. Raven Press, New York.
425. Welsh, J. D., Preiser, H., Woodley, J. F., and Crane, R. K. (1972). An enriched microvillus membrane preparation from frozen specimens of human small intestine. Gastroenterology 62:572.
426. Wilson, F. A., and Dietschy, J. M. (1974). The intestinal unstirred layer: its surface area and effect on active transport kinetics. Biochim. Biophys. Acta 363:112.
427. Wilson, F. A., and Treanor, L. L. (1975). Characterization of the passive and active transport mechanisms for bile acid uptake into rat isolated intestinal epithelial cells. Biochim. Biophys. Acta 406:280.
428. Wojnarowska, F., and Gray, G. M. (1975). Intestinal surface peptide hydrolases: identification and characterization of three enzymes from rat brush border. Biochim. Biophys. Acta 403:147.
429. Woodley, J. F., and Keane, R. (1972). Enterokinase in normal intestinal biopsies and those from patients with untreated coeliac disease. Gut 13:900.
430. Yeh, K., and Moog, F. (1974). Intestinal lactase activity in the suckling rat: influence of hypophysectomy and thyroidectomy. Science 183:77.
431. Younoszal, M. K., and Schedl, H. P. (1972). Effect of diabetes on intestinal disaccharidase activities. J. Lab. Clin. Med. 79:579.
432. Zaglak, B., and Curtius, H. Ch. (1975). The mechanism of the human intestinal sucrase action. Biochem. Biophys. Res. Commun. 62:503.

Index

Adenosine triphosphatase, membrane-bound, and gastric secretion, 144–148
Anatomy, functional, of gastrointestinal vascular beds, 2–4
Antidiuretic hormone and intestinal secretion, 273–274

Bacteria, invasive, and intestinal secretion, 271–272
Basolateral membrane, 339–340
Bicarbonate
mechanism of intestinal secretion of, 226
role in ion transport, 293
secretion by exocrine pancreas, 175
Bile, regulation of pancreas, 195
Bile secretion, 223–256
influences on, 235–236
mechanisms involved in, 225–237
morphological basis and general aspects, 224–225
Biliary motility, 60–61, 223–256
actions of cholecystokinin, 238–239
Bilirubin, biliary secretion of, 228–230
Blood flow: *see also* Circulation
gross, indirect study of, 4–5
intestinal, local and hormonal control of, 17–19
total organ, 4–5
volume, direct measurements of, 4
Brush border, 326
components of, 328–334
deletions of functions of, 346–347
development and control of functions of, 342–346
hydrolase-related transport, 341–342
interaction between amino acid and sugar transport, 340
kinetic advantage, 340–341
mechanisms of absorption, 334–337
organization of functions of, 340–342

Calcium
role in pancreatic protein secretion, 184
role of intracellular free, in gastric secretion, 134–135
secretion by exocrine pancreas, 177–178
Carboxylic ester hydrolasesterol ester hydrolase, digestion of lipids, 309
Cations, divalent, secretion by exocrine pancreas, 177–178
Chloride secretion by exocrine pancreas, 176
Cholecystokinin
and biliary motility, 238–239
and intestinal secretion, 273
Cholecystokinin-pancreoszymin, 83–84
actions and localization of, 84
assay, 84
Cholera as model for intestinal secretion, 261–270
Cholesterol control of bile saturation with, 234–237
secretion of, 233–234
Circulation: *see also* Blood flow
colonic, 23–24
gastric, 8–10
adrenergic control of, 9
gastrointestinal, 1–34
current methods of studying, 4–8
morphological study of, 4
intestinal, 10–23
adrenergic control of, 15–17
pancreatic, 24–27
adrenergic control of, 24–25
Colipase, pancreatic, digestion of lipids, 306–307, 310
Colon
circulation of, 23–24

Colon – *continued*
motility of, 57–60
Crypts of Lieberkühn as source of intestinal secretion, 269–270
Cyclic adenosine monophosphate, role in intestinal secretion, 266–269

Diet, pancreatic adaptation to, 197–198
Digestive-absorptive surface for absorption of water-soluble compounds, 326

Electrolytes
and bile secretion, 225–228
secretion by exocrine pancreas, 174–178, 200
transport of
in ileum, 286–294
in jejunum, 294–295
Enterocyte, uptake of lipids into, 311–313
Enteroglucagon, 85–86
physiology, 85–86
Enterotoxin
cholera
and epithelial cell interaction, 261–262
intestinal fluid induced by, 262–270
Escherichia coli, intestinal secretory effects, 270–272
Shigella, intestinal secretory effects, 271
Enzymes
adsorbed pancreatic, role of, 326–328
lipolytic, digestion of lipids, 306–311
Esophagus, motility of, 36–38

Fat digestion, intraluminal phase of, 311
Food, regulation of pancreas, 195–198

Gastric emptying, 47–48
Gastric secretion, 127–171
energy source for, 148–152
inhibitors, 148–149
metabolite measurements, 149–152
substrates, 148
hormone-receptor interaction, 128–132
membrane-bound ATPase, 144–148
membrane isolation, 141–143
membrane turnover, 137
morphology of secretory membrane, 135–137
proton pump mechanisms, 137–141
secretory process, 137–148
subcellular transport, 157–163
tissue properties, 152–157
transduction of stimulus, 132–135
Gastrin, 77–81
chemistry of, 77
and intestinal secretion, 273
localization of, 79
pathology, 81
pattern of release of, 79–80
pharmacology, 77–79
physiological actions, 80
regulation of pancreas, 192
Gastro-duodenal junction, 48–49
Gastrointestinal tract, circulation of, 1–34
Glucagon
and intestinal secretion, 273
pancreatic, 84–85
role in the gut, 85
Glucose, stimulation of sodium absorption, 297–301
Guanylate cyclase and gastric secretion, 134

Hormone-receptor interaction and gastric secretion, 128–132
Hormones
affecting intestinal fluid movement, 272–274
control of ion transport, 295–297
gastrointestinal, 71–103
advances in method of study of, 73–77
importance and complexity of, 73
purification of, 73–74

regulation of exocrine pancreas, 190–195, 203
trophic effects on pancreas, 198
Hydrolase, brush border transport, 341–342

Ileum, electrolyte transport in, 286–294
Immunocytochemistry, 74
Intestinal fluid, hormones affecting movement of, 272–274
Intestinal mucosa in intestinal secretion, 259–260
Intestinal secretion, 257–284
cholera as a model for, 261–270
composition of, 260–261
inhibitors of, 274–276
Intestine
blood flow and function, 19–23
circulation of, 10–23
severed small, resistance measurements in, 7
small, vascular dimensions of, 10–15
Ions
hormone control of transport of, 295–297
mechanisms underlying absorption of, 285–304

Jejunum, electrolyte transport in, 294–295

Lecithin, biliary secretion, 232–233
Lipase
bile salt-stimulated milk, digestion of lipases, 311
intestinal, digestion of lipids, 310–311
pancreatic, digestion of lipids, 306–309
pregastric, digestion of lipids, 306
Lipids
absorbed, intracellular metabolism of, 313
biliary secretion of, 231–232
digestion and absorption of, 305–323
specific, absorption of, 314–317
uptake into enterocyte, 311–313

Lumen, gastrointestinal, gas absorption from, 6–7
Microscopy in vivo of vascular sections, 5
Motilin, 87–89
pharmacology of, 87–88
physiology, 88–89
Motility
of biliary tract, 60–61
colonic, 57–60
hormonal control of, 60
myogenic control of, 57–59
nervous control of, 59–60
esophageal, 36–38
hormonal control of, 38
myogenic control of, 36–37
nervous control of, 37–38
gastric, 38–47
hormonal control of, 45–47
myogenic control of, 38–42
nervous control of, 42–45
gastrointestinal, 35–69
small intestinal, 49
hormonal control of, 55–57
myogenic control of, 49–51
nervous control of, 51–55
Myoepithelial cells
functions of, 119–120
salivary, 105–125

Nerves
adrenergic, regulation of pancreas, 187–190
cholinergic, regulation of pancreas, 187–188
Nervous system, autonomic, action of, in biliary motility, 240–241

Organics, water-soluble
digestion and absorption of, 325–365
mechanisms of absorption, 334–340

Pancreas
blood flow and endocrine secretion, 26–27
blood flow and exocrine secretion, 25–26
circulation of, 24–27

Pancreas – *continued*
exocrine, 173–221
acinar cell, 182–187
biosynthesis and intracellular transport of exportable proteins, 178–181, 200–202
development of function, 198
duct cell, 182
electrolyte secretion, 174–178, 200
endocrine-exocrine interaction in, 192–193
hormonal regulation, 190–195 203
nervous regulation of, 187–190, 202
pancreatic juice, 195–196
pharmacological effects, 198–199
and prostaglandins,-190
regulation by digestive secretion and food, 195–198
relation between water, electrolyte, and protein secretion, 177
site of water and electrolyte transport, 176–177
stimulus-secretion coupling, 181–187
vessels and ducts, 199–200
vascular dimensions of, 24
Peptide
gastric inhibitory, 89–90
actions, 89
and intestinal secretion, 273
physiology, 90
gastrointestinal, and biliary motility, 239–240
vasoactive intestinal, 90–92, 272–273
actions of, 90–91
physiology and pathology, 91–92
Permeability, capillary, determining, 7–8
pH, role in ion transport, 293
Phosphodiesterases and gastric secretion, 134
Phosphokinase, cAMP-activated, and gastric secretion, 134
Phospholipases, digestion of lipids, 309–310
Phospholipids, absorption of, 314–315
Polypeptide, pancreatic, 93–94
physiology and pathology, 94
Postcapillary resistance section, 4
Precapillary resistance section, 3
Precapillary sphincters, 3
Prostaglandins and exocrine pancreas, 190
Proteins
pancreatic secretion of, 178–181, 200–202
secretory, synthesis of, 179

Radioimmunoassay, 74–77

Salivary myoepithelial cells
biology of, 115–116
functions of, 119–120
general configuration of, 109–110
identification of, 107–109
innervation of, 113
morphology and physiology of, 105–125
historical background, 106–107
physiology and pharmacology of, 116–119
pre-and postnatal development, 112–113
ultrastructure of 110–112
Secretin, 81–83
actions of, 82–83
chemistry of, 81–82
and intestinal secretion, 273
localization of, 82
pathology, 83
pattern of release of, 82
regulation of pancreas, 190–191
Sodium
glucose control of absorption of, 297–301
secretion by exocrine pancreas, 175–176
Somatostatin, 92–93
actions of, 93
Sterols, absorption of, 315–317
Stomach: *see also* entries under Gastric
circulation of, 8–10
vascular dimensions of, 8–9

Thyrocalcitonin and intestinal secretion, 273

Vascular beds, gastrointestinal, functional anatomy of, 2–4
Vascular circuits, parallel-coupled, 8
  of small intestine, 10–12
Vascular sections
  consecutive
    plethysmographic and gravimetric studies of, 7
    of stomach, 8–9
    studies of, 7–8
  parallel-coupled
  clearance studies of, 5
  inert gas wash-out studies of, 5–6
    isotope and microsphere fractionation techniques, 5
    local tissue monitoring of indicator dilution curves, 6
    studies of, 5–7
  series-coupled, of small intestine, 12–15

Water
  and bile secretion, 225–228
  mechanisms underlying absorption of, 285–304
  secretion by exocrine pancreas, 174